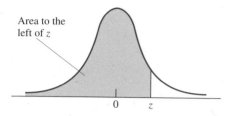

Area to the
left of z

Each table value is the cumulative area to
the left of the specified z-value.

Standard Normal Curve Areas (positive z-values)

z	.00	.01	.02	.03	.04	.05	.06	.07	.08	.09
					The Second Decimal Digit of z					
0.0	.5000	.5040	.5080	.5120	.5160	.5199	.5239	.5279	.5319	.5359
0.1	.5398	.5438	.5478	.5517	.5557	.5596	.5636	.5675	.5714	.5753
0.2	.5793	.5832	.5871	.5910	.5948	.5987	.6026	.6064	.6103	.6141
0.3	.6179	.6217	.6255	.6293	.6331	.6368	.6406	.6443	.6480	.6517
0.4	.6554	.6591	.6628	.6664	.6700	.6736	.6772	.6808	.6844	.6879
0.5	.6915	.6950	.6985	.7019	.7054	.7088	.7123	.7157	.7190	.7224
0.6	.7257	.7291	.7324	.7357	.7389	.7422	.7454	.7486	.7517	.7549
0.7	.7580	.7611	.7642	.7673	.7704	.7734	.7764	.7794	.7823	.7852
0.8	.7881	.7910	.7939	.7967	.7995	.8023	.8051	.8078	.8106	.8133
0.9	.8159	.8186	.8212	.8238	.8264	.8289	.8315	.8340	.8365	.8389
1.0	.8413	.8438	.8461	.8485	.8508	.8531	.8554	.8577	.8599	.8621
1.1	.8643	.8665	.8686	.8708	.8729	.8749	.8770	.8790	.8810	.8830
1.2	.8849	.8869	.8888	.8907	.8925	.8944	.8962	.8980	.8997	.9015
1.3	.9032	.9049	.9066	.9082	.9099	.9115	.9131	.9147	.9162	.9177
1.4	.9192	.9207	.9222	.9236	.9251	.9265	.9279	.9292	.9306	.9319
1.5	.9332	.9345	.9357	.9370	.9382	.9394	.9406	.9418	.9429	.9441
1.6	.9452	.9463	.9474	.9484	.9495	.9505	.9515	.9525	.9535	.9545
1.7	.9554	.9564	.9573	.9582	.9591	.9599	.9608	.9616	.9625	.9633
1.8	.9641	.9649	.9656	.9664	.9671	.9678	.9686	.9693	.9699	.9706
1.9	.9713	.9719	.9726	.9732	.9738	.9744	.9750	.9756	.9761	.9767
2.0	.9772	.9778	.9783	.9788	.9793	.9798	.9803	.9808	.9812	.9817
2.1	.9821	.9826	.9830	.9834	.9838	.9842	.9846	.9850	.9854	.9857
2.2	.9861	.9864	.9868	.9871	.9875	.9878	.9881	.9884	.9887	.9890
2.3	.9893	.9896	.9898	.9901	.9904	.9906	.9909	.9911	.9913	.9916
2.4	.9918	.9920	.9922	.9925	.9927	.9929	.9931	.9932	.9934	.9936
2.5	.9938	.9940	.9941	.9943	.9945	.9946	.9948	.9949	.9951	.9952
2.6	.9953	.9955	.9956	.9957	.9959	.9960	.9961	.9962	.9963	.9964
2.7	.9965	.9966	.9967	.9968	.9969	.9970	.9971	.9972	.9973	.9974
2.8	.9974	.9975	.9976	.9977	.9977	.9978	.9979	.9979	.9980	.9981
2.9	.9981	.9982	.9982	.9983	.9984	.9984	.9985	.9985	.9986	.9986
3.0	.9987	.9987	.9987	.9988	.9988	.9989	.9989	.9989	.9990	.9990
3.1	.9990	.9991	.9991	.9991	.9992	.9992	.9992	.9992	.9993	.9993
3.2	.9993	.9993	.9994	.9994	.9994	.9994	.9994	.9995	.9995	.9995
3.3	.9995	.9995	.9995	.9996	.9996	.9996	.9996	.9996	.9996	.9997
3.4	.9997	.9997	.9997	.9997	.9997	.9997	.9997	.9997	.9997	.9998

STATISTICS IN PRACTICE

Ernest A. Blaisdell
Elizabethtown College
Elizabethtown, Pennsylvania

Saunders College Publishing
Harcourt Brace College Publishers
Fort Worth   Philadelphia   San Diego   New York   Orlando   Austin
San Antonio   Toronto   Montreal   London   Sydney   Tokyo

Text Typeface: Times Roman
Compositor: Progressive Typographers, Inc.
Acquisitions Editor: Robert Stern
Developmental Editor: Donald Gecewicz
Managing Editor: Carol Field
Project Editor: Kimberly A. LoDico
Copy Editor: Zanae Rodrigo
Manager of Art and Design: Carol Bleistine
Art Director: Christine Schueler
Art Assistant: Caroline McGowan
Text Designer: Alan Wendt
Cover Designer: Lawrence R. Didona
Text Artwork: Grafacon, Inc.
Photo Researcher: Dena Digilio-Betz
Director of EDP: Tim Frelick
Production Manager: Bob Butler
Marketing Manager: Monica Wilson
Cover Credit: © 1992 Bishop/Phototake, Inc. NYC

Printed in the United States of America

Statistics in Practice

ISBN: 0-03-032229-4

Library of Congress Catalog Card Number: 92-050412

678901   49   987654

The text and cover were printed on paper made from waste paper, containing 10% post-consumer waste and 40% pre-consumer waste, measured as a percentage of total fibre weight content.

To my loving Mother, Thelma

With fondness, Alf and I recall
those many trips to and fro,
especially when you multiplied
by n, *more or less.*

Now the final journey home remains.

Preface

"The time may not be very remote when it will be understood that for complete initiation as an efficient citizen of one of the great complex world states that are now developing, it is necessary to be able to compute, to think in averages and maxima and minima, as it is now to be able to read and to write."

Though written more than 60 years ago, H. G. Wells' passage from *Mankind in the Making* seems particularly relevant in today's electronic age of global communication. Understanding the uses of statistics and its role in assimilating information contained in reports, scientific journals, political coverage, or even the daily newspaper is a necessary part of modern education. I have attempted to bring my training as a statistician and my years of teaching experience to the shaping of a clear and concisely written text. In addition, there is always the difficulty of persuading students, many of whom have been conditioned into math anxiety, that they might actually enjoy statistics, that statistical concepts are worth learning, and that statistics derives from real problems in the real world. My goal has been to provide a presentation that is pedagogically and mathematically sound, yet sufficiently gentle to minimize math anxiety.

Content Features

- To make data as vivid as possible, traditional methods of summarizing data are blended in Chapter 2 with more recently developed **data analysis techniques** such as dotplots, stem-and-leaf displays, 5-number summaries, and boxplots.

- **Regression analysis,** frequently used in many disciplines, can be introduced much earlier than it usually is in the traditional course. Chapter 3, therefore, is a concise introduction to the descriptive aspects of correlation and regression. It is written so that an instructor can vary its placement within the course syllabus. Furthermore, if the instructor wishes, this coverage can be complemented later with a detailed discussion in Chapter 14 of inferential methods in regression analysis.

- Included in Chapter 4, which introduces probability, is a **concise section on elementary counting techniques.** Students often find this topic difficult because they tend to approach each problem as either ''a permutation or a combination.'' I believe a greater understanding can be achieved by de-emphasizing permutation formulas and stressing the versatility of the multiplication rule.

- The critical topics of **confidence intervals** and **hypothesis testing** merit separate chapters. They are introduced in Chapters 9 and 10, respectively, and in Chapter 11 they are jointly used to discuss two-sample inferences.

- *P*-values are prominent in the research literature of virtually all disciplines. Consequently, after the introduction of hypothesis testing in Chapter 10, the reader is frequently exposed to the use of *P*-values throughout the remainder of the book. This is done, however, only after the student has had adequate opportunity to comprehend the basic concepts of hypothesis testing and rejection regions.

Exercises and Examples

- There is an abundant quantity of **interesting exercises** (1,633) and illustrations based on real-life situations and cited sources from a wide spectrum of disciplines. They are stated concisely, without burying the reader in verbiage.

- The **exercises** have been **carefully selected** and constructed to ensure that they meaningfully contribute to the learning process and enhance an appreciation of how statistics intermingles with our daily lives. The order of presentation progresses from mastering the basics to practical applications. Data used in previous exercise sets are always reproduced when used in subsequent applications. To serve the needs of instructors, odd-numbered problems are frequently paired with even-numbered problems. Answers are given in the book for all review exercises and for all odd-numbered end-of-section exercises.

- **Worked examples** are set up so that students can ''walk through'' them step by step. This approach helps the student understand the rationale of each statistical procedure. **Procedure boxes** that recap in a step-by-step manner what students should understand about a given process are liberally provided.

Pedagogical Features

- Each chapter opens with a preview, ''Looking Ahead.'' The opening photograph and accompanying caption set the theme for the chapter and give the students a foretaste of what they will explore in the pages that follow.

- To enhance the book's appearance and reader friendliness, liberal use is made of photos, marginal notes of interest, newspaper and magazine excerpts, and historical highlights of prominent mathematicians and statisticians.

- To help students master and retain the concepts in each chapter, the end-of-chapter material includes a summary, "Looking Back," a "Key Words" list, a "MINITAB Commands" list, and "Review Exercises," a set of comprehensive problems.

- Important formulas, definitions, procedures, tips, and computer commands are highlighted so that the reader can give them priority on first reading and, later, during review.

- A removable, detailed formula card and separate tables card have been bound into the book for possible use during examinations.

- For a ready reference, the normal, chi-square, and *t* tables are reproduced on the inside covers.

Use of MINITAB

This book is unique in its abundant use of the statistical computer package MINITAB. Instead of being used as a mere appendage to each chapter, MINITAB is woven throughout the text, with each command explained when it is first used in an application. This integration emphasizes the computer's role as a practical tool for relieving much of the drudgery associated with data sets, allowing the user more time to focus on other aspects of the analysis such as selecting a proper procedure, describing and interpreting data, and displaying the results. MINITAB's use also enhances the comprehension of many statistical concepts and techniques presented in a first course.

Although there are many statistical software packages currently available, I chose MINITAB because of its wide acceptance in educational instruction and its extensive use around the world in business and government. It is currently used at more than 2,000 colleges and universities and by 70% of Fortune 500 companies. Of equal importance, MINITAB can be learned quickly and easily, providing students with a powerful data analysis system that can be used in other courses and in their professional careers.

Because not all students have access to MINITAB, I have incorporated it into the text in a manner that affords an instructor considerable flexibility concerning its usage in the course. An instructor can choose any of the following options, each of which has been used by the author during the class-testing of this book.

- **Active Computer Usage.** Sufficient MINITAB instruction is provided so that students can write their own commands for statistical analyses. The book contains a total of 240 MINITAB assignments. They are flagged with the symbol Ⓜ and are placed at the end of an exercise set. MINITAB coverage is so extensive that the usual MINITAB supplement manual is unnecessary.

- **Passive Computer Usage.** Students can be instructed to examine only the output of the MINITAB exhibits and to just look over (or ignore) the commands used to generate the results.

- **No Computer Usage.** An instructor may prefer to have the class skip entirely the MINITAB exhibits. Implementation of this option is facilitated by the fact that all MINITAB output is prominently highlighted.

Flexibility in Topical Coverage

The book is designed so that an instructor has a great deal of flexibility in topical coverage. The diagram below displays several possibilities.

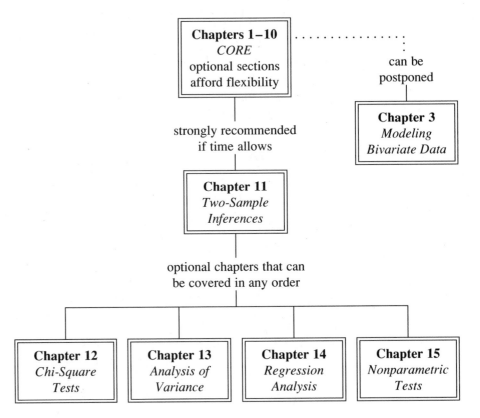

Chapters 1 through 10 form the core of the text, and Chapter 11 is strongly recommended if time allows. For instructors who want to spend less time covering these chapters, optional sections 6.4, The Hypergeometric Probability Distribution; 6.5, The Poisson Probability Distribution; and 11.3, Small-Sample Inferences for Two Means: Independent Samples and Unequal Variances, can be excluded. Additional time can be gained by also excluding sections 9.7, Chi-Square Probability Distributions; 9.8, Confidence Interval for a Variance; 10.5, Hypothesis Test for a Variance; 11.6, F Probability Distributions, and 11.7, Inferences for Two Variances.

Supplements

The following supplements have been prepared to enhance the use of this book. They are available, free of charge, to instructors who adopt the text.

- **Student Solutions Manual.** This supplement was prepared by Ronald L. Shubert of Elizabethtown College and is available to students for purchase. It contains detailed solutions for all review exercises and all odd-numbered, end-of-section exercises.

- **Instructor's Manual.** Also prepared by Ronald L. Shubert, this manual contains detailed solutions to all exercises, and a sample course syllabus with helpful suggestions.

- **ExaMaster™ Computerized Test Bank.** Available for IBM-compatible and Macintosh computers, this test bank contains more than 1,200 questions, each written especially for this text. A virtually unlimited number of tests can be custom designed by an instructor. Tests can contain a mixture of multiple-choice and free-response questions sorted by several different categories. An instructor can also add and edit questions, and grading keys can be generated. Full documentation accompanies the test bank.

- **Printed Test Bank.** A printed version of the Computerized Test Bank is also available. This supplement was prepared by Amy F. Relyea of Miami University of Ohio.

- **Professor's Grade Book.** Computer software for managing student records accompanies the Computerized Test Bank.

- **Data Disk.** A computer diskette is available containing the data sets for 384 exercises, including all data sets for the MINITAB problems. The data sets are stored as ASCII files, and they may be freely duplicated for student use at adopting institutions. The data disk will be distributed to instructors free of charge upon adoption.

Acknowledgments

My sincere thanks are extended to the many individuals who contributed to the development of this book. The following people served as reviewers.

Graydon Bell
Northern Arizona Univeristy

Patricia M. Buchanan
Pennsylvania State University

Chris Burditt
Napa Valley College

Darrell F. Clevidence
Carl Sandburg College

Pat Deamer
Skyline College

William D. Ergle
Roanoke College

Bryan V. Hearsey
Lebanon Valley College

William E. Hinds
Midwestern State University

Kermit Hutcheson
University of Georgia

Marlene J. Kovaly
Florida Junior College at Jacksonville

Mike Orkin
California State University, Hayward

Larry Ringer
Texas A & M University

Gerald Rogers
New Mexico State University

Adele Shapiro
Palm Beach Community College

George Sturm
Grand Valley State University

Mary Sue Younger
University of Tennessee

Douglas A. Zahn
Florida State University

Considerable effort has been devoted to make the text as accurate as possible. I am grateful to both Patricia Buchanan of the Pennsylvania State University and David Mathiason of the Rochester Institute of Technology for serving as accuracy reviewers of all examples and all answers at the back of the text. Any errors that might remain are, however, the sole responsibility of the author.

For their assistance in class-testing the manuscript at Elizabethtown College, I would like to thank Donald E. Koontz, John E. Koontz, Robert K. Morse, Larry Polin, and Laurie Showers. Special thanks are extended to Ronald Shubert, who also participated in the class-testing and wrote the Student Solutions Manual and the Instructor's Manual. I also want to thank Amy F. Relyea for preparing the Printed Test Bank.

I am grateful to Minitab, Inc. for providing a copy of the latest release of MINITAB available at the time of this writing—Release 8.2 for the MS-DOS standard version. Information about MINITAB can be obtained by contacting

Minitab, Inc.
3081 Enterprise Drive
State College, PA 16801
Phone: (814) 238-3280
Fax: (814) 238-4383

MINITAB is a registered trademark of Minitab, Inc.

I am greatly indebted to the staff at Saunders College Publishing. I wish to thank Robert B. Stern (Senior Acquisitions Editor), Kimberly A. LoDico (Project Editor), Donald J. Gecewicz (Developmental Editor), Karyn Valerius (Editorial Assistant), Christine Schueler (Art Director), and Dena Digilio-Betz (Photo Research Editor).

Finally, I wish to extend special thanks to my wife, Judith, for her support and encouragement during the preparation of this work.

E.A.B.
Lebanon, Pennsylvania
November, 1992

Contents

Chapter 3
Modeling Bivariate Data: Describing the Relationship Between Two Variables   96

Chapter 4
Probability: Measuring Uncertainty   134

Chapter 9
Estimating Means, Proportions, and Variances: Single Sample   324

Chapter 10
Tests of Hypotheses: Single Sample   374

Index of Applications

STATISTICS
IN PRACTICE

(© Julian Baum and David Angus/Science Photo Library/Photo Researchers, Inc.)

1

A First Look at Statistics and MINITAB

◀ *1990 Census Reveals United States Population Nears 250 Million*
Based on its 1990 census, the Census Bureau reported 249,632,692 as the official count of the nation's population. This figure represents an increase of 10.2 percent over the 1980 census count of 226,545,805 (Figure 1.1). As mandated by the Constitution, a census of the population is conducted every 10 years. The results are used in reassigning the number of seats held by each state in the House of Representatives. The population figures are also instrumental in determining funding levels for federal programs.

Figure 1.1
U.S. population during the years 1900 to 1990. (Source: U.S. Census Bureau.)

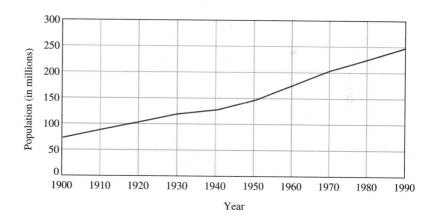

1.1 Why Study Statistics?

As you begin your statistics course, perhaps you are wondering why this subject should be a part of your educational program. Some reflection reveals that statistics influences nearly all facets of our society. Decisions in government, education, business, sports, politics, and many other fields are often based on statistical considerations.

Since the nation's first census in 1790, the science of statistics has played an important role in the governmental affairs of the United States. The Census Bureau is the government agency principally responsible for compiling data on the characteristics of the U.S. population. In addition to its census every 10 years, the bureau conducts monthly and other periodic surveys that influence a multitude of federal policies. Also active in the collection and analysis of governmental data are many other federal agencies, such as the Bureau of Labor Statistics; the National Science Foundation; the Internal Revenue Service; and the Departments of Agriculture, Commerce, Defense, Education, Veterans Affairs, and Health and Human Services. See Table 1.1.

In addition to governmental applications, statistics has assumed increasing importance as a decision-making tool in business. Statistical analyses influence the availability of a wide range of goods and services, from what we watch on television to the choice of treatments for medical ailments.

In recent years, the role of statistics in the educational curriculum has rapidly grown, and this increase has not been confined just to higher education. Statistical instruction has become more prevalent in high school programs, and it has even filtered down to the elementary level.

Because statistics pervades so much of our daily activities through a barrage of economic figures and product superiority claims, statistical literacy needs to be a part of one's educational training. A first course in statistics should aid the student in separating fact from fiction by developing the ability to look at statistical claims with a critical eye and thus strip away the hyperbole.

The course that you are now taking will also familiarize you with the language of

Table 1.1

Ten Most Populous States in 1990 and the Changes in House Seats

State	Population (millions)	Change in Seats
California	29.8	+7
New York	18.0	−3
Texas	17.0	+3
Florida	13.0	+4
Pennsylvania	11.9	−2
Illinois	11.5	−2
Ohio	10.9	−2
Michigan	9.3	−2
New Jersey	7.8	−1
N. Carolina	6.7	+1

Source: *Census Bureau.*

statistics and its basic concepts. It will help you make generalizations about random events that occur during your daily activities, and it will give you an increased awareness of the critical role that statistics plays in the scientific method. Further, the course will introduce you to statistical techniques that are applicable in the physical, life, and social sciences; in business; and in many other disciplines. The following examples illustrate some of the types of problems to which this methodology will be applied in following chapters.

- A manufacturer of laptop computers needs to estimate the average length of time that a battery charge will power the system. This information is essential, since some consumers consider charge life an important factor in the selection of a portable computer.

- In planning the amount of food that must be prepared for a weekend, the manager of a college cafeteria needs to estimate the proportion of the student body that will vacate the campus during that period.

- A food manufacturer wants to investigate the variation in the number of raisins placed in boxes of raisin bran cereal. Too few or too many raisins will negatively affect the cereal's sales.

- The manager of a presidential campaign wishes to estimate the percentage of registered voters who would favor a tax increase to reduce the federal budget deficit.

- A university's director of career development wants to know if a difference exists in the mean starting salaries of ultrasound technicians and radiologic technologists.

- A pharmaceutical company that produces a quit-smoking drug wants to show that the proportion of its successful users exceeds the proportion of smokers who quit with the aid of a competitor's drug.

- A manufacturing company maintains counts of the number of sick days used by each employee during the year. It wants to determine if a dependency exists between the age of an employee and the number of sick days used.

- A social scientist is concerned with how a city's crime rate is related to its unemployment rate.

- A nutritionist wants to determine if there is a relationship between the amount of cereal eaten by children and the cereal's percentage of sugar.

- An economist is interested in modeling the relationship between Christmas season sales and sales on the Friday after Thanksgiving.

- A consumer would like to know whether a significant difference exists in the average amounts of fat contained in four brands of low-fat frozen yogurt.

- A financial analyst wishes to investigate how the share price of IBM's common stock is related to its price-earnings ratio.

1.2   What Is Statistics?

For many people, the first memorable encounter with statistics came as a child through an association with an athletic activity. Perhaps this involved listening to the play-by-play of the Los Angeles Dodgers and the occasional references to the announcer's

Some baseball managers compile detailed records of their players' performances versus opposing pitchers. These statistics aid managers in selecting a starting lineup or a pinch hitter. *(© Walter Iooss, Jr. / The Image Bank)*

statistician. It might have come by serving as score keeper for the seventh grade basketball team. Perhaps it consisted of keeping tabs on the weekly rushing performances of a favorite football player. For some, the recollection will pertain to a nonathletic activity. It might consist of having maintained sales records during a scouts' fund-raising campaign. It could have been keeping track of one's average grade in algebra. Perhaps the first recollection of statistics was watching the returns of a presidential election.

All of the preceding are illustrations of one branch of statistics. This field is called **descriptive statistics,** and refers to describing the results of collected data.

> **Descriptive statistics** is the process of collecting data, summarizing it, and describing its characteristics.

The descriptive aspect is probably the image of statistics that most people conjure up when the term is used. In descriptive statistics, interest is focused on describing only the set of data that has been collected, with no attempt to generalize to other groups of data.

If statistics were only descriptive, then a first course in this field would be a short and less exciting one. Today, however, most course time is spent on **inferential statistics,** which is interested in obtaining information about a large collection of elements, referred to as the **population.**

> The **population** is the entire collection of elements of interest in a study.

Usually the population will contain a very large number of elements; thus, because of time and cost considerations, it would not be feasible to examine each. Consequently, a **sample** of the population is selected.

> The **sample** is a portion of the population.

The sample is analyzed and, based on this information, generalizations (inferences) are made concerning the population (Figure 1.2).

> **Inferential statistics** is the process of selecting a sample and using its information to make a generalization about a characteristic of a population.

Information contained in the sample is usually used to make an inference concerning a **parameter,** which is a numerical characteristic of the population. For instance, the parameter might be the average value of the population, or perhaps its largest value, or possibly the percentage of the population possessing a particular attribute. A parameter is estimated by computing a similar characteristic of the sample, called a **statistic.**

Population

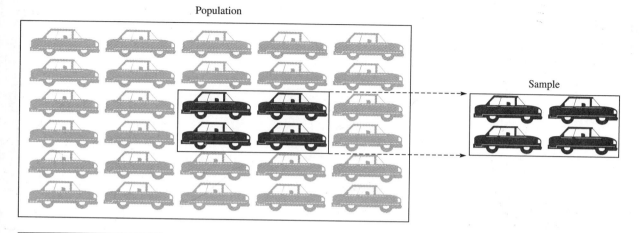

Sample

Figure 1.2
Sample versus population.

> A **parameter** is a numerical characteristic of the population.
>
> A **statistic** is a numerical characteristic of the sample.

Example 1.1   Six million people now attend two-year colleges in the United States (*American Association of Community and Junior Colleges,* 1990). To estimate the average yearly income of these students, an educational organization plans to conduct a national survey and interview 2,000 people from this group. The **population** of interest consists of the six million students who are currently attending two-year colleges in the United States. The **parameter** of interest is the average income of the six million students. The 2,000 students who will be interviewed constitute a **sample,** and the average income of this sample is a **statistic** that could be used to estimate the parameter of interest.

Example 1.2   Many mail order distributors of personal computer supplies sell generic computer diskettes at prices lower than brand names. Suppose one such company is considering the purchase of 100,000 diskettes from the manufacturer. It will make the purchase if no more than 1 percent of the diskettes in the lot are defective. It is not realistic for the company to test each diskette in the **population** of 100,000. Instead, it will select a **sample** of 800 diskettes and test these for defects. The results of this sample will then be used to estimate the percentage of defective diskettes in the entire lot. This characteristic of the population, the percentage of the 100,000 diskettes that are defective, is the population **parameter** of interest. Most likely, the **statistic** that will be used to estimate it will be the percentage of defectives found in the sample of 800 diskettes.

As you progress through the course, try to keep in mind that the main application of statistics is the making of inferences. The basic concept is quite simple: information contained in a sample is used to draw a conclusion concerning a population. An

The entire contents of the book you are now reading can be stored on two 3.5″ diskettes.
(© Murray Alcosser / The Image Bank)

inferential problem, however, involves more complexity than is captured in this summary. Each statistical inference problem must deal with the following six components.

Components of an Inferential Problem
1. Specify the population of interest.
2. Choose an appropriate sampling method.
3. Collect the sample data.
4. Analyze the pertinent information in the sample.
5. Use the results of the sample analysis to make an inference about the population.
6. Provide a measure of the inference's reliability.

Example 1.3   To illustrate the six aspects of an inferential problem, consider again the mail order company in Example 1.2 that is contemplating the purchase of 100,000 computer diskettes from a manufacturer.

1. Specify the population of interest:
The population of interest is the entire lot of 100,000 diskettes. This is the collection of elements on which interest is focused.

2. Choose an appropriate sampling method:
The sampling method involves the decision of how many diskettes to sample, as well as how these diskettes should be selected from the population. These are considerations in an area of statistics known as sample design. The choice of one sampling method over another could result in considerable savings in time and expenses.

3. Collect the sample data:

For each diskette sampled, the company is only interested in whether or not it is defective. Consequently, the data collected will consist of recording whether or not the diskette is defective for each of the 800 sampled diskettes.

4. Analyze the pertinent information in the sample:

The analysis of the sample data will involve computing the percentage of the sampled diskettes that were found to be defective.

5. Use the results of the sample analysis to make an inference about the population:

The statistical inference will be in the form of providing an estimate of the percentage of defective diskettes in the entire lot of 100,000.

6. Provide a measure of the inference's reliability:

Because the estimate provided (the sample percentage of defectives) is not expected to be exactly equal to the population percentage of defectives, a bound on the estimate's error should be given. For example, we might be able to say that we are 95% confident that the estimate is off by no more than 0.7% (seven-tenths of one percent).

The preceding components of inferential problems will become more clear as we progress through the text.

1.3   An Introduction to MINITAB

MINITAB is a general-purpose statistical analysis software package that is used for the organization, analysis, and presentation of data. It was developed in 1972 for introductory statistics courses at The Pennsylvania State University. MINITAB is now available for use on a wide variety of mainframe, mini, and personal computers, and it is used extensively around the world in businesses, governments, and institutions of higher education. Like other statistical software, MINITAB can be used to perform calculations that vary from the tedious to the near impossible. The user is freed to employ more complex techniques than are possible with hand calculations and has additional time to focus on other aspects of the analysis such as describing the data, interpreting it, and displaying the results.

In 1990 MINITAB was used in the teaching of statistics at more than 2,000 colleges and universities. It is also utilized at 70% of Fortune 500 companies.

MINITAB's vast popularity can be attributed to its ability to perform sophisticated analyses of data using simple commands that require little time to learn. In the few minutes needed to complete this section, you will be substantially on the way to discovering the power, yet simplicity, of this software package. Additional capabilities of MINITAB will be introduced as you progress through the text, increasing your comprehension of the system. Not only can MINITAB relieve the burden of performing many laborious calculations, but it can also aid in understanding many of the basic statistical concepts and techniques presented in a first course. Perhaps of equal importance, you will learn a powerful data analysis system that can be used in other courses and later in your professional career.

The understanding of a few basic concepts is needed to begin using MINITAB. These will be illustrated by going through a MINITAB session. This will allow you to see how data are entered, how commands are given, how MINITAB responds, and how output is obtained. If it is feasible, we recommend that you perform this session on your computer as you read this section.

The first step is to load the MINITAB program. How this is done will depend on the type of computer being used. If it is a personal computer, you will either insert the MINITAB program diskettes in the disk drive(s) or load the program from a hard drive. This may be as simple as typing

```
MINITAB
```

If the starting procedure is more complicated, you should either consult your course instructor or follow the directions given in the manual accompanying the software.

If you are using a mainframe computer or some other multiuser system, you will probably first have to identify yourself to the system by typing an account number, name, or password. Your instructor can provide you with the exact details, and you may want to copy or attach them to the inside cover of this book.

Next, the MINITAB software is loaded by typing the command **MINITAB** and then pressing the **Enter** (or **Return**) key. You can use either upper- or lowercase letters to type commands. In response to your typed command, MINITAB will load, sometimes provide a message, and then display the following.

```
MTB >
```

This is the prompt that MINITAB uses to indicate that it is waiting for a command to be typed. (There are over 200 commands in Release 8.2 of MINITAB.)

Usually, the next step is to enter the data to be organized or analyzed. Data are stored in columns labeled as C1, C2, C3, and so on. However, the data do not have to be entered in column format. To illustrate, we will form a column of data that will be called C1 and contain the following 10 odd integers: 1, 3, 5, 7, 9, 11, 13, 15, 17, 19. This can be accomplished by using the **SET** command. Type the following and press the Enter (Return) key after completing each line.

```
SET C1
1 3 5 7 9 11 13 15 17 19
END
```

The above will actually appear on your screen or paper as

```
MTB > SET C1
DATA> 1 3 5 7 9 11 13 15 17 19
DATA> END
MTB >
```

In the above, **MTB >** and **DATA>** are prompts that MINITAB prints after the Enter (Return) key is pressed.

Do not be concerned if you commit an error and type a misspelling or a meaningless command. MINITAB will warn you with an error message, and you can then retype a corrected statement. If you enter the wrong data, just repeat the above procedure.

Another way that data can be entered is with the **READ** command. This command is often used when entering more than one column of data, all of which contain the same number of values. For example, let's put the 10 even integers 2 through 20 in column C2, and in column C3 let's place the numbers 5, 10, 15, 20, 25, 30, 35, 40, 45, and 50. To accomplish this, type the following.

```
READ C2 C3
2 5
4 10
6 15
8 20
10 25
12 30
14 35
16 40
18 45
20 50
END
```

Note that each row of data consists of two values separated by a space (they may also be separated by a comma). The first value in each row is stored in column C2, and the second is placed in C3. If data were being read into eight columns such as C2, C3, C4, C5, C6, C7, C8, and C9, we would have to enter eight values for each row of data. Furthermore, the command could be

```
READ C2 C3 C4 C5 C6 C7 C8 C9
```

Or, we could simply type

```
READ C2-C9
```

In the latter command a hyphen (-) is used to designate a string of consecutively numbered columns.

The contents of one or more columns can be examined at any time by using the **PRINT** command. Let's look at what is stored in columns C1, C2, and C3 by typing the following:

```
PRINT C1-C3
```

MINITAB will respond with the following screen display.

```
MTB > PRINT C1-C3

 ROW    C1    C2    C3

   1     1     2     5
   2     3     4    10
   3     5     6    15
   4     7     8    20
   5     9    10    25
   6    11    12    30
   7    13    14    35
   8    15    16    40
   9    17    18    45
  10    19    20    50
```

In addition to columns, MINITAB stores constants that are labeled K1, K2, K3, and so on. As an illustration, we will determine the total of the values in column C3 and store this result in K1. Arithmetic operations can be performed by using the **LET** command. With this and the **SUM** command, we will now form the sum of the numbers in C3 and have the result called K1. This is done by typing

```
LET K1 = SUM(C3)
```

Note that **SUM** is applied to column C3, which must be enclosed in parentheses. To check on the value of K1, let's obtain it by typing the command **PRINT K1.**

```
MTB > PRINT K1
K1        275.000
```

From MINITAB's response to the **PRINT** command, we have that the total of the values in column C3 is 275.

As we progress with the use of MINITAB in this book, there will be occasions when an algebraic expression needs to be evaluated. This can be accomplished with the **LET** command. To illustrate, we will instruct MINITAB to:

1. multiply each number in column C1 by the number in the same position of C2,
2. divide each number in C3 by 5, and
3. add each result from (1) to the corresponding result from (2) and store these values in column C4.

To execute these steps, type

```
LET C4 = C1*C2 + C3/5
```

Observe that MINITAB uses an asterisk (*) to denote **multiplication** and a slash (/) for **division.** To check on the contents of C4, let's display it, along with C1, C2, and C3, by applying the **PRINT** command to columns C1 through C4.

```
MTB > PRINT C1-C4
```

ROW	C1	C2	C3	C4
1	1	2	5	3
2	3	4	10	14
3	5	6	15	33
4	7	8	20	60
5	9	10	25	95
6	11	12	30	138
7	13	14	35	189
8	15	16	40	248
9	17	18	45	315
10	19	20	50	390

In addition to performing arithmetic operations and evaluating algebraic expressions, the **LET** command can be used to make corrections and changes in the stored values of a column. To demonstrate this, we will replace the 33 in column C4 with an 8. Since 33 is the third value in C4, type

```
LET C4(3) = 8
```

To change the last (tenth) value in C4 to 1,528, type

```
LET C4(10) = 1528
```

Observe that 1,528 was typed without a comma. With MINITAB, commas are never used in numbers. The new contents of C4 can be displayed with the **PRINT** command. They appear below.

```
MTB > PRINT C4

  3   14    8   60   95  138  189  248  315  1528
```

Only the first four letters of a MINITAB command are required. Consequently, we could have instructed MINITAB to display the contents of C4 by typing **PRIN C4.**

To enhance the appearance and comprehension of the MINITAB worksheet, there are several features that one can use. These are listed and illustrated below.

1. Text can be added between command names and the arguments to which the commands apply. For example, consider the following command.

```
READ the following data into columns C2 and C3
```

MINITAB will treat the above command the same as

```
READ C2 C3
```

2. Informative statement lines can be typed by beginning them with the command **NOTE** or the # symbol.

```
NOTE   Whitney Lehman                    Math 151
#      MINITAB assignment 5        November 27, 1993
```

3. Comments can be typed at the end of a command or data line by preceding them with the # symbol.

```
LET C4(10) = 1528 # Changes 10th value of C4 to 1528
```

4. A column can be assigned a name. This is particularly helpful in remembering its contents.

```
NAME C1 'Odd Nos'  C2 'Even Nos'
```

A name must be enclosed in single quotes (apostrophes), and it may contain up to eight characters, except #, an apostrophe, or a beginning or ending blank. After naming C1 'Odd Nos', it can subsequently be referred to as either C1 or 'Odd Nos'. However, all output produced by MINITAB that involves C1 will utilize its name. For example,

```
LET C10 = 100*SQRT('Odd Nos') + 9
```

MINITAB will respond to this command by taking the square root of each number in column 'Odd Nos', multiplying each result by 100, adding 9 to each resulting value, and then placing the final results in column C10.

When you have completed your session with MINITAB, you may want to save your results so that they can be retrieved later for additional work. A file can be created that will store the worksheet for later recall. To accomplish this, assign the file a name (up to 8 characters). To illustrate, let's assign the name 'Chapter1' to this session, and we will save it by typing

```
SAVE 'Chapter1'
```

If you are using a personal computer, you may want to specify a particular drive in the **SAVE** command. For instance, if you want to save to drive B the file named Chapter1, then type

```
SAVE 'B:\Chapter1'
```

The **SAVE** command will preserve all columns, constants, and names that were created in the worksheet. To recall them for use in a future session, load MINITAB as discussed at the beginning of this section and type

```
RETRIEVE 'Chapter1'
```

For a personal computer, you may have to specify the drive as explained with the **SAVE** command. For instance, if the file is located in drive B, then type

```
RETRIEVE 'B:\Chapter1'
```

When your session with MINITAB is completed, you must inform MINITAB that you want to exit it. This is accomplished by typing

```
STOP
```

With this command, you will exit MINITAB and return to the computer's operating system. You must then log off the computer. The instructions for accomplishing this will vary for different systems. Consult your instructor for these, and then copy them in a convenient place, such as on the inside cover of this text.

Usually you will want a "hard copy" of your work with MINITAB so that it can be referred to later or turned in to your instructor as part of an assignment. Of course, you will have this if you are using a computer terminal that prints the results on paper as you type and as commands are executed. However, you may be using a system with a VDT (video display terminal). If there is a printer attached, your work can be duplicated on paper by typing the following command at the *start* of your session (at the first appearance of the MINITAB prompt **MTB** >).

```
PAPER
```

When you no longer want output from the printer, type

```
NOPAPER
```

If your system does not have a printer attached, you may be able to create and save a file and have it sent to a remote printer after you have exited from MINITAB. This can be accomplished by using the **OUTFILE** command. Choose a file name (we will use 'Chapter1'), and type the following at the *beginning* of your MINITAB session (at the first occurrence of **MTB** >).

```
OUTFILE 'Chapter1'
```

After typing the above, all commands and output displayed on your screen will be saved in the file 'Chapter1'. At the end of your MINITAB session, before typing **STOP**, type

```
NOOUTFILE
```

Check with your instructor for information on where and how the created file can be printed.

You can now begin experimenting with MINITAB. Do not worry about making mistakes. If you make one, MINITAB will usually issue a warning, and you can simply retype the correct information. If at any time you need information concerning a command, just type **HELP** and then the command. For example, if you are not sure about the **SQRT** command, type

```
HELP SQRT
```

To find out more about the capabilities of the **HELP** command, type

```
HELP HELP
```

A new version of MINITAB for Macintosh and DOS personal computers, Release 8, became available in 1991. In addition to retaining the command structure that you will use in this book, Release 8 provides the option of using menus and dialog boxes to construct the commands. If you own a computer, you may want to consider purchasing the Student Version of Release 8. It is very inexpensive and can probably be obtained through your campus bookstore or by contacting Addison-Wesley Publishing Company, Inc., 1 Jacob Way, Reading, Massachusetts, 01867.

Looking Back

Statistics consists of two main branches: descriptive and inferential. **Descriptive statistics** is concerned with collecting data, summarizing it, and describing its results. No attempt is made to generalize the results to other groups of data.

Inferential statistics is concerned with making generalizations that extend beyond the collected data to a larger group. To learn something about a large collection of elements (the **population**), a relatively small portion of it (a **sample**) is selected. Information contained in the sample is analyzed, and then a generalization pertaining to the population is made. Often these inferences will take the form of estimating some numerical characteristic of the population (a **parameter**). This is done by computing a similar characteristic of the sample (a **statistic**).

Today, the primary application of statistics is making inferences. Each statistical inference problem involves six main components:

1. Specify the population of interest.
2. Choose an appropriate sampling method.
3. Collect the sample data.
4. Analyze relevant information in the sample.
5. Use the results of the sample analysis to make an inference about the population.
6. Provide a measure of the inference's reliability.

Key Words

In reviewing this chapter, you should be able to define, explain, and illustrate each of the following.

descriptive statistics *(page 4)*   parameter *(page 4)*

population *(page 4)*   statistic *(page 4)*

sample *(page 4)*   components of an inferential

inferential statistics *(page 4)*   problem *(page 6)*

ⓂMINITAB Commands

SET _ *(page 8)*   END *(page 8)*

READ _ *(page 8)*   PRINT _ *(page 9)*

LET _ *(page 10)*   SUM( ) *(page 10)*

NOTE *(page 11)*   # *(page 11)*

NAME _ _ *(page 11)*   SQRT( ) *(page 12)*

SAVE _ *(page 12)*   RETRIEVE _ *(page 12)*

STOP *(page 12)*   HELP *(page 13)*

PAPER *(page 13)*   NOPAPER *(page 13)*

OUTFILE _ *(page 13)*   NOOUTFILE *(page 13)*

SORT _ _ *(page 19)*   RANK _ _ *(page 21)*

Review Exercises

NOTE: Exercises that require the use of MINITAB are indicated by Ⓜ before the problem number.

1.1   Steeped coffee is made by stirring coffee grounds in boiling water and then filtering. *The Wall Street Journal* (November 24, 1989) reported on a Dutch experiment that indicated that coffee brewed by the steeping method tends to raise blood cholesterol levels. More than 100 volunteers participated in the study. Does this group constitute a sample or a population?

1.2   At the commencement of the 1992 presidential election year, *USA TODAY* had the Gallup Organization conduct a poll of 1,433 adults nationwide (January 3–6, 1992). One of its findings revealed that 45% of poll participants feared that they might not be able to pay their medical or health-care costs within the next 12 months. Identify the sample and the population of interest for this survey.

1.3   The numbers of recorded cases of tuberculosis in the Mountain states during 1988 were as follows:

Arizona	202
Colorado	89
Idaho	18
Montana	15
Nevada	48
New Mexico	89
Utah	18
Wyoming	5

Source: *United States Centers for Disease Control.*

For each of the following, determine if the statement is descriptive or inferential. If it is inferential, state whether you believe the data justify the conclusion.
a. Arizona had the largest number of recorded cases in 1988.
b. There is little risk of contracting tuberculosis in Wyoming.
c. Two Mountain states tied for the second largest number of recorded cases in 1988.
d. For 1988, more recorded cases occurred in Nevada than in Utah.
e. A person living in Arizona is more likely to contract tuberculosis than one living in Colorado.

1.4   The following gives the composition of the 113 members of a college faculty.

	Highest Degree Held	
	Doctorate	**Master**
Men	52	43
Women	13	5

Determine for each of the following if the statement is descriptive or inferential. If it is inferential, state whether you believe the data justify the conclusion.

a. A larger percentage of women than men at this college possess a doctorate.

b. College female faculty members in general are more likely to possess a doctorate.

c. This college's administration probably prefers to hire male faculty members.

d. The average salary of women at this college exceeds that of men, since a larger percentage of women hold the doctorate.

−1.5 Give an example of a situation in which the 113 faculty members of the college referred to in Exercise 1.4 would be considered (a) as a sample and (b) as a population.

1.6 In a nationwide telephone poll of 1,030 adults conducted for the J. M. Smucker Co. (*Bruskin Associates,* Nobember 3–5, 1989) a large percentage of respondents said that they intend to say grace or a prayer before a holiday meal. The geographical breakdown of respondents was the following:

Region	**% Intending to Give Thanks**
South	90
Midwest	86
Northeast	77
West	72

For each of the following, determine if the statement is descriptive or inferential. For each inference, state whether you believe it is justified based on the given data.

a. A larger percentage of southerners than westerners in the survey said that they have intentions of giving thanks before a holiday meal.

b. Of those surveyed, more southerners than westerners have intentions of giving thanks before a holiday meal.

c. Southerners are more religious than westerners.

−1.7 A politician is campaigning for the mayoral office of her city. To assess her chances of winning, a political poll will be conducted.

a. If the sample is selected from a list of all registered city voters, explain why this is probably not the population of interest to the candidate.

b. Explain why it is impossible to actually select before the election a sample from the population of interest.

1.8 In a January issue of a city newspaper, there were five supermarket advertisements that featured onions and potatoes. The prices per pound are given below.

Store	Onions	Potatoes
1	$0.69	$0.39
2	$0.69	$0.39
3	$0.65	$0.39
4	$0.75	$0.45
5	$0.49	$0.29

Determine for each of the following if the statement is descriptive or inferential in nature.

a. Two of the stores had the same prices for onions and potatoes.

b. Of the five stores, the one with the least expensive onions also had the least expensive potatoes.

c. In this city, onions are more expensive than potatoes.

d. The difference in the highest and lowest cost of onions for the five stores surveyed was $0.26.

e. Store 5 had the least expensive produce.

1.9 A sardine processor wants to estimate the average weight of sardines in a fishing trawler's catch. To accomplish this, a sample of 278 sardines is selected, and the average weight of the sample is found to be 38.9 grams.

a. Identify the population of interest.

b. What is the parameter of interest?

c. What statistic is used to estimate the parameter?

d. If the sample consisted of the first 278 sardines caught by the trawler, explain why this sample might not be representative of the population.

1.10 In 1990, the Carnegie Foundation surveyed 10,000 college and university faculty members in the United States. One of the questions asked was, "Do your interests lie primarily in research or teaching?" Seventy percent said that teaching was their main interest.

a. With what population was the Carnegie Foundation concerned?

b. The purpose of this question was to estimate what parameter of the population?

c. What statistic was used to estimate the parameter?

1.11 In planning for an expansion of its sewage treatment facility, a large city wants to estimate the percentage of city homes that have an in-sink garbage disposal. Eight hundred and eighty homes were surveyed, and 35 percent were found to have disposals.

a. Identify the population of interest.

b. What is the parameter of interest?

c. What statistic is used to estimate the parameter?

1.12 *Newsweek* (November 13, 1989) reported that a study of 16,936 Harvard graduates determined that moderate exercise could extend a person's life by up to 2 years. Although the sample was from a population of Harvard graduates, do you think that this is the population of primary interest to the researchers?

MINITAB Assignments

Ⓜ 1.13 The 10 most actively traded issues on the New York Stock Exchange during 1991 appear below. For each issue, the high and low share prices (in dollars) are given, as well as the number of shares (in millions) traded during the year (The *Wall Street Journal,* January 2, 1992).

Issue	High	Low	Shares
RJR Nabisco	12.88	5.50	762
Philip Morris	81.75	48.25	498
PepsiCo	35.62	23.50	489
AT&T	40.38	29.00	457
IBM	139.75	83.50	439
American Express	30.38	18.00	421
Citicorp	17.50	8.50	410
General Motors	44.38	26.75	364
General Electric	78.12	53.00	360
Wal-Mart	59.88	28.00	335

Use MINITAB to determine the total number of shares traded for these issues during 1991.

Ⓜ 1.14 With reference to the data in Exercise 1.13, the price range for an issue is the difference between the high and the low price for the year. Have MINITAB create a column that contains the price range for each issue in the order listed in the exercise.

Ⓜ 1.15 The formula for converting temperatures in degrees Fahrenheit to degrees Celsius is

$$\text{Celsius} = \frac{5(\text{Fahrenheit} - 32)}{9}.$$

Use MINITAB to convert from degrees Fahrenheit to Celsius the temperatures from 0 to 100 in increments of 10. (The integers 0, 10, 20, . . . 100 are equally spaced, and consequently can be quickly placed in column C1 by typing the following.)

```
SET C1
0:100/10
END
```

Ⓜ 1.16 Create a table of square roots, squares, and reciprocals for the first 25 positive integers. Use MINITAB to assign suitable labels to each column of output. (See Exercise 1.15 for an easy way to place the integers from 1 to 25 in column C1.)

Ⓜ 1.17 In MINITAB the symbol for exponentiation is ****** (a double asterisk). For example, suppose we want to add 5 to each of the numbers in column C1, raise each result to the 9th power, and store these values in C2. We would use the command

```
LET C2 = (C1 + 5)**9
```

Use MINITAB to create a column of cubes and a column of cube roots for the positive integers from 1 to 25.

Ⓜ 1.18 If an amount of money A is invested at an interest rate i (in decimal form) and compounded annually, at the end of n years A will grow to an amount F, where F is given by

$$F = A(1 + i)^n.$$

Use MINITAB to determine the value at the end of each year of $1,000 that is invested for 20 years at an annual rate of 8 percent.

Ⓜ 1.19 In an advanced statistics class of 10 students, the following scores were obtained on the final examination.

Student Number	Final Exam Score
1	73
2	85
3	72
4	69
5	87
6	94
7	86
8	81
9	58
10	98

Store the student numbers and scores in two columns, assign the columns appropriate names, and use MINITAB to print their contents.

Ⓜ 1.20 For the final examination scores in Exercise 1.19, have MINITAB compute the average score and print its value.

Ⓜ 1.21 In MINITAB the numbers in a column can be arranged in ascending numerical order by employing the **SORT** command. For example, suppose we are working with two columns of data, C1 and C2, and we want to sort the numbers in C1 and preserve the row-by-row correspondence with C2. The following command will accomplish this.

```
SORT C1, carry along C2, place results in C1 and C2
```

Or more concisely, just type

```
SORT C1 C2 C1 C2
```

For the final examination scores in Exercise 1.19, use MINITAB to create a listing of the scores arranged in numerical order with the corresponding student numbers.

Ⓜ 1.22 Suppose the scores of students 3 and 4 were erroneously switched in Exercise 1.19. Use the **LET** command to make the necessary corrections, and print the revised examination scores.

Student	Recorded Grade	Actual Grade
3	72	69
4	69	72

Ⓜ 1.23 Each student's course grade in the advanced statistics class referred to in Exercise 1.19 was based on three regular examinations in addition to the final examination. The scores were:

Student	Exam 1	Exam 2	Exam 3	Final Exam
1	69	78	54	73
2	89	87	72	85
3	75	81	47	69
4	91	86	58	72
5	93	91	69	87
6	99	95	83	94
7	72	82	68	86
8	84	89	70	81
9	65	69	39	58
10	94	94	82	98

In determining each student's course average, the final examination counted as two regular exams. Use MINITAB to create a list of each student's four examination scores and course average. Assign appropriate names to the columns of output.

Ⓜ 1.24 The following command can be used to rank the numerical contents of a column.

```
RANK C1 C2
```

This will rank the numbers in column C1 by assigning 1 to the smallest value, a rank of 2 to the next value in size, and so on. The assigned ranks for the numbers in C1 are then stored in column C2. For the examination scores in Exercise 1.23, create a table that will give each student's rank on each exam.

Ⓜ 1.25 The figures below are the per capita amounts (in dollars) that the 50 states spent for public libraries in 1989. *(National Center for Education Statistics.)*

AL	AK	AZ	AR	CA	CO	CT	DE	FL	GA
8.13	23.64	14.53	5.98	15.89	18.01	22.10	8.22	11.93	10.06
HI	ID	IL	IN	IA	KS	KY	LA	ME	MD
17.23	9.84	19.25	17.95	11.85	15.85	7.04	11.53	12.66	24.45
MA	MI	MN	MS	MO	MT	NE	NV	NH	NJ
19.18	13.83	18.62	6.58	13.37	7.56	13.16	9.65	14.46	21.16
NM	NY	NC	ND	OH	OK	OR	PA	RI	SC
11.34	29.48	10.55	8.01	23.34	10.24	14.74	10.99	11.71	8.72
SD	TN	TX	UT	VT	VA	WA	WV	WI	WY
13.09	7.70	8.94	16.64	12.37	16.29	19.81	7.46	16.53	19.83

Use the procedure described in Exercise 1.24 to obtain a table that shows each expenditure and its corresponding rank.

(© Stuart Dee/ The Image Bank)

2 Describing Data: Graphical and Numerical Methods

29, 24, 27

◀ ***Women Will Soon Earn More Ph.D.s than Men***
Although men still receive more Ph.D.s than women, the gap between the sexes is quickly closing. A Department of Education study projects that by 2001, more Ph.D.s will be awarded to women than men. As evidence of this trend, the study cites that men earned 77 percent of the 34,064 Ph.D.s awarded in 1976, while in 1989 only 64 percent of the 35,379 Ph.D. recipients were men (Source: The Wall Street Journal, *December 27, 1990).*

Looking Ahead

In this chapter you will learn several methods for describing and presenting data in a clear and concise manner. Two general types of techniques will be discussed: **graphical methods** and **numerical methods.**

Figure 2.1 shows a graphical method (pie chart) used to convey the distributions of Ph.D.s between the sexes for the years 1976 and 1989. Graphical methods

Figure 2.1
Ph.D.s earned by men and women during the years 1976 and 1989.

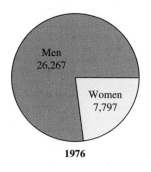

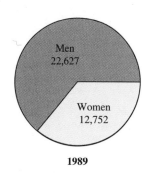

23

Table 2.1

1989 Mean Salary Offers to New Ph.D.s in Selected Fields

Field of Study	Annual Salary
Electrical Engineering	$48,666
Chemical Engineering	$47,853
Mechanical Engineering	$45,893
Chemistry	$43,215
Physics	$42,632
Mathematics	$37,500
Civil Engineering	$37,214

Source: *Statistical Abstract of the United States,* 1990.

summarize and display data in a manner that affords the viewer a quick impression of their distribution. We will see, however, that for many situations graphical methods are inadequate, and it is necessary to use numerical methods.

Several numerical techniques will be introduced and illustrated. Among the numerical methods to be discussed is the **mean,** which is used in Table 2.1 to describe salary offers extended to graduating Ph.D. students in selected fields of study. The mean is one of several numerical techniques that measure the center of a distribution. In this chapter, you will also learn how to describe the variation in the values of a distribution and how to describe the relationship of a single value to an entire distribution.

2.1 Graphical Methods for Quantitative Data: Frequency Distributions and Histograms

The methods discussed in this section and most of the chapter are designed for describing **quantitative data.** These data are numerical, such as the cost of a frozen yogurt serving, its weight, the time required to serve it, and the number of servings sold during a particular month. In contrast, **qualitative data** are categorical or attributive in nature, such as the flavor of the yogurt; its color; the serving size as measured by small, medium, or large; and the month in which it was bought. This section and the next discuss graphical methods for quantitative data, whereas Section 2.3 considers qualitative data.

Frequency Distributions

We will begin our consideration of graphical methods for summarizing quantitative data by discussing a frequency distribution. Table 2.2 lists the 1988 maximum monthly AFDC (aid for dependent children) benefit for a family of three for each of the 50 states. Table 2.3 contains the amounts only. All figures are in dollars.

Table 2.2

1988 Maximum Monthly AFDC Benefit for a Family of Three

State	Benefit	State	Benefit	State	Benefit
Alabama	118	Louisiana	190	Ohio	309
Alaska	779	Maine	416	Oklahoma	310
Arizona	293	Maryland	359	Oregon	412
Arkansas	202	Massachusetts	510	Pennsylvania	402
California	633	Michigan	528	Rhode Island	503
Colorado	356	Minnesota	532	South Carolina	200
Connecticut	601	Mississippi	120	South Dakota	366
Delaware	319	Missouri	282	Tennessee	159
Florida	275	Montana	359	Texas	184
Georgia	263	Nebraska	350	Utah	376
Hawaii	515	Nevada	325	Vermont	603
Idaho	304	New Hampshire	486	Virginia	354
Illinois	342	New Jersey	424	Washington	492
Indiana	288	New Mexico	264	West Virginia	249
Iowa	381	New York (City)	539	Wisconsin	517
Kansas	409	North Carolina	266	Wyoming	360
Kentucky	207	North Dakota	371		

Source: House Ways and Means Committee.

A quick inspection of the 50 amounts appearing in Table 2.3 reveals that they range from a low of 118 to a high of 779. How are the other 48 values distributed? A quick impression of their distribution can be obtained by constructing either a tabular display, such as a **frequency distribution,** or a pictorial exhibit, such as a **histogram.** Regardless of which we choose, the first step is to divide the interval from 118 to 779 into a number of subintervals, called **classes.** There is no unique way of selecting the subdivisions. In fact, we do not have to start with 118 and end with 779. The choice is subjective, made for informative purposes and clarity of presentation. For these data,

Table 2.3

Benefit Amounts in Dollars

118	356	515	409	510	359	264	310	366	354
779	601	304	207	528	350	539	412	159	492
293	319	342	190	532	325	266	402	184	249
202	275	288	416	120	486	371	503	376	517
633	263	381	359	282	424	309	200	603	360

Table 2.4

A Frequency Distribution for the 50 AFDC Benefits

Benefit	Number of States
100–199	5
200–299	11
300–399	16
400–499	7
500–599	7
600–699	3
700–799	1

such a choice might be the classes consisting of 100s, 200s, 300s, and so forth. In the next example, we will see that there is sometimes a "natural choice" of classes. Usually, you will use from 5 to 15 classes, with the high end of this range reserved for very large data sets.

Let's use the classes suggested above. To construct a frequency distribution, we count the number of measurements that fall within each subinterval, and then list the classes and their associated frequencies. These are given in Table 2.4.

The frequency distribution provides a quick impression of how the 50 amounts are distributed. For example, it shows that slightly more than half (27) of the values are in the 200s and 300s. This characteristic would be much more difficult to discern by only looking at the 50 values before they had been grouped.

The following important points should be noted about frequency distributions.

> The numbers that define the classes are called the **class limits.**

For the first class of our example, the **lower class limit** is 100, and 199 is the **upper class limit.**

> The midpoint of a class is called the **class mark.**

For the first class, this is $(100 + 199)/2 = 149.5$. The class mark for the second class is $(200 + 299)/2 = 249.5$.

> The distance between consecutive class marks is the **class width** (or **class increment**).

For our example, the class width is $(249.5 - 149.5) = 100$.

> The numbers midway between the upper limit of one class and the lower limit of the next class are called the **class boundaries.**

For the frequency distribution in Table 2.4, we designated the classes by using the class limits. They could have been denoted by using class boundaries as illustrated below.

Classes (using limits)	Classes (using boundaries)
100–199	99.5–199.5
200–299	199.5–299.5
300–399	299.5–399.5
400–499	399.5–499.5
500–599	499.5–599.5
600–699	599.5–699.5
700–799	699.5–799.5

Whether one uses class limits or class boundaries is a personal choice. Class boundaries have the advantage of providing continuity in the sense that there are no gaps between the classes; where one class ends, the next class begins. This feature makes boundaries better suited for use in graphical displays in which the classes are represented by intervals on the x-axis. This will be illustrated when we discuss histograms.

> Class limits are numbers that can occur in the sample data, while class boundaries are often impossible values.

For example, the first class has limits of 100 and 199, which are possible values for the sample. The sample will not have any values equal to the boundaries 99.5 and 199.5.

Example 2.1   Twenty-five students are enrolled in an advanced mathematical statistics course. Their scores on the final examination are given in Table 2.5. Construct a frequency distribution of the scores.

Table 2.5

Final Exam Scores for a Mathematical Statistics Class

86	52	69	74	64
83	71	78	77	79
56	88	64	73	71
98	75	78	90	83
72	91	81	85	64

Solution

The first step is to select the classes. Because there are only 25 values, we will use a small number of classes. Although there are several possibilities, for many instructors a "natural choice" might be the 50s, 60s, 70s, 80s, and 90s, corresponding to the letter grades F, D, C, B, and A. We'll use these in Table 2.6.

Notice that class limits have been used to designate the classes. If you prefer, denote the classes using the class boundaries 49.5–59.5, 59.5–69.5, 69.5–79.5, 79.5–89.5, and 89.5–99.5.

Note also that the class marks are $(50 + 59)/2 = 54.5, 64.5, 74.5, 84.5$, and 94.5. Thus, the class width is 10, the difference between consecutive class marks. It can also be obtained by taking the difference between the upper and lower boundaries of a class; for the first class this would give $(59.5 - 49.5) = 10$.

TIP: The class width can *not* be obtained by taking the difference between the upper and lower limits of a class.

Suppose in Table 2.6 one attempts to determine the width of the first class by subtracting the lower limit 50 from the upper limit 59. This would result in the erroneous value $(59 - 50) = 9$.

Table 2.6

A Frequency Distribution for the 25 Final Exam Scores

Score	Frequency
50–59	2
60–69	4
70–79	10
80–89	6
90–99	3

Relative Frequency Distributions

A frequency distribution shows how many of the sample values fall within each class. Often we're more interested in what proportion (or percentage) is represented by this figure. Such is usually the case when comparing distributions based on different sample sizes. A table that shows the proportion (or percentage) for each class is called a **relative frequency distribution.**

Example 2.2 Construct a relative frequency distribution for the frequency distribution of final examination scores repeated below.

Score	Frequency
50–59	2
60–69	4
70–79	10
80–89	6
90–99	3

Solution

The relative frequency for a class is obtained by dividing its frequency by the total number of values in the sample. For the five classes, the relative frequencies are $\frac{2}{25} = 0.08$, $\frac{4}{25} = 0.16$, $\frac{10}{25} = 0.40$, $\frac{6}{25} = 0.24$, and $\frac{3}{25} = 0.12$. Except for round-off error, the relative frequencies must sum to one. If you prefer, you can use the percentages 8, 16, 40, 24, and 12. When percentages are used, they must add to 100. The relative frequency distribution is given in Table 2.7.

Table 2.7

A Relative Frequency Distribution for the 25 Final Exam Scores

Score	Relative Frequency
50–59	0.08
60–69	0.16
70–79	0.40
80–89	0.24
90–99	0.12

Histograms

A frequency (or relative frequency) distribution can be displayed pictorially by constructing a **histogram.** For a distribution with classes of equal width, each class is denoted by an interval on the horizontal axis, and a rectangle is constructed having this interval as its base and having a height equal to its frequency (or relative frequency). To denote the intervals on the horizontal axis that correspond to the classes, class boundaries are used, because they provide continuity with no gaps between the classes.

Example 2.3 Construct a histogram for the frequency distribution of the 50 AFDC benefits given on the next page.

**A Frequency Distribution
for the 50 AFDC Benefits**

Benefit	Number of States	Relative frequencies
100–199	5	1
200–299	11	.22
300–399	16	.32
400–499	7	.14
500–599	7	.14
600–699	3	.06
700–799	1	.02

(handwritten annotations: "frequencies" above "Number of States"; "Relative frequencies" heading; relative frequency values .22, .32, .14, .14, .06, .02)

Solution

The classes are represented along the horizontal axis by the intervals 99.5–199.5, 199.5–299.5, . . . , 699.5–799.5. Next, over each interval a rectangle is constructed with a height equal to the class frequency. The resulting histogram appears in Figure 2.2.

From the histogram it is easy to see how the 50 amounts are distributed. More than half the values are in the 200s and 300s, the amounts are evenly distributed in the 400s and 500s, and they drop off sharply in the 600s and 700s. Since a picture has a more rapid impact and lasting impression than a table, a histogram is often preferred over a frequency distribution for displaying data.

TIP: A **relative frequency histogram** can be constructed by making the height of each rectangle equal to the relative frequency of the class.

Figure 2.2
Histogram for the 50 AFDC benefits.

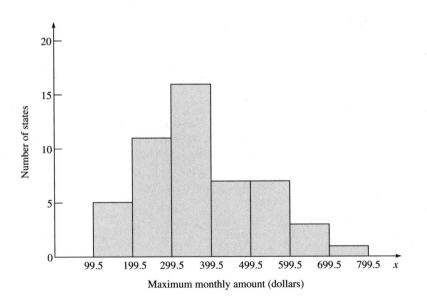

Maximum monthly amount (dollars)

Figure 2.3
Relative frequency histogram
for the 50 AFDC benefits.

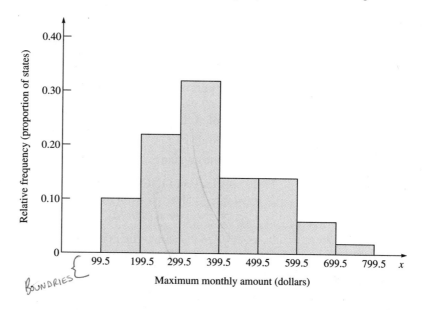

A relative frequency histogram for the AFDC amounts is displayed in Figure 2.3.

Ⓜ**Example 2.4**   Use MINITAB to construct a frequency distribution and histogram for the 50 AFDC amounts below.

118	356	515	409	510	359	264	310	366	354
779	601	304	207	528	350	539	412	159	492
293	319	342	190	532	325	266	402	184	249
202	275	288	416	120	486	371	503	376	517
633	263	381	359	282	424	309	200	603	360

Solution

First we will store the amounts in column C1. Since data are entered in only one column, the **SET** command will be used.

```
SET C1
118 779 293 202 633 356 601 319 275 263 515 304 342 288 381
409 207 190 416 359 510 528 532 120 282 359 350 325 486 424
264 539 266 371 309 310 412 402 503 200 366 159 184 376 603
354 492 249 517 360
END
```

To obtain both a frequency distribution and a histogram, type

```
HISTOGRAM C1;
START 149.5;
INCREMENT 100.
```

The first line in the above commands contains:

1. the main command **HISTOGRAM**,
2. the column that contains the data,

3. an ending semicolon that informs MINITAB that another command (subcommand) will follow on the next line.

The second line contains three essential items:

1. the subcommand **START** that is used to indicate where the histogram is to begin,
2. the starting value, which is the first class mark (the midpoint of the first class),
3. a semicolon that tells MINITAB that another subcommand will appear on the following line.

The third line contains:

1. the subcommand **INCREMENT** that is used to indicate the class width,
2. the desired value for the class width,
3. a period that indicates to MINITAB the end of the series of commands.

In response to the above commands, MINITAB will produce the following output:

Exhibit 2.1

```
MTB > HISTOGRAM C1;
SUBC> START 149.5;
SUBC> INCREMENT 100.

Histogram of C1    N = 50

Midpoint    Count
      150        5    *****
      250       11    ***********
      350       16    ****************
      450        7    *******
      550        7    *******
      650        3    ***
      750        1    *
```

Output from MINITAB will sometimes differ from conventional displays. Note that MINITAB does not designate the classes by using class limits or boundaries. Rather, it gives the class midpoints (class marks) in the first column, and these have been rounded to a whole number. In the second column, under the heading ''Count,'' appear the class frequencies. These first two columns constitute a frequency distribution.

To the right of these columns appears a histogram with asterisks (*) used for the rectangles. In addition, the asterisks are drawn horizontally rather than vertically as in a conventional histogram. If Exhibit 2.1 is turned 90 degrees to the left, then the MINITAB histogram will more closely resemble that drawn earlier in Figure 2.2.

The subcommands **START** and **INCREMENT** are optional with the **HISTOGRAM** command. If one just types

```
HISTOGRAM C1
```

then MINITAB will automatically choose a set of classes.

An obvious disadvantage of using frequency distributions or histograms to describe a data set is that they do not retain the individual values. In the next section we will consider a graphical method without this limitation and with many of the attributes of a frequency distribution and a histogram.

Section 2.1 Exercises
Graphical Methods for Quantitative Data: Frequency Distributions and Histograms

2.1   The following figures were obtained from a sample of 16 private colleges in Pennsylvania. They are the percentages of applicants accepted at the colleges for the 1987–1988 school year.

$$58 \quad 81 \quad 38 \quad 75 \quad 73 \quad 57 \quad 70 \quad 70$$
$$64 \quad 74 \quad 47 \quad 55 \quad 74 \quad 60 \quad 43 \quad 50$$

Construct a frequency distribution for the percentages. Use classes of equal width and begin with the class 30–39.

2.2   Forty employees were donors during a recent visit of a bloodmobile unit at their company. The ages of the donors are given below.

35	53	61	43	21	42	23	29	35	37
39	58	27	64	27	31	36	48	41	22
37	35	42	32	43	34	59	50	38	43
31	30	41	37	29	45	23	56	46	41

Construct a frequency distribution for the ages. Use classes of equal width and a first class of 20–29.

2.3   Obtain a relative frequency histogram for the percentages of accepted applicants that are given in Exercise 2.1.

2.4   Construct a frequency histogram for the 40 ages in Exercise 2.2.

2.5   A frequency distribution with five classes was constructed for a set of integers. The class width is 8, and the lower limit of the first class is 20. Determine the limits of the five classes.

2.6   A frequency distribution for a set of integers has seven classes with a class width of 20. If the upper class limit of the first class is 49, find the limits of the seven classes.

2.7   The lower class limits of a frequency distribution for a set of integers are 100, 140, 180, 220, 260, and 300. Determine the class boundaries for the six classes.

2.8   The upper class limits of a frequency distribution of sales are $2.70, $2.90, $3.10, $3.30, $3.50, $3.70, and $3.90. Use class boundaries to designate the classes.

2.9   The class marks of a frequency distribution are 74, 83, 92, 101, 110, 119, 128, and 137. Determine the class boundaries.

2.10 Use class boundaries to designate the classes of a frequency distribution whose class marks are 32.5, 36.5, 40.5, 44.5, 48.5, 52.5, and 56.5.

2.11 The figure below is a relative frequency histogram of the ages of 20 participants in a special education program. List the age of each student.

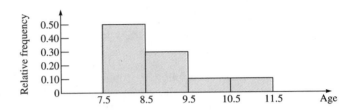

2.12 The following is a frequency distribution of the times required to complete the final examination for a class of 33 students in an introductory statistics course. The figures are in minutes.

Time	Number of Students
81–85	1
86–90	1
91–95	2
96–100	3
101–105	3
106–110	7
111–115	10
116–120	6

For this frequency distribution, determine:
a. the class width,
b. the boundaries of the third class,
c. the lower boundary of the first class,
d. the upper boundary of the last class,
e. the lower limit of the second class,
f. the upper limit of the fourth class,
g. the number of students who required 100 minutes or less to complete the exam,
h. the number of students who took more than 110 minutes to finish, and
i. the number of students who required anywhere from 86 to 105 minutes to complete the test.

2.13 The distribution of the ages of the full-time faculty at a certain college is given below.

Mark	Age	**Number of Faculty**
27	25–29	4
32	30–34	14
37	35–39	19
	40–44	28
	45–49	26
	50–54	23
	55–59	20
	60–64	16
	65–69	8

Determine for this distribution of ages:
a. the class width,
b. the boundaries of the fifth class,
c. the lower boundary of the first class,
d. the upper boundary of the last class,
e. the lower limit of the seventh class,
f. the upper limit of the third class,
g. the number of faculty younger than 40,
h. the number of faculty older than 59, and
i. the number of faculty older than 39 but younger than 60.

2.14 During the 1988 Summer Olympics in Seoul, South Korea, more than half the gold medals were awarded to the USSR (55), to East Germany (37), and to the United States (36). Twenty-eight other countries received gold medals. The numbers awarded to these countries are given below.

```
12  11  11  10  7  6  6  5  5  5  4  3  3  3
    3   3   2   2  2  2  1  1  1  1  1  1  1  1
```

Construct a relative frequency distribution for the data.

2.15 Construct a relative frequency distribution for the 1987 unemployment rates (percent) that appear below for the 50 states (*U.S. Government*).

AL	AK	AZ	AR	CA	CO	CT	DE	FL	GA
7.8	10.8	6.2	8.1	5.8	7.7	3.3	3.2	5.3	5.5
HI	ID	IL	IN	IA	KS	KY	LA	ME	MD
3.8	8.0	7.4	6.4	5.5	4.9	8.8	12.0	4.4	4.2
MA	MI	MN	MS	MO	MT	NE	NV	NH	NJ
3.2	8.2	5.4	10.2	6.3	7.4	4.9	6.3	2.5	4.0
NM	NY	NC	ND	OH	OK	OR	PA	RI	SC
8.9	7.9	4.5	5.2	7.0	7.4	6.2	5.7	3.8	5.6
SD	TN	TX	UT	VT	VA	WA	WV	WI	WY
4.2	6.6	8.4	6.4	3.6	4.2	7.6	10.8	6.1	8.6

2.16 Obtain a relative frequency histogram for the 28 numbers of gold medals that are given in Exercise 2.14.

— 2.17 Construct a relative frequency histogram for the 50 unemployment rates that appear in Exercise 2.15.

MINITAB Assignments

Ⓜ 2.18 Use MINITAB to construct a frequency distribution and histogram of the 50 unemployment rates that are given in Exercise 2.15. Use a class width of 1 and 2.45 as the first class mark.

Ⓜ 2.19 The following are the SAT math scores of 79 students in a first year calculus course.

580	490	560	550	670	510	540	610	590	600
480	510	520	540	580	490	600	620	680	590
570	530	580	560	540	560	530	580	520	540
610	580	570	590	530	560	570	560	540	590
590	600	560	500	510	600	580	540	470	690
580	710	680	600	580	540	540	570	580	590
520	540	580	550	550	560	640	670	560	680
570	540	470	480	530	490	490	680	720	

Use MINITAB to construct a histogram of the scores, where the first class mark is 470 and the class width is 20.

Ⓜ 2.20 The figures below are the per capita amounts (in dollars) that the 50 states spent for public libraries in 1989 (*National Center for Education Statistics*).

AL	AK	AZ	AR	CA	CO	CT	DE	FL	GA
8.13	23.64	14.53	5.98	15.89	18.01	22.10	8.22	11.93	10.06
HI	ID	IL	IN	IA	KS	KY	LA	ME	MD
17.23	9.84	19.25	17.95	11.85	15.85	7.04	11.53	12.66	24.45
MA	MI	MN	MS	MO	MT	NE	NV	NH	NJ
19.18	13.83	18.62	6.58	13.37	7.56	13.16	9.65	14.46	21.16
NM	NY	NC	ND	OH	OK	OR	PA	RI	SC
11.34	29.48	10.55	8.01	23.34	10.24	14.74	10.99	11.71	8.72
SD	TN	TX	UT	VT	VA	WA	WV	WI	WY
13.09	7.70	8.94	16.64	12.37	16.29	19.81	7.46	16.53	19.83

Use MINITAB to obtain a histogram of the 50 amounts. Allow MINITAB to select the classes.

Ⓜ 2.21 The following are the median purchase prices of existing single family houses in 32 metropolitan areas for the years 1980 and 1987. The figures are in thousands of dollars. Use MINITAB to construct a histogram of the prices for each of the 2 years. Allow MINITAB to select each set of classes, and observe the shift in prices that has occurred from 1980 to 1987.

Metropolitan Area	1980	1987
Atlanta, GA	81.8	131.1
Baltimore, MD	69.5	125.4
Boston-Lawrence-Lowell, MA-NH	70.5	186.2
Chicago-Gary, IL-IN	78.0	116.9
Cleveland-Akron-Lorain, OH	63.8	89.4
Columbus, OH	65.7	92.9
Dallas, TX	89.4	128.5
Denver-Boulder, CO	80.0	139.8
Detroit-Ann Arbor, MI	67.9	93.6
Greensboro-Winston Salem-High Point, NC	63.6	110.0
Honolulu, HI	119.3	177.6
Houston-Galveston, TX	85.7	106.8
Indianapolis, IN	60.3	94.7
Kansas City, MO-KS	64.2	105.8
Los Angeles-Long Beach-Anaheim, CA	110.7	167.7
Louisville, KY-IN	58.5	85.6
Miami-Fort Lauderdale, FL	75.3	103.9
Milwaukee-Racine, WI	84.9	95.7
Minneapolis-St. Paul, MN-WI	80.9	139.4
New York-Newark-Jersey City, NY-NJ-CT	93.8	181.8
Philadelphia-Wilmington-Trenton, PA-DE-NJ-MD	62.6	113.7
Phoenix, AZ	92.7	132.9
Pittsburgh, PA	64.6	82.8
Portland, OR-WA	75.5	108.5
Rochester, NY	57.5	79.6
St. Louis, MO-IL	56.5	92.1
Salt Lake City-Ogden, UT	73.2	121.0
San Diego, CA	105.7	156.7
San Francisco-Oakland-San Jose, CA	122.9	179.9
Seattle-Tacoma, WA	82.2	127.7
Tampa-St. Petersburg, FL	64.0	100.3
Washington, DC-MD-VA	100.1	159.5

Source: Federal Home Loan Bank Board, *Savings and Home Financing Source Book,* annual.

2.2   Graphical Methods for Quantitative Data: Dotplots and Stem-and-Leaf Displays

This chapter is concerned with different methods for describing data. Given a set of measurements (raw data), how can these be summarized and presented in a manner that quickly conveys a sense of the data's distribution? The previous section discussed two graphical methods: a frequency distribution and its pictorial representation—a histogram. In these, the raw data are grouped into intervals, called classes. There is a pictorial display without this grouping but similar in appearance to a histogram. It is called a **dotplot** and is illustrated in the next example.

Dotplots

Example 2.5   In Example 2.1 the final examination scores were given for 25 students enrolled in an advanced mathematical statistics course. The scores are repeated below.

$$
\begin{array}{ccccc}
86 & 52 & 69 & 74 & 64 \\
83 & 71 & 78 & 77 & 79 \\
56 & 88 & 64 & 73 & 71 \\
98 & 75 & 78 & 90 & 83 \\
72 & 91 & 81 & 85 & 64 \\
\end{array}
$$

To construct a dotplot, draw a number line that covers the range of the data; in this case, the line must include values from 52 through 98. Each value in the data is now plotted as a dot above its representation on the real number line—hence the name dotplot. The completed dotplot is given in Figure 2.4.

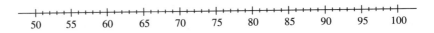

Usually dotplots are used only with small sample sizes. For large sets of data, dotplots tend to produce several plots of only one or two dots, and consequently, they do not give as clear a presentation of data clustering as a histogram or similar graphical displays.

Figure 2.4
Dotplot for the 25 final exam scores.

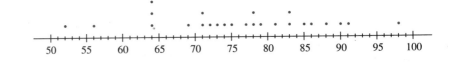

ⓂExample 2.6   Use MINITAB to construct a dotplot for the 25 final exam scores in Example 2.5.

Solution

The data are first entered and stored in column C1.

```
SET C1
86 83 56 98 72 52 71 88 75 91 69 78 64
78 81 74 77 73 90 85 64 79 71 83 64
END
```

To generate a dotplot of the data in C1, type

```
DOTPLOT C1
```

MINITAB will respond with the dotplot given in Exhibit 2.2.

Exhibit 2.2

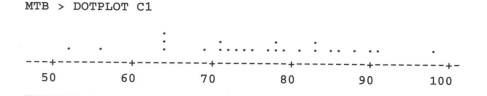

Stem-and-Leaf Displays

A relatively new method for summarizing and describing data was popularized about two decades ago by John Tukey. It is called a **stem-and-leaf display** and is one of several exploratory data analysis tools that have become popular recently. A stem-and-leaf combines characteristics of a frequency distribution and a histogram with other desirable features to produce a display that is both graphical and tabular. Specifically, it has the grouping of a frequency distribution, presents a pictorial display similar to a histogram, sorts the data according to size, and lists the individual values.

Example 2.7   We will construct a stem-and-leaf display for the 25 final examination scores that were used in Example 2.6 and repeated below.

86	52	69	74	64
83	71	78	77	79
56	88	64	73	71
98	75	78	90	83
72	91	81	85	64

Each number is split into digits consisting of a **stem** and a **leaf.** For this example the tens digit is the stem, and the units digit is the leaf. The first step in constructing the stem-and-leaf display is to list the stems in increasing order in a vertical column, followed by a vertical line (see Figure 2.5). Next, for each data value, list its leaf in the row of its stem. For example, for the exam score of 86 the stem is 8 (representing 80) and the leaf is 6. This score is recorded in the display by placing a 6 to the right of the vertical line in the row with a stem of 8.

The stem-and-leaf display provides considerable information about the data. It possesses the grouping of a frequency distribution with the stems corresponding to classes consisting of the 50s, 60s, 70s, 80s, and 90s. The leaves present a pictorial

Figure 2.5
Stem-and-leaf display of the
25 final exam scores.

Stem	Leaf
5	6 2
6	9 4 4 4
7	2 1 5 8 8 4 7 3 9 1
8	6 3 8 1 5 3
9	8 1 0

display similar to the rectangles in a histogram. The low and high values of the data are quickly discerned, and each individual value is displayed. Moreover, if MINITAB is used to construct the display, then the leaves will be arranged in increasing order, thus providing a sorting of the data according to size.

In constructing a stem-and-leaf display, one should include all possible stems between the largest and smallest stems, even if some of them do not have leaves. For instance, in the last example the stem of 6 would be listed even if the data had no values in the 60s.

> **TIP:** Sometimes each stem value is used twice and will have two rows in the display. This is often done when there are several data values with the same stem, and thus the row for that stem would be very long.

When a stem value is used twice, one of its rows is used for the leaves 0, 1, 2, 3, and 4, while the other stem row is for the leaves 5, 6, 7, 8, and 9. This is illustrated in the following example.

Ⓜ Example 2.8 Use MINITAB to construct a stem-and-leaf display for the 25 final examination scores.

Solution

The scores are stored in column C1.

```
SET C1
86 83 56 98 72 52 71 88 75 91 69 78 64
78 81 74 77 73 90 85 64 79 71 83 64
END
```

To have MINITAB construct a stem-and-leaf display, the command **STEM-AND-LEAF** is used. Since only the first four letters of a command are needed, we can just type **STEM.**

```
STEM C1
```

MINITAB will respond with the output given in Exhibit 2.3.

The stem-and-leaf display created by MINITAB places the stems in the second column, with the leaves appearing to the right. The first column gives a cumulative count of the number of leaves on that line and on those lines from either the beginning

Exhibit 2.3

```
MTB > STEM C1

Stem-and-leaf of C1          N = 25
Leaf Unit = 1.0

     1      5 2
     2      5 6
     5      6 444
     6      6 9
    11      7 11234
    (5)     7 57889
     9      8 133
     6      8 568
     3      9 01
     1      9 8
```

or the end of the displ   example, consider the third line for which 5 appears in the first column. The fiv   es that there are 5 values from the beginning line through the third line of the   . The 9 in the first column of the seventh row shows that there are 9 values f   t line through the last line. The line that contains the value midway from the   ig and the end of the display is highlighted with parentheses in the first colum   umber enclosed indicates how many values are represented on that line. The   umn is usually not included when a stem-and-leaf display is constructed by

In Exhib   ch stem value is used twice and has two rows in the display. To create a disp   ilar to Figure 2.5 in which each stem appears once, the subcommand **INCF   T** can be used to specify the distance between lines. The following commands   oduce the stem-and-leaf display in Exhibit 2.4.

```
STEM C
INCREM        10.
```

Exhibit 2.4

```
MTB >    M C1;
SUBC    CREMENT 10.

Stem    1-leaf of C1      N = 25
Lea     .it = 1.0

        5 26
        6 4449
    )   7 1123457889
        8 133568
    }   9 018
```

The first step in making a stem-and-leaf display is deciding how to split each data value into a stem and a leaf. Often the leaf will be just the last digit, as was the case with the final examination scores. However, for some data sets this procedure would create a very large number of stems and render such a choice impractical. For instance, earlier we considered the maximum monthly AFDC benefits paid by the 50 states to a family of 3. These values are repeated below.

118	356	515	409	510	359	264	310	366	354
779	601	304	207	528	350	539	412	159	492
293	319	342	190	532	325	266	402	184	249
202	275	288	416	120	486	371	503	376	517
633	263	381	359	282	424	309	200	603	360

The values range from a low of 118 to a high of 779. If the units digit is used as the leaf, then we would have 67 stems from 11 through 77. This would create too large a display and defeat the objective of condensing and summarizing the data. A better choice would be to use the 100s digit for the stem, the 10s digit for the leaf, and drop the units digit. Such a stem-and-leaf display created by MINITAB appears in Exhibit 2.5.

Exhibit 2.5

```
MTB > STEM C1;
SUBC> INCREMENT 100.

Stem-and-leaf of C1        N = 50
Leaf Unit = 10

    5     1 12589
   16     2 00046667889
  (16)    3 0011245555566778
   18     4 0011289
   11     5 0111233
    4     6 003
    1     7 7
```

In the display of Exhibit 2.5, we are told that the leaf unit equals 10. For instance, the data value of 528 has a leaf of 2 because this is its 10s digit. Because its stem is 5, it appears in stem row 5 and is represented by a 2. Note that the units digit is dropped and not rounded. The number 528 is treated by MINITAB as if it were 520.

Section 2.2 Exercises
Graphical Methods for Quantitative Data: Dotplots and Stem-and-Leaf Displays

2.22 The following dotplot was constructed to show the numbers of United States flags sold by a gift shop during the first 2 weeks of the Persian Gulf War in 1991.

List the numbers of sales for the 14 days.

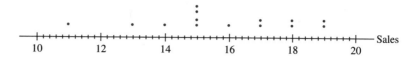

2.23 The dotplot below shows the times required by 18 laboratory animals to perform a complex task during a psychology experiment. The times were recorded to the nearest tenth of a second. List each of the 18 required times.

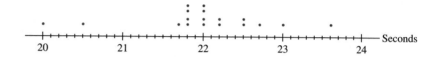

2.24 The following is a stem-and-leaf display of the weights of the players for a college baseball team. List the 25 individual weights.

Stem	Leaf
15	3
16	5 8 9
17	0 3 6 6 8 9 9
18	0 5 5 8 8
19	2 5 7
20	5 5 9
21	3
22	0
23	
24	4

2.25 The hourly wage rates of 21 part-time employees of a fast food restaurant are summarized in the following stem-and-leaf display. List the individual wage rates.

Stem	Leaf
42	5 5 5
43	
44	0 0 0 5
45	0 0 0 0 5
46	0 5
47	5 5 5
48	
49	
50	0 0 0 0

2.26 For its report on campus crime, *USA TODAY* surveyed 549 four-year colleges/
universities and published several statistics, one of which was the number of
assault cases during 1987 (*USA TODAY,* October 6, 1988). The numbers of
assaults for 34 colleges in New York State are given below.

7	2	3	6	8	1	2	16	5	5	3	11	17	0	3	14	19
9	2	18	8	0	3	2	3	3	2	1	1	28	2	7	38	5

Construct a stem-and-leaf display of the data.

2.27 Included in its report on campus crime (see Exercise 2.26), *USA TODAY* also
gave the students-per-crime ratio for each school. These figures, rounded to an
integral value, are given below for 37 schools in California.

17	26	45	39	13	35	26	50	49	47	47	31	66	19	26
104	177	11	10	16	29	37	27	194	5	10	9	15	17	8
14	12	11	17	14	13	6								

Construct a stem-and-leaf display of these figures.

2.28 Construct a dotplot of the numbers of assault cases given in Exercise 2.26.

2.29 During the 1988 Summer Olympics, over half the gold medals were awarded to
4 countries. The numbers received by 28 other countries are given below.

12	11	11	10	7	6	6	5	5	5	4	3	3	3
3	3	2	2	2	2	1	1	1	1	1	1	1	1

Construct a dotplot of these values.

2.30 The following figures are from a sample of 16 private colleges in Pennsylvania
and represent the percentages of applicants accepted for the 1987–1988 school
year.

58	81	38	75	73	57	70	70
64	74	47	55	74	60	43	50

Construct a dotplot of these values.

2.31 The ages of 40 blood donors were given in Exercise 2.2 and are repeated below.

35	53	61	43	21	42	23	29	35	37
39	58	27	64	27	31	36	48	41	22
37	35	42	32	43	34	59	50	38	43
31	30	41	37	29	45	23	56	46	41

Construct a dotplot of these ages.

2.32 Construct a stem-and-leaf display of the 16 percentages given in Exercise 2.30.

2.33 Construct a stem-and-leaf display of the 40 ages that appear in Exercise 2.31.

2.34 The following are the weights in ounces of 20 packages of bulk cheddar cheese
featured in a supermarket display.

19.7	21.4	22.8	19.1	22.7	20.9	22.3	20.5	20.4	21.1
18.8	19.5	21.7	18.8	19.5	22.1	23.0	20.5	22.2	19.8

a. Construct a stem-and-leaf display of the data.
b. Construct a stem-and-leaf display for which the leaves are arranged in order of magnitude.

2.35 The cumulative grade point averages are given below for 18 students in an honors section of probability and statistics.

3.67	3.53	3.29	3.71	3.60	3.45	3.74	3.41	3.39
3.50	3.64	3.88	3.75	4.00	3.90	3.73	4.00	3.62

a. Construct a stem-and-leaf display of the data.
b. Construct a stem-and-leaf display for which the leaves are arranged in order of magnitude.

MINITAB Assignments

Ⓜ 2.36 In the 1989 *Gas Mileage Guide* published by the U.S. Department of Energy, estimates of miles-per-gallon (mpg) are given for the following 2-wheel drive small pickup trucks equipped with a manual 5-speed transmission. All mpg estimates are for highway driving.

Truck	mpg
Chevrolet S10 (2.5 liter engine)	27
Chevrolet S10 (2.8 liter engine)	26
Dodge Ram 50 (2.0 liter engine)	28
Dodge Ram 50 (2.6 liter engine)	25
Ford Courier	26
Ford Ranger (2.3 liter engine)	27
Ford Ranger (2.9 liter engine)	23
GMC S15 (2.5 liter engine)	27
GMC S15 (2.8 liter engine)	26
Isuzu (2.3 liter engine)	24
Isuzu (2.6 liter engine)	24
Mazda B2200	26
Mazda B2600	26
Mitsubishi (2.0 liter engine)	28
Mitsubishi (2.6 liter engine)	25
Nissan (2.4 liter engine)	26
Nissan (3.0 liter engine)	24
Toyota (2.4 liter engine)	26
Toyota (3.0 liter engine)	24

Use MINITAB to construct a dotplot of the mileage estimates.

Ⓜ **2.37** The following are the cumulative grade point averages of the 47 members of the Math Club at a liberal arts college. Use MINITAB to construct a stem-and-leaf plot of these values.

3.29	3.67	3.28	2.89	3.01	2.94	3.31	2.80	3.54	3.15
3.08	3.01	3.17	3.36	3.57	3.39	3.32	3.19	2.95	3.19
2.84	3.16	3.21	3.26	3.36	3.37	3.39	3.75	3.41	3.97
2.87	2.99	3.02	3.09	3.87	3.14	3.18	3.24	3.41	3.61
2.94	3.06	2.94	3.66	3.77	3.72	3.51			

Ⓜ **2.38** In MINITAB the subcommand **INCREMENT** can be used with the **DOT-PLOT** command to specify the distance between the numerical markings on the axis. For example, the following commands will produce a dotplot for data in column C1, where the distance is 50. Since there are 10 spaces between numerical markings, each of the 10 spaces will denote 5 units.

```
DOTP C1;
INCR 50.
```

Use MINITAB to construct a dotplot of the following 79 SAT math scores, where the numerical markings are 100 units apart.

580	490	560	550	670	510	540	610	590	600
480	510	520	540	580	490	600	620	680	590
570	530	580	560	540	560	530	580	520	540
610	580	570	590	530	560	570	560	540	590
590	600	560	500	510	600	580	540	470	690
580	710	680	600	580	540	540	570	580	590
520	540	580	550	550	560	640	670	560	680
570	540	470	480	530	490	490	680	720	

Ⓜ **2.39** For the SAT scores in Exercise 2.38, use MINITAB to construct a stem-and-leaf plot with four stems representing 400, 500, 600, and 700.

2.3 Graphical Methods for Qualitative Data: Bar and Pie Charts

The methods discussed in Sections 2.1 and 2.2 were designed for describing quantitative data, that is, numerical data. A person's age, weight, and height are quantitative, as are the distance between two towns, the speed of a passing train, the cost of a college education, the time required to adjust the valves in a Porsche 911 engine, and the number of home runs hit during the 1992 World Series.

Contrasted with quantitative data are qualitative data that are categorical or attributive in nature. Some examples are the color of a person's hair, the classification system for grading the size of eggs, the country in which a person was born, the manufacturer of an automobile, a person's favorite flavor of ice cream, and the days of the week. In this section we will discuss two methods for describing and presenting data with a qualitative aspect.

Figure 2.6
1988 Summer Olympics gold medals awarded.

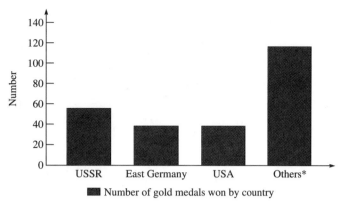

* "Others" consists of 28 countries

Bar Charts

Example 2.9

The 1988 Summer Olympics was held in Seoul, South Korea. Thirty-one countries received a total of 241 gold medals. Over half of these were awarded to just three countries: the United States (36), the former USSR (55), and the former East Germany (37). One way of displaying this information is by means of a **bar chart** in which the qualitative data (countries) are represented along a horizontal axis (*x*-axis). Above each country a vertical bar is drawn with a height equal to the number of gold medals received. The bar chart is shown in Figure 2.6.

Notice that the appearance of a bar chart is similar to a histogram, except that the bars are usually separated in a bar chart, whereas the rectangles in a histogram are connected. Also, the bar chart has qualitative values along the horizontal axis, while a histogram has quantitative values.

Table 2.8

1988 Summer Olympics

Country	Number of Gold Medals
U.S.S.R.	55
E. Germany	37
U.S.	36
Others	113

Pie Charts

The information presented in the bar chart of Figure 2.6 can also be displayed in the form of a **pie chart.** While both convey the same information, the pie chart more effectively shows the relationship of each qualitative category to the totality. This is accomplished by dividing a circle (the "pie") into sectors ("slices") and allocating these to the different categories of the qualitative variable. The procedure for constructing a pie chart is illustrated in Example 2.10.

Example 2.10

Construct a pie chart to display the information given in Table 2.8.

Solution

Since the qualitative variable (country) has four categories, the pie will be divided into four slices. To determine the size of the slice for each country, we must first determine the percentage of medals won by each. For example, the United States received 36 of the 241 medals awarded, which is

$$\frac{36}{241} = 14.9\%.$$

The sector of the pie allocated to the United States will be 14.9 percent of the total. A sector of a circle is measured by its central angle, and there are 360 degrees in the entire circle. Consequently, the slice of the pie associated with the United States will be

$$(14.9\%)(360 \text{ degrees}) = 54 \text{ degrees}.$$

The size of the slice for each of the other countries is computed in the same way, and these are given in Table 2.9. The pie chart appears in Figure 2.7.

Table 2.9

1988 Summer Olympics

Country	Number of Gold Medals	Percent of Medals	Size of Slice
U.S.S.R.	55	22.8%	82 degrees
E. Germany	37	15.4%	55 degrees
U.S.	36	14.9%	54 degrees
Others	113	46.9%	169 degrees
	241		

Figure 2.7
1988 Summer Olympics gold medals awarded.

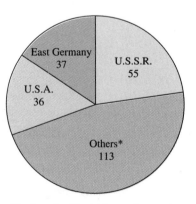

Number of gold medals won by country

* "Others" consists of 28 countries

(© Jay Freis/The Image Bank)

Example 2.11   Although U.S. field soldiers in the Persian Gulf War were generally quite satisfied with their ready-to-eat meals, there was one selection that they strongly disliked—chicken á la king (*Newsweek,* February 11, 1991). A major complaint lodged against this particular meal was its high salt content. The chicken á la king selection contained 1,446 milligrams of salt—nearly one-half the maximum daily amount (3,000 mg) recommended by the American Heart Association. The meal's components and their respective salt content can be displayed effectively with the bar chart in Figure 2.8 or, using the slice sizes given in Table 2.10, with the pie chart in Figure 2.9.

Figure 2.8
Bar chart of salt content by item. (Source: U.S. Defense Department.)

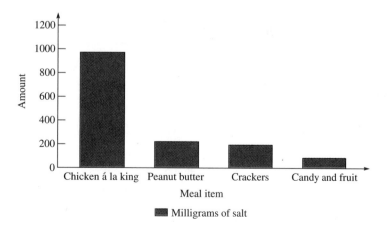

Table 2.10

Salt Content — Chicken á la king

Item	Milligrams Salt	Percent of Total	Size of Slice
Chicken á la king	965	66.7%	240 degrees
Peanut butter	218	15.1%	54 degrees
Crackers	184	12.7%	46 degrees
Candy and fruit	79	5.5%	20 degrees
	1,446		

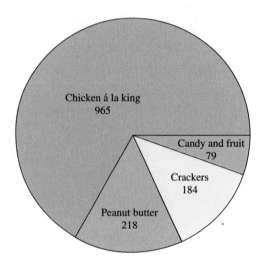

Figure 2.9
Pie chart of salt content by item. (Source: U.S. Defense Department.)

Section 2.3 Exercises

Graphical Methods for Qualitative Data: Bar and Pie Charts

2.40 In recent years the medical profession has intensified its efforts to inform the public that a high blood cholesterol level increases one's risk of heart disease. For someone in their 20s, the recommended level is under 180 mg/dl. Those with levels from 200 to 220 are considered at moderate risk, and those with levels above 220 are considered at high risk. Levels between 180 and 200 are borderline. Suppose a sample of 100 college students were tested and classified as follows:

Cholesterol Level	Number of Students
Recommended	25
Borderline	10
Moderate Risk	50
High Risk	15

a. Construct a bar chart to display the distribution.
b. Use a pie chart to present the distribution.

2.41 A history professor gave a final examination to a class of 20 students. The grade distribution consisted of 3 As, 5 Bs, 8 Cs, and 4 Ds. Use a bar chart to display this distribution.

2.42 The distribution of course grades by the history professor referred to in Exercise 2.41 consisted of 4 As, 5 Bs, 7 Cs, 3 Ds, and 1 F. Use a bar chart to present the grade distribution.

2.43 Use a pie chart to present the grade distribution in Exercise 2.41.

2.44 Display the distribution of course grades in Exercise 2.42 by using a pie chart.

2.45 *USA TODAY* (January 17, 1992) gave the following breakdown as to how workers in the United States spend their pay checks. The figures show the average amounts per $100 allocated to the various categories.

Housing	$41.36	Miscellaneous	$6.35
Transportation	17.80	Clothing	6.07
Food	17.71	Entertainment	4.32
Health care	6.39		

Display the distribution with a pie chart.

2.46 Ottis Anderson, the Most Valuable Player of Super Bowl XXV, was the leading rusher in the 1991 National Football League title game. His rushing performance of 102 yards was only the tenth time in Super Bowl history that the century mark has been broken. The table below shows the players who have accomplished this feat. Use a bar chart to display each player and his rushing yards.

Player	Yards
Tim Smith (Washington Redskins)	204
Marcus Allen (Los Angeles Raiders)	191
John Riggins (Washington Redskins)	166
Franco Harris (Pittsburgh Steelers)	158
Larry Csonka (Miami Dolphins)	145

(*Continued on next page*)

(*Continued*)

Player	Yards
Clarence Davis (Oakland Raiders)	137
Matt Snell (New York Jets)	121
Tom Matte (Baltimore Colts)	116
Larry Csonka (Miami Dolphins)	112
Ottis Anderson (New York Giants)	102

2.47 Sometimes a bar graph is displayed with horizontal bars drawn from the *y*-axis. This is usually more convenient when the categorical descriptions of the qualitative variable are lengthy, as is the case in Exercise 2.46. Construct a horizontal bar graph to display the information in that exercise.

2.48 Use a horizontal bar chart (see Exercise 2.47) to display the distribution of cholesterol levels given in Exercise 2.40.

2.49 In 1988, approximately $8 billion were spent in the United States on candies and chocolates. The following gives the market share by company (*Business Week,* October 30, 1989).

Company	% of Market Share
Hershey	20.5
M&M/Mars	18.5
Jacobs-Suchard	6.7
Nestle	6.7
Leaf	5.6
Others	42.0

Use a horizontal bar chart to display this information (see Exercise 2.47).

2.50 IBM had the largest share of the personal computer market in 1988. The following table shows the market share by manufacturer during that year (The *Wall Street Journal,* January 18, 1989).

Company	% of Market Share
IBM	27.9
Apple	12.2
Compaq	8.3
Zenith	5.4
Tandy	5.0
Others	41.2

Use a pie chart to display the distribution of market share.

2.51 The average monthly basic rates for cable TV and pay TV are given below for the years 1982 through 1986 (*Statistical Abstract of the United States,* 1990).

Year	Cable TV	Pay TV
1982	$8.46	$10.36
1983	$8.76	$10.25
1984	$9.20	$10.32
1985	$10.24	$10.53
1986	$11.09	$10.40

Display this information by constructing a double vertical bar chart. Place the years along the horizontal axis and the rates along the vertical axis. For each year draw vertical bars for cable TV and for pay TV.

2.4   Numerical Measures of Central Location

Graphical methods are useful for describing a set of data because they summarize the measurements in a form that affords the viewer a quick impression of the data's distribution. However, graphical methods are not always convenient or desirable. For example, suppose you take a biology exam and then encounter your instructor the next day in the bookstore. If you inquire how the class performed, you cannot expect the professor to produce a frequency distribution or histogram on the spot. Rather, you can probably expect a numerical value that describes the middle of the grade distribution. Descriptions of the center of a distribution are called **measures of central location** (sometimes called **averages**). Although there are many such measures, the

"The time may not be very remote when it will be understood that for complete initiation as an efficient citizen of one of the great complex world states that are now developing, it is necessary to be able to compute, to think in averages and maxima and minima, as it is now to be able to read and to write." H. G. Wells, *Mankind in the Making,* 1929. *(Photo courtesy of UPI/Bettmann Archives)*

mean is the most popular, and we will use it almost exclusively throughout the book. However, we will briefly look at a few other measures.

The Mean

The **mean** is the most commonly used measure of central location. It is easy to find, most people are familiar with it, and one usually thinks of the mean when the term **average** is mentioned. The mean has been more thoroughly researched than any other measure of central location, and it possesses several desirable properties. Consequently, many of the statistical methods discussed in subsequent chapters are based on the mean.

To illustrate the mean, consider again the biology student referred to earlier. Suppose she completed the course and obtained the following scores on the 5 exams that were administered during the semester.

$$91 \quad 93 \quad 94 \quad 91 \quad 99$$

The **mean** of the scores is obtained by summing all the scores and dividing by 5, the number of scores. To describe this more concisely with a formula, we need to introduce some notation. We will let the variable x denote an examination score. In this example, there are 5 values of x, namely, the 5 scores. Subscripts are used on a variable in order to represent its particular values. Here, we will let x_1 denote the first score, x_2 the second, x_3 the third, x_4 the fourth, and x_5 the fifth score.

$$x_1 = 91, x_2 = 93, x_3 = 94, x_4 = 91, x_5 = 99$$

Σ, the Greek letter sigma, is used to denote the sum of the values of a subscripted variable. In particular,

$$\Sigma x = x_1 + x_2 + x_3 + x_4 + x_5$$
$$= 91 + 93 + 94 + 91 + 99$$
$$= 468.$$

We can now give a formula for the mean of the exam scores. It is symbolized by $\bar{x}$ (read as "x bar") and equals

$$\bar{x} = \frac{\Sigma x}{5} = \frac{468}{5} = 93.6.$$

The **mean** of a sample of n measurements $x_1, x_2, x_3 \ldots x_n$ is denoted by $\bar{x}$ and equals

$$\bar{x} = \frac{\Sigma x}{n}. \tag{2.1}$$

Example 2.12 In Example 2.1 the final examination scores were given for 25 students enrolled in an advanced mathematical statistics course. The scores are repeated below. Find the mean score.

$$
\begin{array}{ccccc}
86 & 52 & 69 & 74 & 64 \\
83 & 71 & 78 & 77 & 79 \\
56 & 88 & 64 & 73 & 71 \\
98 & 75 & 78 & 90 & 83 \\
72 & 91 & 81 & 85 & 64 \\
\end{array}
$$

Solution

From Formula 2.1, the mean score equals

$$\bar{x} = \frac{\Sigma x}{n}$$

$$= \frac{1,902}{25}$$

$$= 76.08.$$

Ⓜ**Example 2.13**   Use MINITAB to obtain the mean of the final examination scores in Example 2.12.

Solution

The scores are first stored in column C1, and then the command **MEAN** is applied to the column.

```
SET C1
86 83 56 98 72 52 71 88 75 91 69 78 64
78 81 74 77 73 90 85 64 79 71 83 64
END
MEAN C1
```

In response, MINITAB will produce the output in Exhibit 2.6.

Exhibit 2.6

```
MTB > MEAN C1
   MEAN     =        76.080
MTB >
```

Numerical characteristics of a population are called parameters. When the population is large, parameters are usually unknown and must be estimated by computing similar characteristics of a sample. These are called statistics. For example, the sample mean (a statistic) is often used to estimate the population mean (a parameter). Parameters are usually denoted by Greek letters. For this reason we will use μ (read as "mu") to represent a population mean. It is not always clear whether a set of data constitutes a sample or a population. For instance, the 25 final exam scores in Example 2.12 would be a population if that class of students were the only one of interest. If, however, these scores were going to be the basis for making generalizations about a

A recent study by the ad agency D'Arcy Masius Benton & Bowles estimated that the mean number of cents-off coupons distributed yearly by manufacturers was 2,910 per household. The mean household income was $29,000, and the mean weekly grocery bill of $74 was reduced an average of $6 by using coupons (The *Wall Street Journal,* February 20, 1991).

larger group of students, then the 25 scores would be viewed as a sample. Unless stated otherwise, assume that all examples in this chapter pertain to samples.

The **sample mean** is a statistic and is denoted by $\bar{x}$.

The **population mean** is a parameter and is denoted by μ.

The Weighted Mean

Sometimes when computing the mean, we do not want all the measurements to carry the same level of importance. For example, consider again the five examination scores of the biology student referred to earlier.

$$91 \quad 93 \quad 94 \quad 91 \quad 99$$

We found the mean score to be 93.6. This was determined under the premise that all five scores are equally important. However, suppose the first score was obtained on a short quiz, the next three scores on regular exams, and the last score was from a comprehensive final examination. In this case, the instructor would probably assign a weight to each exam that would indicate its importance relative to the other exams. For instance, suppose the three regular exams are equally important and should count twice as much as the first exam. Further assume that the final exam is to count twice as much as each of the regular exams. We can take these considerations into account by computing a **weighted mean.** The first step is to assign a weight of 1 to the least importance score (exam one); then assign a weight of 2 to each of the three regular exams; finally, a weight of 4 is assigned to the final exam, since this counts twice as much as each regular exam.

Exam Scores: $x_1 = 91, x_2 = 93, x_3 = 94, x_4 = 91, x_5 = 99$

Weights: $w_1 = 1, \ w_2 = 2, \ w_3 = 2, \ w_4 = 2, \ w_5 = 4$

The weighted mean is obtained by multiplying each score by its weight, summing these, and dividing the result by the sum of the weights.

$$
\begin{aligned}
\bar{x} &= \frac{\Sigma xw}{\Sigma w} \\[6pt]
&= \frac{91 \cdot 1 + 93 \cdot 2 + 94 \cdot 2 + 91 \cdot 2 + 99 \cdot 4}{1 + 2 + 2 + 2 + 4} \\[6pt]
&= \frac{1{,}043}{11} \\[6pt]
&= 94.82
\end{aligned}
$$

In essence, the formula treats the data as though there were one score of 91, two scores of 93, two scores of 94, two additional scores of 91, and four scores of 99.

> The **weighted mean** of a set of measurements $x_1, x_2, x_3 \ldots x_n$ with relative weights $w_1, w_2, w_3 \ldots w_n$ is given by
>
> $$\bar{x} = \frac{\Sigma xw}{\Sigma w}.$$ (2.2)

The weighted mean formula has many applications, some of which will be seen in subsequent chapters. In this section, we will show how it can be used to approximate the mean of data grouped into a frequency distribution.

Example 2.14   In the 1988 *Forbes* Four Hundred issue of the 400 richest people in America, the following distribution of their ages was given.

Age	Number
Thirties	8
Forties	46
Fifties	88
Sixties	134
Seventies	78
Eighties	42
Nineties	4

To approximate the mean age of the 400 richest people, the class mark is used to represent the age of each person falling within that class. A weighted mean is then calculated, where the xs are the class marks and the weights are the corresponding class frequencies.

Age	Class Mark x	Frequency f	$x \cdot f$
30 up to 40	35	8	280
40 up to 50	45	46	2,070
50 up to 60	55	88	4,840
60 up to 70	65	134	8,710
70 up to 80	75	78	5,850
80 up to 90	85	42	3,570
90 up to 100	95	4	380
		400	25,700

$$\bar{x} = \frac{\Sigma xw}{\Sigma w} = \frac{\Sigma xf}{\Sigma f}$$

$$= \frac{25,700}{400}$$

$$= 64.25 \text{ years}$$

The Median

The **median** of a set of measurements is the middle value after the measurements have been ranked in order of size. Earlier we considered the five examination scores of a student enrolled in a biology course.

$$91 \quad 93 \quad 94 \quad 91 \quad 99$$

To find the median, we first arrange the numbers in an **array,** that is, in order of size.

$$91 \quad 91 \quad 93 \quad 94 \quad 99$$

Since 93 is the middle number, the median is

$$Md = 93.$$

Notice that the median is located in the third position of the array. In general, if there are n numbers, and they have been arranged in order, then the position of the median in the array is $(n + 1)/2$. For example, if there are 35 measurements, then the median is found by first arranging them in order, and then selecting the number that occupies the $(35 + 1)/2 = 18$th position in the array. The following example considers a situation for which n, the number of measurements, is even.

Example 2.15 Suppose the biology student referred to previously had taken one additional exam for which her score was 89. Then her six examination scores would be as follows.

$$91 \quad 93 \quad 94 \quad 91 \quad 99 \quad 89$$

To find the median score, first arrange the scores in numerical order.

$$89 \quad 91 \quad 91 \quad 93 \quad 94 \quad 99$$

Second, locate the **position** of the median in the array.

$$\text{Position} = \frac{n + 1}{2} = \frac{6 + 1}{2} = 3.5$$

Position 3.5 is interpreted to mean that the median is halfway between the values in the 3rd and 4th positions of the array. Thus, the median is the midpoint of 91 (the 3rd value) and 93 (the 4th value).

$$Md = \frac{91 + 93}{2} = 92$$

The **median** of a set of n measurements is the middle value when the measurements are ranked according to size. After the numbers have been arranged in order, the **position** of the median in the array is given as follows.

$$\text{Position} = \frac{n + 1}{2}$$

ⓂExample 2.16   Using MINITAB, obtain the median of the 25 final examination scores considered in Example 2.12.

Solution

After storing the scores in column C1, the **MEDIAN** command is used.

```
SET C1
86 83 56 98 72 52 71 88 75 91 69 78 64
78 81 74 77 73 90 85 64 79 71 83 64
END
MEDIAN C1
```

MINITAB will produce the output shown in Exhibit 2.7.

Exhibit 2.7

```
MTB > MEDIAN C1
   MEDIAN =        77.000
MTB >
```

Although the mean is the most popular measure of central location, the median may be preferred when the measurements contain an **outlier,** an unusually small or large value. The median is less sensitive to these extreme values. For instance, suppose the biology student's five exam scores had been 91, 93, 94, 20, and 99. The 20 is an outlier because it is atypical with respect to the others. The mean is greatly affected by this and has a value of only 79.4, while the median is 93. Many would feel that the median instead of the mean would be the more appropriate measure to use in determining the student's course grade. Because the 20 is so far out of line, it probably can be attributed to exceptional circumstances such as illness or a serious personal problem.

When the mean and the median of a data set are the same, the distribution of the measurements is **symmetric.** Otherwise the distribution is said to be **left skewed** or **right skewed** as illustrated in Figure 2.10.

Figure 2.10
A, Left-skewed; *B,* symmetric; and *C,* right-skewed distributions.

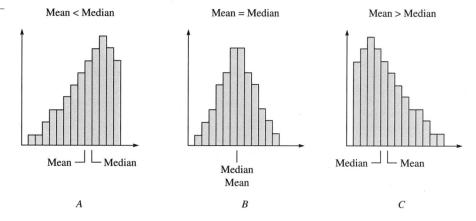

The Mode

The last measure of central location that we will consider is the **mode.** Unlike the mean and the median, the mode does not always exist.

> If a data set has a value that occurs more often than any of the others, then that value is said to be the **mode.**
>
> If two values occur the same number of times and more often than the others, the data set is said to be **bimodal.**
>
> The data set is **multimodal** if there are more than two values that occur with the same greatest frequency.

At the beginning of this section we gave the following scores for the five exams taken by a student in a biology course.

91   93   94   91   99

The modal score is 91 since this occurs twice.

> **TIP:**   Unlike the mean and the median, the mode is applicable to qualitative as well as quantitative data.

This versatility, and the fact that the mode is an indication of the value at which a distribution is most concentrated, make it particularly useful in marketing and inventory considerations. A small convenience store might be limited to stocking only one size of workmen's gloves, only one brand of paper towels, either white or brown eggs, and so on. In reaching a decision concerning the stocking of these items, the owner would probably be guided by the modal choices of his or her customers, that is, those products chosen most frequently.

Section 2.4 Exercises
Numerical Measures of Central Location

2.52 During his most recent visit to a supermarket, a shopper redeemed manufacturers' cents-off coupons with the following values.

$0.75 $0.40 $1.25 $0.50 $0.35 $0.25 $1.00 $0.75 $0.60

Determine the total amount saved and the mean redemption value of the coupons.

2.53 Two weeks ago, the shopper's wife in Exercise 2.52 redeemed 10 coupons that had a mean value of 75 cents. How much did she save with these coupons?

2.54 Only five managers have won 20 or more World Series games. Casey Stengel, Joe McCarthy, John McGraw, Connie Mack, and Walter Alston won 37, 30, 26, 24, and 20 games, respectively. Determine the mean number of World Series victories by this group.

2.55 Consider the sample 8, 3, 6, 1, 8. Find the (a) mean, (b) median, and (c) mode.

2.56 For the sample 7, 11, 8, − 6, 4, 0, 4, 12, find the (a) mean, (b) median, and (c) mode.

2.57 A mathematics department has 7 members, and their ages are 36, 49, 58, 27, 63, 32, and 41.
   a. Find the mean age.
   b. Give an example of a situation for which the data would be regarded as a sample. In this case, is the mean a parameter or a statistic, and what symbol should be used to denote it?
   c. Give an example of a situation for which the data would be considered a population. Would the mean be a parameter or statistic, and what symbol should be used to denote it?

2.58 The heights in inches of the members of a basketball team are 76, 69, 77, 80, 77, 71, 75, 76, 74, and 78.
   a. Find the mean height.
   b. For what situation could the data be considered as a sample? Would the mean be a parameter or a statistic, and what symbol should be used to denote it?
   c. For what situation could the data be considered as a population? In this case, is the mean a parameter or a statistic, and what symbol should be used to denote it?

2.59 A law firm consists of 6 attorneys, and their present salaries are as follows (in dollars).

$$\begin{array}{cccccc} 5\,4 & 3 & 2 & 4 & & 1 \\ 55{,}000 & 52{,}000 & 49{,}000 & 54{,}000 & 198{,}000 & 45{,}000 \end{array}$$

   a. Determine the mean salary.
   b. Determine the median salary.
   c. Which measure of central location do you believe more accurately describes the "typical" salary at this firm?
   d. Which average do you think the firm's president would use if he wishes to create an image of prosperity concerning his staff?

2.60 Last year the mean number of cars sold per month by a dealership was 25. For the months of January through November, the number of sales were as follows.

$$13 \quad 11 \quad 16 \quad 29 \quad 32 \quad 39 \quad 34 \quad 33 \quad 28 \quad 29 \quad 21$$

How many cars were sold in December by this dealership?

2.61 A dance hall floor can safely support a maximum weight of 30,000 pounds. Recently there were 197 dancers on the floor, and their mean weight was 156 pounds. Determine if the patrons' safety was being jeopardized at that time.

2.62 A restaurant offers a luncheon special that includes a choice of apple, lemon, or custard pie. The dessert choices of the last 30 customers purchasing the special were as follows.

lemon	lemon	apple	lemon	apple	apple
custard	apple	custard	custard	lemon	apple
custard	apple	lemon	apple	custard	custard
lemon	apple	apple	lemon	custard	apple
apple	apple	lemon	lemon	lemon	custard

Find the modal choice.

2.63 During the fall semester, a mathematics department offered three sections of calculus. The mean grades of the sections on the final examination were 78.0, 84.6, and 70.0, and the class sizes were 28, 25, and 21, respectively. Determine the mean exam grade for the 74 students.

2.64 In 1991, a woman purchased 10 shares of IBM stock at a price of $117.50 per share. Later in the year she bought 15 shares at $110.75. When the price dropped to $98, she purchased an additional 25 shares. Find the mean price paid for the 50 shares.

2.65 Last semester, a student took 5 courses and received the following grades:

Course	Credit Hours	Grade
Hist 105	3	C
Math 121	4	A
Phys 111	4	B
Engl 105	3	C
PhEd 151	1	D

Assuming that an A, B, C, and D are 4, 3, 2, and 1 point, respectively, determine the student's grade point average for the semester.

2.66 One hundred families were surveyed, and the number of children, x, was recorded for each family. The results were as follows:

x	Frequency
0	37
1	22
2	19
3	12
4	8
5	2

Determine the mean number of children per family.

2.67 For drivers aged 25 through 49, the following table gives the number of arrests (in hundreds) in the United States during 1986 for driving under the influence. Use the weighted mean formula to approximate the mean age of this group.

	Age	Arrests
27	25–29	220
32	30–34	158
37	35–39	111
42	40–44	72
47	45–49	49

Source: U.S. Bureau of
Justice Statistics, *Drunk
Driving, Special Report.*

2.68 The following is a frequency distribution of the times required to complete the final examination for a class of students in an introductory statistics course. The figures are in minutes.

Time	Number of Students	
81–85	1	83
86–90	1	
91–95	2	
96–100	3	
101–105	3	
106–110	7	
111–115	10	
116–120	6	

Use the weighted mean formula to approximate the mean time required to complete the exam.

MINITAB Assignments

Ⓜ 2.69 *USA TODAY* (May 25, 1990) gave the following state percentages of motorists who use their seat belts.

AL	AK	AZ	AR	CA	CO	CT	DE	FL	GA
31.0	45.0	48.0	30.1	66.0	47.0	54.8	43.0	55.2	38.8
HI	ID	IL	IN	IA	KS	KY	LA	ME	MD
80.5	33.7	40.5	47.4	59.0	52.0	20.5	40.6	34.4	67.0
MA	MI	MN	MS	MO	MT	NE	NV	NH	NJ
28.0	45.6	44.1	17.0	54.1	63.0	32.0	38.4	50.1	44.1
NM	NY	NC	ND	OH	OK	OR	PA	RI	SC
51.7	60.0	62.0	28.0	44.4	36.3	48.0	49.5	23.9	37.9
SD	TN	TX	UT	VT	VA	WA	WV	WI	WY
26.0	41.0	63.0	44.2	35.0	55.0	55.4	42.0	50.3	35.5

a. Use MINITAB to determine the mean of the 50 percentages.
b. The mean percentage obtained in part (a) does not give the percentage of motorists in the 50 states who use their seat belts. Explain how this percentage could be obtained by using the weighted mean formula.

Ⓜ **2.70** Use MINITAB to obtain the median of the 50 percentages given in Exercise 2.69.

Ⓜ **2.71** A certain candy bar has a label weight of 75 grams. A sample of 40 bars from a production run had the following weights in grams. Use MINITAB to obtain the mean and the median weight of the sample.

75.4	68.8	71.3	74.4	77.0	73.0	70.8	75.2	70.4	73.6
73.4	75.2	73.7	77.2	74.9	73.1	68.8	73.5	77.9	76.2
75.2	74.1	73.9	75.8	74.1	74.3	70.5	74.0	73.5	71.6
75.6	76.1	71.9	71.1	75.4	74.0	77.4	76.4	78.8	74.2

2.5   Numerical Measures of Variability

In the previous section, we saw that it is important to use some type of measure of central location when describing a distribution of measurements. The most frequently used measure is the mean, and it is the one that will be employed in the remainder of this text. However, describing only the center of a distribution is incomplete, because it fails to consider another significant characteristic—variability. It is important to know how a set of measurements spreads out or fluctuates. In this section, we will discuss three such measures of variation, namely, the **range, variance,** and **standard deviation.**

The Range

To illustrate the concept of variation and its associated measures, consider the following situation. A chain of drug stores sells canned mixed nuts that it purchases from a distributor and then markets under its own label. The chain is considering two possible suppliers: A and B. It samples five cans from each supplier and counts the number of peanuts in each can. The results are given below:

Supplier A: 21   20   19   20   20

Supplier B: 29   11   10   33   17

Notice that the two sets of measurements have the same mean of 20 peanuts but differ dramatically in terms of their variability. The numbers for supplier A are all close to 20, while those for B fluctuate considerably from the mean (see the dotplots in Figure 2.11).

Figure 2.11
Dotplots of the peanut counts for suppliers *A* and *B*.

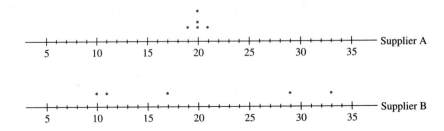

A simple way of measuring this variability is the **range,** which is the difference between the largest and smallest values.

Supplier A: 21   20   19   20   20.   Range $= 21 - 19 = \ 2$

Supplier B: 29   11   10   33   17.   Range $= 33 - 10 = 23$

> For a set of measurements, the **range** equals the difference between the largest and smallest values.
>
> $$\text{Range} = \text{largest} - \text{smallest}$$

Notice that the range gives the amount of spread between the largest and smallest values. The range is simple in concept, easily determined, and is used with industrial quality control charts. Intuitively, however, we would feel that a good measure of variation should involve all the data values. The most commonly used measures of variability are the variance and standard deviation.

The Variance

The **variance** measures how a set of measurements fluctuate relative to their mean. To find the variance, we first determine the amount by which each value x deviates from the mean, $(x - \bar{x})$. We will illustrate this for the data consisting of the number of peanuts in the five cans from supplier B. The deviations from the mean appear in Table 2.11.

One might think that a good way of measuring the variation in the values of x would be to average the deviations from the mean. But note from Table 2.11 that the sum of the deviations is 0. This will always be the case for every possible data set. Numbers greater than the mean will have positive deviations, those smaller than the mean will have negative deviations, and the positive and negative deviations will cancel each other. To circumvent this balancing of the positive and negative deviations, we could average their absolute values and obtain the **mean absolute deviation.** However, this is difficult to work with mathematically and is limited in its usefulness. A better alternative is to work with the **sum of the squared deviations from the**

Table 2.11

Deviations from the Mean

x	$(x - \bar{x})$
29	9
11	-9
10	-10
33	13
17	-3
$\Sigma x = 100$	$\Sigma(x - \bar{x}) = \quad 0$
$\bar{x} = \ 20$	

mean. This quantity will be denoted by $SS(x)$ and is the basis for the definition of **variance** that follows.

For a sample of n measurements $x_1, x_2 \ldots x_n$, the **sample variance** is denoted by s^2 and is given by

$$s^2 = \frac{SS(x)}{n-1},$$

where

$$SS(x) = \Sigma(x - \bar{x})^2. \tag{2.3}$$

Note: If $x_1, x_2 \ldots x_N$ constitute a population of N measurements, then the sum of the squared deviations $SS(x)$ is divided by N, and the **population variance** is denoted by σ^2.

Notice that the sample variance is almost an average of the sum of the squared deviations from the mean. Instead of dividing by n, however, we divide by $n-1$. This is done because the sample variance s^2 is used to estimate the population variance σ^2. Theory shows that one can expect better estimates by using a divisor of $n-1$.

Example 2.17   Find the sample variance of the numbers of peanuts in the five cans of mixed nuts from supplier B.

Supplier B: 29   11   10   33   17

Solution

The calculations are facilitated by constructing Table 2.12.

Table 2.12

Calculations for Obtaining $SS(x)$

x	$(x - \bar{x})$	$(x - \bar{x})^2$
29	9	81
11	-9	81
10	-10	100
33	13	169
17	-3	9
$\Sigma x = 100$		$\Sigma(x - \bar{x})^2 = 440$
$\bar{x} = 20$		

$$SS(x) = \Sigma(x - \bar{x})^2 = 440$$

$$s^2 = \frac{SS(x)}{n-1} = \frac{440}{4} = 110$$

The variance of the numbers of peanuts in the five cans from supplier A can be shown to be $s^2 = 0.5$. The variance of the sample for supplier B is many times this because of the much greater variability in the values (see Figure 2.11). The magnitude of the variance is tied to the variability in the sample, increasing with greater variation.

TIP: Formula 2.3 implies that the smallest possible value for the variance is 0. This will occur when there is no variation in the data, that is, when all the measurements are the same.

$SS(x)$, the sum of the squared deviations, was easy to calculate in Example 2.17 because the mean came out to be a whole number, thus making the deviations from the mean also whole numbers. Often the calculations will not work out this conveniently, and consequently, $SS(x)$ is usually calculated by using the following shortcut formula.

Shortcut Formula for Finding $SS(x)$

$$SS(x) = \Sigma x^2 - \frac{(\Sigma x)^2}{n}$$

(2.4)

Notice that the shortcut formula does not require that one obtain deviations from the mean. Only Σx^2, the sum of the squared measurements, and Σx, the sum of the measurements, are needed.

Example 2.18   Use the shortcut formula to determine the variance of the following sample: 9, 1, 0, -10, 8, 7.

Solution

To find s^2, the sample variance, we first need to obtain $SS(x)$.

x	x^2
9	81
1	1
0	0
-10	100
8	64
7	49
$\Sigma x = 15$	$\Sigma x^2 = 295$

$$SS(x) = \Sigma x^2 - \frac{(\Sigma x)^2}{n} = 295 - \frac{(15)^2}{6}$$

$$= 295 - 37.5 = 257.5$$

Thus, the sample variance is

$$s^2 = \frac{SS(x)}{n-1} = \frac{257.5}{5}$$

$$= 51.5.$$

The Standard Deviation

At this stage, we can calculate the variance of a sample, and we realize that it measures in some way the variation in the data. To obtain a practical interpretation, we need to work with its positive square root, which is called the **standard deviation.** The standard deviation is used frequently throughout the remainder of the text and has the same units of measurement as the data. This is because the variance involves the square of the measurements, and the standard deviation is the square root of the variance. For example, if the measurements involve inches, the standard deviation will also be in inches.

> The **standard deviation** of a set of measurements is equal to the positive square root of the variance.
>
> $$\text{Standard deviation} = \sqrt{\text{variance}}$$
>
> The **standard deviation of a sample** is denoted by s.
> The **standard deviation of a population** is denoted by σ.

Example 2.19   According to the U.S. Department of Agriculture, the per capita consumption of fresh fruits for the years 1980 through 1984 were 86.7, 83.9, 83.9, 88.0, and 87.5 pounds, respectively. Find the standard deviation of these amounts.

Solution

To find s we first need to obtain $SS(x)$ and from this, the variance. Note that s will be in pounds, since the data are given in these units.

x	x^2
86.7	7,516.89
83.9	7,039.21
83.9	7,039.21
88.0	7,744.00
87.5	7,656.25
$\Sigma x = 430.0$	$\Sigma x^2 = 36,995.56$

$$SS(x) = \Sigma x^2 - \frac{(\Sigma x)^2}{n} = 36,995.56 - \frac{(430)^2}{5}$$

$$= 15.56$$

The variance is

$$s^2 = \frac{SS(x)}{n-1} = \frac{15.56}{4} = 3.89$$

Thus, the standard deviation is

$$s = \sqrt{3.89} = 1.97 \text{ pounds.}$$

ⓜExample 2.20 At the beginning of this chapter we gave for each of the 50 states the maximum monthly AFDC (aid for dependent children) benefit in 1988 for a family of three. Use MINITAB to find the mean, median, and standard deviation of these amounts.

118	356	515	409	510	359	264	310	366	354
779	601	304	207	528	350	539	412	159	492
293	319	342	190	532	325	266	402	184	249
202	275	288	416	120	486	371	503	376	517
633	263	381	359	282	424	309	200	603	360

Solution

The data are first stored in column C1.

```
SET C1
118 779 293 202 633 356 601 319 275 263 515 304 342 288 381
409 207 190 416 359 510 528 532 120 282 359 350 325 486 424
264 539 266 371 309 310 412 402 503 200 366 159 184 376 603
354 492 249 517 360
END
```

To obtain the mean, median, and standard deviation, type the following.

```
MEAN C1
MEDIAN C1
STDEV C1
```

MINITAB will produce the output given in Exhibit 2.8.

Exhibit 2.8

```
MTB > MEAN C1
    MEAN    =       368.04
MTB > MEDIAN C1
    MEDIAN =       357.50
MTB > STDEV C1
    ST.DEV. =      139.86
```

In the next section we will consider the practical significance of the standard deviation, and many applications of it will be encountered as we progress through the text.

Section 2.5 Exercises

Numerical Measures of Variability

2.72 The numbers of bicycles and cars in 1985 are given below for four countries (*Bicycle Federation of America, Motor Vehicle Manufacturers Association,* and *International Trade Centre*). The figures are in millions.

Country	Bicycles	Cars
United States	95	132
Japan	58	28
West Germany	45	26
India	45	2

Guess which sample is more variable, and then check by finding the standard deviation for the numbers of bicycles and for the numbers of cars.

2.73 Consider the sample 9, 1, 0, 4, 6. Find the (a) range, (b) variance, and (c) standard deviation.

2.74 For the sample 8, 10, 3, 6, -6, 9, find the (a) range, (b) variance, and (c) standard deviation.

2.75 Consider the following sample:

$$5 \quad 8 \quad 2 \quad 1 \quad 7 \quad 1 \quad 0$$

a. Calculate $SS(x)$ by using its defining equation; that is, by finding the sum of the squared deviations of each value from the mean.
b. Calculate $SS(x)$ by using the shortcut formula.
c. Find the sample standard deviation.

2.76 Use two different methods to calculate $SS(x)$ for the following sample. Then find the standard deviation:

$$10 \quad 9 \quad 11 \quad 8 \quad 6 \quad 10 \quad 12 \quad 9$$

2.77 The average hourly wages (in dollars) in 1988 are given below for five Asian countries. Calculate the range and the standard deviation of the wages.

Country	Hourly Wage
Japan	13.14
Taiwan	2.71
Singapore	2.67
South Korea	2.46
Hong Kong	2.43

Source: *U.S. Bureau of Labor Statistics.*

2.78 The average hourly wages (in dollars) in 1988 are given below for five European countries. Calculate the range and the standard deviation of the wages.

Country	Hourly Wage
Norway	19.43
West Germany	18.07
Switzerland	17.94
Sweden	16.85
Britain	10.56

Source: *U.S. Bureau of Labor Statistics.*

2.79 Consider the following two samples:

Sample A: 5   8   9   1   0   10   2
Sample B: 1   4   4   6   11   4   5

a. Show that the samples have the same mean and range.
b. By visual inspection, which sample do you think is more variable?
c. Check your answer to Part b by determining the standard deviation of each sample.
d. Does the range or the standard deviation provide a better indication of variability? Explain.

2.80 Adding (or subtracting) the same number from each measurement in a sample does not alter the standard deviation. Coding the data in this way can often be used to simplify the calculation of s. To illustrate, consider the following weights for a sample of six male college students.

178   169   163   158   170   183

a. Calculate the standard deviation of the sample.
b. Subtract 173 from each weight and calculate the standard deviation of the resulting numbers. Note that the result is the same as that for Part a.

2.81 If each measurement in a sample is multiplied by the same constant k, the standard deviation of the resulting sample will equal k times the standard deviation of the original sample. To illustrate, consider the following numbers.

0.013   0.009   0.011   0.013   0.009

a. Calculate the standard deviation of the sample.
b. Multiply each measurement by 1,000, and find the standard deviation of the resulting numbers. How could this value be used to obtain the standard deviation of the original sample?

2.82 One hundred families were surveyed, and the number of children x was recorded for each family. The results are as follows.

x	frequency f
0	37
1	22
2	19
3	12
4	8
5	2

a. Determine the range of the 100 numbers.

b. Find the standard deviation of the sample. Hint:

$$SS(x) = \Sigma x^2 f - \frac{(\Sigma xf)^2}{n},$$

where $n = \Sigma f$.

2.83   A flooring specialist has received a large shipment of boxes of ceramic tiles. Each box contains 36 tiles. She samples 42 boxes and finds the following numbers of broken tiles per box.

```
2  5  1  0  2  4  3  6  2  1  8  2  6  3
2  3  0  0  0  4  8  5  0  1  0  2  3  6
2  0  0  0  0  4  2  1  0  2  5  3  1  3
```

First group the data as was done in Exercise 2.82, and then use the hint to find the standard deviation of the sample.

MINITAB Assignments

Ⓜ 2.84   The figures below are the per capita amounts (in dollars) that the 50 states spent for public libraries in 1989 (*National Center for Education Statistics*).

AL	AK	AZ	AR	CA	CO	CT	DE	FL	GA
8.13	23.64	14.53	5.98	15.89	18.01	22.10	8.22	11.93	10.06
HI	ID	IL	IN	IA	KS	KY	LA	ME	MD
17.23	9.84	19.25	17.95	11.85	15.85	7.04	11.53	12.66	24.45
MA	MI	MN	MS	MO	MT	NE	NV	NH	NJ
19.18	13.83	18.62	6.58	13.37	7.56	13.16	9.65	14.46	21.16
NM	NY	NC	ND	OH	OK	OR	PA	RI	SC
11.34	29.48	10.55	8.01	23.34	10.24	14.74	10.99	11.71	8.72
SD	TN	TX	UT	VT	VA	WA	WV	WI	WY
13.09	7.70	8.94	16.64	12.37	16.29	19.81	7.46	16.53	19.83

Use MINITAB to calculate the mean and the standard deviation of the 50 expenditures.

Ⓜ 2.85   The following are the cumulative grade point averages of the 47 members of the Math Club at a liberal arts college. Use MINITAB to obtain the mean, median, and the standard deviation of the grade point averages.

3.29	3.67	3.28	2.89	3.01	2.94	3.31	2.80	3.54	3.15
3.08	3.01	3.17	3.36	3.57	3.39	3.32	3.19	2.95	3.19
2.84	3.16	3.21	3.26	3.36	3.37	3.39	3.75	3.41	3.97
2.87	2.99	3.02	3.09	3.87	3.14	3.18	3.24	3.41	3.61
2.94	3.06	2.94	3.66	3.77	3.72	3.51			

Ⓜ 2.86 The following are the state percentages of motorists who use their seat belts (*USA TODAY,* May 25, 1990). Use MINITAB to calculate the standard deviation of these figures.

AL	AK	AZ	AR	CA	CO	CT	DE	FL	GA
31.0	45.0	48.0	30.1	66.0	47.0	54.8	43.0	55.2	38.8
HI	ID	IL	IN	IA	KS	KY	LA	ME	MD
80.5	33.7	40.5	47.4	59.0	52.0	20.5	40.6	34.4	67.0
MA	MI	MN	MS	MO	MT	NE	NV	NH	NJ
28.0	45.6	44.1	17.0	54.1	63.0	32.0	38.4	50.1	44.1
NM	NY	NC	ND	OH	OK	OR	PA	RI	SC
51.7	60.0	62.0	28.0	44.4	36.3	48.0	49.5	23.9	37.9
SD	TN	TX	UT	VT	VA	WA	WV	WI	WY
26.0	41.0	63.0	44.2	35.0	55.0	55.4	42.0	50.3	35.5

Ⓜ 2.87 The following are the median purchase prices of existing one-family houses in 32 metropolitan areas for the years 1980 and 1987. The figures are in thousands of dollars. Use MINITAB to obtain the mean, median, and standard deviation of the prices for each of the two years.

Metropolitan Area	**1980**	**1987**
Atlanta, GA	81.8	131.1
Baltimore, MD	69.5	125.4
Boston-Lawrence-Lowell, MA-NH	70.5	186.2
Chicago-Gary, IL-IN	78.0	116.9
Cleveland-Akron-Lorain, OH	63.8	89.4
Columbus, OH	65.7	92.9
Dallas, TX	89.4	128.5
Denver-Boulder, CO	80.0	139.8
Detroit-Ann Arbor, MI	67.9	93.6
Greensboro-Winston Salem-High Point, NC	63.6	110.0
Honolulu, HI	119.3	177.6
Houston-Galveston, TX	85.7	106.8
Indianapolis, IN	60.3	94.7
Kansas City, MO-KS	64.2	105.8
Los Angeles-Long Beach-Anaheim, CA	110.7	167.7
Louisville, KY-IN	58.5	85.6
Miami-Fort Lauderdale, FL	75.3	103.9
Milwaukee-Racine, WI	84.9	95.7

(*Continued on next page*)

(*Continued*)

Metropolitan Area	1980	1987
Minneapolis-St. Paul, MN-WI	80.9	139.4
New York-Newark-Jersey City, NY-NJ-CT	93.8	181.8
Philadelphia-Wilmington-Trenton, PA-DE-NJ-MD	62.6	113.7
Phoenix, AZ	92.7	132.9
Pittsburgh, PA	64.6	82.8
Portland, OR-WA	75.5	108.5
Rochester, NY	57.5	79.6
St. Louis, MO-IL	56.5	92.1
Salt Lake City-Ogden, UT	73.2	121.0
San Diego, CA	105.7	156.7
San Francisco-Oakland-San Jose, CA	122.9	179.9
Seattle-Tacoma, WA	82.2	127.7
Tampa-St. Peterburg, FL	64.0	100.3
Washington, DC-MD-VA	100.1	159.5

Source: Federal Home Loan Bank Board, *Savings and Home Financing Source Book,* annual.

2.6 Understanding the Significance of Standard Deviation

The standard deviation is the most common way of measuring the variation in a set of measurements. For two data sets, the one that has the greater variability will have the larger standard deviation. The utility of the standard deviation is not limited just to

Pursuing zero defects using six sigma.

Six Sigma is a strategy devised by Motorola in 1987 to reduce the number of manufacturing defects to nearly zero. Three years after its implementation, the program produced a $500 million savings. Digital Equipment Corporation, Raytheon, Corning, IBM, and several other U.S. companies have since initiated Six Sigma programs to improve their manufacturing quality. The theoretical basis for the Six Sigma strategy is an extension of the Empirical rule to a six-standard deviation interval from a distribution's mean. Sigma is used in the strategy's name because the Greek letter sigma is the symbol for denoting a population's standard deviation. (*The New York Times,* Jan. 13, 1991.) (*Photo courtesy of General Motors Corporation.*)

Figure 2.12
Some mound-shaped distributions.

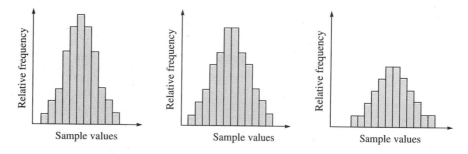

comparisons of data sets. In this section we will see that it is also useful in describing a single distribution of measurements. This description will be accomplished by examining two statements that relate the standard deviation to the percentage of measurements near their mean. First we will consider the **Empirical rule** and then examine **Chebyshev's theorem.**

The Empirical Rule

The **Empirical rule** assumes that the distribution of measurements is **mound-shaped.** By this we mean that the distribution is approximately symmetric, and the measurements tend to cluster near the center and drop off sharply toward the ends of the distribution. Some examples are given in Figure 2.12. Mound-shaped distributions, also called bell-shaped distributions, occur frequently in life and will be discussed in detail in Chapter 7.

The Empirical Rule:
If a set of measurements has a mound-shaped distribution, then

a. The interval from $\bar{x} - s$ to $\bar{x} + s$ will contain approximately 68 percent of the measurements.
b. The interval from $\bar{x} - 2s$ to $\bar{x} + 2s$ will contain approximately 95 percent of the measurements.
c. The interval from $\bar{x} - 3s$ to $\bar{x} + 3s$ will contain approximately all the measurements.

The Empirical rule tells us what percentage of the measurements can be expected to lie within one, two, and three standard deviations of the mean of a mound-shaped distribution.

Example 2.21   A manufacturer of oat bran cereal sampled 300 boxes to check the amount of bran being dispensed by its filling machine. It found a mean weight of 16.5 ounces and a standard deviation of 0.2 ounce. Use the Empirical rule to describe the distribution of weights.

Solution

Weight distributions of containers filled by automated equipment are usually mound shaped. Consequently, it is reasonable to assume that the Empirical rule is applicable in this situation. Therefore, we can expect that:

a. Approximately 68 percent of the boxes have weights falling in the interval 16.5 ± 0.2, i.e., from 16.3 to 16.7 ounces.
b. Approximately 95 percent of the boxes have weights falling in the interval 16.5 ± 0.4, i.e., from 16.1 to 16.9 ounces.
c. Approximately 100 percent of the boxes have weights falling in the interval 16.5 ± 0.6, i.e., from 15.9 to 17.1 ounces.

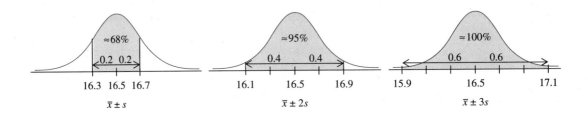

The Empirical rule assumes that a distribution is mound shaped. Even when this is not the case because the distribution is slightly skewed, the Empirical rule often gives good approximations for the percentage of measurements falling within 1, 2, and 3 standard deviations of the mean. This is illustrated in the following example.

Example 2.22   Earlier, MINITAB was used to obtain the mean, median, and standard deviation of the 50 states' maximum AFDC monthly benefits for a family of three. The results are reproduced in Exhibit 2.9, and the histogram of the 50 amounts appears in Figure 2.13.

   The distribution is skewed slightly to the right (positively skewed) because the mean ($368) is greater than the median ($358). In spite of this, Table 2.13 shows that the Empirical rule still provides good approximations for the percentage of measurements falling within 1, 2, and 3 standard deviations of the mean.

Table 2.13

Empirical Rule Applied to the AFDC Benefits

Interval	Number of Values in the Interval	%
$\bar{x} \pm s = 368 \pm 140 = 228$ to 508	32	64
$\bar{x} \pm 2s = 368 \pm 280 = 88$ to 648	49	98
$\bar{x} \pm 3s = 368 \pm 420 = -52$ to 788	50	100

Figure 2.13
Histogram for the 50 AFDC benefits.

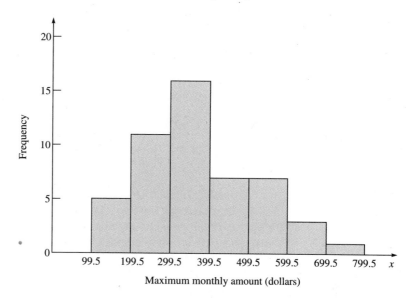

Maximum monthly amount (dollars)

Exhibit 2.9

```
MTB > MEAN C1
    MEAN    =       368.04
MTB > MEDIAN C1
    MEDIAN =        357.50
MTB > STDEV C1
    ST.DEV. =       139.86
```

Chebyshev's Theorem

When the shape of a distribution is such that the Empirical rule is not applicable, one can always resort to **Chebyshev's theorem.** It is more general than the Empirical rule because it applies to *all* data sets.

> **Chebyshev's Theorem**
> For any set of measurements and any number $k \geq 1$, the interval from $\bar{x} - ks$ to $\bar{x} + ks$ will contain at least $\left(1 - \dfrac{1}{k^2}\right)$ of the measurements.

In Chebyshev's theorem the number k can be chosen to be any value that is one or more, and k does not have to be a whole number. In the Empirical rule, k is either one, two, or three.

Example 2.23   What does Chebyshev's theorem say about the percentage of measurements that will fall within 2 standard deviations of their mean? Within 2.5 standard deviations?

Solution

For a choice of $k = 2$, Chebyshev's theorem says at least $\left(1 - \dfrac{1}{2^2}\right)$ of the values will lie within the interval $\bar{x} \pm 2s$. That is to say, *at least* three-fourths (75 percent) will be between $\bar{x} - 2s$ and $\bar{x} + 2s$.

For $k = 2.5$, at least $\left(1 - \dfrac{1}{2.5^2}\right)$ will fall within the interval $\bar{x} \pm 2.5s$, that is, *at least* 84 percent of the values will be from $\bar{x} - 2.5s$ to $\bar{x} + 2.5s$.

It should be emphasized that Chebyshev's theorem gives the *minimum* proportion of the measurements that will lie within k standard deviations of their mean. For instance, in the previous example we saw that *at least* 75 percent will fall within a two–standard deviation distance of the mean. Usually the actual percentage is much greater than this figure. Chebyshev's theorem is very conservative, since it applies to all possible distributions. Because of its extreme conservatism, Chebyshev's theorem is not widely used in practical applications. Its main value is in theoretical considerations.

Section 2.6 Exercises
Understanding the Significance of Standard Deviation

2.88 What kind of data sets can be described by Chebyshev's theorem?

2.89 What kind of data sets can be described with the Empirical rule?

2.90 What does Chebyshev's theorem say about the percentage of measurements that will lie within each of the following intervals?
   a. From $\bar{x} - s$ to $\bar{x} + s$.
   b. From $\bar{x} - 2s$ to $\bar{x} + 2s$.
   c. From $\bar{x} - 3s$ to $\bar{x} + 3s$.

2.91 When the Empirical rule is applicable, what can we say about the percentage of measurements that will lie within each of the intervals given in Exercise 2.90?

2.92 What does Chebyshev's theorem say about the percentage of measurements that will lie within each of the following intervals?
   a. From $\bar{x} - 1.5s$ to $\bar{x} + 1.5s$.
   b. From $\bar{x} - 2.75s$ to $\bar{x} + 2.75s$.
   c. From $\bar{x} - 10s$ to $\bar{x} + 10s$.

2.93 In one state, automobile insurance premiums average $343 with a standard deviation of $35. What can we say about the percentage of policies that have premiums between $203 and $483?

2.94 A taxi company has found that its fares average $7.80 with a standard deviation of $1.40. What can we say about the percentage of fares that are between $5.00 and $10.60 if
   a. the distribution of fares is mound shaped?
   b. the distribution of fares is not mound shaped?

2.95 A bakery makes loaves of rye bread that have an average weight of 28 ounces and a standard deviation of 0.8 ounce. The distribution of weights is mound shaped.

a. About 95 percent of the loaves will have weights that lie within what interval?

b. Nearly all the loaves will have weights within what interval?

c. Approximately what percent of the loaves will weigh more than 28.8 ounces?

2.96 A pharmaceutical company manufactures capsules that contain an average of 507 grams of vitamin C. The standard deviation is 3 grams. According to Chebyshev's theorem, at least 96 percent of the capsules will contain what amount of vitamin C?

2.97 The state unemployment rates for 1987 are given below.

AL	AK	AZ	AR	CA	CO	CT	DE	FL	GA
7.8	10.8	6.2	8.1	5.8	7.7	3.3	3.2	5.3	5.5
HI	ID	IL	IN	IA	KS	KY	LA	ME	MD
3.8	8.0	7.4	6.4	5.5	4.9	8.8	12.0	4.4	4.2
MA	MI	MN	MS	MO	MT	NE	NV	NH	NJ
3.2	8.2	5.4	10.2	6.3	7.4	4.9	6.3	2.5	4.0
NM	NY	NC	ND	OH	OK	OR	PA	RI	SC
8.9	7.9	4.5	5.2	7.0	7.4	6.2	5.7	3.8	5.6
SD	TN	TX	UT	VT	VA	WA	WV	WI	WY
4.2	6.6	8.4	6.4	3.6	4.2	7.6	10.8	6.1	8.6

Using the fact that $\bar{x} = 6.32$ and $s = 2.17$, determine the percentage of the data that have values between

a. $\bar{x} - s$ and $\bar{x} + s$,

b. $\bar{x} - 2s$ and $\bar{x} + 2s$, and

c. $\bar{x} - 3s$ and $\bar{x} + 3s$.

d. Compare the percentages found in Parts a, b, and c with the percentages given by the Empirical rule.

e. Compare the percentages found in Parts b and c with the percentages given by Chebyshev's theorem.

2.98 Chebyshev's theorem and the Empirical rule imply that most of the values in a sample will lie within the interval

$$\bar{x} \pm 2s.$$

Since this interval extends from 2 standard deviations below the mean to 2 standard deviations above it, the interval covers a range R of 4 standard deviations. Consequently, a rough approximation of the standard deviation can be obtained by setting R equal to $4s$. Solving for s, we obtain

$$s \approx \frac{R}{4}.$$

Use this procedure to approximate the standard deviation of the 50 unemployment rates given in Exercise 2.97.

2.99 The ages of 40 blood donors are given below. Make a cursory inspection of the data, and use the procedure described in Exercise 2.98 to obtain a rough approximation of the standard deviation (the actual value of s is 10.96).

$$
\begin{array}{cccccccccc}
35 & 53 & 61 & 43 & 21 & 42 & 23 & 29 & 35 & 37 \\
39 & 58 & 27 & 64 & 27 & 31 & 36 & 48 & 41 & 22 \\
37 & 35 & 42 & 32 & 43 & 34 & 59 & 50 & 38 & 43 \\
31 & 30 & 41 & 37 & 29 & 45 & 23 & 56 & 46 & 41
\end{array}
$$

2.7 Measures of Relative Position

Measures of relative position describe the relationship of a single value to the entire set of measurements. A simple way of comparing one value with the complete distribution is to calculate its **z-score,** which shows how far away, in terms of the number of standard deviations, the value is from the mean of the distribution.

Z-Scores

For a set of measurements with mean $\bar{x}$ and standard deviation s, the **z-score** of a single value x is

$$
z = \frac{x - \bar{x}}{s}. \tag{2.5}
$$

From Formula 2.5, it is seen that the z-score measures how many standard deviations x is from the mean of the distribution. Also note that

1. the z-score is negative when x is smaller than the mean,
2. the z-score is positive when x is larger than the mean, and
3. the z-score equals 0 when x equals the mean.

Furthermore, for data sets whose distribution is mound shaped, the Empirical rule gives the following results:

1. approximately 68 percent of the data have z-scores between -1 and 1,
2. approximately 95 percent of the data have z-scores between -2 and 2, and
3. nearly all the measurements have z-scores between -3 and 3.

Example 2.24   Suppose a student has taken two quizzes in his statistics course. On the first quiz the mean score was 32, the standard deviation was 8, and the student received a 44. The student obtained a 28 on the second quiz, for which the mean was 23 and the standard deviation was 3. On which quiz did the student perform better relative to the rest of the class?

Solution

For the first quiz, $\bar{x} = 32$ and $s = 8$. Therefore, a score of 44 has a z-score of

$$z = \frac{(44 - 32)}{8} = 1.5.$$

For the second quiz, $\bar{x} = 23$ and $s = 3$. The z-score for a score of 28 is

$$z = \frac{(28 - 23)}{3} = 1.67.$$

Since the student scored 1.67 standard deviations above the mean on the second quiz, compared to 1.5 on the first, he performed better on the second.

Ⓜ **Example 2.25** Use MINITAB to obtain the z-scores for the 50 states' maximum monthly AFDC benefits that were stored in column C1 in Example 2.20.

Solution

The following commands will calculate the z-scores for a data set in column C1 and place the results in C2, and print the data and their z-scores.

```
LET C2 = (C1 - MEAN(C1))/STDEV(C1)
PRINT C1 C2
```

The 50 AFDC benefits and the resulting z-scores are given in Exhibit 2.10.

Exhibit 2.10

```
MTB > LET C2 = (C1 - MEAN(C1))/STDEV(C1)
MTB > NAME C1 'AFDC' C2 'Z-SCORE'
MTB > PRINT C1 C2
```

ROW	AFDC	Z-SCORE	ROW	AFDC	Z-SCORE	ROW	AFDC	Z-SCORE
1	118	-1.78776	18	412	0.31431	35	120	-1.77346
2	356	-0.08608	19	159	-1.49461	36	486	0.84340
3	515	1.05075	20	492	0.88630	37	371	0.02116
4	409	0.29286	21	293	-0.53653	38	503	0.96495
5	510	1.01500	22	319	-0.35063	39	376	0.05691
6	359	-0.06463	23	342	-0.18618	40	517	1.06505
7	264	-0.74387	24	190	-1.27296	41	633	1.89443
8	310	-0.41498	25	532	1.17229	42	263	-0.75102
9	366	-0.01459	26	325	-0.30773	43	381	0.09266
10	354	-0.10038	27	266	-0.72957	44	359	-0.06463
11	779	2.93832	28	402	0.24281	45	282	-0.61518
12	601	1.66564	29	184	-1.31586	46	424	0.40011
13	304	-0.45788	30	249	-0.85112	47	309	-0.42213
14	207	-1.15142	31	202	-1.18717	48	200	-1.20147
15	528	1.14369	32	275	-0.66523	49	603	1.67994
16	350	-0.12898	33	288	-0.57228	50	360	-0.05749
17	539	1.22234	34	416	0.34291			
```
```

The z-scores above could also have been obtained by using the **CENTER** command. The following instructs MINITAB to calculate the z-scores for the values in column C1 and place the results in column C2.

```
CENTER C1 C2
```

> **TIP:**   If all the values of a data set are transformed to z-scores, the resulting data (the z-scores) will have a mean of 0 and a standard deviation of 1.

Percentiles

A measure of relative position describes the relationship of a single value to the entire data set. We have seen that one such measure is the z-score, which tells how many standard deviations the value is from the mean. **Percentiles** offer another method of comparing a single value with the complete distribution. Although they can be defined for small data sets, percentiles are only practical when the number of measurements is large.

> For a large set of measurements, the percentiles are denoted by $P_1, P_2, P_3, \ldots, P_{99}$.
>
> P_k is called the **kth percentile** and is the value such that approximately $k\%$ of the measurements are less and $(100 - k)\%$ are more.

Example 2.26   Elevated blood levels of cholesterol are a major risk factor in cardiovascular diseases, the leading cause of death in the United States. Levels that previously were considered acceptable are now being regarded as unsafe. A major reason for this change was a 1983 paper by Rifkind and Segal* in which plasma cholesterol and triglyceride levels were reported for more than 60,000 Americans. Tables were given with percentile levels for men, women, and children. Table 2.14 gives the plasma total cholesterol values in adult males. The table reveals, for example, that for males in the age group 20 to 24, the 5th percentile is $P_5 = 125$. Thus, only about 5 percent of men this age have cholesterol levels below 125. Since $P_{95} = 220$ for this age group, about 95 percent have levels below 220, and approximately 90 percent have cholesterol levels between $P_5 = 125$ and $P_{95} = 220$.

Three frequently used percentiles are P_{25}, P_{50}, and P_{75}. For a set of measurements arranged in order, these three percentiles divide the distribution into 4 groups, each containing approximately 25 percent of the measurements. For instance, the results of national examinations are often reported in percentile scores. Suppose for such a test that $P_{25} = 136, P_{50} = 159$, and $P_{75} = 185$. Then for the distribution of scores on this

---

* Rifkind, BM and Segal P. *"Lipid Research Clinics Reference Values for Hyperlipidemia and Hypolipidemia,"* Journal of the American Medical Association (1983) 250:14, 1869.

Table 2.14

Plasma Total Cholesterol (mg/dl) in Adult Males

Age	Mean	P_5	P_{75}	P_{90}	P_{95}
0–19	155	115	170	185	200
20–24	165	125	185	205	220
25–29	180	135	200	225	245
30–34	190	140	215	240	255
35–39	200	145	225	250	270
40–44	205	150	230	250	270
45–69	215	160	235	260	275
≥ 70	205	150	230	250	270

Adapted from Rifkind, BM and Segal P. *"Lipid Research Clinics Reference Value for Hyperlipidemia and Hypolipidemia."* Journal of the American Medical Association, (1983) 250:14, 1869.

test, approximately 25 percent were below 136, approximately 25 percent were between 136 and 159, approximately 25 percent were between 159 and 185, and approximately 25 percent were above 185.

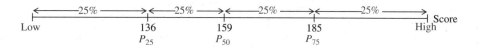

P_{25}, P_{50}, and P_{75} are called the **first, second,** and **third quartiles,** respectively, and are denoted by Q_1, Q_2, and Q_3. The second quartile Q_2 was considered earlier, since it is just the median. Thus,

$$\text{The median} = P_{50} = Q_2.$$

5-Number Summaries

The three quartiles, together with the smallest and largest values, are used to provide a 5-number summary of a data set.

> The *5-number summary* of a set of measurements are the values
>
> $$\text{MIN}, Q_1, Q_2, Q_3, \text{MAX},$$
>
> where Q_1, Q_2, Q_3 are the first, second, and third quartiles, and MIN and MAX are the smallest and largest values.

The 5-number summary is illustrated in the following example.

Exhibit 2.11

```
MTB > DESCRIBE C1

            N      MEAN    MEDIAN    TRMEAN    STDEV    SEMEAN
C1         50     368.0     357.5     363.4    139.9      19.8

          MIN       MAX        Q1        Q3
C1      118.0     779.0     272.7     487.5
```

ⓂExample 2.27

In MINITAB the **DESCRIBE** command can be used to obtain several numerical descriptions of a distribution, including the three quartiles. Exhibit 2.11 gives the MINITAB output for the 50 AFDC amounts stored earlier in column C1. The second quartile is labeled as the median and equals 357.5. The first and third quartiles are $Q_1 = 272.7$ and $Q_3 = 487.5$. As with the median, they are not whole numbers because interpolation has been used to locate their positions in the arrangement of the measurements. The smallest and largest observations are labeled MIN and MAX, respectively.

The 5-number summary for the data set consists of

$$MIN = 118, Q_1 = 272.7, Q_2 = 357.5, Q_3 = 487.5, \text{ and } MAX = 779.$$

The 5-number summary is illustrated in Figure 2.14. Note that from it the range can be determined by calculating

$$Range = MAX - MIN = 661.$$

Furthermore, the distance between the first and third quartiles gives the range of the middle 50 percent of the distribution and is called the **interquartile range (IQR).** It equals

$$IQR = Q_3 - Q_1 = 487.5 - 272.7 = 214.8.$$

The MINITAB output from the **DESCRIBE** command also gives TRMEAN and SEMEAN. The latter is the standard error of the mean and will be discussed in Chapter 8. TRMEAN is a 5 percent trimmed mean calculated by deleting approximately the smallest 5 percent and the largest 5 percent of the data, and then averaging the remaining values.

Figure 2.14
A 5-number summary for the 50 AFDC benefits.

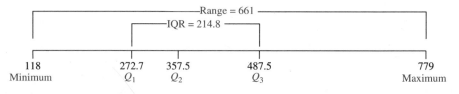

Boxplots

Although there are many variations of a **boxplot,** one of the simplest consists of a pictorial display of a distribution's 5-number summary. The boxplot provides the investigator with a visual impression of how a distribution's values are spread out from their median value Q_2.

Figure 2.15
Boxplot for the 50 AFDC
benefits.

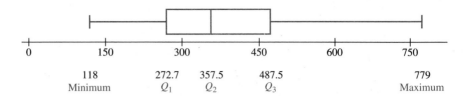

	118	272.7	357.5	487.5	779
	Minimum	Q_1	Q_2	Q_3	Maximum

> **Constructing a Boxplot:**
> A box is drawn above the real number line such that
> a. the ends of the box are at Q_1 and Q_3, called the left and right hinges of the box;
> b. a vertical line is drawn within the box to indicate Q_2; and
> c. a line is drawn from the left end of the box to the MIN value, and a line connects the right end of the box to the MAX value (the lines are called whiskers).

Example 2.28   The 5-number summary in Example 2.27 was used to construct the boxplot in Figure 2.15 for the 50 AFDC amounts. The boxplot provides a snapshot of the distribution's 5-number summary and quickly conveys several characteristics of the distribution. The box portrays the middle half of the data, and the length of the box (distance between the hinges) is the IQR. The left and right whiskers represent the lower and upper fourth of the distribution, and the distance between the extreme ends of the whiskers is the range.

Boxplots are often used in the initial exploratory analysis of data, and some types of boxplots are especially suitable for identifying outliers. The boxplot produced by MINITAB for the 50 AFDC amounts appears in Exhibit 2.12. It is generated by applying the command **BOXPLOT** to C1, the column that contains the data values. Notice that MINITAB uses ''+'' to identify the median Q_2, and the hinges are denoted with an ''I.''

Exhibit 2.12   MTB > BOXPLOT C1

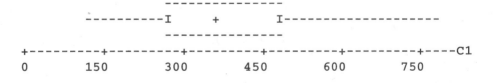

Section 2.7 Exercises
Measures of Relative Position

2.100 Four students took a national test for which the mean score was 480 and the standard deviation was 100. Their scores were 570, 440, 630, and 550. Determine the *z*-score for each student.

2.101 A sample of restaurants in a large city revealed that the average price of a cup of coffee was 58 cents, and the standard deviation was 8 cents. Three of the restaurants sampled charged 50, 45, and 60 cents. Find the z-scores for these restaurants.

2.102 A student obtained a z-score of 1.60 on the national test referred to in Exercise 2.100. What was the student's actual test score?

2.103 A restaurant from the survey referred to in Exercise 2.101 charges a price that has a z-value of -2.25. What is the cost of a cup of coffee at this restaurant?

2.104 A married couple are employed by the same company. The husband works in a department for which the mean hourly rate is $12.80 and the standard deviation is $1.20. His wife is employed in a department where the mean rate is $13.50 and the standard deviation is $1.80. Relative to their departments, which is better paid if the husband earns $14.60 and the wife earns $15.75?

2.105 The following table gives the 1988–1989 resident and nonresident tuition and required fees for the seven professional programs in medicine in Pennsylvania *(Pennsylvania Department of Education)*. All figures are in dollars.

School	Resident	Nonresident
Hahnemann University	17,092	17,092
Medical College of Pennsylvania	14,820	14,820
The Pennsylvania State University	11,508	17,406
Temple University	13,332	18,084
Thomas Jefferson University	16,025	16,025
University of Pennsylvania	16,650	16,650
University of Pittsburgh	13,802	19,262
Mean	14,747	17,048
Standard deviation	2,006	1,430

a. Find the z-score for the resident tuition at the University of Pennsylvania.
b. Find the z-score for the nonresident tuition at Temple University.
c. Compared to the other schools, which of the tuitions referred to in (a) and (b) is relatively cheaper?

2.106 An applicant's score of 215 on a company's employment examination was at the 84th percentile. Approximately what percentage of people taking this test achieve a higher score?

2.107 On the employment test referred to in Exercise 2.106, the first, second, and third quartile scores were 104, 150, and 195, respectively. Approximately what percentage of scores on this test are (a) between 104 and 195, (b) above 104, and (c) above 195?

2.108 Selected percentile values for the plasma total cholesterol levels for males aged 35 to 39 are given below.

Age	Mean	P_5	P_{75}	P_{90}	P_{95}
35–39	200	145	225	250	270

For males in this age group, approximately what percentage have cholesterol values (a) between 145 and 225, (b) between 225 and 270, (c) above 225, and (d) below 270?

2.109 The **DESCRIBE** command in MINITAB was applied to the ages of 40 blood donors. Use the resulting output reproduced below to determine (a) the 5-number summary, (b) the range, and (c) the interquartile range.

```
MTB > DESCRIBE C1

            N       MEAN    MEDIAN    TRMEAN    STDEV    SEMEAN
C1         40      39.10     37.50     38.78    10.96      1.73

          MIN        MAX        Q1        Q3
C1      21.00      64.00     31.00     44.50
```

2.110 Use the results given in Exercise 2.109 to construct a boxplot for the 40 ages.

2.111 MINITAB was used to obtain the following boxplot. Determine the 5-number summary, IQR, and range.

```
MTB > BOXPLOT C1


                    ---------------------
----------I              +          I--------------------
                    ---------------------

+---------+---------+---------+---------+---------+---------+----C1
0        200       400       600       800      1000
```

MINITAB Assignments

Ⓜ 2.112 Use MINITAB to obtain the 5-number summary for the 1987 state unemployment rates given below.

AL	AK	AZ	AR	CA	CO	CT	DE	FL	GA
7.8	10.8	6.2	8.1	5.8	7.7	3.3	3.2	5.3	5.5
HI	ID	IL	IN	IA	KS	KY	LA	ME	MD
3.8	8.0	7.4	6.4	5.5	4.9	8.8	12.0	4.4	4.2
MA	MI	MN	MS	MO	MT	NE	NV	NH	NJ
3.2	8.2	5.4	10.2	6.3	7.4	4.9	6.3	2.5	4.0
NM	NY	NC	ND	OH	OK	OR	PA	RI	SC
8.9	7.9	4.5	5.2	7.0	7.4	6.2	5.7	3.8	5.6
SD	TN	TX	UT	VT	VA	WA	WV	WI	WY
4.2	6.6	8.4	6.4	3.6	4.2	7.6	10.8	6.1	8.6

Ⓜ 2.113 Use MINITAB to obtain the z-scores for the following grade point averages of the members of the Math Club at a liberal arts college.

3.29	3.67	3.28	2.89	3.01	2.94	3.31	2.80	3.54	3.15
3.08	3.01	3.17	3.36	3.57	3.39	3.32	3.19	2.95	3.19
2.84	3.16	3.21	3.26	3.36	3.37	3.39	3.75	3.41	3.97
2.87	2.99	3.02	3.09	3.87	3.14	3.18	3.24	3.41	3.61
2.94	3.06	2.94	3.66	3.77	3.72	3.51			

Ⓜ 2.114 Have MINITAB construct a boxplot for the unemployment rates in Exercise 2.112.

Ⓜ 2.115 Use MINITAB to construct a boxplot for the grade point averages in Exercise 2.113.

Ⓜ 2.116 The population percentage changes for the 50 states during the period from January, 1980 to July, 1989 are given below *(U.S. Census Bureau)*. Use MINITAB to determine the z-score for each state.

AL	AK	AZ	AR	CA	CO	CT	DE	FL	GA
5.8	31.1	30.8	5.2	22.8	14.8	2.0	13.3	30.0	17.8
HI	ID	IL	IN	IA	KS	KY	LA	ME	MD
15.2	7.4	2.0	1.9	−2.5	6.3	1.8	4.2	8.6	11.3
MA	MI	MN	MS	MO	MT	NE	NV	NH	NJ
3.1	−1.0	6.8	4.0	4.9	2.4	2.6	38.9	20.2	2.2
NM	NY	NC	ND	OH	OK	OR	PA	RI	SC
17.3	2.2	11.7	1.1	1.0	6.6	7.1	1.5	5.4	12.5
SD	TN	TX	UT	VT	VA	WA	WV	WI	WY
3.5	7.6	19.4	16.8	11.0	14.0	15.2	5.0	3.4	1.1

Looking Back

Techniques for describing sets of data can be broadly classified as **graphical** and **numerical. Graphical methods** summarize and present the data in a manner that quickly conveys to the viewer a sense of its distribution. For **quantitative data** this might be accomplished by constructing a **frequency distribution, relative frequency distribution, histogram, dotplot,** or a **stem-and-leaf display.** For **qualitative data,** that is, data that are categorical or attributive in nature, a **bar chart** or a **pie chart** could be used.

When graphical methods are neither feasible nor desirable for describing a distribution of measurements, one relies on the use of **numerical techniques.** Descriptions of the center of a distribution are called **measures of central location.** Although the **mean** is the most popular, the **median** is sometimes preferred because it is less sensitive to outliers. The **weighted mean** affords the opportunity of assigning weights to the measurements in accordance to their relative importance. The **mode** is the only average that is applicable to qualitative as well as quantitative data. It indicates the value at which a distribution is most concentrated, making it particularly useful in marketing and inventory control.

The **mean** of a sample of n measurements $x_1, x_2, x_3 \ldots x_n$ is

$$\bar{x} = \frac{\Sigma x}{n}.$$

The **weighted mean** of a set of measurements $x_1, x_2, x_3 \ldots x_n$ with relative weights $w_1, w_2, w_3 \ldots w_n$ is given by

$$\bar{x} = \frac{\Sigma xw}{\Sigma w}.$$

The **median** of a set of n measurements is the middle value when the measurements are ranked according to size. After the numbers have been arranged in order, the position of the median in the array is given by

$$\text{Position} = \frac{(n + 1)}{2}.$$

Describing only the center of a distribution is incomplete, since it fails to consider the important characteristic of **variability.** The **range** gives the spread between the largest and smallest values, thus providing a quick but simplistic measure of variation. The **variance** measures how a set of measurements fluctuates relative to their mean. The **standard deviation** is the most common way of measuring the variation in a set of measurements. It equals the square root of the variance and can be used to compare the variation of two data sets. For two distributions of data, the one that has the larger standard deviation will have the greater variability. The standard deviation is also useful in describing a single distribution of measurements.

For a set of measurements, the **range** equals the difference between the largest and smallest values.

$$\text{Range} = \text{largest} - \text{smallest}$$

For a sample of n measurements $x_1, x_2 \ldots x_n$, **the sample variance** is denoted by s^2 and is given by

$$s^2 = \frac{SS(x)}{n - 1}$$

where

$$SS(x) = \Sigma x^2 - \frac{(\Sigma x)^2}{n}.$$

Standard deviation $= \sqrt{\text{variance}}$

Two statements that describe how the standard deviation relates to a distribution are the **Empirical rule** and **Chebyshev's theorem.** To use the Empirical rule, one must assume that the distribution of measurements is **mound shaped.** This assumption is not necessary with Chebyshev's theorem because it is applicable to all distributions.

Chebyshev's theorem: For any set of measurements and any number $k \geq 1$, the interval from $\bar{x} - ks$ to $\bar{x} + ks$ will contain at least $\left(1 - \frac{1}{k^2}\right)$ of the measurements.

The **Empirical rule:** If a set of measurements has a mound-shaped distribution, then

a. the interval from $\bar{x} - s$ to $\bar{x} + s$ will contain approximately 68 percent of the measurements,

b. the interval from $\bar{x} - 2s$ to $\bar{x} + 2s$ will contain approximately 95 percent of the measurements, and

c. the interval from $\bar{x} - 3s$ to $\bar{x} + 3s$ will contain approximately all the measurements.

Measures of relative position describe the relationship of a single value to the entire set of measurements. A simple measure is the value's **z-score.** This gives the number of standard deviations that the value is from the mean.

For a set of measurements with mean $\bar{x}$ and standard deviation s, the **z-score** of a single value x is

$$z = \frac{x - \bar{x}}{s}.$$

Percentiles are another method of comparing a single value with the complete distribution. Three frequently used percentiles are the 25th, 50th, and 75th. These are called the **first, second,** and **third quartiles,** and they divide a ranked distribution into four groups, each containing approximately 25 percent of the measurements. The distance from the first quartile to the third quartile is called the **interquartile range (IQR),** and it gives the range of the middle 50 percent of the distribution. A distribution's **5-number summary** consists of its smallest value, its three quartiles, and its largest value.

The **5-number summary** of a set of measurements are the values

MIN, Q_1, Q_2, Q_3, MAX,

where Q_1, Q_2, Q_3 are the first, second, and third quartiles, and MIN and MAX are the smallest and largest values.

A **boxplot** is a pictorial display of a 5-number summary.

Key Words

In reviewing this chapter, you should be able to define, explain, and illustrate each of the following.

Ⓜ *MINITAB Commands*

Review Exercises

2.117 A primary factor in the operating speed of a personal computer is the clock speed of its microprocessor chip, the system's central processing unit (CPU). A computer manufacturer tests 10 chips and obtains the following speeds (in megahertz):

3   4

5   9   6   7   3   8   10   8   2   1
13.0   14.3   13.6   13.8   12.8   12.9   15.2   14.0   12.0   11.9

a. Determine the mean speed.
b. Find the median speed.
c. Determine the range of the speeds.
d. Find the standard deviation of the speeds.
e. Construct a stem-and-leaf display of the speeds.

2.118 For a sample of 50 measurements, $\Sigma x = 780$ and $\Sigma x^2 = 17,526$. Find the mean and the standard deviation.

2.119 Five brothers have a mean weight of 172 pounds, and four of them weigh 173, 149, 197, and 126. Determine the weight of the remaining brother.

2.120 Find the standard deviation of the following sample. (Hint: See Exercise 2.80.)

35,514   35,507   35,511   35,501   35,515   35,510

2.121 In a certain city the mean price of a quart of milk is 63 cents and the standard deviation is 8 cents. The mean price of a pound of ground beef is $1.75 and the standard deviation is 12 cents. A late-night convenience store charges 89 cents for a quart of milk and $2.12 for a pound of ground beef. Which of these items is relatively more overpriced?

2.122 A company has offices at 3 sites. The numbers of employees at the 3 locations are 78, 103, and 86, and the mean ages of the workers at these sites are 42, 37, and 35, respectively. Find the mean age of all workers employed by this company.

2.123 The number of hours worked last week by the employees at a manufacturing plant have been grouped in the following frequency distribution. Determine the mean number of hours worked by all employees.

Hours	Number of Workers
26–30	7
31–35	18
36–40	65
41–45	90
46–50	48
51–55	12

2.124 A restaurant employs 36 waitresses. The following show the numbers of hours that each worked during the first week of June.

52   43   51   55   42   41   48   40   25   39   41   49
48   52   51   42   46   27   59   46   61   38   42   30
51   40   32   40   40   42   47   52   34   48   42   43

Construct a stem-and-leaf display of the data.

2.125 Construct a dotplot of the numbers of hours that are given in Exercise 2.124.

2.126 The following gives the numbers of miles since the last oil change for 29 state police cars.

4,215	3,567	5,631	876	7,429	2,395	1,034	4,521
1,112	7,234	3,415	2,689	4,234	765	7,651	7,031
2,031	3,141	5,441	4,222	272	1,012	4,425	1,172
4,447	3,391	8,215	3,919	5,879			

Use a histogram to display the data.

2.127 An outstanding collegiate baseball player obtained the following numbers of hits for the 48 games that he played during his senior year.

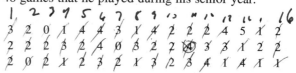

a. Display the data in a frequency distribution for which each class consists of a single value.

b. Use the grouping in Part a to find the mean and the standard deviation of the numbers of hits. (Hint: See Exercise 2.82.)

2.128 A company performs a yearly performance evaluation of its employees. Each employee receives a rating of meritorious (M), acceptable (A), or inadequate (I). The 32 employees in the research and development division received the following ratings.

A A M I M M A M M A A M A M M A
I M I A A M M M A A I I A M A M

Determine the modal rating.

2.129 The three pizza chains with the largest share of total sales are Pizza Hut, Domino's, and Little Caesar's (The *Wall Street Journal,* April 21, 1989). Their percentages of total sales are given below.

Company	% of Total Sales
Pizza Hut	20.7
Domino's	17.0
Little Caesar's	6.7
Others	55.6

Use a bar chart to display the sales distribution.

2.130 Use a pie chart to display the sales distribution that is given in Exercise 2.129.

2.131 Although its use has generated considerable controversy since the Three Mile Island incident in 1979, nuclear energy is still America's number two source of electricity. The following table gives the sources of electricity in the United States. Use a horizontal bar chart to display this information.

Source	Percentage
Coal	55
Nuclear	19
Natural gas	11
Hydropower	8
Oil	5
Others	2

Source: *Edison Electric Institute*, 1988.

2.132 The following are the average automobile insurance premiums for the 50 states during 1986 *(A. M. Best)*. All figures are in dollars.

AL	AK	AZ	AR	CA	CO	CT	DE	FL	GA
278	602	554	434	566	444	466	462	389	450
HI	ID	IL	IN	IA	KS	KY	LA	ME	MD
454	344	419	338	244	345	369	515	333	506
MA	MI	MN	MS	MO	MT	NE	NV	NH	NJ
556	481	417	297	403	373	324	550	330	604
NM	NY	NC	ND	OH	OK	OR	PA	RI	SC
378	522	362	307	327	369	398	512	477	450
SD	TN	TX	UT	VT	VA	WA	WV	WI	WY
256	292	426	397	364	381	394	455	373	384

a. Construct a relative frequency histogram for which the lower limit of the first class is $225 and each class width is $50. Is the distribution symmetric or skewed?

b. The mean and standard deviation of the premiums are $413 and $89, respectively. Determine what percentage of the data falls within 2 standard deviations of the mean. What percentage falls within 3 standard deviations?

c. Compare the percentages found in Part b with those given by the Empirical rule.

d. Compare the percentages found in Part b with those given by Chebyshev's theorem.

2.133 In Exercise 2.98, a method was discussed for obtaining a rough approximation of the standard deviation. This consists of finding the range R and dividing the result by 4. Use this to approximate the standard deviation of the 50 premiums in Exercise 2.132 (the actual value is $s = 89).

2.134 Final examination scores in a calculus course have averaged 74 with a standard deviation of 8 points. Assuming that the distribution of scores is approximately mound shaped, determine the percentage of scores that are (a) between 66 and 82, (b) below 58, and (c) above 90.

2.135 For the distribution of scores in Exercise 2.134, what does Chebyshev's theorem say about the percentage of scores between 54 and 94?

2.136 In Exercise 2.87, the 1987 median purchase prices of existing one-family houses in 32 metropolitan areas were given. The following is the MINITAB output from applying the **DESCRIBE** command to those values. Use the results to determine the 5-number summary, the range, and the interquartile range.

```
MTB > DESCRIBE C1
           N      MEAN    MEDIAN   TRMEAN    STDEV   SEMEAN
C1        32    122.73    115.30   121.32    31.70     5.60

         MIN       MAX       Q1       Q3
C1     79.60    186.20    94.95    139.70
```

2.137 Use the information in Exercise 2.136 to construct a boxplot for the prices.

MINITAB Assignments

Ⓜ 2.138 A small town has a senior citizens club, and the following are the ages of its 72 members.

69	82	68	82	76	74	69	76	59	62
72	58	68	90	58	67	59	56	62	71
63	82	64	59	56	65	67	69	64	55
74	78	75	71	67	74	81	59	67	63
62	64	63	59	68	67	62	64	59	57
67	68	62	64	59	56	57	62	63	64
91	75	87	56	65	64	69	79	57	75
98	58								

Use MINITAB to (a) construct a dotplot, (b) summarize the data in a stem-and-leaf display, and (c) obtain the 5-number summary of the ages.

Ⓜ 2.139 Use MINITAB to construct a boxplot for the ages of the senior citizens in Exercise 2.138.

Ⓜ 2.140 The following are the closing prices of a certain stock for 35 consecutive trading days. The figures are in dollars.

35.75	35.12	35.88	35.50	35.50	36.75	36.25	36.62
37.25	37.75	37.62	38.50	38.25	38.25	38.88	37.88
38.00	36.25	37.75	38.62	38.50	40.50	41.25	43.75
44.50	43.00	42.25	42.75	42.88	44.25	45.75	45.50
46.00	46.50	45.25					

Use MINITAB to obtain a frequency distribution with a class width of $1.00 and $35.00 as the first class mark.

Ⓜ 2.141 A study of professors' salaries at a local college revealed a mean salary of $34,576 and a standard deviation of $4,593. The z-scores for the salaries of the 16 members of the Chemistry Department are given below. Use MINITAB to find the salaries of the chemistry faculty.

1.25	0.69	−1.06	2.97	0.89	1.27	−1.87	0.02
0.98	−1.52	−2.65	−0.92	1.06	−1.22	0.52	3.02

(© Walter Iooss, Jr./The Image Bank)

3

Modeling Bivariate Data: Describing the Relationship Between Two Variables

◀ *Using ERA to Predict Number of Wins?*

Many baseball experts and fans argue that pitching is the key to a team's success. A frequently used indicator of a pitching staff's effectiveness is its earned run average (ERA). The ERA is a measure of the number of earned runs that a pitcher gives up per nine innings, and one strives to keep this as low as possible. Table 3.1 and Figure 3.1 show the number of wins (y) and the ERA (x) for each American League team at the end of the 1990 season. Examining these, one might wonder if x and y are mathematically related. This chapter provides an introduction to the problem of investigating relationships between two variables. We will be concerned with describing the relation by means of a mathematical model.

Table 3.1

Earned Run Average and Number of Wins for the American League's 1990 Season

Team	ERA (x)	Wins (y)
Oakland	3.18	103
Chicago	3.61	94
Seattle	3.69	77
Boston	3.72	88
California	3.79	80
Texas	3.83	83
Toronto	3.84	86
Kansas City	3.93	75
Baltimore	4.04	76
Milwaukee	4.08	74
Minnesota	4.12	74
New York	4.21	67
Cleveland	4.26	77
Detroit	4.39	79

Figure 3.1
Earned run average and number of wins; American League 1990 season.

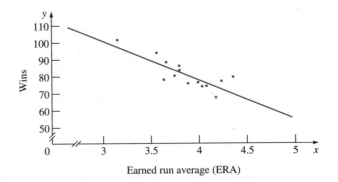

Earned run average (ERA)

Looking Ahead

In 1988, researchers at Harvard University published a 1,000 page study referred to as the RBRVS (Resource-Based Relative Value Scale). The report presented radical changes in the way that doctors are reimbursed for their services by Medicare. Instead of basing payments on historical charges by physicians, the study suggested that a relative value scale be used.

In the formation of their proposal, the Harvard researchers developed a **mathematical model** to measure the relative value associated with a procedure. The model consisted of an equation involving several variables such as time, technical skill, mental effort, stress, training required, and overhead expenses. Knowledge of these variables would be used to determine an appropriate reimbursement for a service.

In this chapter you will be introduced to the problem of investigating the relationship between two variables. Many statistical studies involve **bivariate** (two-variable) data, and the following four examples illustrate the type of problems on which this chapter will focus.

Example 3.1   Companies spend money on advertising (x) with the expectation that there will be a resulting increase in sales (y). One would certainly expect that a relationship exists between these two variables, and a company may want to use knowledge of the amount of advertising expenditures to predict future sales.

Example 3.2   The mathematics departments of some colleges and universities administer placement tests to aid in assigning appropriate math courses to incoming freshmen. They may use a student's score (x) to predict his or her grade (y) in a freshman mathematics course.

Example 3.3   Some contend that vitamin C is helpful in reducing one's susceptibility to the common cold. Is there a relationship between the daily intake of vitamin C (x) and the number of colds (y) that people have each year?

Blood pressure tends to increase as one grows older. It is also affected by many other factors, such as one's weight, physical condition, dietary and smoking habits, level of stress, amount of exercise, and use of drugs and alcohol. *(© Ansell Horn/ Phototake, NYC)*

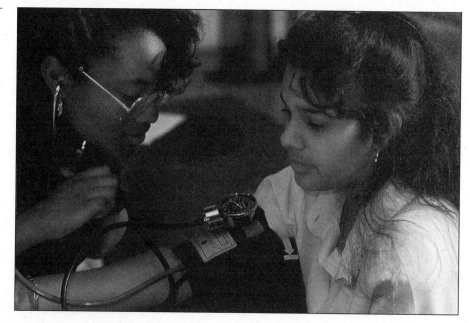

Example 3.4   *X* might denote a woman's age and *y* her systolic blood pressure. Are age and blood pressure related? If so, is it possible to describe approximately this relationship by means of a mathematical model (formula)? If this can be done, one might use the model to predict a woman's systolic pressure (*y*) by substituting her age (*x*) into the formula.

Seldom will an exact mathematical relationship exist between *x* and *y*; thus, the investigator must be content with obtaining a formula that will describe it approximately. Consequently, the researcher will also want to obtain a measure of how well the formula approximates the relationship as exhibited by the collected data. This and additional bivariate concepts will be illustrated in this chapter.

3.1   Straight Lines and the *SS*(   ) Notation

Frequently, when two variables *x* and *y* are related, a complicated mathematical formula is required to model the relationship. In this chapter, however, only problems will be considered for which the relation between *x* and *y* can be approximated with a straight line. Once the line's equation is obtained, it can then be used as a mathematical model to predict the value of *y* for a specified value of *x*. Before beginning this modeling process, we will briefly review some basic concepts of straight lines and introduce notation that will be used in the following sections.

Figure 3.2
Line with slope 3, y-intercept
2.

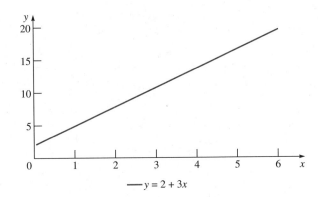

$\qquad\qquad y = 2 + 3x$

Straight Lines

> The graph of the following equation is a **straight line.**
>
> $$y = b_0 + b_1 x$$
>
> Its **slope** is given by b_1, the coefficient of x, and its **y-intercept** is b_0, the constant term in the equation.

For example,

$$y = 2 + 3x$$

is the equation of a straight line with slope $b_1 = 3$ and y-intercept $b_0 = 2$. The y-intercept of 2 indicates that the line crosses the y-axis 2 units above the origin. This is shown in Figure 3.2. The slope of a line indicates how y and x change as one moves from any point to a second point on the line. More precisely, it measures the ratio of the change in y (denoted by Δy) to the change in x (Δx) as movement occurs between any two points on the line. The change in y is called the **rise,** and the change in x is called the **run.** For our example,

$$b_1 = \text{slope} = \frac{\text{rise}}{\text{run}} = \frac{\Delta y}{\Delta x} = 3.$$

Since 3 can be written as

$$3 = \frac{3}{1} = \frac{\Delta y}{\Delta x},$$

we see that each unit change in x ($\Delta x = 1$) will be accompanied by a 3-unit change in y ($\Delta y = 3$). Similarly, since

$$3 = \frac{7.5}{2.5},$$

for every two and one-half units of run ($\Delta x = 2.5$) there will be seven and one-half units of rise ($\Delta y = 7.5$). These examples are illustrated in Figure 3.3.

Figure 3.3

Line with slope 3, y-intercept 2.

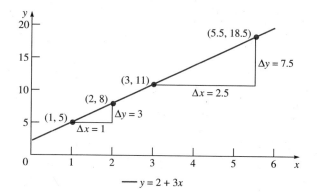

Example 3.5   A manufacturer of hospital diagnostic systems believes that next year's expenses y (in millions) can be modeled by the following equation, where x denotes the number of units produced during the year.

$$y = 4.3 + 1.7x.$$

By how much will expenses increase for each production increase of 5 systems?

Solution

Since the model is an equation of the form

$$y = b_0 + b_1 x,$$

its graph is a straight line with slope $b_1 = 1.7$. Consequently, each unit increase in x will increase expenses by \$1.7 million. Thus, a production increase of 5 units will raise total expenses by

$$\Delta y = b_1 \Delta x = \$1.7(5) = \$8.5 \text{ million.}$$

The graph of the assumed model is shown in Figure 3.4.

Figure 3.4

Modeling manufacturing expenses.

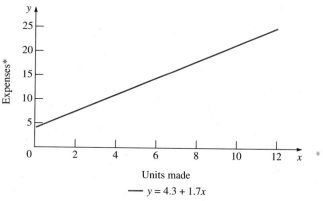

* Costs are in millions

The SS( ) Notation

In defining the concept of variance in Chapter 2, the $SS(x)$ notation was introduced. We defined

$$SS(x) = \Sigma(x - \bar{x})^2.$$

As emphasized in Chapter 2, it is usually easier to calculate $SS(x)$ by means of the following shortcut formula.

$$SS(x) = \Sigma x^2 - \frac{(\Sigma x)^2}{n} \qquad (3.1)$$

In the following sections we will find it convenient to extend this notation to $SS(y)$ and $SS(xy)$. These quantities are defined as follows.

$$SS(y) = \Sigma(y - \bar{y})^2$$

$$SS(xy) = \Sigma(x - \bar{x})(y - \bar{y})$$

As with $SS(x)$, we will use shortcut formulas to compute $SS(y)$ and $SS(xy)$. These are given in Formulas 3.2 and 3.3.

$$SS(y) = \Sigma y^2 - \frac{(\Sigma y)^2}{n} \qquad (3.2)$$

$$SS(xy) = \Sigma xy - \frac{(\Sigma x)(\Sigma y)}{n} \qquad (3.3)$$

Example 3.6   For the following 5 data points, compute $SS(x)$, $SS(y)$, and $SS(xy)$.

x	6	4	8	0	-5
y	2	6	1	5	12

Solution

In Formulas 3.1, 3.2, and 3.3, the basic quantities that need to be calculated are the sum of the xs and their squares, the sum of the ys and their squares, and the sum of the products of x with y. You may have a calculator that accumulates these sums. If not, then you might find it convenient to arrange the sums as in Table 3.2. The required sums for obtaining $SS(x)$, $SS(y)$, and $SS(xy)$ appear in the last row of Table 3.2. Applying Formulas 3.1 through 3.3, we obtain the following.

$$SS(x) = 141 - \frac{(13)^2}{5} = 107.2$$

$$SS(y) = 210 - \frac{(26)^2}{5} = 74.8$$

$$SS(xy) = -16 - \frac{(13)(26)}{5} = -83.6$$

Table 3.2

Quantities Needed in Calculating $SS(x)$, $SS(y)$, and $SS(xy)$

x	y	xy	x^2	y^2
6	2	12	36	4
4	6	24	16	36
8	1	8	64	1
0	5	0	0	25
-5	12	-60	25	144
13	26	-16	141	210

Section 3.1 Exercises
Straight Lines and the SS(   ) Notation

In Exercises 3.1–3.8, determine the slope and the *y*-intercept of the line whose equation is given.

3.1 $y = 3 + 8x$

3.2 $5x - 4y = 20$

3.3 $y = 4x$

3.4 $2x + 3y = 10$

3.5 $y = 10$

3.6 $x + 6y = 12$

3.7 $4x - 3y = 24$

3.8 $1.5x + 0.5y = 3$

3.9 Graph each of the lines in Exercises 3.1 through 3.4.

3.10 Graph each of the lines in Exercises 3.5 through 3.8.

3.11 Determine the slope of the following line.

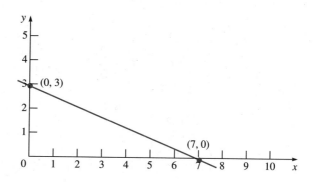

3.12 Give the slope and the y-intercept of the line whose graph appears below.

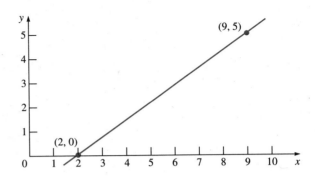

3.13 Write the equation of the line that has a slope of 10 and y-intercept 7.

3.14 Find the equation of the line that intercepts the y-axis at -3 and has a slope of 5.

3.15 Consider the line that passes through the points (4, 0) and (0, 4). Determine its equation and graph the line.

3.16 Find the equation of the line that passes through the points $(-5, 10)$ and $(2, -11)$. Also graph the line.

3.17 For one day's use of a compact car, a rental company charges a fixed fee of $35.00 plus 30 cents per mile. Let x denote the number of miles traveled and y the rental cost for that day.
   a. Find the equation that relates y to x.
   b. What are the slope and y-intercept of the line?
   c. A customer rents the car for 2 days and drives 100 miles further the second day than on the first day. Use the slope of the line to determine the additional cost for the second day.

3.18 The service department of an appliance store charges $50 for a service call plus an hourly rate of $45.00 to repair a dishwasher in the home. Let x denote the number of hours required and y the cost for a home repair.
   a. Obtain the equation that gives the relationship between y and x.
   b. Give the slope and y-intercept of the line.
   c. A technician for the company made two service calls last Monday, and the first call required three more hours of labor than the second. Use the slope of the line to determine the amount by which the cost of the first service call exceeded that of the second.

3.19 Next year's projected sales y for a small company are modeled by the following equation, where x denotes the amount of advertising expenditures during the year.

$$y = 975{,}000 + 1.5x$$

   a. By how much will sales rise for each increase of $1 in advertising expenditures?
   b. What will be the increase in sales if advertising is increased by $1,000?

3.20 The relationship between the demand for a product y and its price x (in dollars) is modeled by the following.

$$y = 35,760 - 200x$$

a. How will a price increase of \$1 affect the demand for the product?
b. How will the demand be affected if the price is lowered by \$25?

3.21 Calculate $SS(x)$, $SS(y)$, and $SS(xy)$ for the following 5 data points.

x	5	0	4	-1	3
y	10	4	7	2	6

3.22 Determine $SS(x)$, $SS(y)$, and $SS(xy)$ for the points (3, 8), (5, 12), (9, 20), (0, 2), $(-5, -12)$, (1, 4).

3.23 The tread depth (in hundredths of an inch) and the number of miles of usage (in thousands) are given below for a sample of 10 tires of the same brand. Calculate $SS(x)$, $SS(y)$, and $SS(xy)$.

Miles (x)	39	37	35	15	25	38	18	60	36	40
Tread (y)	10	15	16	30	21	14	34	3	10	14

3.24 For 100 points, $\Sigma x = 387$, $\Sigma y = 588$, $\Sigma x^2 = 1,981$, $\Sigma y^2 = 5,023$, and $\Sigma xy = 2,530$. Find $SS(x)$, $SS(y)$, and $SS(xy)$.

3.2 Scatter Diagrams

During the cold winter months a restaurant in the Northeast features a hot cereal specially prepared a few hours before opening. The chef wants to anticipate the daily demand, since the cereal is only made once each day, and any that remains after

(© George Haling/Photo Researchers, Inc.)

closing must be thrown away. Based on his observations of sales over a long time, the chef believes that the number of sales each day is dependent on the outside temperature. To develop a more accurate method of forecasting sales, he randomly selects 10 business days from last winter and examines the number of hot cereal sales and the daily temperature low forecasted for his town. These figures are given in Table 3.3.

The objective is to obtain a model that relates sales to temperature. More specifically, the chef would like to obtain an equation that would use knowledge of the temperature low forecasted for that day to predict the number of hot cereal sales. Such a **prediction equation** is called a **regression equation.** The variable that we want to predict is called the **dependent variable (output variable** or **response variable)** and is denoted by y. The variable whose knowledge will be utilized to predict y is called the **independent variable (input variable** or **predictor variable)** and is denoted by x. Thus, in this example the independent variable x is the daily temperature low and the dependent variable y is the number of hot cereal sales.

After the data have been collected, the next step toward obtaining a prediction equation is to decide what type of relationship might exist between x and y. Although in this chapter we will only consider straight line relationships, x and y may require a more sophisticated curve to approximate the relation. Often a **scatter diagram** will indicate what type of curve might be feasible. A scatter diagram is just a graph of the x and y values. Figure 3.5 shows a scatter diagram for the data in Table 3.3. From the scatter diagram, it appears that the number of sales (y) tends to decrease as the temperature (x) increases. Moreover, there seems to be approximately a straight line relationship between the two variables. We will pursue this in more detail in Section 3.3.

Table 3.3

Temperature Lows and Cereal Sales for 10 Randomly Selected Days

Day	Temperature Low	Hot Cereal Sales
1	18°F	64
2	10°	50
3	5°	80
4	24°	60
5	38°	22
6	14°	72
7	20°	58
8	32°	31
9	29°	54
10	35°	48

Figure 3.5
Scatter diagram. Temperature
low versus hot cereal sales.

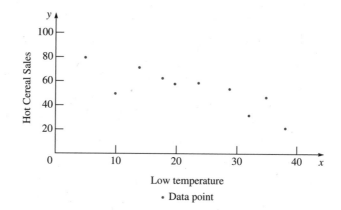

Low temperature

• Data point

Regression was originally
called ''reversion'' and was
introduced by Francis Galton
(1822–1911). A cousin of
Charles Darwin, Galton investigated the relationship
between the heights of parents and their offspring. He
found a ''regression'' of
heights toward the mean.
Tall parents tend to have
children of shorter heights
than they, while short parents
tend to have taller children.
*(Photo courtesy of the Bett-
mann Archive.)*

A scatter diagram can be obtained in MINITAB by using the command **PLOT.** In
the following we have first instructed MINITAB to store the x and y values in columns
C1 and C2, respectively, and then we have named the columns so that the computer
output will have easily identified labels.

```
READ C1 C2
18   64
10   50
 5   80
24   60
38   22
14   72
20   58
32   31
29   54
35   48
END
NAME C1 'TEMP' C2 'SALES'
PLOT C2 C1
```

> **NOTE:**   In the **PLOT** command the column of y values must be given first,
> followed by the column of x values.

In response to the above commands, MINITAB will produce the scatter diagram
shown in Exhibit 3.1.

With the **PLOT** command as used above, MINITAB automatically selects appropriate scales along the x and y axes. However, you can choose something else if
you wish. For example, suppose we want to obtain a graph with the axes labeled as in
Figure 3.5, namely, x starting at 0, increasing in increments of 10, and ending at 40; y

Exhibit 3.1   `MTB > PLOT C2 C1`

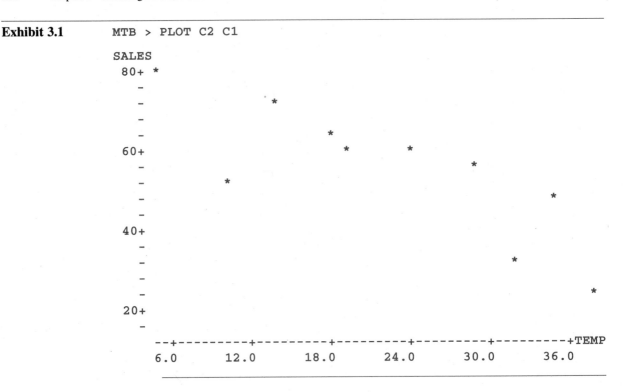

```
SALES
 80+  *
   -
   -                        *
   -                      *
   -                  *
 60+               *        *
   -                         *
   -       *
   -   *
   -
 40+
   -
   -                           *
   -
   -                            *
 20+
   -
      --+---------+---------+---------+---------+---------+TEMP
       6.0       12.0      18.0      24.0      30.0      36.0
```

starting at 0, increasing in increments of 20, and ending at 100. To specify this, type the following commands.

```
PLOT C2 C1;
XSTART 0 stop 40;     # Comment: "stop" is unnecessary
XINCREMENT 10;
YSTART 0 stop 100;    # Comment: "stop" is unnecessary
YINCREMENT 20.
```

If the specifications for the axes are too restrictive for MINITAB to implement, it will make the necessary adjustments and attempt to meet the specifications as closely as possible.

Ⓜ **Example 3.7**   To aid in assigning appropriate math courses to incoming freshmen, the Mathematics Department at a liberal arts college administers a placement test during freshmen orientation. The 25-point test measures one's quantitative and analytical skills. The department believes that the test score is a good predictor of a student's final average in their introductory statistics course. Table 3.4 contains the placement test scores and final averages for a sample of 15 former students in the course. To determine what type of relationship might exist between x and y, we will have MINITAB construct a scatter diagram as in Exhibit 3.2.

Table 3.4

Placement Test Scores and Course Averages

Student	Test Score x	Course Average y
1	21	69
2	17	72
3	21	94
4	11	61
5	15	62
6	19	80
7	15	65
8	23	88
9	13	54
10	19	75
11	16	80
12	25	93
13	8	55
14	14	60
15	17	64

```
READ C1 C2
21   69
17   72
21   94
11   61
15   62
19   80
15   65
23   88
13   54
19   75
16   80
25   93
 8   55
14   60
17   64
END
NAME C1 'TEST' C2 'CRSE AVG'
PLOT C2 C1
```

From the scatter diagram it appears that there is approximately a straight line relationship between x (test score) and y (course average), and y tends to increase as x increases. In the next section we will see how a mathematical model (equation) can be obtained to approximate this relationship.

Exhibit 3.2

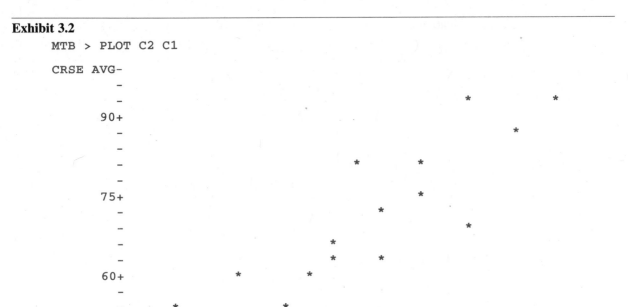

```
MTB > PLOT C2 C1

CRSE AVG-
        -
        -                                                    *              *
     90+
        -                                                 *
        -
        -                              *         *
        -
     75+                                         *
        -                          *
        -                      *
        -                  *         *
     60+          *              *
        -
        -      *              *
        -
        --+---------+---------+---------+---------+---------+----TEST
         7.0      10.5      14.0      17.5      21.0      24.5
```

3.3   Linear Regression and the Method of Least Squares

At the beginning of Section 3.2 we considered a northeastern restaurant that featured a hot cereal on its winter breakfast menu. The chef believes that the number of sales

Table 3.5

Calculations Needed to Compute $SS($ $)$ Terms

Temperature (x)	Sales (y)	xy	x^2	y^2
18	64	1,152	324	4,096
10	50	500	100	2,500
5	80	400	25	6,400
24	60	1,440	576	3,600
38	22	836	1,444	484
14	72	1,008	196	5,184
20	58	1,160	400	3,364
32	31	992	1,024	961
29	54	1,566	841	2,916
35	48	1,680	1,225	2,304
225	539	10,734	6,155	31,809

each day (y) is related to the temperature low (x) that day. He would like to obtain a mathematical model that could predict daily sales by using the forecasted temperature low reported by the weather service. The first step toward this goal is to obtain some data. The chef randomly selected 10 business days from last winter and examined the numbers of hot cereal sales and the forecasted daily temperature lows. These figures are given in Table 3.3.

After collecting the data, the next step is to construct a scatter diagram to determine the type of relationship that might exist between x and y. This was done in Figure 3.5, and the scatter diagram indicates that a straight line might describe quite well the association between the daily temperature low and the number of hot cereal sales. But which straight line should we use to approximate this relationship? Statisticians usually define the "best" choice to be the **least squares line.** Shortly we will explain why this is considered to be the line of best fit, but first we will show how it can be obtained.

The equation of the **least squares line** is

$$y = b_0 + b_1 x,$$

where the slope b_1 of the line is given by

$$b_1 = \frac{SS(xy)}{SS(x)}, \tag{3.4}$$

and the y-intercept b_0 of the line is given by

$$b_0 = \bar{y} - b_1 \bar{x}. \tag{3.5}$$

Let's now obtain the least squares line for the hot cereal example. The data given earlier in Table 3.3 are reproduced in the first two columns of Table 3.5, which also contains the basic calculations needed to compute $SS(x)$, $SS(xy)$, and $SS(y)$. Although $SS(y)$ is not required to obtain the least squares line, it will be needed in the next section for determining the correlation coefficient. Since you will usually compute the correlation coefficient whenever you determine the least squares line, it is suggested that you calculate $SS(y)$ at the same time that you find $SS(x)$ and $SS(xy)$. Using the column totals in Table 3.5 on the previous page, we obtain the following.

$$SS(x) = \Sigma x^2 - \frac{(\Sigma x)^2}{n} = 6{,}155 - \frac{(225)^2}{10} = 1{,}092.5$$

$$SS(y) = \Sigma y^2 - \frac{(\Sigma y)^2}{n} = 31{,}809 - \frac{(539)^2}{10} = 2{,}756.9$$

$$SS(xy) = \Sigma xy - \frac{(\Sigma x)(\Sigma y)}{n} = 10{,}734 - \frac{(225)(539)}{10} = -1{,}393.5$$

To determine the slope of the least squares line, substitute $SS(x)$ and $SS(xy)$ into Formula 3.4.

$$b_1 = \frac{SS(xy)}{SS(x)} = \frac{-1{,}393.5}{1{,}092.5} = -1.2755$$

The y-intercept is given by Formula 3.5.

$$b_0 = \bar{y} - b_1\bar{x} = \frac{539}{10} - (-1.2755)\left(\frac{225}{10}\right) = 82.60$$

Thus, the equation of the least squares line is

$$y = b_0 + b_1 x$$

$$y = 82.6 - 1.28x.$$

Notice that in calculating b_1 we retained 4 decimal places (-1.2755) and rounded this off to -1.28 when placed in the least squares equation. Four places were initially retained because b_1 was subsequently used in an intermediate calculation to obtain b_0.

Also note that, since the slope can be written as

$$b_1 = \frac{\Delta y}{\Delta x} = -1.28 = \frac{-12.8}{10},$$

we see that a temperature increase of 10 degrees ($\Delta x = 10$) will result in a decrease of approximately 13 sales ($\Delta y = -12.8$).

The least squares line $y = 82.6 - 1.28x$ provides us with a model for predicting the number of daily hot cereal sales (y) from a knowledge of the temperature low (x) for the day. For instance, suppose the weather bureau reports that the low temperature for tomorrow is expected to be 28°F. Then the chef could use the regression equation to predict how many bowls of cereal will be sold and, consequently, how much cereal should be prepared for that day. The best prediction of tomorrow's sales is

$$y = 82.6 - 1.28(28)$$

$$= 46.76$$

$$\approx 47$$

It should be noted that we do not expect that exactly 47 bowls of cereal will be sold tomorrow. This is only an estimate of the average number of sales for all winter days having a temperature low of 28 degrees. Moreover, this prediction is based on an estimated model. A more indepth discussion of this will be given in Chapter 14 on Regression Analysis.

You can have MINITAB obtain the least squares line by employing the command **REGRESS.** Suppose you want the least squares equation for a set of data points whose x and y values are stored in columns C1 and C2, respectively. Type the following command.

```
REGRESS C2 1 C1
```

If it is easier to remember, you could type something like

```
REGRESS the y values in C2 on 1 x variable in C1
```

Notice that the **REGRESS** command must be followed by the column of y values, then the number 1, and, last, by the column of x values. Because we are only considering problems involving one independent variable x in this chapter, you may think that the number 1 is redundant in the command. However, the **REGRESS** command is also used in multiple regression that involves k ($k > 1$) independent variables, in which case the 1 would be replaced by k in the command.

The MINITAB output for the cereal and temperature data is given in Exhibit 3.3. The first information produced by MINITAB in Exhibit 3.3 is the equation of the least squares line. This is

$$\text{SALES} = 82.6 - 1.28 \text{ TEMP}.$$

If you want more decimal digits in the coefficients, these can be obtained from the 'Coef' column just below the regression equation. Using these, we obtain

$$\text{SALES} = 82.599 - 1.2755 \text{ TEMP}.$$

MINITAB has also generated considerably more information than we are prepared to interpret at this point. An understanding of the remaining output produced by the **REGRESS** command will be acquired as these topics are encountered in later chapters.

Exhibit 3.3

```
MTB > REGRESS C2 1 C1        # Comment: y values in C2, x in C1

The regression equation is
SALES = 82.6 - 1.28 TEMP

Predictor        Coef        Stdev      t-ratio         p
Constant       82.599        8.305         9.95     0.000
TEMP          -1.2755       0.3348        -3.81     0.005

s = 11.06       R-sq = 64.5%      R-sq(adj) = 60.0%

Analysis of Variance

SOURCE         DF           SS           MS          F         p
Regression      1       1777.4       1777.4      14.52     0.005
Error           8        979.5        122.4
Total           9       2756.9

Unusual Observations
Obs.     TEMP      SALES       Fit Stdev.Fit   Residual    St.Resid
  2      10.0      50.00     69.84      5.45     -19.84      -2.06R

R denotes an obs. with a large st. resid.
```

Example 3.8   The placement test scores (x) and course averages (y) given in Example 3.7 are repeated in the first two columns of Table 3.6.

1. Use Formulas 3.4 and 3.5 to find the least squares line.
2. Predict the course average of a student who receives a placement score of 20.
Ⓜ 3. Use MINITAB to obtain the least squares line.

Table 3.6

Calculations Needed to Compute SS(   ) Terms

Test Score (x)	Course Avg (y)	xy	x^2	y^2
21	69	1,449	441	4,761
17	72	1,224	289	5,184
21	94	1,974	441	8,836
11	61	671	121	3,721
15	62	930	225	3,844
19	80	1,520	361	6,400
15	65	975	225	4,225
23	88	2,024	529	7,744
13	54	702	169	2,916
19	75	1,425	361	5,625
16	80	1,280	256	6,400
25	93	2,325	625	8,649
8	55	440	64	3,025
14	60	840	196	3,600
17	64	1,088	289	4,096
254	1,072	18,867	4,592	79,026

Solution

1. To facilitate the computation of $SS(x)$ and $SS(xy)$, we first construct Table 3.6. From the column totals we have the following.

$$SS(x) = \Sigma x^2 - \frac{(\Sigma x)^2}{n} = 4{,}592 - \frac{(254)^2}{15} = 290.9333$$

$$SS(xy) = \Sigma xy - \frac{(\Sigma x)(\Sigma y)}{n} = 18{,}867 - \frac{(254)(1{,}072)}{15} = 714.4667$$

Using Formula 3.4, the slope of the least squares line is

$$b_1 = \frac{SS(xy)}{SS(x)} = \frac{714.4667}{290.9333} = 2.4558.$$

From Formula 3.5, the y-intercept is

$$b_0 = \bar{y} - b_1\bar{x} = \frac{1{,}072}{15} - 2.4558 \left(\frac{254}{15} \right) = 29.88.$$

The equation of the least squares line is

$$y = 29.88 + 2.46x.$$

Note that since the slope is 2.46, the model indicates that each increase of one point

($\Delta x = 1$) in the placement score results in about a two and one-half point ($\Delta y = 2.46$) increase in the course average.

2. To predict the course average of a student with a test score of 20, we substitute a value of 20 for x in the least squares equation.

$$y = 29.88 + 2.46(20) = 79.08 \approx 79$$

3. In Example 3.7 we had MINITAB store the x and y values in columns C1 and C2, respectively, and the columns were named 'TEST' and 'CRSE AVG'. Proceeding from that point, type the command

```
REGRESS C2 1 C1
```

MINITAB will produce the output in Exhibit 3.4. From this, the least squares equation is given as

$$\text{CRSE AVG} = 29.9 + 2.46 \text{ TEST}.$$

Exhibit 3.4

```
MTB > REGRESS C2 1 C1

The regression equation is
CRSE AVG = 29.9 + 2.46 TEST

Predictor        Coef        Stdev      t-ratio          p
Constant       29.882        7.304         4.09      0.001
TEST           2.4558       0.4175         5.88      0.000

s = 7.121        R-sq = 72.7%       R-sq(adj) = 70.6%

Analysis of Variance

SOURCE          DF           SS           MS          F          p
Regression       1       1754.6       1754.6      34.60      0.000
Error           13        659.2         50.7
Total           14       2413.7
```

In the preceding example, the least squares line was used to predict the course average of a student with a test score of $x = 20$. In Chapter 14 it will be shown that the prediction precision tends to deteriorate as x is chosen farther away from $\bar{x}$ (for this example $\bar{x}$ equals $\frac{254}{15} = 16.93$). Consequently, one should be cautious in using the least squares line to predict at values of x distant from the mean of the x values in the sample. Another reason for caution is that while a straight line might be a good model near $\bar{x}$, it could be a poor choice near the extremes of the x values for which the model was developed. For instance, curvature in the relationship between x and y might be present beyond the end points for x.

Before completing this section, we will briefly discuss why the least squares line is generally considered the line of best fit. What special property does it possess that makes it unique among the infinite number of possible straight line models? To

The Census Bureau's population projections in 1964 illustrate the risks of predicting far into the future. It projected that the U.S. population would rise to 207.5 million in 1970, and this turned out to be very close to the actual figure of 205.1 million. However, its projection of 260.2 million for 1985 overestimated the actual number by 20.9 million. *(© Bonnie Freer/Photo Researchers)*

answer this, consider the four data points (2, 1), (4, 8), (6, 9), and (7, 4). The least squares line for these points can be shown to be $y = 1.7966 + 0.7797x$. In Figure 3.6 we have sketched this line and the 4 data points. Each vertical line represents the difference between the y value for the data point and the predicted y value computed from the least squares line. For instance, for the data point (4, 8) the predicted value of y obtained by substituting $x = 4$ in the least squares equation is

$$y = 1.7966 + 0.7797(4)$$
$$= 4.92.$$

The difference between the collected value of y and this predicted value is $(8 - 4.92) = 3.08$. This is called the **error** (or **residual**) at the data point (4, 8). Table 3.7

Figure 3.6
Data points and their least squares line. (Vertical line denotes the error.)

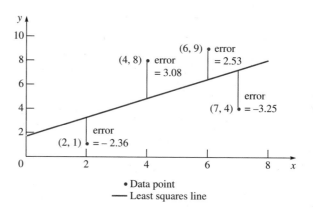

Table 3.7

Errors of the Data Points from the Least Squares Line

x	y	Predicted y	Error	(Error)^2
2	1	3.36	-2.36	5.57
4	8	4.92	3.08	9.49
6	9	6.47	2.53	6.40
7	4	7.25	-3.25	10.56
			0	32.02

gives the error for each of the 4 data points. Notice that points with positive errors lie above the least squares line, and those with negative errors lie below it.

> **NOTE:**   The least squares line is unique in that it is the line for which the sum of the squares of the errors (SSE) is minimized.

No other line would give a smaller value of SSE for the four given data points. This minimum error sum of squares is the total of the last column in Table 3.7, namely, SSE = 32.02. Notice from the total of the error column, the sum of the errors is zero. This is true in general for the least squares line. The positive and negative errors from the least squares line will always cancel each other out.

Sections 3.2 & 3.3 Exercises
Scatter Diagrams and Linear Regression

3.25 Consider the following 5 data points:

x	0	2	4	6	8
y	2	4	9	12	15

   a. Draw a scatter diagram.
   b. From the scatter diagram, does it appear that a straight line might provide a good fit to the data points?

3.26 Construct a scatter diagram for the following 7 data points. From the plot do you think that a straight line should be used to model the relationship between x and y?

x	-3	-2	-1	0	1	2	3
y	0	1	5	4	6	6	10

3.27 Obtain the least squares line for the data points in Exercise 3.25.

3.28 Find the least squares line for the 7 points in Exercise 3.26.

3.29 In Exercise 3.22 the reader was asked to calculate $SS(x)$ and $SS(xy)$ for the following points. Use those results to find the least squares line.

$$(3, 8), (5, 12), (9, 20), (0, 2), (-5, -12), (1, 4)$$

3.30 The tread depth (in hundredths of an inch) and the number of miles of usage (in thousands) are given below for a sample of 10 tires of the same brand.

Miles (x)	39	37	35	15	25	38	18	60	36	40
Tread (y)	10	15	16	30	21	14	34	3	10	14

a. Construct a scatter diagram.
b. Use the values of $SS(x)$ and $SS(xy)$ found in Exercise 3.23 to obtain the least squares line.
c. Graph the least squares line on the scatter diagram. Does the line appear to fit the data points well?
d. Use the least squares line to predict the tread depth of a tire with 50,000 miles of wear.

3.31 Consider the following 4 data points:

x	0	1	3	4
y	1	0	4	5

a. Obtain the least squares line.
b. Construct a table similar to Table 3.7 that shows the predicted value of y, the error, and the squared error for each data point.
c. From the table constructed in Part b, find SSE, the sum of the squared errors, and show that the sum of the errors is zero.
d. Make a scatter diagram and also graph the least squares line. For each data point, show its error by drawing a vertical line from the point to the least squares line.

3.32 The average prices (in dollars) per ounce for gold and silver for the years 1980 through 1987 are given below (Source: U.S. Bureau of Mines, *Minerals Yearbook*).

Year	Gold	Silver
80	613	20.63
81	460	10.52
82	376	7.95
83	424	11.44
84	361	8.14
85	318	6.14
86	368	5.47
87	448	7.01

a. Construct a scatter diagram, where x is the silver price and y is the gold price.
b. Fit a least squares line relating gold price y to silver price x.

c. Predict the price of an ounce of gold when the price of an ounce of silver is $8.00.

3.33 In Wall Street parlance, the "January effect" refers to a historic pattern in which low-priced stocks, such as those traded on the NASDAQ exchange, often outperform the Dow Jones Industrial Average (DJIA) during January. The table below gives the percentage change during January for the DJIA and the NASDAQ composite for the years 1980 through 1989.

Year	DJIA % Change (x)	NASDAQ % Change (y)
80	4.4	7.0
81	−1.7	−2.2
82	−0.4	−3.8
83	2.8	6.9
84	−3.0	−3.6
85	6.2	12.7
86	1.6	3.3
87	13.8	12.4
88	1.0	4.3
89	7.2	4.7

Source: The *Wall Street Journal,* December 4, 1989.

a. Construct a scatter diagram of the data points.
b. Obtain the least squares line to approximate the relationship between the January percentage changes in the DJIA and the NASDAQ composite.
c. Use the model obtained in Part b to predict the percentage change in the NASDAQ composite for a January in which the DJIA increases by 2 percent.

3.34 Can physical exercise extend a person's life? This popular belief was supported in a recent study by Paffenbarger, Hyde, Wing, and Hsieh. They examined the physical activity and other life-style characteristics of 16,936 Harvard alumni for relationships to lengths of life ("Physical Activity, All-Cause Mortality, and Longevity of College Alumni," *New England Journal of Medicine,* 1986, 314). The table below gives estimates of years of added life gained by men expending 2,000 or more kcal per week on exercise, as compared with those expending less than 500 kcal.

Age at the Start of Follow up (x)	Estimated Years of Added Life (y)
37	2.51
42	2.34
47	2.10
52	2.11
57	2.02
62	1.75
67	1.35
72	0.72
77	0.42

a. Construct a scatter diagram of the data points (observe that there appears to be approximately a linear relationship between x and y).
b. Model the relationship between y and x by obtaining the least squares line.
c. Use the least squares line to predict the number of years of added life for a male who was 47 at the start of the follow up. Compare your answer with the y-value for the data point with $x = 47$.

MINITAB Assignments

Ⓜ 3.35 Have MINITAB construct a scatter diagram for the ages and years of added life in Exercise 3.34.

Ⓜ 3.36 Use MINITAB to obtain the least squares line for the points in Exercise 3.34.

Ⓜ 3.37 Use MINITAB to construct a scatter diagram for the following heights (inches) and weights (pounds) of a sample of 12 members of a high school football team.

Player	Height (x)	Weight (y)
1	62	135
2	68	182
3	69	168
4	73	198
5	70	174
6	68	159
7	75	221
8	72	197
9	71	182
10	70	170
11	66	154
12	77	234

Ⓜ 3.38 For the data in Exercise 3.37, use MINITAB to construct a scatter diagram with the y-axis starting at 120 and having an increment of 40. Have the x-axis start at 60 with a spacing of 4.

Ⓜ 3.39 Have MINITAB obtain the least squares line for the sample of heights and weights in Exercise 3.37.

3.4 Correlation: Measuring the Usefulness of the Model

We have seen that the first step in describing the relationship between two variables is to construct a scatter diagram for the data points. If this graph reveals that the points tend to lie close to a straight line, then we proceed to obtain the least squares line for approximating this linear relationship. However, before we adopt the least squares line as our model, we need to measure in some way how well a straight line describes the association between x and y. That is, how "close" do the data points tend to fall to a straight line? The most familiar measure of the strength of a linear relationship is Pearson's product-moment **correlation coefficient** that is defined in Formula 3.6.

Pearson's sample product-moment **coefficient of linear correlation** r is given by the formula

$$r = \frac{SS(xy)}{\sqrt{SS(x)SS(y)}}.$$ (3.6)

A number of important points concerning the correlation coefficient r are noted below.

Properties of the Correlation Coefficient r:
1. The value of r is always between -1 and $+1$ inclusive.
2. r has the same sign as b_1, the slope of the least squares line.
3. r is near $+1$ when the data points fall close to a straight line that is rising; that is, when y tends to increase as x increases (see Figure 3.7).
4. r is near -1 when the data points fall close to a straight line that is falling; that is, when y tends to decrease as x increases (see Figure 3.8).
5. If all the data points fall exactly on a straight line with positive slope, then $r = +1$.
6. If all the data points fall exactly on a straight line with negative slope, then $r = -1$.
7. A value of r near 0 indicates little or no linear relationship between y and x (see Figure 3.9).

Figure 3.7
Large positive correlation;
$r = 0.97$.

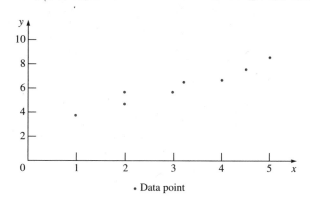

• Data point

Figure 3.8
Large negative correlation;
$r = -0.93$.

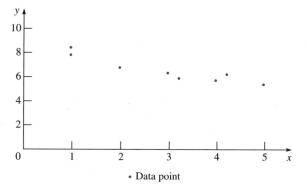

• Data point

Figure 3.9
Lack of correlation; $r = 0.01$.

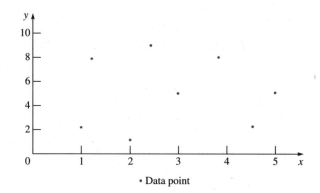

• Data point

Example 3.9   In Section 3.3, Table 3.5 was constructed to obtain the least squares line for the hot cereal sales example. The table is reproduced below.

Temperature (x)	Sales (y)	xy	x^2	y^2
18	64	1,152	324	4,096
10	50	500	100	2,500
5	80	400	25	6,400
24	60	1,440	576	3,600
38	22	836	1,444	484
14	72	1,008	196	5,184
20	58	1,160	400	3,364
32	31	992	1,024	961
29	54	1,566	841	2,916
35	48	1,680	1,225	2,304
225	539	10,734	6,155	31,809

From the column totals we obtained

$$SS(x) = 1,092.5, \quad SS(y) = 2,756.9, \quad SS(xy) = -1,393.5.$$

The correlation coefficient is

$$r = \frac{SS(xy)}{\sqrt{SS(x)SS(y)}} = \frac{-1,393.5}{\sqrt{(1,092.5)(2,756.9)}} = -0.80.$$

This rather large negative value of r suggests that the relationship between hot cereal sales and the daily temperature low is approximately linear, and that sales y tend to decrease as the temperature x increases.

Perhaps you are wondering how close r has to be to ± 1 to be considered "large." An understanding of testing statistical hypotheses is required to answer this definitively. The purpose of the present chapter is to provide you with an introduction to modeling the relationship between two variables—an important and frequently occurring problem in statistics. A more detailed and thorough development of this topic is presented in Chapter 14.

Example 3.10   Find the correlation coefficient for the test scores and course averages in Example 3.8.

Solution

Table 3.6 from Example 3.8 is repeated below.

Test Score (x)	Course Avg (y)	xy	x^2	y^2
21	69	1,449	441	4,761
17	72	1,224	289	5,184
21	94	1,974	441	8,836
11	61	671	121	3,721
15	62	930	225	3,844
19	80	1,520	361	6,400
15	65	975	225	4,225
23	88	2,024	529	7,744
13	54	702	169	2,916
19	75	1,425	361	5,625
16	80	1,280	256	6,400
25	93	2,325	625	8,649
8	55	440	64	3,025
14	60	840	196	3,600
17	64	1,088	289	4,096
254	1,072	18,867	4,592	79,026

Using the column totals of the table, we obtained earlier

$$SS(x) = 290.9333$$

$$SS(xy) = 714.4667$$

To calculate r, we also need the value of $SS(y)$.

$$SS(y) = \Sigma y^2 - \frac{(\Sigma y)^2}{n} = 79,026 - \frac{(1,072)^2}{15} = 2,413.7333$$

Applying Formula 3.6, we obtain

$$r = \frac{SS(xy)}{\sqrt{SS(x)SS(y)}} = \frac{714.4667}{\sqrt{(290.9333)(2,413.7333)}} = 0.85.$$

The fact that r is quite close to $+1$ indicates that the data points tend to lie near the least squares line and that course averages tend to increase with higher test scores.

NOTE: The presence of a strong positive correlation does not imply a causal relationship between two variables.

In the last example, a high placement test score is not causing the course average to be high. The placement test is given before the course begins, and the score on the placement exam is not involved in the computation of the course average. Other factors, such as motivation, native mathematical ability, and previous mathematical experiences, cause placement test scores and course averages to increase together. One must be careful not to conclude that a causal relationship exists whenever two variables are highly correlated.

MINITAB uses the command **CORRELATION** to produce the correlation coefficient. Earlier we stored the test scores and course averages in columns C1 and C2 and assigned the names TEST and CRSE AVG, respectively. To obtain the coefficient of correlation between these two variables, type

```
CORRELATION C1 C2
```

MINITAB will respond with the computer output given in Exhibit 3.5.

Exhibit 3.5

```
MTB > CORRELATION C1 C2    # Can also write C2 C1

Correlation of TEST and CRSE AVG = 0.853
```

Ⓜ**Example 3.11**   At the beginning of this chapter, the number of wins (y) and the earned run average (x) were given for each American League team at the end of the 1990 season. The scatter diagram in Figure 3.1 suggested that the relationship between x and y might be approximated by a straight line. To measure the strength of a linear relation, we had MINITAB calculate the correlation coefficient r. The result appears in Exhibit 3.6, from which we conclude that $r = -0.817$. The closeness of r to -1 indicates that the relationship between the earned run average and the number of wins is described quite well by the least squares line sketched in Figure 3.1, repeated below. In Exercise 3.52, the reader is asked to obtain the equation of this estimated model.

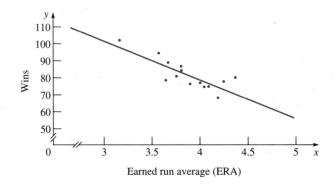

Earned run average (ERA)

Exhibit 3.6

```
MTB > READ C1 C2
DATA> 3.18   103
DATA> 3.61    94
DATA> 3.69    77
DATA> 3.72    88
DATA> 3.79    80
DATA> 3.83    83
DATA> 3.84    86
DATA> 3.93    75
DATA> 4.04    76
DATA> 4.08    74
DATA> 4.12    74
DATA> 4.21    67
DATA> 4.26    77
DATA> 4.39    79
DATA> END
MTB > NAME C1 'ERA' C2 'WINS'
MTB > CORR C1 C2

Correlation of ERA and WINS = -0.817
```

Section 3.4 Exercises
Coefficient of Correlation

In Exercises 3.40–3.47, state whether the variables x and y are likely to be positively correlated, negatively correlated, or uncorrelated.

3.40 X is the weight of an automobile, and y is its average miles per gallon.

3.41 X is an adult's age, and y is the time required to run a mile.

3.42 X is the distance from a student's home to his/her college, and y is the student's grade point average.

3.43 X is a company's advertising expenditures, and y is its sales revenue.

3.44 X is a student's height, and y is his/her SAT math score.

3.45 X is the horsepower of a truck's engine, and y is the load capacity of the truck.

3.46 X is the quantity of alcohol consumed by a laboratory animal, and y is the speed at which it can run a maze.

3.47 X is an adult's age, and y is his/her systolic blood pressure.

3.48 For each of the following correlation coefficients, what can be said about the relationship between the x and y values of the data points?
   a. $r = 0.98$ b. $r = -0.05$
   c. $r = -0.89$ d. $r = 1$

3.49 Construct a scatter diagram and calculate the correlation coefficient for the following data points.

x	0	2	5	8	10
y	0	3	4	5	7

3.50 For the following data points, draw a scatter diagram and obtain the correlation coefficient.

x	-9	-7	-5	-3	-1	1	3	5	7	9
y	0	2	3	5	5	6	7	6	7	8

3.51 In Exercise 3.30 the least squares line was fit to the following data consisting of the tread depth and the number of miles of usage for a sample of 10 tires.

Miles (x)	39	37	35	15	25	.38	18	60	36	40
Tread (y)	10	15	16	30	21	14	34	3	10	14

Calculate the correlation coefficient r and interpret its value.

3.52 Many baseball experts consider pitching the key to a successful season. A frequently used measure of a pitching staff's effectiveness is its earned run average (ERA), which a pitcher strives to keep as low as possible. Table 3.1, reproduced below, gives the number of wins and the earned run averages of the American League teams for the 1990 season.

Team	ERA (x)	Number of Wins (y)
Oakland	3.18	103
Chicago	3.61	94
Seattle	3.69	77
Boston	3.72	88
California	3.79	80
Texas	3.83	83
Toronto	3.84	86
Kansas City	3.93	75
Baltimore	4.04	76
Milwaukee	4.08	74
Minnesota	4.12	74
New York	4.21	67
Cleveland	4.26	77
Detroit	4.39	79

For the above data, $\Sigma x = 54.69$, $\Sigma y = 1{,}133$, $\Sigma x^2 = 214.9087$, $\Sigma y^2 = 92{,}815$, and $\Sigma xy = 4{,}395.19$.

a. Construct a scatter diagram of the 14 data points.

b. Model the relationship between the number of wins and the earned run average by obtaining the least squares line.

c. The earned run average of the Milwaukee Brewers was 4.08. Use the model to predict the number of wins for a team with this ERA. Compare the prediction with the number of wins achieved by Milwaukee.

3.53 The data given in Exercise 3.52 suggest that the number of games won by a major league team is related to the team's earned run average. Is there a relationship between the number of wins and a team's batting average? These figures are given below for the American League teams during the 1990 season. Calculate the correlation between the number of wins and the batting average. Do you expect this relationship to be positively or negatively correlated?

Team	Batting Average (x)	No. of Wins (y)
Oakland	.254	103
Chicago	.258	94
Seattle	.259	77
Boston	.272	88
California	.260	80
Texas	.259	83
Toronto	.265	86
Kansas City	.267	75
Baltimore	.245	76
Milwaukee	.256	74
Minnesota	.265	74
New York	.241	67
Cleveland	.267	77
Detroit	.259	79

For the above, $\Sigma x = 3.627$, $\Sigma y = 1,133$, $\Sigma x^2 = 0.940577$, $\Sigma y^2 = 92,815$, and $\Sigma xy = 293.746$.

3.54 Calculate the correlation coefficient for the gold and silver prices per ounce given in Exercise 3.32 and repeated below. Interpret the resulting value.

Year	Gold	Silver
80	613	20.63
81	460	10.52
82	376	7.95
83	424	11.44
84	361	8.14
85	318	6.14
86	368	5.47
87	448	7.01

3.55 Calculate and interpret the correlation coefficient for the age and years of added life data in Exercise 3.34 and copied below.

Age at the Start of Follow up (x)	Estimated Years of Added Life (y)
37	2.51
42	2.34
47	2.10
52	2.11
57	2.02
62	1.75
67	1.35
72	0.72
77	0.42

MINITAB Assignments

Ⓜ 3.56 Use MINITAB to obtain the correlation coefficient for gold and silver prices given in Exercise 3.54.

Ⓜ 3.57 Use MINITAB to determine the correlation coefficient between batting average and number of wins for the American League teams listed in Exercise 3.53.

Ⓜ 3.58 Use MINITAB to calculate the correlation coefficient for the following heights (inches) and weights (pounds) of a sample of 12 members of a high school football team.

Player	Height (x)	Weight (y)
1	62	135
2	68	182
3	69	168
4	73	198
5	70	174
6	68	159
7	75	221
8	72	197
9	71	182
10	70	170
11	66	154
12	77	234

— Ⓜ 3.59 The owner of an expensive gift shop believes that her weekly sales are related to the performance of the stock market for that week. To explore this possibility, she determines the amount of sales and the mean of the Dow Jones Industrial Average for each of 16 randomly selected weeks. These figures are given below, with the sales figures expressed in units of $1,000.

DJIA	Sales
2215	58.3
2518	62.9
1781	46.3
1823	48.2
2117	58.2
2703	65.8
1423	36.7
1532	32.3
1879	52.7
1713	39.3
2122	58.7
2346	60.9
1629	40.5
2609	70.3
1515	39.1
1687	45.9

Use MINITAB to:

a. construct a scatter diagram with the DJIA on the x-axis and sales as the dependent variable;
b. obtain the least squares line for the purpose of predicting weekly sales;
c. calculate the correlation coefficient r for the DJIA figures and the corresponding sales in order to measure the strength of a linear relationship.

Looking Back

This chapter has been concerned with investigating the relationship between two variables. There are many problems in which one is interested in determining whether x and y are related and, if so, how. If they are related, is it possible to describe the relationship by means of a mathematical model (equation)? If this can be done, then one could use the model to predict values of y for specified values of x. The chapter has provided an introduction to addressing this type of problem, and it is suggested that you proceed as follows:

1. With the collected data, make a scatter diagram to discern what type of relationship might exist between x and y (in this chapter, only straight line relations are considered).
2. If the scatter diagram indicates that the data points tend to lie close to a straight line, then obtain the least squares line for consideration as a possible model.

   The equation of the least squares line is

   $$y = b_0 + b_1 x,$$

   where

   $$b_1 = \frac{SS(xy)}{SS(x)}$$

   and

   $$b_0 = \bar{y} - b_1 \bar{x}.$$

3. Before adopting the model, assess its usefulness. There are several formal ways of determining the utility of the model, and these are discussed in Chapter 14. In this introductory chapter we calculated the correlation coefficient r to measure the strength of the linear relationship between x and y.

   The equation of the correlation coefficient is

   $$r = \frac{SS(xy)}{\sqrt{SS(x)SS(y)}}.$$

   A value of r close to -1 or $+1$ indicates a good fit between the data points and the least squares line.

Key Words

In reviewing this chapter, you should be able to define, explain, and illustrate each of the following.

bivariate data *(page 98)*

mathematical model *(page 99)*

straight line *(page 100)*

slope *(page 100)*

y-intercept *(page 100)*

$SS(x)$, $SS(y)$, $SS(xy)$ *(page 102)*

scatter diagram *(page 106)*

regression equation *(page 106)*

prediction equation *(page 106)*

independent (predictor) variable
(page 106)

dependent (response) variable
(page 106)

least squares line *(page 111)*

predicted value *(page 112)*

error (residual) *(page 116)*

error sum of squares (SSE) *(page 117)*

correlation coefficient *(page 120)*

causal relationship *(page 124)*

Ⓜ *MINITAB Commands*

READ _ _ *(page 107)*   PLOT _ _; *(page 107)*

END *(page 107)*   XSTART _ _;

NAME _ _ *(page 107)*   XINCREMENT _;
                                         YSTART _ _;
REGRESS _ _ _ *(page 112)*   YINCREMENT _.

CORRELATION _ _ *(page 124)*

Review Exercises

3.60 Determine the y-intercept and the slope, and graph the line whose equation is $5x + 2y = 10$.

3.61 Find the equation of the line with slope $-\frac{7}{3}$ and y-intercept $\frac{1}{6}$.

3.62 A local automobile dealership pays each salesperson a weekly salary of $135 plus $75 for each car that he or she sells that week. Let x denote the number of cars sold in a week, and let y denote a salesperson's salary for that week.
a. Find the equation that gives the relation between y and x.
b. Give the slope and y-intercept of the line.

3.63 An experiment is conducted to enhance a child's ability to reason deductively. Each child is tested before and after the experiment. The increase (y) in test scores and the age (x) of a child are modeled by the equation

$$y = 13.2 + 5.0x.$$

If two brothers differ in age by two years, by how much are their scores on this test expected to differ?

3.64 Draw a scatter diagram of the following data points.

x	-5	0	4	5	10
y	10	8	4	0	-5

3.65 Obtain the least squares line for the points in Exercise 3.64.

3.66 Calculate the correlation coefficient of the data in Exercise 3.64, and interpret its value.

3.67 In addition to the slope of the least squares line and its correlation coefficient having the same sign, the two quantities are related by the equation

$$r = b_1 \sqrt{\frac{SS(x)}{SS(y)}}$$

Show that this relation holds for the following data points.

x	0	1	3	4
y	1	0	4	5

3.68 The Jebsen-Taylor Hand Function Test is used to measure the recovery of coordination after traumatic injury. The following are the times after injury (in weeks) and the scores on one subtest for eight patients with similar medial nerve injuries.

Time After Injury	x	3	2	5	6	2	4	10	5
Subtest Score	y	6	8	5	3	7	6	3	4

a. Make a scatter diagram of the data.
b. Model the relationship between x and y by obtaining the least squares line.
c. Find the correlation coefficient for x and y, and interpret its value.
d. Predict the score of a patient who was injured 7 weeks ago.

3.69 The Dow Jones Utility Average uses the prices of 15 stocks to measure the stock market performance of utility companies. The table below gives the percentage change in the price of a share for each company during November, 1989.

Company	Closing Price (x) on 10/31	Percent Change (y) for November
Pacific Gas & Electric	20.00	6.9
Consolidated Edison	25.50	6.4
Consolidated Natural Gas	43.88	5.1
Columbia Gas System	45.50	4.9
Panhandle Eastern	27.63	4.5
SCEcorp.	36.88	3.7
Detroit Edison	24.13	3.6
Public Service Enterprises	27.25	3.2
Peoples Energy	23.75	3.2
Commonwealth Edison	38.50	1.3
Centerior Energy	19.38	1.3
American Electric Power	30.88	0.4
Niagara Mohawk Power	14.13	− 0.9
Houston Industries	35.25	− 1.8
Philadelphia Electric	22.88	− 3.3

Source: *The Wall Street Journal,* December 8, 1989.

For the above data, $\Sigma x = 435.54$, $\Sigma y = 38.5$, $\Sigma x^2 = 13,836$, $\Sigma y^2 = 224.45$, and $\Sigma xy = 1,194.7$.

a. Calculate r to measure the correlation between the percentage change and share price.
b. Does it appear that higher priced stocks tend to have the larger percentage increases?

3.70 The time required for a factory worker to install a certain component in a video camcorder appears to be related to the number of days of experience with this procedure. Installment times (in seconds) and the numbers of days of experience appear below for a sample of 10 workers.

Experience	x	6	8	10	8	1	3	2	5	5	7
Time Required	y	32	30	25	28	39	35	40	30	33	38

a. Construct a scatter diagram of the data points.
b. Obtain the least squares line to approximate the relationship between y and x.
c. Calculate the correlation coefficient r and interpret its value.

MINITAB Assignments

Ⓜ 3.71 Use MINITAB to solve Exercise 3.70.

Ⓜ 3.72 The following figures give the daily attendance and the number of hot dog sales for a sample of 10 games of a minor league baseball team.

Attendance (x)	Hot Dog Sales (y)
8747	6845
5857	4168
8360	5348
6945	5687
8688	6007
4534	3216
7450	5018
5874	4652
9821	7002
5873	3897

Use MINITAB to construct a scatter diagram.

Ⓜ 3.73 Using MINITAB and the data in Exercise 3.72, obtain the least squares line for the purpose of predicting hot dog sales.

Ⓜ 3.74 Measure the strength of the linear relationship in Exercise 3.72 by using MINI-TAB to find the correlation coefficient.

(© Courtesy of the National Optical Astronomy Laboratory)

4 Probability: Measuring Uncertainty

◀ ***The Question Persists: Are There Civilizations Beyond Our Galaxy?***
"Fifty years ago physicist Enrico Fermi posed a question that has intrigued scientists ever since: Are there civilizations beyond our galaxy, and, if so, why have they made no attempt to contact us? Fermi's question was a logical one given the common view of the cosmos at the time. There were, after all, billions of stars in the pin-wheel-shaped disk of stars, dust, and gases making up our Milky Way galaxy. If only a tiny fraction of these had evolved similarly to the sun, there still had to be millions of stars with planetary systems suitable for life. Even if the appearance of life is an extremely rare occurrence, and its evolution to an advanced stage still more rare, there still should be thousands of advanced civilizations scattered throughout the galaxy. Where, asked Fermi, are they?" (From an article by Clair Wood, Science Columnist, the Bangor (Maine) Daily News, September 24–25, 1988. Reprinted with permission from the Bangor Daily News, Bangor, Maine 04401.)

Looking Ahead

Like most sciences, the birth of probability can be traced to the desire to solve problems for which no reasonable method existed at the time. For centuries, scholars had been intrigued by profound issues of uncertainty, such as the existence of life on other planets. Before the seventeenth century, however, relatively little had been accomplished in measuring uncertainty and the likelihood of things happening. In 1654, modern probability theory began, not from a complex issue such as the uncertainty of life, but from investigations of games of chance. A gambler, the Chevalier de Méré, posed some gaming questions to the French mathematician Blaise Pascal. Pascal wrote about his solutions to another mathematician, Pierre Fermat, and their exchanges gave rise to a systematic development of probability and launched an extremely fertile period in this field. In 1657 Christian Huygens, a Dutch scientist, wrote what is generally regarded as the first published textbook

Blaise Pascal (1623–1662).
(Courtesy of The Bettmann Archive)

on probability. The period around 1660 resulted in several contributions of probability applications in a variety of disciplines.

Today, the term probability and similar words are frequently used in one's writings and conversations:

I'll *probably* go to the movies tonight.

Drawing a flush is less *likely* than a straight in five-card poker.

The *odds* are slim that the Orioles will be able to sweep their home series with the Yankees.

The *chance* of winning the state's daily lottery this evening is only 1 in 1,000.

Over the course of a year, the *likelihood* of an accident at a particular intersection is 10 percent.

Rain tomorrow is *probable.*

In this chapter, you will learn how the use of probability can be made more precise by quantifying the concept. We will begin the chapter by looking at some different ways of assigning probabilities. This will result in the need to count the number of possible outcomes for an experiment. Consequently, we will consider some basic counting techniques, including combinations and permutations. Once it has been determined what is *possible* when an experiment is performed, we can then determine what is *probable* by employing the concepts of probability and its various rules that are discussed in the remaining sections.

4.1   Probability, Sample Spaces, and Events

Statistics involves the analysis of data, and the process of collecting data is called an **experiment.**

> An **experiment** is the process of obtaining observations or measurements.

An experiment can represent a variety of activities. It might consist of:

1. playing 15 hands of blackjack at a Las Vegas casino;
2. surveying 300 students at a local university to determine if they are left- or right-handed;
3. measuring the maximum sustainable pressure of a new beverage container;
4. counting the number of houses that have a smoke detector.

Before conducting an experiment, one is usually interested in the likelihood of possible outcomes occurring. For example, how likely is it that:

1. a player will win at least 10 of 15 blackjack hands?
2. fewer than 30 of 300 students will be left-handed?
3. in a test of 100 new beverage containers, all can withstand a pressure of 75 pounds per square inch?
4. a majority of houses in a neighborhood will have a smoke detector?

Probability is used to measure the likelihood that a particular outcome will occur when an experiment is performed. Before defining probability, we need to illustrate the concept of a **sample space** for an experiment.

> A **sample space** is the collection of all possible outcomes for an experiment.

Example 4.1 Consider the experiment in which a nickel and a dime are tossed together once. A listing of the possible outcomes is the following, where the first letter indicates whether the nickel lands heads (H) or tails (T), and the second letter shows the results for the dime. This listing of all the possible outcomes for the experiment is a sample space. It consists of four elements (HH, HT, TH, TT), each of which is called a **sample point,** and every possible outcome of the experiment is represented by exactly one sample point. That is, when the experiment is performed, one, and only one, of the four sample points will result.

   H   H   H   T   T   H   T   T

Example 4.2 For the experiment in Example 4.1 suppose we are only interested in how many heads show when the nickel and the dime are tossed. Then another possible sample space for the experiment is the following set of integers.

$$0 \quad 1 \quad 2$$

This is a sample space because each possible outcome of the experiment is represented by exactly one sample point in the listing. For example, if both coins show heads, then this outcome is represented by the sample point 2.

The preceding examples illustrate that an experiment can have more than one possible sample space. However, when determining probabilities, we often try to obtain a sample space for which the points are equally likely to occur, since this will facilitate the assignment of probabilities. Equally likely points occur in Example 4.1, but in Example 4.2 the sample point 1 is twice as likely as either 0 or 2.

Example 4.3 An experiment consists of tossing together a black and a white die once. A sample space for which the points are equally likely appears in Table 4.1. It consists of 36 ordered pairs, where the first position indicates the number of spots tossed on the black die, and the second position gives the number of spots for the white die.

Table 4.1

A Sample Space for the Toss of Two Dice

(1, 1)	(2, 1)	(3, 1)	(4, 1)	(5, 1)	(6, 1)
(1, 2)	(2, 2)	(3, 2)	(4, 2)	(5, 2)	(6, 2)
(1, 3)	(2, 3)	(3, 3)	(4, 3)	(5, 3)	(6, 3)
(1, 4)	(2, 4)	(3, 4)	(4, 4)	(5, 4)	(6, 4)
(1, 5)	(2, 5)	(3, 5)	(4, 5)	(5, 5)	(6, 5)
(1, 6)	(2, 6)	(3, 6)	(4, 6)	(5, 6)	(6, 6)

After a sample space has been constructed, interest is usually focused on a certain portion of it, which is called an **event.**

An **event** is a portion of a sample space, that is, a collection of sample points.

For example, we might be interested in obtaining a sum of 7 when the 2 dice in Example 4.3 are tossed. Then we could represent the event by E and define it as the collection of the following sample points.

$$(1, 6) \quad (2, 5) \quad (3, 4) \quad (4, 3) \quad (5, 2) \quad (6, 1)$$

When the experiment is performed, exactly one of the 36 points in the sample space will result. If the resulting point is one of the 6 points in event E, then we say that event E has occurred.

For centuries, several different variations of dice have been used in games of chance. In ancient times, gamblers tossed tali, which were polished heel bones of certain animals. These were shaped so that, when thrown, they could land upright in only 4 possible ways.

The Relative Frequency Definition of Probability

Having discussed sample spaces and events, we will now present a definition of probability and consider how probabilities can be assigned to the points of a sample space.

> **The Relative Frequency Definition of Probability:**
> The **probability of event** E is denoted by $P(E)$ and is the proportion of the time that E can be expected to occur in the long run.

Note that because $P(E)$ is a relative frequency of occurrences, we then have that $0 \leq P(E) \leq 1$, that is, a probability is always a number between 0 and 1 inclusive. Also observe that the relative frequency concept of probability involves repetitions of the experiment an unlimited number of times. Of course this is impossible. In practice, the definition is applied by repeating the experiment a large number of times and using the relative frequency of event E to approximate its actual probability. The following examples illustrate this concept.

Example 4.4

A company manufactures steel cable that is guaranteed to support a weight of 10,000 pounds. As a safety factor, the company strives to have the cable capable of holding 14,000 pounds. To estimate the probability that a roll of cable can support this weight, 500 rolls were tested, and it was found that 465 rolls could hold 14,000 pounds. If we let E be the event

E: a roll of cable can support 14,000 pounds,

then the probability of E is estimated to be

$$P(E) \approx \frac{465}{500} = 0.93.$$

Example 4.5

A soft drink bottler is considering the addition of a new flavor to its line. To assess the potential popularity of this flavor, preliminary marketing tests were conducted. In 4 test areas, 4,857 people were given samples of the product, and 3,215 said that they liked the flavor. Estimate the probability that a randomly selected consumer will like the drink.

Solution

Letting E be the event

E: a consumer likes the flavor,

we estimate the probability of E to be

$$P(E) \approx \frac{3,215}{4,857} = 0.66.$$

Ⓜ**Example 4.6**   MINITAB can be used to simulate a variety of experiments. For instance, to simulate the tossing of a fair coin 100 times, we could have MINITAB generate a sequence of 100 equally likely digits, where each is either a 1 (a head) or a 0 (a tail). This can be accomplished by using the command **RANDOM** and the subcommand **INTEGER.**

The first two commands below generate 100 digits (0 or 1) and place the results in column C1. The **SUM** command is then applied to C1 to determine the number of 1s (heads) that have been generated.

```
RANDOM 100 C1;      # Need semicolon since a subcommand follows
INTEGERS 0 to 1.    # Need period since this is last subcommand
SUM C1
```

To illustrate the relative frequency definition of probability, we had MINITAB perform the above simulation of 100 tosses of a fair coin. This was repeated 10 times for a total simulation of 1,000 tosses. The relative frequency of the number of heads was computed after each 100 tosses. The results are summarized in Table 4.2 and plotted in Exhibit 4.1.

From Table 4.2 and Exhibit 4.1 it is seen that the relative frequencies fluctuate considerably for the first few hundred tosses, but then they begin to stabilize and tend toward a probability of 0.5. As the number of tosses n increases, we expect the relative frequencies to converge to this value.

Exhibit 4.1

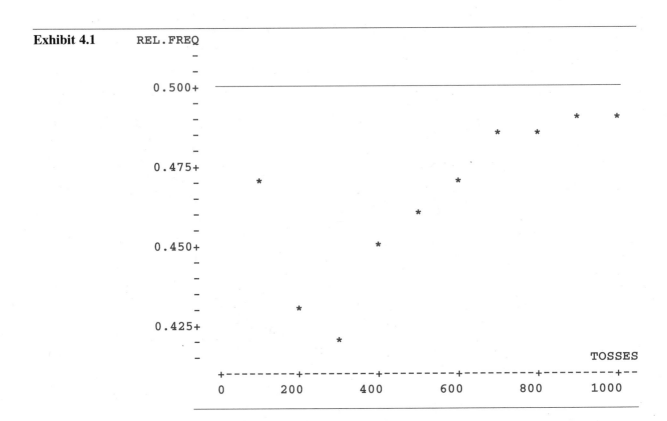

Table 4.2

Results for 1,000 Simulations of a Coin Toss

Number of Tosses n	Number of Heads x	Relative Frequency x/n
100	47	0.470000
200	86	0.430000
300	126	0.420000
400	180	0.450000
500	231	0.462000
600	283	0.471667
700	339	0.484286
800	387	0.483750
900	442	0.491111
1,000	491	0.491000

The Theoretical Method of Assigning Probabilities

The last three examples illustrate how probabilities of events can be approximated by repeating the experiment a large number of times. Instead of using these empirical approximations, it is sometimes possible to apply an alternative method of assigning probabilities. For instance, we mentioned earlier that it is advantageous to use a sample space for which the points are equally likely. When this is possible, the following method of assigning probabilities can be utilized.

The Theoretical Method of Assigning Probabilities:
Suppose a sample space has s equally likely points, and E is an event consisting of e points. Then

$$P(E) = \frac{e}{s}.$$

Example 4.7

In Example 4.1, we considered the experiment in which a nickel and a dime are tossed together once. The sample space is given below, where the first letter indicates whether the nickel shows a head (H) or a tail (T), and the second letter denotes the result on the dime.

HH HT TH TT

Assuming that each coin is fair (heads and tails are equally likely), find the probability that only one of the coins will show heads.

Solution

The sample space has $s = 4$ equally likely points, and the event of interest is

$$E: \text{only one coin shows heads.}$$

Event E contains $e = 2$ points (HT and TH). Therefore, by the theoretical method of assigning probabilities,

$$P(E) = \frac{e}{s} = \frac{2}{4} = 0.5.$$

In Example 4.2, the following sample space was given for the experiment in Example 4.1.

$$0 \quad 1 \quad 2$$

This gives the number of heads that show when the two coins are tossed. Note that the theoretical method of assigning probabilities cannot be applied to this sample space, since its points are not equally likely.

Example 4.8   Consider again the experiment that consists of tossing together a black and a white die once. Assuming that each die is fair, a sample space for which the points are equally likely is given by the following 36 ordered pairs. The first position indicates the number of spots tossed on the black die, and the second position gives the number of spots for the white die.

$$
\begin{array}{cccccc}
(1, 1) & (2, 1) & (3, 1) & (4, 1) & (5, 1) & (6, 1) \\
(1, 2) & (2, 2) & (3, 2) & (4, 2) & (5, 2) & (6, 2) \\
(1, 3) & (2, 3) & (3, 3) & (4, 3) & (5, 3) & (6, 3) \\
(1, 4) & (2, 4) & (3, 4) & (4, 4) & (5, 4) & (6, 4) \\
(1, 5) & (2, 5) & (3, 5) & (4, 5) & (5, 5) & (6, 5) \\
(1, 6) & (2, 6) & (3, 6) & (4, 6) & (5, 6) & (6, 6)
\end{array}
$$

Find the probability of tossing

1. 7 or 11 spots;
2. fewer than 5 spots.

Solution

The sample space consists of $s = 36$ equally likely points, and therefore the theoretical method of assigning probabilities can be used.

1. The event of interest is

$$A: \text{a sum of 7 or 11 is tossed.}$$

There are $a = 8$ points in event A. These include the following.

$$(1, 6) \quad (2, 5) \quad (3, 4) \quad (4, 3) \quad (5, 2) \quad (6, 1) \quad (5, 6) \quad (6, 5)$$

$$P(A) = \frac{a}{s} = \frac{8}{36} = 0.22$$

2. In this case the event of interest is

   B: fewer than 5 spots are tossed.

   There are $b = 6$ points in event B. These include the following.

$$(1, 1) \quad (1, 2) \quad (1, 3) \quad (2, 1) \quad (2, 2) \quad (3, 1)$$

$$P(B) = \frac{b}{s} = \frac{6}{36} = 0.17$$

The Subjective Method of Assigning Probabilities

It is important to keep in mind that the theoretical method of assigning probabilities is only applicable when the sample space has equally likely points. In many situations, this is not the case. For example, in tossing a pair of coins or dice, one or both may be unbalanced. When the points of a sample space are not equally likely, we try to approximate $P(E)$ empirically by using the relative frequency of E based on a large number of repetitions of the experiment. There are, however, situations that cannot be repeated several times and do not have sample spaces with equally likely points. Consequently, neither a relative frequency approximation nor the theoretical method of assigning probabilities is applicable. In such situations one can resort to a **subjective method of assigning probabilities.**

> **The Subjective Method of Assigning Probabilities:**
> Probabilities are assigned to the points of a sample space on the basis of one's personal assessments. The assignment of probabilities must satisfy:
>
> 1. $P(E_i) \geq 0$ for each sample point E_i;
> 2. $\Sigma P(E_i) = 1$.

The subjective approach allows considerable flexibility in the assignment of probabilities, requiring only that each probability is nonnegative and all the probabilities sum to one. However, the assignment should be made so that it reflects the experimenter's degree of belief that the sample points will occur.

Example 4.9   Suppose you want to assign a probability to the event that you will pass Calculus next semester. A sample space for the experiment is

$$E_1 \quad \text{and} \quad E_2$$

where E_1 and E_2 denote the events *pass* and *not pass,* respectively. Since the sample points are not equally likely, you cannot use the theoretical method of assigning probabilities. Also, you cannot repeat the course a large number of times, so a relative frequency approximation must be ruled out. Only the subjective method is applicable for assigning a probability to your passing the course. This approach relies on using your experience and expertise to assess the situation and all pertinent facts to arrive at your best estimate of the probability. Before obtaining this, you would probably weigh such factors as your mathematical aptitude and abilities, the time of day at which the course is offered and how well you function then, your course load next semester, the instructor who is teaching the course, the level of rigor of the course, and anything else that you feel would be relevant to passing the course. After considerable reflection, you might decide to make the following subjective assignment of probabilities.

$$P(E_1) = 0.8 \quad \text{and} \quad P(E_2) = 0.2$$

Perhaps you do not find the subjective approach very appealing or satisfying. In practice, however, there are situations for which an assignment of probabilities is desired, and the subjective approach is the only method applicable. The chief of planning for a corporation that is considering the construction of a new facility might need to provide its board of directors with an estimated probability of success. Before adding a new menu item, a fast-food franchise may want to estimate the probability that its total market share of sales will increase. A patient who is contemplating a newly developed experimental treatment may request an estimate of the probability that the procedure will be successful. Subjective probabilities such as the above are personal measures of the evaluator's belief that an event will occur. In order to provide meaningful estimates of subjective probabilities, considerable expertise, experience, and sound judgement are required on the part of the evaluator.

Section 4.1 Exercises

Probability, Sample Spaces, and Events

4.1   A test contains two true-false questions. Construct a sample space to show the different ways in which the two questions can be answered.

4.2   For the experiment described in Exercise 4.1, suppose a student has no knowledge of the material and answers each question by guessing. Using the theoretical method of assigning probabilities, find the probability of answering both questions correctly.

4.3   List a sample space showing the different ways that three true-false questions can be answered.

4.4   Construct a sample space that lists the different ways of answering two multiple-choice questions, if each question has three alternatives (a, b, and c).

4.5   Assume that a student has a complete lack of knowledge of the two multiple-choice questions referred to in Exercise 4.4. If each question is answered by guessing, what is the probability that the student will answer both correctly?

4.6   A die and a quarter are tossed together once.
   a. Construct a sample space.
   b. Find $P(A)$, where A is the event that the quarter shows heads.
   c. Find $P(B)$, where B is the event that more than 4 spots show on the die.
   d. Find the probability that events A and B both occur when the experiment is performed.

4.7   An experiment consists of selecting 1 card from a standard deck of 52 and tossing a coin if the card is red; if it is black, then a die is tossed.
  a. List a sample space for the experiment.
  b. Are the points in the sample space equally likely?

4.8   A computer is programmed to generate a sequence of 4 digits, where each digit is either 0 or 1, and each of these is equally likely to occur.
  a. Make a sample space that shows all possible 4-digit sequences of 0s and 1s.
  b. What is the probability that such a sequence will contain 3 zeros? At least 3 zeros?

4.9   In 1986, there were 3,757,000 live births in the United States, and 906,000 of these were by caesarean section (U.S. National Center for Health Statistics). Use these data to estimate the probability that a birth in the United States will be by caesarean section.

4.10 Of the 3,757,000 live births in the United States during 1986, 1,832,000 were female. Use these results to estimate the probability of a female birth in the United States.

4.11 For a *Newsweek* Poll, the Gallup Organization interviewed a national sample of 757 adults to assess how Americans feel about the status of the family. Based on the sample results, the estimated probability is 0.49 that an adult believes the American family is worse off than it was 10 years ago. Which method of assigning probabilities was used to obtain this estimate?

4.12 The federal government requires airlines to report flight cancellations, and it periodically publishes the cancellation rates for each airline. During a recent period one airline canceled 2,148 of 165,312 flights. Use these results to estimate the probability of a flight being canceled by this airline.

4.13 As part of an examination, a student was asked to find the probability of obtaining no heads on 2 tosses of a fair coin. The student's answer was $\frac{1}{3}$ based on the following reasoning: "A sample space consists of the values 0, 1, 2, which are the number of heads that are possible on the 2 tosses. Since 1 of the 3 sample points represents no heads, the probability is $\frac{1}{3}$." Find the flaw in his reasoning.

4.14 Sometimes the probability of an event is expressed in **odds.** The odds that event E will occur are given by the ratio of $P(E)$ to $[1 - P(E)]$. For example, if $P(E) = 0.80$, then the odds in favor of E are $\frac{80}{100}$ to $\frac{20}{100}$, or in the most simplified terms, 4 to 1 (odds are usually given in terms of integers that have no common factors). Express each of the following probabilities of event E in terms of odds in favor of the event.
  a. $P(E) = 0.70$ b. $P(E) = 0.75$
  c. $P(E) = 0.5$ d. $P(E) = 0.35$

4.15 An automobile muffler franchiser will assign one franchise to a particular city. Four locations (A, B, C, and D) are under consideration. An investor believes that locations A, B, and C are equally appealing, and that each has twice the chance of being selected compared to location D. Determine the subjective probability that the investor is assigning to each location.

4.16 With reference to Exercise 4.15, suppose a second investor subjectively assigns the four locations the following probabilities.

$$P(A) = 0.2 \qquad P(B) = 0.2 \qquad P(C) = 0.3 \qquad P(D) = 0.5$$

Is this a valid subjective assignment of probabilities? Explain.

4.17 Take a coin and estimate the probability of it landing heads by tossing it a total of
   a. 20 times,
   b. 40 times,
   c. 60 times,
   d. 80 times,
   e. 100 times.
   f. Do the estimated probabilities obtained in Parts (a) through (e) appear to be stabilizing around some value?

MINITAB Assignments

Ⓜ 4.18 Use MINITAB to simulate the tossing of a fair coin a total of
   a. 200 times,
   b. 400 times,
   c. 600 times,
   d. 800 times,
   e. 1,000 times.
   f. For each of the above parts, calculate the relative frequency of the number of heads. Do they appear to be stabilizing around 0.5?

Ⓜ 4.19 Use MINITAB to simulate the recording of the sexes of 200 births. Have MINITAB display the number of male births. Assume that each sex is equally likely to occur.

Ⓜ 4.20 In Example 4.6 the command **RANDOM** and the subcommand **INTEGER** were used to simulate the tossing of a coin. Use MINITAB to simulate 300 tosses of a fair die by typing the following.

```
RANDOM 300 C1;
INTEGERS 1 to 6.
TALLY C1
```

The **TALLY** command will display the number of occurrences of 1, 2, 3, 4, 5, and 6. Use the results to estimate the probability of each value.

Ⓜ 4.21 A national clothier purchases 25 percent of its blouses from each of four manufacturers, A, B, C, and D. Use MINITAB to simulate the selection of a sample of 400 blouses, and display the number made by each manufacturer. (See Exercise 4.20.)

4.2 Basic Techniques for Counting Sample Points

In Section 4.1, we saw that the theoretical method of assigning probabilities is often used to determine $P(E)$, the probability of event E. This method however is only applicable for a sample space with equally likely points. Frequently such a sample space exists, but it may be so large that it is not feasible to construct the sample space and list its points. However, a listing of the sample points is actually not necessary to determine $P(E)$. We only need to *count* the number of points in the sample space and the number in event E. Determining the number of points can often be accomplished by using only two basic, yet very potent, counting techniques: the **Multiplication rule** and the **Combination formula.**

The Multiplication Rule for Counting

To discuss the Multiplication rule, consider again the experiment in Example 4.1 in which a nickel and a dime are tossed once. The experiment can be thought of as two actions (steps).

Action 1: tossing the nickel.

Action 2: tossing the dime.

One action (tossing the nickel) has 2 possible outcomes; the other action (tossing the dime) has 2 possible outcomes. The Multiplication rule for counting states that both actions together will have

$$2 \cdot 2 = 4 \text{ possible outcomes.}$$

(Recall that 4 is the number of points obtained when the sample space was listed in Example 4.1.)

The Multiplication Rule for Counting:
If an action has m possible outcomes, and a second action has n possible outcomes, then both actions together have $m \cdot n$ possible outcomes.

Example 4.10   In Example 4.8, a black and a white die were tossed together once.

1. Use the Multiplication rule to show that there are 36 points in the sample space.
2. In how many ways can the dice fall with different sides landing upright?
3. Find the probability that the dice will show different sides.

Solution

1. The black die can fall in 6 ways and the white die in 6 ways. By the multiplication rule they can land together

$$6 \cdot 6 = 36 \text{ ways.}$$

2. For the dice to show different sides, two actions must occur:

   Action 1: the black die shows any of its 6 sides.

   Action 2: the white die shows any of the 5 sides that are different from that showing on the black die.

   The first action has 6 possible outcomes, and the second action has 5. Therefore, by the multiplication rule the dice can show different sides

$$6 \cdot 5 = 30 \text{ ways.}$$

3. Let E be the event that the two dice show different sides. By the theoretical method of assigning probabilities,

$$P(E) = \frac{e}{s},$$

where s is the number of points in the sample space, and e is the number of points in event E. From part 1, $s = 36$, and from part 2, $e = 30$. Thus,

$$P(E) = \frac{30}{36} = 0.83.$$

The Multiplication rule can be extended to a sequence of more than two actions. This is illustrated in the following example.

Example 4.11   A college bookstore offers a personal computer system consisting of a computer, a monitor, and a printer. A student has a choice of two computers, three monitors, and two printers, all of which are compatible. In how many ways can a computer system be bundled?

Solution

The computer can be selected in 2 ways, then the monitor chosen in 3 ways, and then the printer in 2 ways. Together, all 3 components can be chosen

$$2 \cdot 3 \cdot 2 = 12 \text{ ways.}$$

If a listing of the points in a sample space is desired, it can be obtained in a systematic way by constructing a **tree diagram.** Figure 4.1 gives a tree diagram for the above example. In the figure, the two computers, three monitors, and two printers are denoted by C_1, C_2, M_1, M_2, M_3, P_1, P_2. The construction of the tree diagram is started by drawing a set of main branches, one for each of the two ways that the computer can be selected. Next, off each of the two main branches draw three secondary branches to show the different ways that the monitor can be chosen. Finally, from each of the secondary branches construct a third set of two branches to show the different ways that the printer can be selected.

By following all possible paths from the beginning point to the ending points, and marking the branches transversed, one obtains a listing of the 12 possible ways of selecting a computer, a monitor, and a printer. These possibilities appear in the right margin of Figure 4.1.

Example 4.12   A small diner has a lunch counter with four stools. A family of four enter and sit at the counter. In how many different ways can they be seated at the four stools?

Solution

The first stool can be occupied by any of the four members, and then the second stool can be occupied by any of the remaining three, then the third stool by either of the

Figure 4.1
Tree diagram to show the possible computer systems.

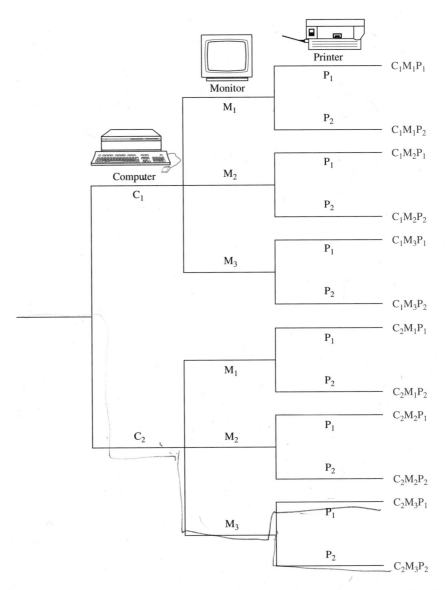

remaining two, and the fourth stool must be taken by the remaining person. By the Multiplication rule, there are $4 \cdot 3 \cdot 2 \cdot 1 = 24$ different ways.

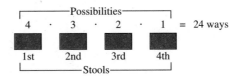

The 24 possible seating arrangements are listed on the next page, where F, M, D, and S denote father, mother, daughter, and son.

FMDS	FMSD	FDMS	FDSM	FSMD	FSDM
MFDS	MFSD	MDFS	MDSF	MSFD	MSDF
DFMS	DFSM	DMFS	DMSF	DSFM	DSMF
SFMD	SFDM	SMFD	SMDF	SDFM	SDMF

An ordered arrangement of distinct elements is called a **permutation.** Each of the 24 arrangements in the preceding example is a permutation of the letters F, M, D, and S. For the interested reader, a formula for the number of permutations of k items selected from n items is given in Exercise 4.55 (see page 155). It is not essential, however, since one can always use the Multiplication rule, as was done in Example 4.12.*

Some of the remaining work in this section involves the product of consecutive descending integers that terminate with 1. Such a product occurred in Example 4.12, namely, $4 \cdot 3 \cdot 2 \cdot 1$. This is called **"four factorial,"** and is denoted by 4!. Other examples of factorials include the following.

$$8! = 8 \cdot 7 \cdot 6 \cdot 5 \cdot 4 \cdot 3 \cdot 2 \cdot 1 = 40{,}320$$
$$2! = 2 \cdot 1 = 2$$
$$1! = 1$$

Mathematicians have found it convenient to define zero factorial. By definition,

$$0! = 1.$$

As an application of factorials, we will soon use the fact that $k!$ gives the number of permutations of k distinct elements. This follows directly from the multiplication rule.

The Combination Formula

In many applications, one must determine the number of different ways that k elements can be selected, where the order in which they are chosen is not important. Such a selection is called a **combination** (see Figure 4.2). The formula for determining the number of possible combinations appears in Equation 4.1.

Figure 4.2
Combination $C(n,k)$.

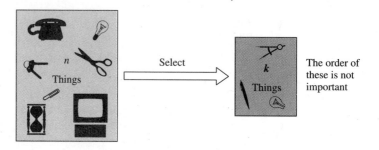

---

* It is the author's belief that students often find this section difficult because they tend to approach each problem as either "a permutation or a combination." It is believed that a greater understanding can be achieved by deemphasizing permutation formulas and stressing the versatility of the multiplication rule.

> **The Combination Formula:**
> $C(n, k)$ denotes the number of combinations of k selected from n. It gives the number of ways that k items can be selected from n items, without regard to the order of selection.
>
> $$C(n, k) = \frac{n!}{k!(n - k)!} \qquad\qquad \textbf{(4.1)}$$

Before explaining its derivation, let's illustrate the formula.

Example 4.13   A pizza shop offers a combination pizza consisting of a choice of any three of the four ingredients: pepperoni (P), mushrooms (M), sausage (S), and anchovies (A). Determine the number of possible combination pizzas.

Solution

We need to determine the number of ways that three ingredients can be chosen from the four possibilities. Since the order of selection is not relevant, the solution is just the number of combinations of 3 things selected from 4. Applying Formula 4.1, we have

$$C(4, 3) = \frac{4!}{3!(4 - 3)!} = \frac{4!}{3!1!} = \frac{24}{6 \cdot 1} = 4.$$

Thus, there are 4 different combination pizzas, namely, the following.

$$PMA \qquad PMS \qquad PSA \qquad MSA$$

To derive Formula 4.1, note the following, each of which is obtained directly from the Multiplication rule.

1. Each combination of k items can be ordered (arranged) in
   $k(k - 1)(k - 2) \ldots 1 = k!$ ways.
2. The number of ordered ways of selecting k from n items is
   $n(n - 1)(n - 2) \ldots (n - k + 1)$.

$C(n,k)$ denotes the number of *unordered* ways of selecting k from n items. If this is multiplied by the number of ways that each combination can be *ordered* (Note 1), the result must equal the number given in Note 2, thus yielding the following.

$$C(n, k) \cdot k! = n(n - 1)(n - 2) \cdots (n - k + 1)$$

Dividing each side by $k!$, we have

$$C(n, k) = \frac{n(n - 1)(n - 2) \cdots (n - k + 1)}{k!}.$$

Multiplying the numerator and denominator by $(n - k)!$ we obtain

$$C(n, k) = \frac{n(n - 1)(n - 2) \cdots (n - k + 1)(n - k)!}{k!(n - k)!} = \frac{n!}{k!(n - k)!}.$$

Example 4.14   A bridge hand consists of 13 cards dealt from a standard deck of 52. Determine the number of possible hands.

Solution

The order in which the cards are selected is not important. Therefore, the solution to this problem is simply the number of ways that 13 things can be selected from 52, that is, the number of combinations of 13 selected from 52.

$$C(52, 13) = \frac{52!}{13!(52 - 13)!} = \frac{52 \cdot 51 \cdots 40 \cdot 39 \cdot 38 \cdots 1}{(13 \cdot 12 \cdots 1)(39 \cdot 38 \cdots 1)}$$

$$= \frac{52 \cdot 51 \cdots 40}{13 \cdot 12 \cdots 1} = 635{,}013{,}559{,}600$$

Thus, there are more than 635 billion different bridge hands, making it highly improbable that a bridge player would ever see the same hand twice in the course of a lifetime.

This section has focused on two basic counting techniques for determining the number of points in a sample space: the Multiplication rule and the Combination formula. Armed with just these two tools, you can solve a wide range of counting problems. We close this section with an example that requires using both techniques.

Example 4.15   A manufacturer of mainframe computers is trying to win a large contract with a client. The manufacturer is going to send to the client's facilities 5 salespersons, 7 technicians, and 3 software engineers. In how many ways can these 15 people be selected if the manufacturer has available 10 salespersons, 9 technicians, and 7 software engineers?

Solution

The 5 salespersons can be selected from the available 10 as follows.

$$C(10, 5) = \frac{10!}{5!(10 - 5)!} = 252 \text{ ways.}$$

The 7 technicians can be selected from the 9 in

$$C(9, 7) = \frac{9!}{7!(9 - 7)!} = 36 \text{ ways.}$$

Three software engineers can be chosen from the available 7 in

$$C(7, 3) = \frac{7!}{3!(7 - 3)!} = 35 \text{ ways.}$$

By the Multiplication rule, the number of ways that 5 salespersons, 7 technicians, and 3 software engineers can be chosen together is

$$C(10, 5) \cdot C(9, 7) \cdot C(7, 3) = 252 \cdot 36 \cdot 35 = 317{,}520.$$

Section 4.2 Exercises
Basic Techniques for Counting Sample Points

In Exercises 4.22–4.33, evaluate the given expression.

4.22 5!	4.23 7!	4.24 10!/4!	4.25 100!/99!
4.26 9!1!/0!	4.27 $C(10, 3)$	4.28 $C(5, 4)$	4.29 $C(12, 9)$
4.30 $C(88, 1)$	4.31 $C(88, 87)$	4.32 $C(88, 88)$	4.33 $C(88, 0)$

4.34   Show that:
   a. $C(25, 5) = C(25, 20)$.
   b. $C(n, k) = C(n, n - k)$.

4.35   A quiz contains five multiple-choice questions, each of which has four alternatives. In how many different ways can the five questions be answered?

4.36   A quiz consists of five true-false questions. In how many different ways can all of these be answered?

4.37   If a quiz contains 5 true-false questions and 5 multiple-choice questions, each with 4 alternatives, in how many different ways can one answer the 10 questions?

4.38   A national motel chain has replaced the key lock for each room with a key card entry system. A door is unlocked by inserting a plastic card into a slot above the door knob. Each key's unique identity is determined by a grid of 63 cells, each of which is either solid or punched.

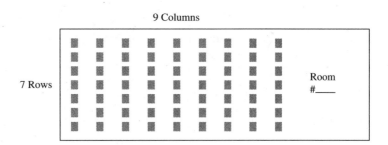

   a. Determine the number of different key cards possible with this system.
   b. How many are possible if each key card must have at least one punched cell?

4.39   Each October the champions of the American and the National Leagues meet in the World Series, with the winner being the first team to win four games. Suppose the series is tied after four games. Use a tree diagram to show the different ways that the two teams can complete the remaining games.

4.40   A restaurant offers a luncheon special consisting of a soup, sandwich, and beverage. If a choice of three soups, two sandwiches, and four beverages is available, in how many different ways can one select the luncheon special?

4.41   The Chemistry Department at a university has four rooms available for three courses that are scheduled at the same time. Determine the number of possible room assignments that can be made to the three courses.

4.42 For the luncheon special mentioned in Exercise 4.40, the choice of soups is chicken, vegetable, and tomato. The sandwich can be tuna fish or beef, and the beverage can be milk, coffee, tea, or lemonade. Draw a tree diagram showing the various ways that one can select the special.

4.43 Let C_1, C_2, and C_3 denote the three courses referred to in Exercise 4.41, and denote the four rooms by R_1, R_2, R_3, and R_4. Use a tree diagram to display the different possible room assignments to the three courses.

4.44 To satisfy the core requirements at a certain college, students must select one of four math courses, one of five science courses, one of four English courses, and one of seven freshman seminar courses. In how many different ways can a student satisfy the core requirements at this college?

4.45 A mail order company sells vitamin C in bottles of 100, 250, and 500 tablets. Vitamin C can also be ordered in potencies of 100 mg, 250 mg, 500 mg, and 1,000 mg. The tablets can also be obtained coated or uncoated. In how many different ways can a customer order a bottle of vitamin C from this company?

4.46 A college baseball team consists of the following 24 players: 6 pitchers, 2 catchers, 2 first basemen, 3 second basemen, 4 players who can play either shortstop or third base, and 7 people who can play any of the three outfield positions. In how many different ways can the team manager select 9 starting players?

4.47 Assume the manager mentioned in Exercise 4.46 has chosen nine starting players. In how many different ways can he assign a batting order?

4.48 License plates in a certain state consist of 5 digits and an ending letter of the alphabet. Determine the number of possible plates if:
a. all digits and letters can be used, and digits can be repeated;
b. all digits and letters are possible, with the exception that the first digit cannot be 0;
c. the first digit cannot be 0, and all digits must be different.

4.49 How many different 5-card poker hands can be dealt from a standard deck of 52 cards?

4.50 In how many ways can a 5-card hand consisting of 3 aces and a pair of kings be dealt from a deck of 52 cards?

4.51 Use the results of Exercises 4.49 and 4.50 to find the probability of being dealt a 5-card hand consisting of 3 aces and a pair of kings.

4.52 A college math club consists of 16 men and 20 women.
a. In how many different ways can a committee of 5 members be chosen?
b. In how many different ways can a committee of 5 be chosen and have it consist of 2 men and 3 women?
c. If the 5 committee members are randomly selected, what is the probability that the committee will consist of 2 men and 3 women?

4.53 Many states now conduct weekly lotteries. In one such lottery, a player selects six different numbers from 1 to 40. Determine the number of possible lottery tickets.

4.54 A college faculty consists of 30 professors, 38 associate professors, 30 assistant professors, and 5 instructors. A president's search committee is to be formed consisting of 4 professors, 3 associate professors, 2 assistant professors, and 1 instructor. In how many different ways can this committee be selected?

4.55 The number of permutations of k things selected from n distinct things is sometimes denoted by $P(n, k)$. Use the multiplication rule for counting to show that

$$P(n, k) = n(n - 1)(n - 2) \cdot \ldots \cdot (n - k + 1) = \frac{n!}{(n - k)!}.$$

4.56 Eight runners are entered in a 100-yard dash. Determine the number of different ways they can place first, second, and third by using:
a. the multiplication rule.
b. the permutation formula given in Exercise 4.55.

4.3   The Addition Rule of Probability

We have seen that probability is used to measure the likelihood that an event will occur when an experiment is performed. $P(E)$ is a number between 0 and 1 inclusive.

Every event E has a probability of at least 0 but not more than 1.

$$0 \leq P(E) \leq 1$$

If event E can never occur when an experiment is performed, then $P(E) = 0$; if event E always occurs, then $P(E) = 1$. Usually an event will have a probability that lies somewhere between these two extremes, and the closer $P(E)$ is to 1, the more likely E will occur.

"Once in a blue moon" is a phrase that is sometimes applied to a rare event. When a month has two full moons, the second is called a blue moon. A blue moon occurs every 2.8 years. In 1990, there was one on New Year's Eve. We'll have to wait until 2009 for the next New Year's Eve occurrence of this unusual event.

Figure 4.3
Venn diagram of event *A* or *B*.

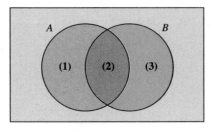

(1): *A* and not *B*.
(2): *A* and *B*.
(3): *B* and not *A*.

At this point, the types of events for which we can calculate probabilities are quite limited. However, the remainder of this chapter is concerned with expanding these capabilities to include probability calculations of **compound events,** such as:

1. the probability that *A* **or** *B* will occur,
2. the probability that *A* **and** *B* will occur,
3. the probability that *A* **will occur if it is known that** *B* **has occurred,** and
4. the probability that **something other than** *A* will occur.

The present section considers the first of these compound events, namely, obtaining $P(A \text{ or } B)$. The formula for finding this is called the **Addition rule** and is given in Equation 4.2.

The Addition Rule of Probability:
The probability that event *A* or *B* will occur when an experiment is performed is given by

$$P(A \text{ or } B) = P(A) + P(B) - P(AB) \qquad (4.2)$$

where $P(AB)$ is the probability that both events, *A* and *B*, will occur.

In the Addition rule the conjunction *or* is used in the inclusive sense. Event *A* **or** *B* will occur if at least one occurs. This includes the possibility that both may occur. To clarify the meaning of event *A* or *B*, refer to Figure 4.3. A **Venn diagram** is drawn in which the rectangle represents all the points in the sample space, and the interiors of the two circles represent the points in event *A* and in event *B*. Event *A* or *B* is denoted by the three regions marked (1), (2), and (3). When the experiment is performed, if the resulting sample point lies in any of the regions (1), (2), or (3), then event *A* or *B* has occurred (in set notation, event *A* or *B* is $A \cup B$, the union of *A* with *B*).

Example 4.16   An office supply shop has a sale on 3.5 inch computer diskettes. These are normally purchased in packages of 10, but it has a bin of 200 mixed high density and double density diskettes that were manufactured either in the United States, Japan, or Korea. An exact breakdown appears in Table 4.3. If a customer randomly selects one diskette from the box, find the probability that the diskette is

1. either high density or made in Korea
2. either made in the United States or Japan.

Table 4.3

Distribution of the 200 Diskettes

Type of Diskette	Korean Made	Japanese Made	American Made	Total
H (high)	20	26	24	70
D (double)	40	14	76	130
Total	60	40	100	200

Solution

1. Let H denote the event that the diskette is high density, and let K be the event that it was made in Korea. Then

$$P(H \text{ or } K) = P(H) + P(K) - P(HK).$$

Since 70 of the 200 diskettes are high density, there are 70 chances out of 200 that H will occur, and thus $P(H) = \frac{70}{200}$. Similarly, $P(K) = \frac{60}{200}$, because 60 of the 200 diskettes were made in Korea. $P(HK) = \frac{20}{200}$, since 20 are both high density and made in Korea. Thus,

$$P(H \text{ or } K) = \frac{70}{200} + \frac{60}{200} - \frac{20}{200}$$

$$= \frac{110}{200}$$

$$= 0.55.$$

2. Let U denote the event that a diskette was made in the United States and J signify that it was made in Japan.

$$P(U \text{ or } J) = P(U) + P(J) - P(UJ)$$

$P(U) = \frac{100}{200}$ and $P(J) = \frac{40}{200}$, since 100 and 40 diskettes were made in the United States and in Japan, respectively. $P(UJ)$ is the probability that a diskette was made in both the United States and in Japan. Of course this is impossible because U and J are **mutually exclusive events.** Consequently $P(UJ) = 0$ and

$$P(U \text{ or } J) = P(U) + P(J) - P(UJ)$$

$$= \frac{100}{200} + \frac{40}{200} - 0$$

$$= 0.70.$$

A and B are **mutually exclusive events** if both cannot occur together in the same experiment. Consequently,

$$P(AB) = 0.$$

The Addition rule can be stated for three or more events, but the formula is rather long (see Exercise 4.77). When the events are mutually exclusive however, the formula simplifies greatly to that in Equation 4.3.

The Addition Rule for Mutually Exclusive Events:
If A, B, C, . . . are mutually exclusive events, then

$$P(A \text{ or } B \text{ or } C \text{ or } \ldots) = P(A) + P(B) + P(C) + \ldots \qquad \textbf{(4.3)}$$

Example 4.17   A lumber yard stocks 2 × 4s (boards that have a nominal depth of 2 inches and a width of 4 inches) in a choice of pine, fir, spruce, and hemlock. Thirty-five percent of purchases are for pine, 27 percent for fir, 22 percent for spruce, and 16 percent for hemlock. What is the probability that the next purchase will be either pine, spruce, or hemlock?

Solution

Define P, F, S, and H as the events that the next 2 × 4 purchased will be pine, fir, spruce, or hemlock, respectively. Since these are mutually exclusive events, Equation 4.3 can be applied.

$$
\begin{aligned}
P(P \text{ or } S \text{ or } H) &= P(P) + P(S) + P(H) \\
&= 0.35 + 0.22 + 0.16 \\
&= 0.73
\end{aligned}
$$

Section 4.3 Exercises
The Addition Rule of Probability

4.57 Are the following paired events, receiving a final grade of A and receiving a final grade of B in your present statistics course, mutually exclusive?

4.58 Are the following paired events, obtaining a masters degree and obtaining a doctorate, mutually exclusive?

4.59 Are the following paired events, a professor teaching a statistics course and teaching a physics course during a particular semester, mutually exclusive?

4.60 Are the following paired events, a baseball player leading the league in home runs and winning the batting title, mutually exclusive?

4.61 Are the following paired events, receiving a speeding ticket and receiving a ticket for going through a stop sign on the same day, mutually exclusive?

4.62 Are the following paired events, observing a total of 10 spots and observing a total of 8 spots in one toss of a pair of dice, mutually exclusive?

4.63 Are the following paired events, observing a total of 10 spots and observing a total of at least 8 spots in one toss of a pair of dice, mutually exclusive?

4.64 Are the following paired events, a retiree drawing a pension from the federal government and drawing a pension from IBM, mutually exclusive?

4.65 In a recent survey of government workers, the *News Digest,* a federal employee's newsletter, received more than 10,000 responses. The highest levels of education completed by the respondents are given below.

Highest Education Level	Proportion of Workers
High school graduate	0.146
Some college	0.292
College graduate	0.243
Some postgraduate	0.138
Postgraduate degree	0.181

What is the probability that a randomly selected respondent has at least a college degree?

4.66 One card is randomly selected from a standard 52-card deck. Find the probability that the selected card is
a. an ace or a black card,
b. an ace or a king,
c. an ace or a king or a queen.

4.67 For the experiment of randomly selecting one card from a standard 52-card deck, give an example of an event E for which
a. $P(E) = 0$,
b. $P(E) = 1$.

4.68 A die is tossed once. Find the probability of tossing 4 or more spots.

4.69 A multinational corporation employs 500 engineers whose distribution is summarized in the following table.

U.S. Citizen?	Job Location	
	U.S.	Europe
Yes	280	120
No	70	30

Determine the probability that an engineer for this corporation is either a U.S. citizen or located in the United States.

4.70 In 1965 the composition of U.S. dimes, quarters, and half dollars was changed from silver to a copper clad. A woman has a bank in which she has accumulated several coins since childhood. The bank's contents are the following:

Date	Dimes	Quarters	Halves
Before 1965	155	113	32
1965 and after	345	287	68

Find the probability that a randomly selected coin is

a. a silver coin,
b. a quarter,
c. a silver quarter,
d. either a silver coin or a quarter,
e. either a dime or a quarter.

4.71 Forty percent of the sales force at a large insurance company have a laptop computer, 65 percent have a desktop computer, and 24 percent have both. What percent of the salespeople have either a laptop or a desktop computer?

4.72 The owner of a sub shop claims that one of three sub sales is with onions, one of two is with meat, and one of five is with both onions and meat. Find the probability that a sub purchase will contain either onions or meat.

4.73 An automobile dealership has found that 37 percent of its new car sales have been dealer financed, 45 percent have been financed by another institution, and 18 percent have been cash sales. Find the probability that the next purchase of a new car at this dealership will be either a cash sale or dealer financed.

4.74 Records of a computer repair facility indicate that 53 percent of the equipment it repairs are disk drives, 22 percent are printers, 14 percent are monitors, and 11 percent involve other hardware. Find the probability that the next item repaired will be

a. a disk drive or a printer,
b. a disk drive or a printer or a monitor.

4.75 Last year 413 students completed the introductory statistics course at a certain college. The proportions of students who received grades of A, B, C, D, and F were 0.15, 0.24, 0.32, 0.20, and 0.09, respectively. Find the probability that a randomly selected student from this group received a grade of C or better.

4.76 At a national convention, 63 percent of the registrants are mathematicians, 42 percent are statisticians, and 97 percent are either a mathematician or a statistician. What percentage of registrants are both a mathematician and a statistician?

4.77 The Addition rule for three mutually exclusive events is given in Formula 4.3. When the events are not mutually exclusive, the following general formula can be used:

$$P(A \text{ or } B \text{ or } C) = [P(A) + P(B) + P(C)] - [P(AB) + P(AC) + P(BC)] + P(ABC)$$

In the above, $P(ABC)$ is the probability that all of the events A and B and C will occur. Use this generalized addition formula to find the probability that a randomly selected card from a 52-card deck will be either an ace or black or a heart.

4.4 The Multiplication and Conditional Rules of Probability

Frequently events A and B are related in such a way that the occurrence of one affects the probability of the other occurring. For instance, the probability of your obtaining an A on the statistics final exam is dependent on whether you prepare for the test. As a second illustration, you might assign a very low probability to the following event.

B: your best friend's car will have a flat tire tomorrow.

However, the assigned probability would likely be much higher if you knew that the following event has occurred.

A: the treads have worn completely off the tires.

The probability of B, knowing that A has occurred, is called the **conditional probability of B given A,** and is denoted by $P(B|A)$. A conditional probability is generally involved in determining $P(AB)$, the probability that both A and B will occur. Specifically, the following formula relates the two probabilities.

The Multiplication Rule of Probability

> **The Multiplication Rule of Probability:**
> The probability that both A **and** B will occur when an experiment is performed is given by
>
> $$P(AB) = P(A) \cdot P(B|A), \qquad (4.4)$$
>
> where $P(B|A)$ is the probability of B if A has occurred, and it is called the conditional probability of B given A.

Example 4.18

In the previous section we considered an office supply shop that has a box of 200 computer diskettes. Some are high density, some double density, and they have been manufactured either in the United States, Japan, or Korea. The breakdown was given in Table 4.3 and is repeated below.

Type of Diskette	Korean Made	Japanese Made	American Made	Total
H (high)	20	26	24	70
D (double)	40	14	76	130
Total	60	40	100	200

Suppose a customer randomly selects 2 diskettes from the box. Find the probability that both diskettes are double density.

Solution

Denote by D_1 the event that the first diskette selected is double density, and let D_2 signify the event that the second diskette chosen is double density. By the Multiplication Rule of Probability, the probability that D_1 and D_2 will both occur is

$$P(D_1 D_2) = P(D_1) \cdot P(D_2|D_1).$$

Since 130 of the 200 diskettes are double density, $P(D_1) = \frac{130}{200}$. $P(D_2|D_1)$ is the probability of obtaining a double density diskette on the second selection, *if* the first selection was double density. If the first selection was double density, then there would be 129 double density diskettes left in the remaining 199, and therefore, $P(D_2|D_1) = \frac{129}{199}$. Thus,

$$P(D_1D_2) = \left(\frac{130}{200}\right)\left(\frac{129}{199}\right)$$

$$= 0.42.$$

While the generalization of the Addition rule is complicated, the Multiplication rule can easily be extended to more than two events. For three events, the extension of Equation 4.4 is the following:

The Multiplication Rule for Three Events:
The probability that A **and** B **and** C will all occur when an experiment is performed is given by

$$P(ABC) = P(A) \cdot P(B|A) \cdot P(C|AB), \tag{4.5}$$

where $P(B|A)$ is the conditional probability of B if A has occurred, and $P(C|AB)$ is the conditional probability of C if both A and B have occurred.

The Multiplication rule can be generalized in a similar way to any number of events.

Example 4.19   Suppose for the diskette example, a customer randomly selects three. Find the probability that all three selections will be double density diskettes.

Solution

Let D_1, D_2, D_3 denote, respectively, the events that the first, second, and third selections are double density. We need to find the probability that all three of these events will occur, that is, $P(D_1D_2D_3)$.

$$P(D_1D_2D_3) = P(D_1) \cdot P(D_2|D_1) \cdot P(D_3|D_1D_2)$$

As was seen in Example 4.18, since 130 of the 200 diskettes are double density, $P(D_1) = \frac{130}{200}$ and $P(D_2|D_1) = \frac{129}{199}$. $P(D_3|D_1D_2)$ is the probability of obtaining a double density diskette on the third selection, *if* selections one and two were double density. If the first two selections were double density, then there would remain 198 diskettes, with 128 of these double density. Therefore $P(D_3|D_1D_2) = \frac{128}{198}$, and

$$P(D_1D_2D_3) = P(D_1) \cdot P(D_2|D_1) \cdot P(D_3|D_1D_2)$$

$$= \left(\frac{130}{200}\right)\left(\frac{129}{199}\right)\left(\frac{128}{198}\right)$$

$$= 0.27.$$

For many events A and B, the probability of B does not depend on whether or not A has occurred. For instance, suppose a fair penny and a fair die are tossed and events A and B are defined as follows.

A: The die shows 6 spots.

B: The penny lands heads.

What happens with the die is irrelevant as far as the penny's chances of landing heads. The probability of B is $\frac{1}{2}$, regardless of whether or not event A occurs. When the probability of B is not affected by the occurrence of A, we say that A and B are **independent events**. Mathematically, $P(B|A) = P(B)$, and this implies that $P(AB) = P(A) \cdot P(B)$. When the probabilities are known, these two equations can be used to check if events A and B are independent. In practice, however, one will often assume that two events are independent, using as a basis their knowledge of the phenomenon that is being modeled.

Independent Events:
A and B are **independent events** if the probability of each event is not affected by whether or not the other event has occurred.

Test for Independent Events
A and B are independent if either $P(A|B) = P(A)$ or $P(AB) = P(A) \cdot P(B)$. Otherwise, the events are **dependent**.

For independent events, the generalized Multiplication rule simplifies to Formula 4.6 below.

The Multiplication Rule for Independent Events:
If A, B, C, . . . are independent events, then

$$P(ABC \ldots) = P(A) \cdot P(B) \cdot P(C) \cdot \ldots \qquad \textbf{(4.6)}$$

Example 4.20   Find the probability of obtaining all heads if a fair penny is tossed

1. twice,
2. four times.

Solution

We will let H_k denote the event that a head shows on the kth toss, where $k = 1, 2, 3, 4$.

1. We need to find $P(H_1 H_2)$, the probability that a head shows on tosses one and two. H_1 and H_2 are independent events, since the outcome on the first toss does not affect what happens on the second toss. Thus,

$$P(H_1 H_2) = P(H_1) \cdot P(H_2)$$
$$= \left(\frac{1}{2}\right)\left(\frac{1}{2}\right)$$
$$= 0.25.$$

Trash the lowly penny? According to *Newsweek* (January 7, 1991), many American businesses would like a "cents less" society. Workers waste an estimated 5.5 million hours a year counting the coins. The public, however, seems to have an attachment to them, with each household having an average hoard of 1,000 cents.

2. As in Part 1, the events H_1, H_2, H_3, and H_4 are independent since the outcome on a toss is not affected by any other toss. Consequently, by the Multiplication rule for independent events,

$$P(H_1 H_2 H_3 H_4) = P(H_1) \cdot P(H_2) \cdot P(H_3) \cdot P(H_4)$$
$$= \left(\frac{1}{2}\right)\left(\frac{1}{2}\right)\left(\frac{1}{2}\right)\left(\frac{1}{2}\right)$$
$$= 0.0625.$$

The Conditional Rule of Probability

The Multiplication rule for two events that was given at the beginning of this section allows one to find the probability that both events will occur when an experiment is performed.

$$P(AB) = P(A) \cdot P(B|A)$$

This formula assumes that one knows the values of $P(A)$ and $P(B|A)$, the conditional probability of B, given that event A has occurred. In some situations it is this conditional probability that is sought. By rewriting the Multiplication rule explicitly for $P(B|A)$, one obtains the following conditional probability formula.

The Conditional Rule of Probability:
The conditional probability of B, given that A has occurred, is given by

$$P(B|A) = \frac{P(BA)}{P(A)} \tag{4.7}$$

> **TIP:**   The Conditional rule might be easier to remember if you think of it in words. It says that the conditional probability of B, given that A has occurred, equals the probability that both events will occur, divided by the probability of the event that is known to have occurred. For instance, if it is known that event R has occurred, then the conditional probability of event T is as follows.
>
> $$P(T \mid R) = \frac{P(\text{both events})}{P(\text{known occurred event})} = \frac{P(TR)}{P(R)}$$

Example 4.21   At a liberal arts college in the South, 60 percent of all freshmen are enrolled in a mathematics course, 73 percent are enrolled in an English course, and 49 percent are taking both. A freshman is randomly selected from this college.

1. What is the probability that the student is taking an English course, if it is known that he/she is enrolled in a mathematics course?
2. If the student is taking an English course, what is the probability that he/she is also enrolled in a mathematics course?
3. If M is the event that a freshman is taking a mathematics course, and E the event that a freshman is taking an English course, are M and E independent events?

Solution

We are told that $P(M) = 0.60$, $P(E) = 0.73$, and $P(ME) = 0.49$.

1. We are given that the student is enrolled in a mathematics course. Given this information, we want the conditional probability that he/she is also in an English course.

$$P(E \mid M) = \frac{P(EM)}{P(M)} = \frac{0.49}{0.60} = 0.82$$

2. Now we are given that the student is taking an English course, and we want the conditional probability that he/she is also in a mathematics course.

$$P(M \mid E) = \frac{P(ME)}{P(E)} = \frac{0.49}{0.73} = 0.67$$

3. We were given that $P(M) = 0.60$, and in Part 2 we determined that $P(M \mid E) = 0.67$. Since $P(M \mid E) \neq P(M)$, E and M are not independent events, that is, they are dependent. One could also reach this conclusion from the fact that $P(E \mid M) \neq P(E)$, or from the fact that $P(ME) \neq P(M) \cdot P(E)$.

As the last example shows, one cannot always rely on common sense to determine if two events are independent. Sometimes independence has to be checked mathematically. The next example provides a second illustration of this.

Example 4.22   Earlier we considered the experiment that consists of tossing together a black and a white die once. A sample space for which the points are equally likely is given by the following 36 ordered pairs, where the first position indicates the number of spots tossed on the black die, and the second position gives the number of spots for the white die.

$$
\begin{array}{cccccc}
(1, 1) & (2, 1) & (3, 1) & (4, 1) & (5, 1) & (6, 1) \\
(1, 2) & (2, 2) & (3, 2) & (4, 2) & (5, 2) & (6, 2) \\
(1, 3) & (2, 3) & (3, 3) & (4, 3) & (5, 3) & (6, 3) \\
(1, 4) & (2, 4) & (3, 4) & (4, 4) & (5, 4) & (6, 4) \\
(1, 5) & (2, 5) & (3, 5) & (4, 5) & (5, 5) & (6, 5) \\
(1, 6) & (2, 6) & (3, 6) & (4, 6) & (5, 6) & (6, 6)
\end{array}
$$

If a sum of 7 spots was tossed, what is the probability that the black die showed an ace (1 spot)?

Solution

Let S denote the event that a sum of 7 is tossed, and let A be the event that the black die shows an ace. We are asked to find $P(A|S)$.

$$
P(A|S) = \frac{P(AS)}{P(S)}
$$

$P(AS) = \frac{1}{36}$, since one of the 36 points, (1, 6), has a sum of 7 with a black ace. There are 6 points in the sample space with a sum of 7, and therefore $P(S) = \frac{6}{36}$. Thus,

$$
P(A|S) = \frac{\dfrac{1}{36}}{\dfrac{6}{36}} = \frac{1}{6}.
$$

Notice from the sample space that $P(A) = \frac{6}{36}$, and therefore $P(A|S) = P(A)$. Thus, A and S are independent events.

TIP: Students often have difficulty recognizing when a conditional probability is given or sought. The tip-off is usually a key word or phrase such as *if, when, given that, knowing that, provided that, we are told that, subject to the fact that, on the condition that,* etc. What follows the key phrase will be the given (known) event. For example, the sentence, "*Knowing that B* has happened, what's the probability of *A*?" requires that we find $P(A|B)$, since B is the given event.

Example 4.23   A grocery store in Virginia occasionally receives a shipment of live Maine lobsters. Seventy-five percent of the time the store will place an ad and sell out its shipment. The store advertises the availability of the lobsters for 80 percent of the shipments. What is the probability that the store will sell an entire shipment if it advertises?

Solution

Define A as the event that the store advertises the availability of the lobsters, and let S denote the event that it sells an entire shipment. We are asked to determine the conditional probability $P(S|A)$.

$$P(S|A) = \frac{P(SA)}{P(A)} = \frac{0.75}{0.80} = 0.94$$

Before ending this section, we need to emphasize that the concepts of **independence** and **mutually exclusive events** are not the same. Mutual exclusiveness refers to the fact that two events cannot both occur; independence indicates that the probability of one event occurring does not depend on the occurrence of the other. For example, suppose a friend will soon give birth to one child, and we define events G and B as follows.

B: the child is a boy.

G: the child is a girl.

B and G **are mutually exclusive events,** since they cannot both occur. To determine if B and G are independent events, we need to ask if $P(B|G) = P(B)$. The probability of a boy, $P(B)$, is about $\frac{1}{2}$. $P(B|G)$ is the probability that the child will be a boy if it is known that it will be a girl. But $P(B|G) = 0$, since the occurrence of G precludes the possibility of B occurring. Consequently, because $P(B|G) \neq P(B)$, events B and G **are not independent events.**

> Mutually exclusive events are always dependent events; independent events are never mutually exclusive events.*

Section 4.4 Exercises

The Multiplication and Conditional Rules of Probability

4.78 Are the following paired events, a professor's car having a flat tire on his way to work and the professor being late for his 8:00 A.M. class that morning, likely to be independent?

4.79 Are the following paired events, a student receiving an A in Calculus I and her obtaining an A in Calculus II, likely to be independent?

4.80 Are events C and D independent, if $P(C) = 0.5$, $P(D) = 0.7$, and $P(CD) = 0.3$?

4.81 Are events E and F independent, if $P(E) = 0.3$ and $P(E|F) = \frac{3}{10}$?

4.82 Are the following paired events, owning an automobile and being left handed, likely to be independent?

* This is true for events with positive probability.

4.83 Are the following paired events, tossing a tail on the first flip of a coin and tossing a tail on the second toss, likely to be independent?

4.84 Are the following paired events, being a smoker and contracting a lung disease, likely to be independent?

4.85 Are the following paired events, buying a lottery ticket and winning the lottery, likely to be independent?

4.86 If $P(A) = 0.7$, $P(B) = 0.2$, and $P(AB) = 0.14$, are A and B independent events?

4.87 If $P(T) = 0.21$ and $P(T|S) = 0.24$, are T and S independent events?

4.88 Let T be the event that a tune up was done correctly on a Porsche 911, and let A denote the event that the tune up was performed by mechanic A. Express in symbols the probability that
a. the tune up was performed by mechanic A;
b. the tune up was done correctly by mechanic A;
c. the tune up was done correctly, if mechanic A did it;
d. if the tune up was performed correctly, then it was done by mechanic A.

4.89 In the following, D is the event that December sales will be good, and Y is the event that a company will show a profit for the year. Explain in practical terms what probabilities are represented by the following:
a. $P(Y|D)$ b. $P(D|Y)$ c. $P(DY)$

4.90 Events A and B are independent, and $P(A) = 0.8$, $P(B) = 0.3$. Find $P(AB)$.

4.91 We are given that $P(A) = 0.5$, $P(B) = 0.4$, where A and B are independent events. Find $P(A|B)$ and $P(B|A)$.

4.92 Events A and B are such that $P(A) = 0.8$, $P(B) = 0.7$, and $P(AB) = 0.6$. Find the following.
a. $P(B|A)$ b. $P(A|B)$
c. Are events A and B independent?

4.93 A and B are independent events with probabilities $P(A) = 0.8$ and $P(B) = 0.3$. Determine $P(A \text{ or } B)$.

4.94 Fifty percent of all 1988 domestic cars were equipped with power windows *(Ward's Automotive Yearbook)*. If a shopping center's parking lot contains five 1988 domestic cars, what is the probability that all are equipped with power windows?

4.95 A supermarket has ten 1-pound packages of ground turkey on display in its meat counter. Six packages are underweight. If a shopper selects 4 packages from the display, what is the probability that
a. all will be underweight?   b. none will be underweight?

4.96 A multinational corporation employs 500 engineers whose distribution is summarized in the following table:

U.S. Citizen?	Job Location	
	U.S.	Europe
Yes	280	120
No	70	30

One employee is randomly selected from the 500 engineers.
a. Find the probability that the employee is a U.S. citizen employed in Europe.
b. Find the probability that the employee is a U.S. citizen.
c. If the employee selected is a U.S. citizen, what is the probability that he/she is employed in Europe?

4.97   The following table gives by sex and age the number of persons in the United States living alone in 1987 (U.S. Bureau of the Census). All figures are in thousands.

Sex	Age					
	15–24	**25–44**	**45–64**	**65–74**	**≥75**	**Total**
Male	661	3,881	1,903	937	865	8,247
Female	591	2,619	2,963	3,223	3,486	12,882
	1,252	6,500	4,866	4,160	4,351	21,129

Suppose one person is chosen at random from this population.
a. Find the probability that a woman is chosen.
b. Find the probability of selecting a woman who is 75 years of age or older.
c. If a woman is chosen, what is the probability that she is 75 years of age or older?
d. Given that a man is chosen, what is the probability that he is 75 years of age or older?

4.98   A random number generator produces a sequence of digits, where each can be any of the 10 integers from 0 through 9 with equal probability. Each digit is generated independently of the others. Find the probability that a sequence of five digits will
a. consist of all ones,
b. consist of all odd integers.

4.99   A statistics professor has found that the probability is 0.80 that a student will pass the final examination. If a student passes the final, the probability is 0.95 that he/she will pass the course. Find the probability that a student will pass both the final exam and the course.

4.100   Thirteen percent of the employees of a large company are female technicians. Forty percent of its workers are technicians. If a technician has been assigned to a particular job, what is the probability that the person is female?

4.101   Last semester 31 percent of a college's enrollment were freshmen, and 16 percent of the student body utilized the services of the college's tutoring center. Of students who used these services, 58 percent were freshmen. For last semester's freshmen, what percentage used the tutoring center?

4.102   At a busy city intersection the probability is 0.08 that an accident will occur on any given day. However, if it rains that day, then the probability of an accident is 0.24. Weather records show that it rains about 10 percent of the days in this city. Suppose you read in the paper that an accident did occur at the intersection yesterday. Knowing this, what is the probability that it rained yesterday?

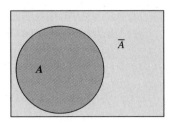

4.5   The Complement Rule of Probability

The **Complement rule** is the last of the four rules of probability. Compared to the other rules, it is the simplest, but the Complement rule can often be a time-saving alternative for solving certain types of probability problems.

In what follows, reference is made to **the complement of event A.** This consists of all the points in a sample space that are not in A, and is illustrated in Figure 4.4. We will use the notation $\overline{A}$ to denote the complement of event A.

By its definition, a sample space is a listing of all the possible outcomes of an experiment. When the experiment is performed, exactly one point in the sample space will result. Consequently, the total probability associated with a sample space is one. This is the basis for the Complement rule given in Equation 4.8.

> **The Complement Rule of Probability:**
> For event A and its complement $\overline{A}$ defined on a sample space,
>
> $$P(A) + P(\overline{A}) = 1. \tag{4.8}$$
>
> Equivalently,
>
> $$P(A) = 1 - P(\overline{A}). \tag{4.9}$$

Usually the Complement rule is used in the form given in Equation 4.9. In words it says the probability that an event will occur equals one minus the probability that the event will not occur. Sometimes it is much easier to find $P(A)$ by using this indirect method of first obtaining the probability that event A will not occur, and then subtracting this probability from one.

Example 4.24   For several years, about 51 percent of births in the United States have been boys (Source: *Statistical Abstract of the United States,* 1990). If a family in this country is going to have five children, what is the probability that they will have *at least one* girl? Assume that consecutive births are independent.

Solution

Let A be the event that at least 1 of the 5 births is a girl. At least 1 means 1 or 2 or 3 or 4 or 5. The probability of obtaining these numbers of girls could be calculated directly by using the Addition and Multiplication rules, but it is a laborious process. It is much

Although males have a slightly higher birth rate, women outnumber men by a wide margin at institutions of higher education. For the fall semester, 1990, they represented 55% of campus enrollments. Since 1979, women have constituted the majority of students, and the difference has been steadily increasing (The *Wall Street Journal,* March 13, 1991).

easier to first find the probability that event A will not occur, $P(\overline{A})$, and then subtract this value from 1. In the following, B_k is used to indicate that the kth birth is a boy.

$$P(\overline{A}) = P \text{ (there is not at least 1 girl in the 5 births)}$$
$$= P \text{ (all 5 births are boys)}$$
$$= P(B_1B_2B_3B_4B_5)$$
$$= P(B_1) \cdot P(B_2) \cdot P(B_3) \cdot P(B_4) \cdot P(B_5)$$
$$= (0.51)(0.51)(0.51)(0.51)(0.51)$$
$$= 0.0345$$

Therefore, the probability of at least 1 girl in 5 births is

$$P(A) = 1 - P(\overline{A})$$
$$= 1 - 0.0345$$
$$= 0.9655.$$

TIP: Always use the Complement rule to find the probability of obtaining **at least one** of some occurrence.

$$P(\text{at least one occurrence}) = 1 - P(\text{no occurrences})$$

The next example is an application of the Complement rule that has a rather surprising result. A well-known problem in probability, usually referred to as the birthday problem, raises the following question. What is the smallest group size so that there is about a 50 percent chance that at least 2 people in the group have the same birthday? The problem is interesting because the size is much smaller than most people would guess. With a group of only 23 people, the probability is about 0.51 that 2 or more will have a birthday match. This is shown in the next example.

Example 4.25

The following table shows the probabilities of at least one birthday match for some other group sizes n.

n	Probability
10	0.12
20	0.41
30	0.71
40	0.89

Let A denote the event that at least 2 people in a group of 23 have the same birthday. We will ignore leap years and assume that each of the 365 days of the year are equally probable for being born. $P(A)$ will be found by using the Complement rule. The complement of A is the event that there are not at least 2 with the same birthday, that is, $\overline{A}$ is the event that all 23 people have different birthdays.

To find $P(\overline{A})$ we will use the theoretical method of assigning probabilities. By the Multiplication rule for counting, there are 365^{23} points in the sample space for the possible birthdays of 23 people.

The number of sample points for which all birthdays are different is $365 \cdot 364 \cdot 363 \cdot 362 \cdot \ldots \cdot 344 \cdot 343$.

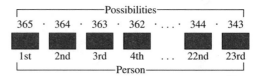

Therefore, by the theoretical method of assigning probabilities, the probability that all 23 have different birthdays is

$$P(\overline{A}) = \frac{365 \cdot 364 \cdot 363 \cdot 362 \cdot \ldots \cdot 344 \cdot 343}{365^{23}} = 0.49.$$

Thus, the probability that at least 2 will have the same birthday is

$$P(A) = 1 - P(\overline{A}) = 0.51.$$

4.6 Combining the Rules of Probability

Having completed our discussion of the rules of probability, we will now focus on problems that involve the application of more than one rule.

Example 4.26 According to the American Academy of Family Physicians, about 30 percent of all doctor visits in 1987 were to physicians specializing in family practice. Find the probability that for two randomly selected people, at least one will see a specialist in family practice on their next doctor visit.

Solution

Let A denote the event that the first person selected will visit a doctor in family practice, and let B denote the event that the second person will visit such a specialist. We need to find the probability that A or B will occur. By the Addition rule,

$$P(A \text{ or } B) = P(A) + P(B) - P(AB).$$

Since the two people were randomly selected, A and B are independent events, and the Multiplication rule gives the following.

$$\begin{aligned} P(AB) &= P(A) \cdot P(B) \\ &= (0.30)(0.30) \\ &= 0.09 \end{aligned}$$

Thus,
$$\begin{aligned} P(A \text{ or } B) &= 0.30 + 0.30 - 0.09 \\ &= 0.51. \end{aligned}$$

Example 4.27 Solve Example 4.26 by using the Complement and Multiplication rules.

Solution

A denotes the event that the first person selected will visit a doctor in family practice, and B is the event that the second person will visit such a specialist. We want to find the probability that at least one of these two events occurs. By the Complement rule,

Figure 4.5
Venn diagram for Examples
4.26 and 4.27.

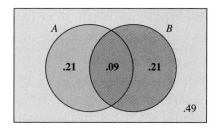

$$P(A \text{ or } B) = 1 - P(A \text{ does not occur } \textbf{and } B \text{ does not occur})$$
$$= 1 - P(\overline{A}\,\overline{B})$$
$$= 1 - P(\overline{A}) \cdot P(\overline{B})$$
$$= 1 - (0.70)(0.70)$$
$$= 0.51.$$

These results are illustrated in the Venn diagram that appears in Figure 4.5.

Example 4.28   In a recent sample of Harvard freshmen, 61 percent said that they were usually up past 1 A.M. Suppose this figure applies to all Harvard freshmen, and that 3 are randomly selected. Find the probability that 2 of the 3 usually stay up past 1 A.M.

Solution

Let U_k for $k = 1, 2, 3$ denote the event that the kth student selected usually stays up past 1 A.M. Then the probability that two of the three stay up after 1 A.M. is $P(U_1 U_2 \overline{U}_3 \text{ or } U_1 \overline{U}_2 U_3 \text{ or } \overline{U}_1 U_2 U_3)$. Since the events $U_1 U_2 \overline{U}_3$, $U_1 \overline{U}_2 U_3$, and $\overline{U}_1 U_2 U_3$ are mutually exclusive, application of the Addition rule gives the following.

$$P(U_1 U_2 \overline{U}_3 \quad \text{or} \quad U_1 \overline{U}_2 U_3 \quad \text{or} \quad \overline{U}_1 U_2 U_3)$$
$$= P(U_1 U_2 \overline{U}_3) + P(U_1 \overline{U}_2 U_3) + P(\overline{U}_1 U_2 U_3)$$
$$= P(U_1) \cdot P(U_2) \cdot P(\overline{U}_3) + P(U_1) \cdot P(\overline{U}_2) \cdot P(U_3) + P(\overline{U}_1) \cdot P(U_2) \cdot P(U_3)$$
$$= (0.61)(0.61)(0.39) + (0.61)(0.39)(0.61) + (0.39)(0.61)(0.61)$$
$$= 3(0.61^2 0.39^1)$$
$$= 0.44$$

Note that the solution equals the product of 3 and $0.61^2 0.39^1$. The 3 can be thought of as the combination $C(3, 2)$, the number of ways of selecting 2 students from 3. Thus, the solution is of the form

$$C(3, 2)0.61^2 0.39^1$$

In Chapter 6, a simple method that utilizes this fact will be discussed for solving problems of this type.

The work in the above example could have been organized in the structure of a tree diagram. The technique is particularly useful for dependent events and is illustrated in the next example.

Figure 4.6
Tree diagram for three cards
of only aces and kings.

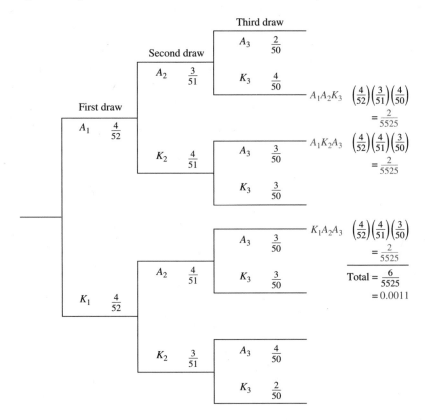

Example 4.29   If 3 cards are randomly selected from a standard deck of 52, what is the probability of obtaining 2 aces and 1 king?

Solution

The tree diagram in Figure 4.6 shows the different ways that the 3 selected cards can result in the selection of aces and kings. In the figure, A_i and K_j $(i, j = 1, 2, 3)$ denote respectively the selection of an ace and the selection of a king on the ith and j^{th} draws.

Starting at the left of the tree diagram, the first set of main branches shows the possible outcomes (A_1 and K_1) and their probabilities for the first draw. Off each of the 2 main branches are secondary branches that show the possibilities (A_2 and K_2) and their probabilities for the second draw. From each of the secondary branches there are branches that show the possible outcomes (A_3 and K_3) for the third draw. To find the probability that 2 aces and 1 king will result, begin at the initial point of the tree and find the paths for which exactly 2 of the 3 selected cards are aces. These are labeled in the diagram. The probability of a particular path is obtained by multiplying the probabilities of the branches transversed. These appear at the far right of the diagram, and are summed to obtain the desired probability of 0.0011.

Example 4.30   The manufacturer of a premium ice cream uses 2 machines to fill pint containers of its best seller—super chocolate almond fudge. Each pint is supposed to have 38 almonds, but 25 percent of the time the older machine will dispense fewer, while this occurs only 10 percent of the time with the newer machine. Eighty percent of all packages are filled by the newer machine.

1. What proportion of super chocolate almond fudge pints contain fewer than 38 almonds?
2. Suppose after a difficult exam you reward yourself and purchase a pint of this ice cream, and you find that it contains fewer than 38 almonds. What is the probability that the container was filled by the older machine?

Solution

First we need to define some events and determine what probabilities are given. Let's define the following.

N: a pint is filled by the newer machine

O: a pint is filled by the older machine

L: a pint contains less than 38 almonds

We are told that 25 percent of the time the older machine dispenses fewer than 38 almonds. Although it may not be readily apparent, this is a conditional probability. It tells us that *if* a container is filled by the older machine, then the probability is 0.25 that it will contain fewer than 38 almonds. Therefore, $P(L|O) = 0.25$. Similarly, $P(L|N) = 0.10$, since 10 percent of pints filled by the newer machine contain fewer than 38 almonds. We are also told that the newer machine fills 80 percent of the containers. Therefore, $P(N) = 0.80$ and $P(O) = 1 - 0.80 = 0.20$.

1. To find the proportion of pints that contain fewer than 38 almonds, we need to find $P(L)$. Event L can occur in the following 2 ways:

   NL: a pint was filled by the newer machine and it contains less than 38 almonds,

   OL: a pint was filled by the older machine and it contains less than 38 almonds.

   $$P(L) = P(NL \text{ or } OL)$$
   $$= P(NL) + P(OL), \text{ since } NL \text{ and } OL \text{ are mutually exclusive events}$$
   $$= P(N) \cdot P(L|N) + P(O) \cdot P(L|O)$$
   $$= (0.80)(0.10) + (0.20)(0.25)$$
   $$= 0.13$$

   Thus, 13 percent of all pints contain fewer than 38 almonds.
2. We are asked to find the probability that a pint was filled by the older machine if we know that it contains fewer than 38 almonds.

   $$P(O|L) = \frac{P(OL)}{P(L)}$$
   $$= \frac{P(O) \cdot P(L|O)}{P(L)}$$
   $$= \frac{(0.20)(0.25)}{0.13}$$
   $$= 0.38$$

Why Study Probability in a Statistics Course?

Before closing this chapter on probability, we should point out why it is necessary to briefly study probability in an introductory statistics course. In Chapter 1, we emphasized that the main application of statistics is statistical inference. To determine something about a large collection of elements, called the population, we resort to sampling. A relatively small portion (sample) of the population is scientifically selected and analyzed. Then, based on information contained in the sample, generalizations (inferences) are made concerning the population. The **mechanism** that allows us to make these inferences is **probability.** For example, suppose the purchasing agent for a television manufacturer is considering the purchase of 1 million transistors. He will buy the lot if the manufacturer's claim is true that at most 2 percent of the transistors are defective. To check on this claim, a sample of 1,000 transistors is tested, and 22 are found defective. Since 2.2 percent of the sample are defective, should the purchasing agent conclude that more than 2 percent of the population are defective, or can the excess amount reasonably be attributed to expected sampling variation? A knowledge of probability is required to provide the answer. A foundation for this knowledge was established in the present chapter and will be employed later in the inferential process.

Sections 4.5 and 4.6 Exercises

The Complement Rule of Probability and
Combining the Rules of Probability

For Exercises 4.103–4.112, describe in words the complement of the given event.

4.103 The birth of a wire-haired fox terrier is a male.

4.104 A compact disk player is defective.

4.105 All components in an optical scanner function properly.

4.106 None of 12 Christmas tree lights are operative.

4.107 At least 1 of 7 department faculty is single.

4.108 More than 1 of 7 department faculty are single.

4.109 Less than 10 in a class of 10 took the exam.

4.110 An elevator is overloaded.

4.111 Eight or more cards in a 13-card bridge hand are red.

4.112 More than 8 cards in a 13-card bridge hand are red.

4.113 Find the probability that 2 tosses of a fair die will result in at least 1 ace (1 spot).

4.114 A fair coin is tossed 4 times. Find the probability of obtaining at least 1 head.

4.115 Ninety percent of 1988 domestic cars were equipped with air conditioning *(Ward's Automotive Yearbook)*. Find the probability that at least 1 of 4 randomly selected 1988 domestic cars has air conditioning.

4.116 A fair coin is tossed 10 times. Use the Complement rule of probability to find the probability of obtaining less than 10 heads.

4.117 Two cards are randomly selected from a 52-card deck. Find the probability of obtaining an ace and a king (in any order).

4.118 A NASA official estimated that there is a probability of $\frac{1}{78}$ that any single space shuttle flight would result in a catastrophic failure (*Lancaster Sunday News*, April 9, 1989). Assume that 100 shuttle missions are planned for the 1990s (present plans are for more than 100 missions). Find the probability that there will be at least 1 shuttle disaster.

4.119 One of the most unusual events in golf history occurred on June 16, 1989, at the Oak Hill Country Club in Rochester, New York. During the second round of the U.S. Open, 4 golfers (not playing together) each made a hole-in-one on the sixth green. Professional golfers average an ace once every 3,750 shots. Suppose that 4 pros are playing a round of golf together, and that the probability of making a hole-in-one on the sixth hole is $\frac{1}{3,750}$. Find the probability that
a. at least 1 will make a hole-in-one on the sixth green,
b. all 4 will make a hole-in-one on the sixth green.

4.120 Although the birth of a boy is slightly more probable than that of a girl, it is usually assumed that they are equally likely. Under this assumption that the probability of each sex is 0.5, determine the probability that a family of 5 children will consist of at least 1 girl. Compare your answer to that obtained in Example 4.24, in which 0.51 was used for the probability of a boy.

4.121 Consider the Venn diagram below in which $P(A) = 0.55$, $P(B) = 0.60$, and $P(AB) = 0.25$. Determine the following probabilities.
a. $P(\overline{A})$ b. $P(A \text{ or } B)$ c. $P(A\overline{B})$
d. $P(A \text{ or } \overline{B})$ e. $P(A|B)$ f. $P(A|\overline{B})$

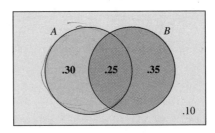

4.122 C and D are independent events for which $P(C) = 0.7$ and $P(D) = 0.2$. Find the following probabilities. (Hint: make a Venn diagram.)
a. $P(C\overline{D})$ b. $P(\overline{C}D)$ c. $P(\overline{C} \text{ or } D)$

4.123 A college senior is going to take the Medical College Admission Test. If he prepares for the test, the probability is 0.7 that he will achieve a satisfactory score. If he doesn't prepare, the probability of obtaining a satisfactory score is only 0.2. If there is a 75 percent probability that he will prepare for the test, what is the probability that the student will achieve a satisfactory test score?

4.124 A tire manufacturer produces tires at three plants, A, B, and C. Fifty, 30, and 20 percent of its tires are made at plants A, B, and C, respectively. Two percent of the tires made at plant A have blemishes, four percent of the tires at plant B have blemishes, and five percent of the tires produced at plant C have blemishes. Find the probability that a tire manufactured by this company will have a blemish. (Hint: Draw a tree diagram.)

4.125 Suppose the student referred to in Exercise 4.123 took the test and achieved a satisfactory score. What is the probability that he prepared for the test?

4.126 A consumer purchases a tire made by the manufacturer mentioned in Exercise 4.124 and it has a blemish. Find the probability that it was made at plant *C*.

Looking Back

An **experiment** in statistics is a process of obtaining observations or measurements. A **sample space** for an experiment is a set of elements such that every possible outcome of the experiment is represented by exactly one element in the sample space. An **event** is a portion of a sample space.

To measure the likelihood that an event will occur when an experiment is performed, the concept of **probability** is used. The **relative frequency definition** of probability was given, and two other methods of assigning probabilities were considered: the **theoretical method** and the **subjective method.** When feasible, one uses the theoretical method to assign probabilities, but this requires a sample space with equally likely points. Its use also requires the ability to count the number of points in the sample space and in some event *E*. Consequently, in Section 4.2 some basic counting techniques were presented.

Armed with just the **Multiplication rule for counting** and the **Combination formula,** a wide variety of counting problems can be solved.

> **The Multiplication Rule for Counting:**
> If an action has *m* possible outcomes, and a second action has *n* possible outcomes, then both actions together have $m \cdot n$ possible outcomes.
>
> **The Combination Formula:**
> $C(n, k)$ denotes the number of combinations of *k* things selected from *n*. It gives the number of ways that *k* items can be selected from *n* items, without regard to the order of selection.
>
> $$C(n, k) = \frac{n!}{k!(n-k)!}$$

Four rules of probability were discussed to obtain probabilities of compound events such as:

1. the probability that *A* **or** *B* will occur,
2. the probability that *A* **and** *B* will occur,
3. the probability that *A* **will occur if it is known that** *B* **has occurred,**
4. the probability that **something other than** *A* **will occur.**

These probabilities can be obtained by using the following rules of probability that were presented in Sections 4.3 through 4.5.

Addition Rule:	$P(A \text{ or } B) = P(A) + P(B) - P(AB)$
Multiplication Rule:	$P(AB) = P(A) \cdot P(B|A)$
Conditional Rule:	$P(A|B) = \dfrac{P(AB)}{P(B)}$
Complement Rule:	$P(A) = 1 - P(\overline{A})$

Key Words

In reviewing this chapter, you should be able to define, explain, and illustrate each of the following.

probability *(page 136)*

experiment *(page 136)*

sample space *(page 137)*

event *(page 138)*

relative frequency definition of probability *(page 139)*

theoretical method of assigning probabilities *(page 141)*

subjective method of assigning probabilities *(page 143)*

odds *(page 145)*

Multiplication rule for counting *(page 147)*

tree diagram *(page 148)*

permutation *(page 150)*

factorial *(page 150)*

combination *(page 150)*

Combination formula *(page 151)*

Addition rule of probability *(page 156)*

Venn diagram *(page 156)*

mutually exclusive events *(page 157)*

Multiplication rule of probability *(page 161)*

independent events *(page 163)*

dependent events *(page 163)*

test for independent events *(page 163)*

Conditional rule of probability *(page 164)*

Complement rule of probability *(page 170)*

Ⓜ MINITAB Commands

RANDOM _ _; *(page 140)*
INTEGERS _ _.

SUM _ *(page 140)*

TALLY _ *(page 146)*

SAMPLE _ _ _ *(page 183)*

SET _ *(page 183)*

END *(page 183)*

Review Exercises

4.127 *A* and *B* are mutually exclusive events for which $P(A) = 0.2$ and $P(B) = 0.4$. Find $P(A \text{ or } B)$.

4.128 *A* and *B* are independent events for which $P(A) = 0.2$ and $P(B) = 0.4$. Find $P(A \text{ or } B)$.

4.129 Evaluate each of the following combinations.
   a. $C(5, 3)$ b. $C(15, 13)$
   c. $C(15, 2)$ d. $C(1,000, 999)$

4.130 A pizza shop sells a super combo that consists of any 5 different ingredients. If the shop offers a choice of 12 ingredients, in how many ways can one order a super combo?

4.131 A sample of 586 adults were asked if they like the taste of liver, and 204 responded positively. Use these results to estimate the probability that an adult likes liver.

4.132 Events A and B are such that $P(A) = 0.83$, $P(B) = 0.56$, and $P(AB) = 0.49$. Find $P(B|A)$ and $P(A|B)$.

4.133 In Exercise 4.132, are A and B independent events? Are they mutually exclusive events?

4.134 Identical twins occur when a single sperm fertilizes a single ovum, which splits during the developmental process. Identical twins constitute one-third of all twin births. Suppose the maternity ward of a hospital has two sets of twins. Find the probability that
   a. both sets are identical twins,
   b. at least one set is identical twins.

4.135 Orbiting debris in space poses a risk to space vehicles. NASA claims that a piece the size of a postage stamp has the potential of destroying a space shuttle. NASA estimates that for each flight the probability is $\frac{1}{30}$ that an impact with this potential will occur (*Bangor Daily News,* July 29, 1989). For the next 4 space shuttles, what is the probability that at least 1 will receive a potentially dangerous impact?

4.136 A couple is going to purchase a lot on which it will eventually build a summer cottage. The couple can select either a shorefront or offshore lot, and it can choose from three different sizes: one acre, two acres, or five acres. Construct a sample space to show the possible selections.

4.137 For each of the following pairs, determine if the events are mutually exclusive:
   a. a company experiencing a decrease in sales for 1992 and the company enjoying an increase in profits for 1992;
   b. in 10 tosses of a coin, obtaining 6 heads and obtaining more than 4 heads;
   c. in 10 tosses of a coin, obtaining 6 heads and obtaining fewer than 4 heads;
   d. a family purchasing an RCA television and purchasing a Toshiba video cassette recorder (VCR).

4.138 A landscaper is going to plant four evergreens that will be located in the front, in the back, and on each side of a house. On his truck there are a Douglas fir, a Frazier fir, a blue spruce, and a scotch pine. In how many different arrangements can the four trees be planted?

4.139 Consider the following Venn diagram where $P(A) = 0.35$, $P(B) = 0.45$, and $P(AB) = 0.20$. Determine the following probabilities.
   a. $P(A \text{ or } B)$ b. $P(\overline{A}B)$
   c. $P(\overline{A} \text{ or } B)$ d. $P(\overline{A}|B)$

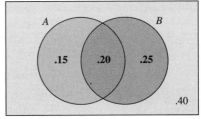

4.140 Three couples have purchased six adjoining seats at a football game. In how many different ways can they be seated if there are no restrictions on their placement?

4.141 For the three couples and six seats referred to in Exercise 4.140, in how many different ways can they be seated if each couple must sit together?

4.142 In 1989, Wade Boggs, third baseman for the Boston Red Sox, led the American League in on-base percentage. He reached base in 43.0 percent of his plate appearances. For a game in which he was at bat 5 times, what is the probability that he reached base at least once? (Assume that plate appearances are independent.)

4.143 A home builder offers its buyers several choices in designing their kitchen. A buyer can choose from four ranges, six refrigerators, two dishwashers, and five microwaves. In how many different ways can a buyer select these appliances?

4.144 Determine if the pair of events in each of the following are apt to be independent:
   a. during the same semester, a Florida State student receiving an A in statistics, and his girlfriend at the University of Alaska receiving an A in statistics;
   b. two close friends in the same statistics class each receiving an A for the course;
   c. selecting 1 card from a standard 52-card deck and obtaining a black card and obtaining an ace.

4.145 A kennel has a litter of 8 retriever pups, 2 of which will be given to the Seeing Eye Association. In how many ways can the pups be selected?

4.146 In addition to the litter of 8 retrievers, the kennel also has a litter of 10 shepherds. The kennel will donate 2 pups from each litter to the Seeing Eye Association. In how many different ways can the pups be chosen?

4.147 An investment club consists of 11 men and 9 women. Five members will be selected at random from the 20.
   a. In how many ways can the sample be selected?
   b. In how many ways can the sample be selected so that it consists of 3 men and 2 women?
   c. Find the probability that the sample of 5 will consist of 3 men and 2 women.

4.148 In situations where it is critical that a system function properly, additional backup systems are usually provided. Suppose a switch is used to activate a component in a satellite. If the switch fails, then a second switch takes over and activates the component. If each switch has a probability of 0.002 of failing, what is the probability that the component will be activated?

4.149 *Newsweek* (December 25, 1989) estimated that the most often requested toy by children for Christmas in 1989 was a Nintendo game pak, with 50 percent requesting a pak. For 3 randomly chosen children, what is the probability that exactly 1 requested a Nintendo game pak?

4.150 A liquidator of surplus goods has purchased four hundred 2-person and 4-person rubber lifeboats. Some boats have manufacturing defects. The distribution of the 400 boats appears below.

Manufacturing Defects?	Type of Boat	
	2-person	4-person
No	202	108
Yes	48	42

  a. Find the probability that a boat is defective.
  b. Find the probability that a boat is defective if it is a 4-person boat.

4.151 If a boat is randomly selected from the 400 lifeboats described in Exercise 4.150, what is the probability that it will either be a 2-person boat or defective?

4.152 Exposure to lead poses a serious environmental health problem to children, since it can cause brain damage even at trace levels. *Newsweek* reported that 60 percent of young children have blood lead levels that may impair their neurological development (March 27, 1989). If 3 young children are randomly selected, what is the probability that at least 1 will have a blood lead level that may impair their neurological development?

4.153 Thirty percent of all residential students at a university have a refrigerator in their room. Four students are randomly chosen. Find the probability that
  a. all have a refrigerator in their room,
  b. at least one has a refrigerator,
  c. fewer than four have a refrigerator.

4.154 Twenty percent of a company's employees are in management. If an employee is in management, there is a probability of 0.88 that he/she is a participant in the company's stock purchase plan. Find the probability that an employee at this company is in management and participates in the stock purchase plan.

4.155 Traffic entering an intersection can continue straight ahead or turn right. Eighty percent of the traffic flow is straight ahead. If a car continues straight, the probability of a collision is 0.0004; if a car turns right, the probability of a collision is 0.0036. Find the probability that a car entering the intersection will have a collision.

4.156 An instructor is being considered for promotion and also for tenure. She believes the probability is 0.8 that she will obtain tenure, and she believes there is a probability of 0.6 that she will receive both. What probability does she assign to being promoted if she is granted tenure?

4.157 Three machines are used to fill bottles of soda. Machines A, B, and C fill 60 percent, 30 percent, and 10 percent of the bottles, respectively. Of those bottles filled by machines A, B, and C, 1 percent, 2 percent, and 5 percent, respectively, are underfilled. Find the probability that a bottle of soda will be underfilled. (Hint: Make a tree diagram.)

4.158 Concerning Exercise 4.157, suppose a bottle is purchased and it is found to be underfilled. What is the probability that it was filled by machine B?

4.159 Along with the Complement rule, the theoretical method of assigning probabilities and the Multiplication rule for counting were used in Example 4.25 to find the probability that at least 2 people in a group of 23 have the same birthday. Redo this problem by using the Complement and the Multiplication rules of probability.

4.160 Find the probability that at least 2 people in a group of 15 have the same birthday.

MINITAB Assignments

Ⓜ 4.161 Use the command **RANDOM** and the subcommand **INTEGER** to simulate the tossing of a fair die a total of
a. 180 times, b. 360 times, c. 540 times, d. 720 times.
e. For each of the above parts, use the **TALLY** command to display the number of occurrences of 1, 2, 3, 4, 5, and 6 spots. Then calculate the relative frequency of the number of ones. Do they appear to be stabilizing around $\frac{1}{6}$?

Ⓜ 4.162 Simulate with MINITAB the selection of 30 cards from a standard deck of 52, where each card is selected singly, observed, and replaced before selecting the next card (sampling with replacement). Assume that you are only interested in whether the card is a club, diamond, heart, or spade.

Ⓜ 4.163 A food processing plant uses three similar machines with equal frequency to fill jars of peanut butter. Use MINITAB to simulate the selection of a sample of 40 jars, where you are only interested in which of the 3 machines filled each jar.

Ⓜ 4.164 MINITAB can be used to simulate an experiment in which elements are *not* returned to the population after being selected (sampling without replacement). For instance, the following command instructs MINITAB to select randomly 5 of the numbers stored in column C1 and then place the 5 values in column C2:

```
SAMPLE 5 C1 C2
```

Place the integers from 1 through 10 in column C1, and have MINITAB randomly select 3. Print the results.

Ⓜ 4.165 Sixty-four patients with high blood pressure agree to participate in a clinical trial to test the effectiveness of a new drug. Thirty-two patients will be randomly selected and given the new drug, while the remaining 32 will be given a standard medication. Assume that the 64 patients are numbered from 1 through 64. Use MINITAB to select the 32 patients who will receive the new drug. (The integers 1 through 64 can easily be placed in column C1 by typing the following commands.)

```
SET C1
1:64
END
```

(© Robert Phillips/The Image Bank)

5

Random Variables and Their Distributions

◄ *How Can They Afford It?*
Some recreational land developers use giveaways to entice prospective purchasers to visit their facilities. Since some of the gifts are very expensive and do not require a purchase, one might wonder how a developer can afford this promotional method. The answer can be determined from information that is often buried in the fine print, namely, the minuscule probabilities associated with the expensive gifts. For instance, consider a company that guarantees to give a visitor one of the following prizes: a $14,500 minivan, a $2,500 large screen television, a $2,000 sailboat, a $1,400 camcorder, or a $69 pair of binoculars. The probabilities of receiving these prizes are, respectively, 0.00001, 0.00002, 0.00002, 0.00004, and 0.99991. Over the long run, what is the average cost per visitor for an offer with these enticements? Techniques will be introduced in this chapter that will allow us to show that the average retail cost is $69.28, scarcely more than the price of the binoculars. Since the gifts are probably purchased at wholesale, the actual cost is likely to be considerably less.

Looking Ahead

In the preceding chapters we have used words to describe events. Sometimes this can be rather cumbersome and lengthy. In this chapter we will learn how to describe many events in a quantitative manner. The vehicle that will allow us to accomplish this is the concept of a **random variable.** We will see that a random variable is involved in showing that the average cost of a promotional gift is $69.28 for the land development problem discussed in this chapter's opener. We will also see that closely associated with a random variable is its **probability distribution.** These serve as mathematical models for describing various physical phenomena.

A distinction will be made between two types of random variables: **discrete** and **continuous.** This chapter will provide a general discussion of discrete random

185

variables, their probability distributions, and their properties. In the next chapter we will look at three particular discrete distributions that play especially important roles in present-day applications of statistics. We will also discuss how to measure the long-run average value and the variability of a discrete random variable.

5.1   Random Variables

In Chapter 4 we saw that an experiment can represent a variety of diverse activities such as:

- tossing a coin 10 times,
- counting the number of bacteria in a vial of yogurt,
- measuring the weights of five pumpkins entered in a county fair competition,
- recording the number of paint defects on a newly manufactured car.

Before conducting an experiment, one is often interested in the probability of various events occurring. To ease the description of these events, statisticians will define one or more random variables and then use the values of these to represent possible outcomes of the experiment.

(© Mark D. Phillips/Photo Researchers, Inc.)

Example 5.1

An official for the state department of weights and measures checks the accuracy of meters on gasoline pumps by filling a one gallon container and comparing this with the quantity indicated on the pump. A particular station has three pumps: regular (R), unleaded (U), and premium (P). The official is going to inspect each pump and record whether it is measuring accurately. A sample space for this experiment is given in Table 5.1, where Y denotes yes and N indicates no.

Table 5.1

Sample Space for the Accuracy of Three Gasoline Pumps

R	U	P
Y	Y	Y
Y	Y	N
Y	N	Y
Y	N	N
N	Y	Y
N	Y	N
N	N	Y
N	N	N

R: regular gas   Y: yes
U: unleaded   N: no
P: premium

Suppose our interest in this experiment is focused on the number of pumps that are measuring accurately. For example, we might want to know the following: What's the probability that two pumps are accurate? What's the probability that at least two are accurate? What's the probability that not more than one is accurate? To facilitate the description of these events, we will let the variable x denote the number of pumps that are measuring accurately. X is called a **random variable** because its value is determined by chance. By this we mean that when the experiment is performed, we do not know which value of x will be observed. For example, if the sample point (Y Y N) results, then x will have a value of 2; if the sample point (N Y N) occurs when the experiment is performed, then x will equal 1. In Table 5.2 we have listed the points in the sample space and the corresponding values of the random variable x. With the introduction of x, we can now replace word descriptions of events with values of the random variable.

Word Description of the Event	**Description Using X**
Two pumps are accurate	$x = 2$
At least two of the pumps are accurate	$x \geq 2$
Not more than one pump is accurate	$x \leq 1$

Table 5.2

Sample Space and the Values of the Random Variable x

R	U	P	
Y	Y	Y	$x = 3$
Y	Y	N	$x = 2$
Y	N	Y	$x = 2$
Y	N	N	$x = 1$
N	Y	Y	$x = 2$
N	Y	N	$x = 1$
N	N	Y	$x = 1$
N	N	N	$x = 0$

R: regular gas   Y: yes
U: unleaded   N: no
P: premium x: the number of accurate pumps (occurrences of Y)

> A **random variable** is a rule that assigns exactly one value to each point in a sample space for an experiment.

Example 5.2   In Example 4.3 we considered an experiment in which a black and a white die were tossed together once. Our sample space consisted of 36 ordered pairs, where the first position indicated the number of spots tossed on the black die and the second position gave the number of spots for the white die. Suppose we are interested in the total

Table 5.3

Sample Space and the Values of the Random Variable x

$(1, 1)\ x = 2$	$(2, 1)\ x = 3$	$(3, 1)\ x = 4$	$(4, 1)\ x = 5$	$(5, 1)\ x = 6$	$(6, 1)\ x = 7$
$(1, 2)\ x = 3$	$(2, 2)\ x = 4$	$(3, 2)\ x = 5$	$(4, 2)\ x = 6$	$(5, 2)\ x = 7$	$(6, 2)\ x = 8$
$(1, 3)\ x = 4$	$(2, 3)\ x = 5$	$(3, 3)\ x = 6$	$(4, 3)\ x = 7$	$(5, 3)\ x = 8$	$(6, 3)\ x = 9$
$(1, 4)\ x = 5$	$(2, 4)\ x = 6$	$(3, 4)\ x = 7$	$(4, 4)\ x = 8$	$(5, 4)\ x = 9$	$(6, 4)\ x = 10$
$(1, 5)\ x = 6$	$(2, 5)\ x = 7$	$(3, 5)\ x = 8$	$(4, 5)\ x = 9$	$(5, 5)\ x = 10$	$(6, 5)\ x = 11$
$(1, 6)\ x = 7$	$(2, 6)\ x = 8$	$(3, 6)\ x = 9$	$(4, 6)\ x = 10$	$(5, 6)\ x = 11$	$(6, 6)\ x = 12$

number of spots tossed on the two dice. Then for each sample point we could let the random variable x indicate the sum of the spots. The resulting values of x are given in Table 5.3. Events can now succinctly be defined in terms of the random variable x. For example, the wordy description of the event

A: tossing a sum of at least 7 spots but fewer than 11,

can be replaced by

$$7 \leq x < 11.$$

Two kinds of random variables will be used in this course: **discrete** and **continuous.** A **discrete random variable** is usually a count variable that indicates the number of times that something is observed. It might be:

- the number of defectives in a shipment of smoke detectors,
- the number of new cars sold last month at your local Honda dealership,
- the number of exams this semester on which you obtain an A,
- the number of houses in your neighborhood that have dangerously high levels of radon gas.

A **discrete random variable** is limited in the number of possible values that it can assume. It can never have more values than the counting numbers 1, 2, 3, 4. . . .

Most of the discrete random variables in this book are count variables. However, not all discrete variables are counts. For example, x might denote the cost in dollars for a roast beef sandwich in a randomly selected New York City restaurant. X could have such values as 2.75, 5.95, 3.99, and so on.

In addition to discrete variables, there are **continuous random variables.**

A **continuous random variable** can assume all the values in some interval on the real number line.

An example of a continuous variable is weight. During a given week someone might weigh anywhere between, say, 150 and 154 pounds. The possibilities include not only the whole numbers 150, 151, 152, 153, and 154, but also the infinity of numbers between these such as 150.235688983, 153.746904531234, 153.1212123, 152.17396, 152.3421543216794532, and so on. Other examples of continuous random variables are:

- the time required to fly from Boston to Atlanta,
- the distance between two sail boats on Lake Ontario,
- the volume of a chunk of ice,
- the pressure at which water is being drawn from a well into a summer cottage.

The remainder of this chapter and all of Chapter 6 will be concerned only with discrete random variables, while Chapter 7 will focus on continuous variables.

Section 5.1 Exercises

Random Variables

5.1   Is the following random variable discrete or continuous: the weight of a newly born kitten?

5.2   Is the following random variable discrete or continuous: the amount of fat in a tenderloin steak?

5.3   Is the following random variable discrete or continuous: the length of a gray squirrel?

5.4   Is the following random variable discrete or continuous: the number of lobsters in a fishing trawler's net?

5.5   Is the following random variable discrete or continuous: a college student's grade on a final examination?

5.6   Is the following random variable discrete or continuous: the number of snowflakes on the roof of a building?

5.7   Is the following random variable discrete or continuous: the depth of snow on a cottage roof?

5.8   Is the following random variable discrete or continuous: the number of fish in the Pacific Ocean?

5.9   Is the following random variable discrete or continuous: the temperature of a wood stove?

5.10 Is the following random variable discrete or continuous: the number of grains of sand in a cinder block?

5.11 Is the following random variable discrete or continuous: the volume of ice in a frozen lake?

5.12 Is the following random variable discrete or continuous: the number of homeless people in the United States?

5.13 Is the following random variable discrete or continuous: the time required to travel from one's home to one's place of employment?

5.14 Is the following random variable discrete or continuous: the height of a building?

5.15 Is the following random variable discrete or continuous: the number of stories in a building?

5.16 Is the following random variable discrete or continuous: the pressure in the passenger compartment of an airplane?

5.17 Is the following random variable discrete or continuous: the number of hairs on a Scottish Terrier?

5.18 Is the following random variable discrete or continuous: the length of one hair on a Scottish Terrier?

5.19 Is the following random variable discrete or continuous: the width of a laptop computer?

5.20 Is the following random variable discrete or continuous: the tensile strength of an airplane rivet?

5.21 Is the following random variable discrete or continuous: the distance traveled by a hiker climbing Mt. Katahdin?

5.22 Is the following random variable discrete or continuous: the cost of a Chevrolet Corvette?

5.23 Is the following random variable discrete or continuous: the weight of a wedge of cheddar cheese?

5.24 Is the following random variable discrete or continuous: the volume of a wedge of cheddar cheese?

5.25 Is the following random variable discrete or continuous: the retail price of a wedge of cheddar cheese?

5.26 Is the following random variable discrete or continuous: the moisture content of a wedge of cheddar cheese?

5.27 Is the following random variable discrete or continuous: the number of display days of a wedge of cheddar cheese?

5.28 Is the following random variable discrete or continuous: the number of mold spots on a wedge of cheddar cheese?

For each of the Exercises 5.29–5.35, identify the random variable of interest, give its possible values, and state whether it is discrete or continuous.

5.29 The publisher of a monthly magazine prints 725,000 copies of its year-end issue, and the publisher is concerned with the number of copies sold.

5.30 Fifty-five satellite drifters are released from the western end of Lake Ontario in May, 1992. During the month of June, a research crew records the number of drifters spotted on the eastern shore.

5.31 An aerial survey of 13 miles of coastline is conducted to determine the length that has been damaged by an oil spill.

5.32 A government safety agency crashes a $10,000 car to measure the amount of damage sustained.

5.33 The 97 students enrolled in Spanish I are surveyed to determine the enrollment for Spanish II.

5.34 To measure the amount of rainfall during April, the height of water in a 500 mm rain gauge is measured.

5.35 To investigate the reliability of a newly developed computer switch, a switch of this type is activated repeatedly until it fails.

5.2 Discrete Probability Distributions

Closely associated with a random variable is its **probability distribution.** As we will see in this and later chapters, probability distributions serve as mathematical models for describing various physical phenomena. In this section we will discuss in general probability distributions of discrete random variables. In the next chapter we will focus on three particular distributions that have special importance in present-day applications of statistics.

Example 5.3

The produce section of a supermarket often features a plentiful selection of Granny Smith apples. The apples are separated in two large bins. One bin is for purchases of six or more, and these are prepackaged at a reduced price. The other bin contains loose

(© David R. Frazier/Photo Researchers, Inc.)

apples for customers who want to purchase five or fewer. For the past year the market has kept records of the number of loose apples per purchase, because it might pre-package smaller quantities if there is sufficient demand for a particular number.

The store's records indicate that during the last year 25 percent of the purchases were for one apple, 40 percent were for two, 20 percent were for three, 10 percent were for four, and 5 percent were for five apples. If we let x denote the number of apples that a customer purchases from the loose bin, then x is a random variable with a probability distribution given in Table 5.4.

The first row of Table 5.4 shows the possible values for x, and the second row gives the values of $P(x)$, the corresponding probabilities. For instance, below $x = 3$ appears 0.20, the value of $P(3)$. $P(3)$ is read as "the probability of 3." Because the table gives the possible values of x and their corresponding probabilities, it is called the **probability distribution** of the random variable x.

The **probability distribution** of a discrete random variable is any device (table, graph, formula) that gives

1. the possible values of the variable,
2. the probability associated with each of these values.

Table 5.4

Probability Distribution of x: Number of Apples Purchased

x	1	2	3	4	5
$P(x)$	0.25	0.40	0.20	0.10	0.05

Sometimes the term **probability function** is used in place of probability distribution.

Example 5.4

In Example 5.2, a black and a white die were tossed together once, and the random variable x denoted the sum of the spots that showed on the two dice. The sample space was given in Table 5.3 and is repeated below. In each ordered pair the first number shows what occurred on the black die, and the second number gives the result for the white die.

Sample Space and the Values of the Random Variable x

$(1, 1)\ x = 2$	$(2, 1)\ x = 3$	$(3, 1)\ x = 4$	$(4, 1)\ x = 5$	$(5, 1)\ x = 6$	$(6, 1)\ x = 7$
$(1, 2)\ x = 3$	$(2, 2)\ x = 4$	$(3, 2)\ x = 5$	$(4, 2)\ x = 6$	$(5, 2)\ x = 7$	$(6, 2)\ x = 8$
$(1, 3)\ x = 4$	$(2, 3)\ x = 5$	$(3, 3)\ x = 6$	$(4, 3)\ x = 7$	$(5, 3)\ x = 8$	$(6, 3)\ x = 9$
$(1, 4)\ x = 5$	$(2, 4)\ x = 6$	$(3, 4)\ x = 7$	$(4, 4)\ x = 8$	$(5, 4)\ x = 9$	$(6, 4)\ x = 10$
$(1, 5)\ x = 6$	$(2, 5)\ x = 7$	$(3, 5)\ x = 8$	$(4, 5)\ x = 9$	$(5, 5)\ x = 10$	$(6, 5)\ x = 11$
$(1, 6)\ x = 7$	$(2, 6)\ x = 8$	$(3, 6)\ x = 9$	$(4, 6)\ x = 10$	$(5, 6)\ x = 11$	$(6, 6)\ x = 12$

Find the probability distribution of the random variable x.

Solution

Since there are 36 equally likely points in the sample space, each has a probability 1/36 of occurring when the experiment is performed. Because one point has a value of $x = 2$, $P(2) = 1/36$. Since two points have $x = 3$, then $P(3) = 2/36$. Proceeding in this way, the following probability distribution is constructed.

x	2	3	4	5	6	7	8	9	10	11	12
$P(x)$	$\frac{1}{36}$	$\frac{2}{36}$	$\frac{3}{36}$	$\frac{4}{36}$	$\frac{5}{36}$	$\frac{6}{36}$	$\frac{5}{36}$	$\frac{4}{36}$	$\frac{3}{36}$	$\frac{2}{36}$	$\frac{1}{36}$

We emphasize that a probability distribution can be given by means of any device, such as a table, a graph, or a formula, as long as it gives the possible values of x and the associated probabilities. This is illustrated in the next example.

Example 5.5

Radon, the second largest cause of lung cancer after smoking, is a radioactive gas produced by the natural decay of radium in the ground. In a 1988 study of seven states (Arizona, North Dakota, Massachusetts, Minnesota, Missouri, Pennsylvania, and Indiana), the Environmental Protection Agency (EPA) found that one-third of the houses tested had dangerous levels of the invisible gas. Suppose two houses are randomly selected from those tested, and we define the random variable x to be the number of houses with dangerous levels. Give the probability distribution of x by means of

1. a table,
2. a graph,
3. a formula.

Solution

1. Since two houses are selected, the possible values for x are 0, 1, and 2. To find their probabilities, we can use the rules of probabilities discussed in Chapter 4.

$$P(2) = P(\text{1st house is unsafe } and \text{ 2nd house is unsafe})$$
$$= P(\text{1st house is unsafe}) \cdot P(\text{2nd house is unsafe})$$
$$= \left(\frac{1}{3}\right)\left(\frac{1}{3}\right)$$
$$= \frac{1}{9}$$

$$P(0) = P(\text{1st house is safe } and \text{ 2nd house is safe})$$
$$= P(\text{1st house is safe}) \cdot P(\text{2nd house is safe})$$
$$= \left(\frac{2}{3}\right)\left(\frac{2}{3}\right)$$
$$= \frac{4}{9}$$

$$P(1) = 1 - [P(0) + P(2)]$$
$$= 1 - \left[\frac{4}{9} + \frac{1}{9}\right]$$
$$= \frac{4}{9}$$

Thus, the probability distribution of x is as follows.

x	0	1	2
$P(x)$	$\frac{4}{9}$	$\frac{4}{9}$	$\frac{1}{9}$

2. Any type of graph can be used to give the probability distribution as long as it shows the possible values of x and the corresponding probabilities. Figure 5.1 shows two possibilities: a **probability spike graph** and a **probability histogram.** Note that in the former, a probability is graphically displayed as the height of a ''spike'' (line), while in the latter a probability is displayed as the height of a rectangle. Moreover, in the probability histogram, the rectangle corresponding to each value of x has an area equal to the probability $P(x)$. The association of a probability with an area in the probability histogram will be seen to be of primary importance when continuous random variables are discussed in Chapter 7.

Figure 5.1
A, Probability spike graph; *B*, Probability histogram.

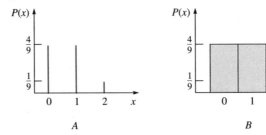

3. The probability distribution of x can also be given by the following formula. Don't be concerned now with where this formula comes from. You will understand it when you get to the next chapter. The only reason it is given here is to illustrate that a formula can sometimes be used to give the probability distribution.

$$P(x) = C(2, x)(\tfrac{1}{3})^x(\tfrac{2}{3})^{2-x}, \qquad \text{for } x = 0, 1, 2 \qquad \textbf{(5.1)}$$

Notice that when $x = 0$ is substituted in Equation 5.1, we obtain

$$P(0) = C(2, 0)(\tfrac{1}{3})^0(\tfrac{2}{3})^2$$
$$= 1 \cdot 1 \cdot (\tfrac{4}{9})$$
$$= \tfrac{4}{9}.$$

Similarly, substituting 1 and 2 for x in Equation 5.1 yields 4/9 and 1/9, respectively.

Every discrete probability distribution must satisfy two basic properties stated below.

Two Basic Properties of a Discrete Probability Distribution:

1. $P(x) \geq 0$ for each value of x.   **(5.2)**

2. $\Sigma P(x) = 1$.   **(5.3)**

Property 5.2 follows from the fact that $P(x)$ is a probability and, consequently, cannot be negative. Equation 5.3 says that the sum of all values of $P(x)$ must be one. This is because exactly one value of x is assigned to each point in the sample space, and the sum of the probabilities attached to the sample points is one.

We close this section by illustrating how MINITAB can be used to simulate the selection of a sample from a probability distribution.

Ⓜ **Example 5.6**   In Example 5.3 the following probability distribution was given, where the random variable x denoted the number of Granny Smith apples per purchase by supermarket shoppers.

x	1	2	3	4	5
$P(x)$	0.25	0.40	0.20	0.10	0.05

We will simulate 1,000 purchases by using MINITAB to generate 1,000 values of x, where the probabilities of selecting the values 1, 2, 3, 4, and 5, are 0.25, 0.40, 0.20, 0.10, and 0.05, respectively. To accomplish this, we first place the possible values of x in column C1 and their corresponding probabilities in C2.

```
READ C1 C2
1 .25
2 .40
3 .20
4 .10
5 .05
END
```

To simulate the 1,000 purchases, the command **RANDOM** is used with the subcommand **DISCRETE**. After entering columns C1 and C2, type the following.

```
RANDOM  1000 values and place in C3;
DISCRETE sample of values from C1 with probs. in C2.
```

Or more concisely, just type the following.

```
RAND 1000 C3;
DISC C1 C2.
```

The results of the simulation are contained in column C3 and can be examined by applying the command **HISTOGRAM** to C3. The output appears in Exhibit 5.1, which reveals that 1, 2, 3, 4, and 5 apples were purchased 258, 407, 181, 108, and 46 times, respectively. Thus, the resulting relative frequencies for the values of x (0.258, 0.407, 0.181, 0.108, 0.046) are quite close to the theoretical relative frequencies of 0.25, 0.40, 0.20, and 0.10, and 0.05.

Exhibit 5.1

```
MTB > RANDOM 1000 C3;
SUBC> DISC C1 C2.
MTB > HIST C3

Histogram of C3    N = 1000
Each * represents 10 obs.

Midpoint   Count
      1      258   ************************
      2      407   ******************************************
      3      181   ******************
      4      108   **********
      5       46   *****
```

Section 5.2 Exercises
Discrete Probability Distributions

5.36 X is a random variable with the following probability distribution. Determine the value of the missing probability, $P(3)$.

x	0	1	2	3	4
$P(x)$	0.2	0.1	0.1		0.2

5.37 The probability distribution of a random variable x is given below. Find the value of the missing probability, $P(4)$.

x	3	4	5	6	7
$P(x)$	0.04		0.16	0.38	0.15

5.38 Use a probability spike graph to display the probability distribution in Exercise 5.36.

5.39 Use a probability spike graph to display the probability distribution that appears in Exercise 5.37.

5.40 Construct a probability histogram for the probability distribution in Exercise 5.36.

5.41 Construct a probability histogram for the probability distribution in Exercise 5.37.

5.42 Explain why the following is or is not a valid probability distribution.

x	0	2	4	6
$P(x)$	0.2	0.1	0.1	0.7

5.43 Explain why the following is or is not a valid probability distribution.

x	-4	-2	0	4	8
$P(x)$	0.15	0.21	0.34	0.18	0.12

5.44 Determine if the following is a valid probability distribution. If it is not, explain why.

x	1	2	3	4	5
$P(x)$	0.1	0.2	-0.4	0.8	0.3

5.45 Determine if the following is a valid probability distribution. If it is, then display it in the form of a probability spike graph.

$$P(x) = \frac{(x-3)^2}{30}, \qquad \text{for } x = -1, 0, 1, 2$$

5.46 Determine if the following is a valid probability distribution. If it is, then display it in the form of a probability spike graph.

$$P(x) = \frac{x^2 - 1}{26}, \quad \text{for } x = 1, 2, 3, 4$$

5.47 Determine if the following is a valid probability distribution. If it is, then display it in the form of a table.

$$P(x) = C(2, x)(.7)^x(.3)^{2-x}, \quad \text{for } x = 0, 1, 2$$

− **5.48** The random variable x has the following probability distribution.

x	0	1	2	3	4	5
$P(x)$	0.05	0.11	0.19	0.25	0.24	0.16

Determine the following probabilities.
a. $P(x \leq 2)$ b. $P(x > 3)$
c. $P(1 \leq x \leq 3)$

5.49 The probability distribution of the random variable x is

$$P(x) = C(3, x)(.4)^x(.6)^{3-x}, \quad \text{for } x = 0, 1, 2, 3.$$

Find the following probabilities.
a. $P(x \leq 1)$ b. $P(x \geq 2)$

5.50 A test contains two true-false questions. The correct answer to each question is true, but a student has no knowledge of the material and answers each question by guessing.
a. Construct a sample space that shows the different ways that the two questions can be answered.
b. Assign appropriate probabilities to the sample points.
c. Let the random variable x denote the number of questions answered correctly. Assign the appropriate value of x to each sample point.
d. Construct the probability distribution of the random variable x.

5.51 A die and a quarter are tossed together once.
a. Construct a sample space that shows the different ways that the die and the quarter can fall.
b. Assign appropriate probabilities to the sample points.
c. Let the random variable x equal the sum of the number spots and the number of heads that show. Assign the appropriate value of x to each sample point.
d. Construct the probability distribution of the random variable x.

− **5.52** *Hippocrates* magazine reported that 20 percent of cat owners talk to their cats as if they were another adult (March/April, 1989). Suppose a sample of 2 cat owners is randomly selected. Let the random variable x be the number in the sample who talk to their cat in this way. Find the probability distribution of x.

5.53 According to the American Cancer Society, 82 percent of Americans know that a poor diet can increase their risk of cancer. Let the random variable x denote the number of people in a sample of 2 who are aware of this fact. Obtain the probability distribution of x.

5.54 A gift shop has found that the probability is 0.70 that a customer will use a charge card to make a purchase. Let the random variable x denote the number of the next 3 customers who use a charge card. Find the probability distribution of x.

5.55 A computer supply shop has five batteries in stock for a particular laptop computer. Unknown to anyone, three of the batteries are defective. Suppose a customer purchases two batteries, and we let x denote the number of defectives selected. Determine the probability distribution of x.

MINITAB Assignments

5.56 Several brands of candy are available in "fun size" boxes. For a particular brand, the number of pieces x per box varies. The probability distribution of x is as follows.

x	6	7	8	9	10
$P(x)$	0.04	0.12	0.65	0.13	0.06

Use MINITAB to simulate 200 purchases, and use a histogram to display the resulting values of x.

5.57 A professional baseball player has gone hitless in 34 percent of the games that he has played during his career. He has obtained 1 hit in 31 percent of the games, 2 hits in 21 percent of the games, 3 hits in 11 percent, and 4 hits 3 percent of the time. Simulate with MINITAB his number of hits per game for a 162 game season, and display the results in a histogram.

5.58 A psychologist has recorded the results of several thousand repetitions of a laboratory experiment. She has determined that the probability distribution of x, the number of errors made by participants, is the following.

x	0	1	2	3	4	5	6
$P(x)$	0.32	0.24	0.13	0.10	0.09	0.07	0.05

Use MINITAB to simulate 500 repetitions of the experiment. Display the resulting values of x in a histogram.

5.59 A coin is bent in such a way that the probability of tossing a head is twice as likely as obtaining a tail. Simulate, using MINITAB, 600 tosses, and use a histogram to display the results.

5.3 The Mean of a Discrete Random Variable

In Example 5.3, we considered a supermarket whose produce section features a bin of Granny Smith apples that can be bought loose in quantities of 1, 2, 3, 4, and 5. The store's records indicate that these numbers are bought 25 percent, 40 percent, 20 percent, 10 percent, and 5 percent of the time, respectively. If we let the random variable x denote the number of apples that a customer purchases, then x has the probability distribution given earlier in Table 5-4 and repeated below.

Probability Distribution of x:
The Number of Apples Purchased

x	1	2	3	4	5
$P(x)$	0.25	0.40	0.20	0.10	0.05

The probability distribution tells the produce manager with what frequency the quantities 1 through 5 are purchased, but she would probably also want to know the average number per purchase. How can this quantity be obtained? Should we add the numbers 1, 2, 3, 4, 5, and divide by 5? After some reflection on this, you'll conclude that it will not give us the correct result, since the values do not occur with equal frequency. For example, 3 apples are purchased twice as often as 4, and 4 times as often as 5 apples. What we need to do is compute a *weighted average* of the values of x, where the weights are the associated probabilities. This weighted average is called the **mean of the random variable** and is denoted by μ, the same symbol that was used earlier to denote the mean of a population.

The **mean of the discrete random variable** x with probability distribution $P(x)$ is a weighted average of the values of x, where the weights are the associated probabilities.

$$\mu = \Sigma x P(x) \tag{5.4}$$

The mean of the number of apples purchased is calculated as follows.

$$\mu = \Sigma x P(x)$$
$$= 1(0.25) + 2(0.40) + 3(0.20) + 4(0.10) + 5(0.05)$$
$$= 0.25 + 0.80 + 0.60 + 0.40 + 0.25$$
$$= 2.3$$

Thus, the mean of the random variable x is 2.3. You can think of this number as the long-run average value of x. By this is meant that over the long run, the average number of apples per purchase is 2.3.

The **mean of a random variable** x is its **long-run average value.** It is also called the **mathematical expectation of x,** or the **expected value of x.**

Example 5.7   An insurance company is willing to issue a policy on a sailboat subject to the following conditions:

1. the replacement cost, $5,000, will be paid for a total loss;
2. if it is not a total loss, but the damage is more than $2,000, then $1,500 will be paid;
3. nothing will be paid for a loss of $2,000 or less.

Suppose the probabilities of (1), (2), and (3) occurring during a policy year are 0.02, 0.10, and 0.35, respectively. How much should the insurance company charge for this type of policy if it wants to receive $50 more than its expected payment?

Solution

Let the random variable x denote the insurance company's payment for a claim. The possible values of x are

1. $5,000 if the boat is totaled; the probability of this is 0.02.
2. $1,500 if the damage is more than $2,000 but not a total loss; this has a probability of 0.10.
3. $0 if the damage is $2,000 or less; this will occur with a probability of 0.35.
4. $0 if there is no damage sustained; the probability of this is

$$1 - (0.02 + 0.10 + 0.35) = 0.53.$$

Therefore, the probability distribution of x, the company's payment for a claim, is as follows.

x	0	1,500	5,000
$P(x)$	0.88	0.10	0.02

The mathematical expectation of x is

$$\mu = \Sigma x P(x)$$
$$= 0(0.88) + 1,500(0.10) + 5,000(0.02)$$
$$= \$250.$$

Thus, the insurance company should charge $300 in order to receive $50 in excess of its expected payment.

It should be emphasized that even though its expected payment is $250, in no year will the company ever pay this amount on a policy. The $250 is just the average amount over the long run that it can expect to pay for similar policies.

Pascal's Famous Wager

"The 'expectation' in a gamble is the value of the prize multiplied by the probability of winning the prize. According to Pascal the value of eternal happiness is infinite. He reasoned that even if the probability of winning eternal happiness by leading a religious life is very small indeed, nevertheless, since the expectation is infinite (any finite fraction of infinity is itself infinite) it will pay anyone to lead such a life."
(E. T. Bell, *Men of Mathematics,* Simon and Schuster, New York, 1937, pp. 88–89.)

At the beginning of this chapter, reference was made to a recreational land company that uses giveaways to entice prospective buyers to visit its facilities. Although some of the gifts are very expensive, we claimed that over the long run the average cost per visitor was only $69.28. The next example shows how this figure is obtained.

Example 5.8

A company guarantees to give a visitor to its facilities one of the following prizes: a $14,500 minivan, a $2,500 large screen television, a $2,000 sailboat; a $1,400 camcorder, or a $69 pair of binoculars. The probabilities of receiving these prizes are, respectively, 0.00001, 0.00002, 0.00002, 0.00004, and 0.99991. Over the long run, what is the average cost per visitor for an offer with these enticements?

Solution

If we let the random variable x denote the cost of a gift, then the probability distribution of x is as follows.

x	69	1,400	2,000	2,500	14,500
$P(x)$	0.99991	0.00004	0.00002	0.00002	0.00001

The mean of the random variable x, that is, its long-run average value, equals

$$\mu = \Sigma x P(x)$$
$$= 69(0.99991) + 1,400(0.00004) + 2,000(0.00002)$$
$$+ 2,500(0.00002) + 14,500(0.00001)$$
$$= \$69.28.$$

Ⓜ Example 5.9   The probability distribution of the random variable x is given by the formula

$$P(x) = \frac{x}{20,100}, \quad \text{for } x = 1, 2, 3, \ldots 200.$$

Use MINITAB to find the mean of x.

Solution

We need to place the values of x and their corresponding probabilities into two columns. This can be accomplished with the following commands:

```
SET C1
1:200/1     #  /1 denotes an increment of 1. It's
END         #  optional since the default value is 1
LET C2 = C1/20100
```

The first three lines above will place in column C1 the numbers $1, 2, 3, \ldots 200$. The **LET** command divides each number in C1 by 20,100 and puts the results in C2. Thus, C2 will contain the following probabilities:

$$\frac{1}{20,100}, \frac{2}{20,100}, \frac{3}{20,100}, \frac{4}{20,100}, \ldots \frac{200}{20,100}$$

Next, we use the **LET** command to form a column C3 that contains the values of $xP(x)$, the product of each x with its probability. Then we'll employ the **SUM** command to add the values of $xP(x)$ that are contained in C3.

```
LET C3 = C1*C2
SUM C3
```

MINITAB will respond with the output in Exhibit 5.2. From the MINITAB output, we
have that the mean of the random variable x is $\mu = 133.67$.

Exhibit 5.2

```
MTB > SET C1
DATA> 1:200/1
DATA> END
MTB > LET C2 = C1/20100
MTB > LET C3 = C1*C2
MTB > SUM C3
   SUM    =        133.67
```

5.4 Measuring the Variability of a Random Variable

In Chapter 2, numerical methods for describing a data set were discussed. We indicated that a minimum description should consist of a measure of central location and a
measure of the variation in the data. Usually these two measures are the sample mean $\bar{x}$
and the sample standard deviation s. Similarly, in applications involving a random
variable, we normally measure its central location and its variability by determining
its mean μ and its standard deviation σ.

> For a discrete random variable x with mean μ and probability distribution $P(x)$, the
> **variance of the random variable** is
> $$\sigma^2 = \Sigma(x - \mu)^2 P(x). \qquad (5.5)$$
> The **standard deviation of the random variable** is
> $$\sigma = \sqrt{\text{variance}}. \qquad (5.6)$$

Equation 5.5 reveals that the variance of a random variable is just a weighted
average of the squared deviations of each x value from the mean; the weights are the
probabilities associated with the values of x. The standard deviation of a random
variable is the square root of its variance.

Example 5.10   For the example involving Granny Smith apples in Section 5.3, the random variable x
denoted the number of loose apples that a customer purchases. Its probability distribution is repeated below. Find the variance and the standard deviation of the random
variable x.

x	1	2	3	4	5
$P(x)$	0.25	0.40	0.20	0.10	0.05

Solution

The calculations are facilitated by constructing Table 5.5 below. The variance of the random variable x appears in the last row of Table 5.5, namely, $\sigma^2 = 1.21$. The standard deviation of x is

$$\sigma = \sqrt{\sigma^2}$$
$$= \sqrt{1.21}$$
$$= 1.1.$$

Table 5.5

x	$P(x)$	$xP(x)$	$x - \mu$	$(x - \mu)^2$	$(x - \mu)^2 P(x)$
1	0.25	0.25	-1.3	1.69	0.4225
2	0.40	0.80	-0.3	0.09	0.0360
3	0.20	0.60	0.7	0.49	0.0980
4	0.10	0.40	1.7	2.89	0.2890
5	0.05	0.25	2.7	7.29	0.3645
		$\mu = 2.30$			$\sigma^2 = 1.2100$

In Chapter 2, after giving the definition of the variance of a sample, we pointed out that it can be obtained more easily by using a formula that does not require deviations from the mean. This is also true for finding the variance of a random variable. We now give a shortcut formula for obtaining σ^2, and henceforth it will be used in place of Equation 5.5.

Shortcut Formula for the Variance of a Random Variable:

$$\sigma^2 = \Sigma x^2 P(x) - \mu^2 \qquad\qquad (5.7)$$

Example 5.11   Use the shortcut formula to find the variance and standard deviation of the random variable x in Example 5.10.

Solution

We need to find the variance and standard deviation of the random variable x whose probability distribution is

x	1	2	3	4	5
$P(x)$	0.25	0.40	0.20	0.10	0.05

We will construct Table 5.6 to perform the calculations.

Table 5.6

x	$P(x)$	$xP(x)$	$x^2P(x)$
1	0.25	0.25	0.25
2	0.40	0.80	1.60
3	0.20	0.60	1.80
4	0.10	0.40	1.60
5	0.05	0.25	1.25
		$\mu = 2.30$	6.50

$$\sigma^2 = \Sigma x^2 P(x) - \mu^2$$
$$= 6.5 - (2.3)^2$$
$$= 1.21$$

The standard deviation of the random variable x is

$$\sigma = \sqrt{1.21}$$
$$= 1.1.$$

Ⓜ Example 5.12   Use MINITAB to find the variance and the standard deviation of the random variable whose probability distribution was given in Example 5.9 and is repeated below.

$$P(x) = \frac{x}{20,100}, \qquad \text{for } x = 1, 2, 3, \ldots 200$$

Solution

We will use the shortcut equation to find the variance.

$$\sigma^2 = \Sigma x^2 P(x) - \mu^2$$

In Example 5.9 we used the following commands to place the x values and their probabilities in columns C1 and C2, respectively, and then have μ, the mean of x, computed. The value of μ is given by **SUM C3**, the sum of the numbers stored in column C3.

```
SET C1
1:200/1
END
LET C2 = C1/20100    # The probs. are placed in C2.
LET C3 = C1*C2       # Each x is multiplied by its prob.
SUM C3               # This is the mean of x.
```

The first part of the shortcut formula, $\Sigma x^2 P(x)$, requires that we create the column (C1**2)*C2, which contains the values of each x squared (C1**2) multiplied by its probability (C2). Then the command **SUM** is used to add the column elements. From this sum we subtract the square of the mean in order to obtain the variance.

```
LET K1 = SUM((C1**2)*C2) - (SUM(C3))**2
PRINT K1
```

To obtain the standard deviation, we instruct MINITAB to take the square root of the variance $K1$.

```
LET K2 = SQRT(K1)
PRINT K2
```

MINITAB will produce the output given in Exhibit 5.3, from which we obtain

$$\sigma^2 = 2{,}233.22$$

$$\sigma = 47.26.$$

Exhibit 5.3

```
MTB > SET C1
DATA> 1:200/1
DATA> END
MTB > LET C2 = C1/20100    # The probs. are placed in C2.
MTB > LET C3 = C1*C2       # Each x is multiplied by its prob.
MTB > SUM C3               # This is the mean of x.
   SUM        =       133.67
MTB > LET K1 = SUM((C1**2)*C2) - (SUM(C3))**2
MTB > PRINT K1                # This is the variance of x.
K1        2233.22
MTB > LET K2 = SQRT(K1)
MTB > PRINT K2               # This is the standard deviation.
K2        47.2570
```

The emphasis in this section has been on how to obtain the variance and the standard deviation of a random variable. If you feel somewhat ambivalent about their usefulness, don't be concerned. At this stage you should be able to compute the variance and standard deviation of a random variable, and you should realize that they are measures of how a random variable fluctuates. Keep in mind that the variance σ^2 is just a weighted average of the squared deviations of the values of x from its mean; the weights are the associated probabilities. The standard deviation σ is the square root of the variance. For nearly all applications in this book, we will be interested in the variance only as a means to obtaining the standard deviation. As you work through the text, many applications of standard deviation will be encountered, and you will progressively feel more comfortable with this concept.

Before closing, we point out that Chebyshev's theorem and the Empirical rule discussed in Chapter 2 are also applicable to random variables and their probability distributions. For instance, we are often concerned about variability within 2 standard deviations of the mean.

> **Chebyshev's theorem** tells us that for a random variable with mean μ and standard deviation σ,
>
> $$P(\mu - 2\sigma \le x \le \mu + 2\sigma) \ge 0.75.$$

In words, this says that at least 75 percent of the time the value of a random variable will fall within 2 standard deviations of its mean (see Figure 5.2).

> The **Empirical rule** says that for a random variable whose probability distribution is approximately mound shaped,
>
> $$P(\mu - 2\sigma \le x \le \mu + 2\sigma) \approx 0.95.$$

In words, for a random variable with a mound-shaped probability distribution, approximately 95 percent of the time it will have a value within 2 standard deviations of its mean (see Figure 5.3).

Many applications will be encountered in subsequent chapters that will lend more meaning to the Empirical rule and to Chebyshev's theorem.

Figure 5.2
Chebyshev's theorem for 2 standard deviations.

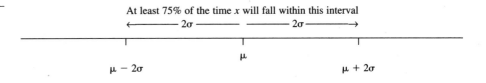

Figure 5.3
The Empirical rule for 2 standard deviations.

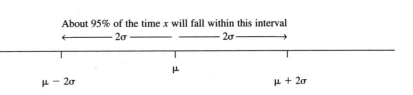

Sections 5.3 and 5.4 Exercises

The Mean of a Discrete Random Variable and Measuring the Variability of a Random Variable

5.60 The random variable x has the following probability distribution. Find its mean, variance, and standard deviation.

x	0	2	4	6
$P(x)$	0.1	0.3	0.1	0.5

5.61 X is a random variable with the following probability distribution. Find its mean, variance, and standard deviation.

x	0	1	2	3	4
$P(x)$	0.2	0.1	0.1	0.4	0.2

5.62 The probability distribution of a random variable x is given below. Determine its mean, variance, and standard deviation.

x	3	4	5	6	7
$P(x)$	0.04	0.27	0.16	0.38	0.15

5.63 For the following probability distribution, find μ, σ^2, and σ.

x	-4	-2	0	4	8
$P(x)$	0.15	0.21	0.34	0.18	0.12

5.64 X is a random variable with a probability distribution given in the following table. Calculate μ, σ^2, and σ.

x	10	20	30	40	50
$P(x)$	0.05	0.20	0.50	0.20	0.05

5.65 Find the mean and the standard deviation of the random variable whose probability distribution is given below.

$$P(x) = \frac{x^2 + 1}{32}, \quad \text{for } x = 2, 3, 4$$

5.66 Calculate μ and σ for the random variable with the following probability distribution.

$$P(x) = \frac{(x-3)^2}{30}, \qquad \text{for } x = -1, 0, 1, 2$$

5.67 Determine the mean and standard deviation of the random variable x whose probability distribution is as follows.

$$P(x) = C(2, x)(.7)^x(.3)^{2-x}, \qquad x = 0, 1, 2$$

5.68 Find the mean and standard deviation of the random variable whose probability distribution is given by the following probability histogram.

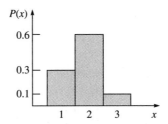

5.69 Determine the mean and standard deviation of the random variable whose probability distribution is given by the following probability spike graph.

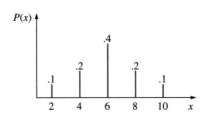

5.70 If a fair coin is tossed 5 times, the probability distribution of x, the number of heads, can be shown to be the following.

x	0	1	2	3	4	5
$P(x)$	$\frac{1}{32}$	$\frac{5}{32}$	$\frac{10}{32}$	$\frac{10}{32}$	$\frac{5}{32}$	$\frac{1}{32}$

Find the mean and standard deviation of x.

5.71 For the probability distribution in Exercise 5.69, find the probability that the random variable x will have a value that is within 2 standard deviations of its mean, that is, determine the following probability.

$$P(\mu - 2\sigma \leq x \leq \mu + 2\sigma)$$

5.72 For the random variable in Exercise 5.70, determine the probability that it will have a value that is within 2 standard deviations of its mean, that is, find the following probability.

$$P(\mu - 2\sigma \le x \le \mu + 2\sigma)$$

5.73 A shop specializes in ice cream cones. One can purchase a cone with 1, 2, 3, or 4 scoops, and shop records reveal that these amounts are bought 50 percent, 28 percent, 17 percent, and 5 percent of the time, respectively. If we let x denote the number of scoops that a customer purchases, then x is a random variable with a probability distribution as follows.

x	1	2	3	4
$P(x)$	0.50	0.28	0.17	0.05

Over the long run, what is the average number of scoops per cone purchased?

5.74 Several brands of candy are available in "fun size" boxes. For a particular brand, the number of pieces x per box varies. The probability distribution of x is as follows.

x	6	7	8	9	10
$P(x)$	0.04	0.12	0.65	0.13	0.06

Find the long-run average number of pieces per box.

5.75 Find the standard deviation of the random variable x whose probability distribution is given in Exercise 5.73.

5.76 Find the standard deviation for the probability distribution in Exercise 5.74.

5.77 An insurance company issues a policy on a $15,000 diamond necklace. If the necklace is stolen, then the company will pay its full value; if it is lost, then the owner will receive $5,000. The company estimates that in a given year the probability of a theft is 0.001, and the probability of it being lost is 0.02. Determine how much the company should charge for this policy if it wants to receive $30 more than its expected payment.

5.78 A company conducted a national contest with a grand prize of $100,000, a second place prize of $10,000, and a third place prize of $1,000. The probabilities of a person winning these amounts were 0.000002, 0.000009, and 0.00004, respectively. Determine the expected winnings for a person who entered this contest.

MINITAB Assignments

Ⓜ 5.79 Use MINITAB to find the mean of the random variable whose probability distribution is given by the following formula.

$$P(x) = \frac{x}{1,560}, \qquad \text{for } x = 2, 4, 6, \ldots 78$$

Ⓜ **5.80** Use MINITAB to obtain the variance and standard deviation of the random variable whose probability distribution is given in Exercise 5.79.

Ⓜ **5.81** The random variable x has the following probability function. Use MINITAB to find its mean and variance.

x	54	56	65	66	69	74	76	79	81	87	98
$P(x)$	0.03	0.11	0.15	0.08	0.18	0.21	0.02	0.09	0.01	0.05	0.07

Ⓜ **5.82** If the random variable x denotes the number of spots that show when a fair die is tossed, then the probability distribution of x is

$$P(x) = \tfrac{1}{6}, \qquad \text{for } x = 1, 2, 3, 4, 5, 6$$

a. Show that the mean and standard deviation of x are $\mu = 3.5$ and $\sigma = 1.708$.

b. Using the command **RANDOM** and the subcommand **INTEGER,** have MINITAB simulate 300 tosses of a fair die and thereby generate 300 values of the random variable x.

c. Use the command **HISTOGRAM** to construct a histogram of the 300 values of x that were obtained from Part b, and compare this with the theoretical probability distribution.

d. Use the commands **MEAN** and **STDEV** to calculate the mean and standard deviation of the 300 values of x obtained from Part b. Compare these with the theoretical values of $\mu = 3.5$ and $\sigma = 1.708$ obtained in Part a.

Looking Back

A **random variable** is a rule that assigns exactly one value to each point in a sample space for an experiment. Random variables allow one to conveniently and succinctly describe many events in a quantitative manner. Two types of random variables were discussed. A **discrete random variable** is usually a count variable that indicates the number of times that something occurs. It might be

• the number of classes that you will miss next semester,

• the number of days next year with temperatures above 90°,

• the number of auto accidents that will occur nationally next year.

A discrete variable can never have more values than the counting numbers. A **continuous random variable** is one that can assume the infinite number of values in some interval on the real number line. These are often measurement variables such as weight, distance, time, and speed. This and the next chapter are concerned with discrete variables; continuous variables will be discussed in Chapter 7.

Closely associated with a random variable is its **probability distribution.** These serve as mathematical models for describing various physical phenomena. The probability distribution of a discrete random variable is any device (table, graph, formula) that gives the possible values of the variable and their corresponding probabilities.

The basic properties of a probability distribution are:

1. $P(x) \geq 0$ for each value of x,
2. $\Sigma P(x) = 1$

The **mean of a random variable** was defined. It gives the long-run average value of the variable.

The **mean of the discrete random variable** x with probability distribution $P(x)$ is a weighted average of the values of x, where the weights are the associated probabilities.

$$\mu = \Sigma x P(x)$$

The **variance of a random variable** is a weighted average of the squared deviations of each value of x from the mean. As with μ, the weights are the associated probabilities. The variance is usually computed using the following shortcut formula.

Shortcut Formula for the Variance of a Random Variable:
For a random variable x with mean μ and probability distribution $P(x)$, the **variance of the random variable** is

$$\sigma^2 = \Sigma x^2 P(x) - \mu^2.$$

In applications, one often uses the square root of the variance. This is called the **standard deviation of the random variable** and is denoted by σ. Chebyshev's theorem and the Empirical rule tell us the following about variability within 2 standard deviations of a random variable's mean:

Chebyshev's Theorem:
At least 75 percent of the time the value of x will fall within 2 standard deviations of its mean.

Empirical Rule:
If the probability distribution of x is nearly mound shaped, then about 95 percent of the time the value of x will fall within 2 standard deviations of its mean.

Key Words

In reviewing this chapter, you should be able to define, explain, and illustrate each of the following.

random variable *(page 187)*

discrete random variable *(page 188)*

continuous random variable *(page 188)*

probability distribution *(page 191)*

probability function *(page 192)*

probability spike graph *(page 194)*

probability histogram *(page 194)*

basic properties of a probability distribution *(page 195)*

mean of a random variable *(page 200)*

mathematical expectation of x *(page 200)*

expected value of x *(page 200)*

variance of a random variable *(page 203)*

standard deviation of a random variable *(page 203)*

shortcut formula for σ^2 *(page 204)*

Chebyshev's theorem *(page 207)*

Empirical rule *(page 207)*

Ⓜ MINITAB Commands

READ _ _ *(page 196)*

SET _ *(page 202)*

END *(page 196)*

LET _ *(page 202)*

SUM _ *(page 203)*

PRINT _ *(page 206)*

RANDOM _ _; *(page 196)*
DISCRETE _ _ .

RANDOM _ _; *(page 211)*
INTEGER _ _ .

HISTOGRAM _ *(page 196)*

MEAN _ *(page 211)*

STDEV _ *(page 211)*

Review Exercises

5.83 Some Japanese consumers are willing to pay premium prices for Maine lobsters. One factor in the high cost is the mortality rate during the long-distance shipping from Maine to Japan. Suppose the number of deaths x per 10 lobsters is given by the following probability distribution.

x	0	1	2	3	4	5
$P(x)$	0.48	0.28	0.14	0.06	0.03	0.01

Over the long run, what is the mean number of deaths per 10 lobsters?

5.84 Each monitor in a lot of 100 computer monitors is tested to determine if it is operating satisfactorily.

a. Identify a random variable of likely interest.
b. Is the variable discrete or continuous?
c. Give the possible values for the random variable.

5.85   A manufacturer of canteen equipment measures the amount of beverage in an 8 ounce cup dispensed by a soft drink machine.
a. Identify a random variable of likely interest.
b. Is the variable discrete or continuous?
c. Give the possible values for the random variable.

5.86   Is the variable, the velocity of a falling comet, discrete or continuous?

5.87   Is the variable, the number of outpatient visits during 1995 to a Veterans Administration clinic, discrete or continuous?

5.88   Is the variable, the number of tosses of a silver dollar until the first head shows, discrete or continuous?

5.89   Is the variable, the weight of pepperoni on a pizza, discrete or continuous?

5.90   Is the variable, the number of pieces of pepperoni on a pizza, discrete or continuous?

5.91   Is the variable, the down time of a mainframe computer, discrete or continuous?

5.92   Is the variable, the amount of fructose in a can of soda, discrete or continuous?

5.93   Is the variable, the systolic blood pressure of an athlete at the completion of a 100 yard dash, discrete or continuous?

5.94   Is the variable, the volume of air inhaled in one breath of a horse, discrete or continuous?

5.95   Is the variable, the daily cost of stay at a hospital, discrete or continuous?

5.96   The probability distribution of a random variable x is given below. Find the value of the missing probability $P(18)$.

x	15	16	17	18	19
$P(x)$	0.11	0.23	0.09		0.13

5.97   Explain why the following is or is not a valid probability distribution.

x	-3	-2	-1	0	1	2	3
$P(x)$	0.36	0.30	0.18	0.10	0.03	0.02	0.01

5.98   Determine if the following is a valid probability distribution. If it is not, explain why.

x	$1	$4	$8	$9
$P(x)$	0.2	0.3	0.4	0.2

5.99   Is the following a valid probability function? Explain.

$$P(x) = \frac{x-1}{15}, \qquad \text{for } x = 2, 3, 4, 5, 6$$

5.100 The probability distribution of a random variable is given by the following table.

x	1	2	3	4	5
$P(x)$	$\frac{1}{55}$	$\frac{4}{55}$	$\frac{9}{55}$	$\frac{16}{55}$	$\frac{25}{55}$

Use a formula to specify the probability distribution.

5.101 Construct a probability spike graph to display the probability distribution in Exercise 5.100.

5.102 Use a probability histogram to give the probability distribution in Exercise 5.100.

5.103 The probability distribution of the random variable x is

$$P(x) = C(4, x)(.2)^x(.8)^{4-x}, \qquad \text{for } x = 0, 1, 2, 3, 4$$

Find the following probabilities.
a. $P(x < 2)$
b. $P(x \geq 3)$

5.104 Find the mean and standard deviation of the random variable whose probability distribution is given in Exercise 5.103.

5.105 Three-quarters of all American houses now have microwave ovens (*Hippocrates,* May/June, 1989). Two houses are randomly selected, and the random variable x is defined to be the number that have a microwave oven. Find the probability distribution of x.

5.106 The last two questions on a quiz are multiple-choice, each with four alternatives (A, B, C, D). The correct answer to each question is alternative B. A student has no knowledge of the material and answers each question by guessing.
a. Construct a sample space that shows the different ways that the two questions can be answered.
b. Assign appropriate probabilities to the sample points.
c. Let the random variable x denote the number of questions answered correctly. Assign the appropriate value of x to each sample point.
d. Construct the probability distribution of the random variable x.
e. Determine the mean and standard deviation of x.

5.107 A college bookstore has found that its supply of the Sunday *New York Times* will sell out about 80 percent of the time. Suppose sales on three randomly selected Sundays are examined. The random variable x is defined to be the number of the three Sundays on which the paper sold out. Determine the probability distribution of x.

5.108 The probability distribution of a random variable x is given below. Determine its mean, variance, and standard deviation.

x	1	3	5	7	9
$P(x)$	0.2	0.1	0.1	0.3	0.3

5.109 For the following probability distribution, find μ, σ^2, and σ.

x	0	1	2	3
$P(x)$	0.27	0.21	0.34	0.18

5.110 Find the mean and standard deviation of the random variable whose probability distribution is given by the following probability histogram.

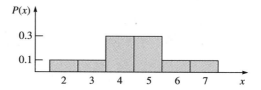

5.111 For the random variable in Exercise 5.110, determine

$$P(\mu - 2\sigma \le x \le \mu + 2\sigma).$$

5.112 Calculate μ and σ for the random variable with the following probability distribution.

$$P(x) = \frac{x + 2}{9}, \qquad \text{for } x = 0, 1, 2$$

5.113 The number of pecans x used to decorate a certain brand of fruit cake varies according to the following probability distribution.

x	7	8	9	10
$P(x)$	0.03	0.42	0.46	0.09

Over the long run, what is the average number of pecans per cake?

5.114 Find the standard deviation of the random variable x whose probability distribution is given in Exercise 5.113.

5.115 The promoter of an outdoor concert wants to insure the event against cancellation. If the concert is cancelled due to inclement weather, the company will cover the entire expenses of $30,000, while it will pay only half this amount if it is cancelled for any other reason. The insurance company assigns a probability of 0.15 to the former and 0.05 to the latter. If the insurance company will charge $1,000 more than its expected payment, what should be the cost of this policy?

MINITAB Assignments

Ⓜ **5.116** Use MINITAB to simulate 360 tosses of a loaded die for which each of the sides with an odd number of spots is twice as probable as each of the sides with an even number. Display the results in a histogram.

Ⓜ **5.117** The random variable x has the following probability distribution. Use MINITAB to find the mean of x.

x	125	135	145	155	165	175	185	195	205	215	225	235
$P(x)$	.10	.04	.14	.07	.17	.20	.01	.08	.01	.04	.06	.08

Ⓜ **5.118** Use MINITAB to find the variance and standard deviation of the random variable x in Exercise 5.117.

Ⓜ **5.119** The probability function of the random variable x is given by the following formula.

$$P(x) = \frac{x}{5050}, \qquad \text{for } x = 1, 2, 3, \ldots 100$$

Use MINITAB to determine μ.

Ⓜ **5.120** For the probability distribution in Exercise 5.119, use MINITAB to find the standard deviation.

Ⓜ **5.121** If a penny, a nickel, a dime, and a quarter are tossed together once, the probability distribution of x, the number of heads, can be shown to be the following.

x	0	1	2	3	4
$P(x)$	$\frac{1}{16}$	$\frac{4}{16}$	$\frac{6}{16}$	$\frac{4}{16}$	$\frac{1}{16}$

a. Show that the mean and standard deviation of x are $\mu = 2$ and $\sigma = 1$.

b. Use MINITAB to simulate 400 repetitions of the tossing of the coins. This will produce 400 values of the random variable x.

c. Using MINITAB, construct a histogram of the 400 values of x that were obtained in Part b, and compare this histogram with the theoretical probability distribution.

d. Use MINITAB to calculate the mean and standard deviation of the 400 values of x obtained in Part b. Compare these with the theoretical values of $\mu = 2$ and $\sigma = 1$ obtained in Part a.

(© Joseph Nettis/Photo Researchers, Inc.)

6 Discrete Probability Distributions

◀ *Lottery Lunacy*

On April 26, 1989, and the 6 days preceding it, Pennsylvania found itself in the grip of gambling fever as 87 million wagers were placed on the state's Super 7 lottery drawing. At stake was a record $115.6 million dollar top prize to be shared by those who could pick 7 of the 11 winning numbers to be drawn at 7:00 P.M. that evening. The next day lottery officials announced that the purchasers of 14 winning tickets would each receive $8,255,641 to be paid in 26 annual installments of $317,525. The 11 winning numbers drawn on April 26 were randomly selected from the integers 1 through 80, and each wager consisted of choosing 7 of these numbers. In addition to the 14 tickets with 7 winning selections, there were 816 tickets with 6 correct picks, 28,776 with 5, and 465,746 with 4 matches. Tickets with 6, 5, and 4 winning numbers each were worth $4,842.50, $274.50, and $7.00, respectively.

Looking Ahead

In the preceding chapter, the concept of a **probability distribution** was introduced. We saw that the probability distribution of a discrete random variable gives the possible values of the variable and their corresponding probabilities. Probability distributions are extremely important in statistics because they can be used as mathematical models for investigating various problems and situations encountered in life. For instance, the Pennsylvania Super 7 lottery drawing can be modeled by the **hypergeometric probability distribution** that will be introduced in Section 6.4. This distribution will be given by a formula that one can use to calculate the probabilities of correctly picking 4, 5, 6, or 7 of the 11 winning numbers. We will show that the chances of winning the first prize jackpot are only 1 in 9,626,413.

There are many different probability distributions. The particular one used in a given situation is dictated by the assumptions that can be made about the phenomenon being studied. In this chapter, we will emphasize one of the most frequently used distributions—the **binomial probability function.** You will see that this can be used to model a multitude of problems, and you will learn how to use the binomial formula to calculate probabilities of interest.

6.1   Binomial Experiments

Many situations involve the repetition of a number of trials where each trial can result in some event of interest occurring **(success)** or not occurring **(failure).** For example, the tossing of a coin 15 times can be thought of as the repetition of 15 **trials** where each trial (single toss) results in either a success (head) or a failure (tail). When the probability of a success, p, remains the same from trial to trial, and the trials are independent, then we have a **binomial experiment.**

A **binomial experiment** is an experiment that satisfies the following properties:

1. The experiment consists of the repetition of n identical trials.
2. Each trial can result in one of two possibilities: success or failure.
3. The probability of a success on a trial is denoted by p, and p remains unchanged from trial to trial.
4. The trials are independent.
5. The random variable x of interest is the total number of successes observed on the n trials.

Note that in a binomial experiment, the random variable x is a count of the successes.

Example 6.1   Many states conduct a daily lottery in which a 3-digit number is randomly selected from the 1,000 possibilities 000 through 999. If an individual purchases a lottery ticket on each of 10 days during the month of September, is this a binomial experiment?

Solution

We must check whether the 5 conditions of a binomial experiment are satisfied.

1. The experiment consists of the repetition of $n = 10$ identical trials. One trial is the purchase of a single ticket.
2. Each trial results in either a success (ticket wins) or a failure (ticket loses).
3. The probability of a success on each trial is $p = 0.001$. Since there are 1,000 possible 3-digit numbers, each ticket has 1 chance in 1,000 of being the winning number.
4. The trials are independent. The chances of a particular ticket winning are unaffected by what happens to any other ticket.
5. If we define the random variable x to be the number of winning tickets among the 10 purchased, then all the required conditions are satisfied, and this is a binomial experiment.

Example 6.2   To help pass time during the airplane flight to the next opponent's city, a professional baseball player and his roommate engage in a game. The game involves 200 tosses of a balanced die and counting the number of aces (1s) obtained. Determine if the 200 tosses constitute a binomial experiment.

Solution

We will check if the 5 necessary conditions hold.

1. The experiment consists of the repetition of $n = 200$ trials, where 1 trial is a single toss of the die.
2. On first inspection it may appear that each trial does not result in either a success or a failure, since a toss can show a 1, 2, 3, 4, 5, or 6. However, because we are only interested in the number of 1s tossed, we can call a success an ace and anything else a failure.
3. The probability of a success on each trial remains the same value. This is $p = \frac{1}{6}$, since the die is balanced.
4. The trials are independent. What happens on a particular toss does not depend on what occurs on any other toss.
5. If we let the random variable x be the number of aces obtained on the 200 tosses, then this is a binomial experiment, since the 5 necessary conditions are satisfied.

Example 6.3   During the return flight home, the baseball player in Example 6.2 and his roommate again play the game in which a die is tossed 200 times and the number of aces is counted. Now, however, the first 100 throws are with the balanced die used earlier, and the remaining 100 tosses are with the roommate's die. This die is unbalanced in such a way that the probability of obtaining an ace is 0.2. Can the 200 tosses be regarded as a binomial experiment?

Solution

For each of the first 100 tosses, the probability of obtaining a success (an ace) is $p = \frac{1}{6}$, but the probability of a success changes to $p = 0.2$ for the remaining 100 throws. Since p does not remain the same for each of the 200 trials, condition 3 fails to hold, and thus, this is not a binomial experiment. (Note, however, that the trials are independent.)

Example 6.4   A liquidator of surplus and salvage goods has a shelf containing 20 automobile fuel pumps that were obtained from a fire sale. These are being sold at 25 percent of their retail value, but on an ''as is'' basis. Unknown to purchasers, 6 pumps are defective. A mechanic purchases at random 8 fuel pumps and is concerned about how many will be in working order. Determine if this is a binomial experiment.

Figure 6.1
Tree diagram of the first two selections.

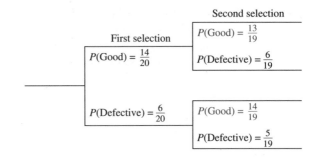

Solution

The experiment can be thought of as the repetition of $n = 8$ identical trials, where a trial is the selection of 1 fuel pump. Because the mechanic is interested in how many of his selections are good, we can consider each trial as resulting in either a success (pump is good) or a failure (pump is defective).

The trials are *not* independent, however, since the conditional probability of obtaining a success for the second selection is dependent on the results of the first selection. This probability is either $\frac{13}{19}$ or $\frac{14}{19}$, depending on whether or not a good pump was chosen on the first selection (see Figure 6.1). Because the trials are not independent, condition 4 fails to hold, and the experiment is not binomial.

Section 6.1 Exercises
Binomial Experiments

For each of the following exercises, determine if the experiment is binomial. If it is binomial, then define a success and give the values of n and p.

6.1   A coin is damaged so that the probability of tossing a head is only 0.4. The coin is tossed 7 times.

6.2   Fifty-eight percent of married women in the United States wear a wedding ring (*Health,* February/March, 1992). One hundred married women are randomly surveyed to determine the number who wear such a ring.

6.3   In 1989, 15 percent of U.S. households had a personal computer (*Newsweek,* April 8, 1991). During that year, 500 households were randomly selected, and a count was made of the number with a personal computer.

6.4   A card is randomly selected from a standard 52-card deck, checked to see if it is red, and then returned to the deck. This is repeated 5 times.

6.5   Five cards are randomly selected without replacement from a 52-card deck. Each card is checked to determine if it is red.

6.6   A computer supply shop has five batteries in stock for a particular notebook computer. Unknown to anyone, three of the batteries are defective. Suppose a customer purchases two batteries, and interest is focused on the number of defectives selected.

6.7   A quiz consists of 10 true-false questions. A student has no knowledge of the material and answers each question strictly by guessing. The student is interested in the number correctly answered.

6.8   According to the American Cancer Society, 82 percent of Americans know that a poor diet can increase their risk of cancer. Fifty people are randomly selected, and the interviewer is interested in the number of people in the sample who are aware of this fact.

6.9   A gift shop has found the probability to be 0.70 that a customer will use a charge card to make a purchase. One is interested in how many of the next 3 customers will use a charge card.

6.10 Ten jumbo shrimp are selected from a fish market's display of 87 jumbo shrimp to determine the number that fail to meet a state's health standards for seafood.

6.11 Three-quarters of all American households now have microwave ovens (*Hippocrates,* May/June, 1989). Twenty households are randomly selected, and the number having a microwave oven is recorded.

6.2   The Binomial Probability Distribution

In this section, we will present and illustrate a formula that can be used to calculate probabilities for a binomial experiment. To see how such a formula can be obtained, consider the binomial experiment in which a fair die is tossed $n = 3$ times. Suppose we want $P(2)$, the probability of obtaining $x = 2$ aces. Two aces in three tosses can occur in the following ways:

$$A_1 A_2 \overline{A}_3 \qquad A_1 \overline{A}_2 A_3 \qquad \overline{A}_1 A_2 A_3,$$

where A_k and $\overline{A}_k$ respectively denote an ace and a nonace on trial k. The desired probability can be obtained by using the Addition and Multiplication rules of probability.

$$
\begin{aligned}
P(2) &= P(A_1 A_2 \overline{A}_3 \text{ or } A_1 \overline{A}_2 A_3 \text{ or } \overline{A}_1 A_2 A_3) \\
&= P(A_1 A_2 \overline{A}_3) + P(A_1 \overline{A}_2 A_3) + P(\overline{A}_1 A_2 A_3) \\
&= \left(\frac{1}{6}\right)\left(\frac{1}{6}\right)\left(\frac{5}{6}\right) + \left(\frac{1}{6}\right)\left(\frac{5}{6}\right)\left(\frac{1}{6}\right) + \left(\frac{5}{6}\right)\left(\frac{1}{6}\right)\left(\frac{1}{6}\right) \\
&= 3 \left(\frac{1}{6}\right)^2 \left(\frac{5}{6}\right)
\end{aligned}
\tag{6.1}
$$

Note the form of $P(2)$ in Equation 6.1. The factor 3 is the number of ways that 3 trials can result in 2 successes, namely, the combination $C(3, 2)$. For the general case of n trials and x successes, this would be $C(n, x)$.

Equation 6.1 also involves the factor $(\frac{1}{6})^2(\frac{5}{6})$, which is the probability of each possible sequence of 2 successes and $3 - 2 = 1$ failure. For the general case of n

Two Bernoulli brothers discussing a geometrical problem. Much of the initial work in the development of the bionomial distribution was due to Jacob Bernoulli (1654–1705). Jacob was one of several Bernoullis who were instrumental in the development of probability theory and calculus. *(Courtesy of The Bettmann Archive.)*

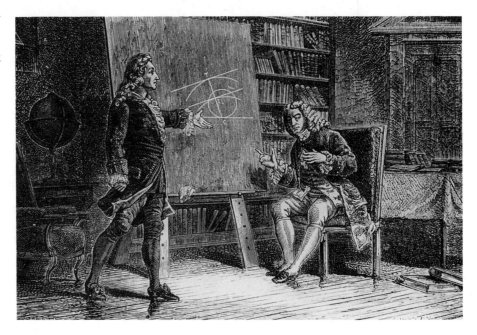

trials, x successes, $n - x$ failures, and success probability p, $(\frac{1}{6})^2(\frac{5}{6})$ would be replaced by $p^x(1 - p)^{n-x}$, resulting in $P(x)$ being given by Formula 6.2 below.

The **binomial probability distribution** is given by the formula

$$P(x) = C(n, x)p^x q^{n-x}, \qquad \text{for } x = 0, 1, 2, 3, \ldots n, \qquad \textbf{(6.2)}$$

where

n is the number of trials,

x is the number of successes on the n trials,

p is the probability of a success on a trial,

q is the probability of a failure on a trial, and $q = 1 - p$.

Example 6.5

Families save an average of $6 per week on their supermarket bills by redeeming manufacturers' coupons (*The Wall Street Journal*, February 20, 1991).

A dog food manufacturer is promoting a new brand with a rebate offer on its 25 pound bag. Each package is supposed to contain a mail-in coupon for $4.00. However, the company has found that the machine dispensing these fails to place a coupon in 10 percent of the bags. Enticed by the large rebate offer, an owner of 2 Airedales purchases 5 bags. Find the probability that

1. 1 of the bags will not contain a coupon,
2. at least 1 bag will fail to have a coupon.

Solution

We can think of this problem as a binomial experiment consisting of $n = 5$ trials, where a trial is the purchase of 1 bag. Because we are interested in finding probabilities concerning bags without a coupon, we will consider the absence of a coupon to be a success. Then $p = 0.10$, since 10 percent of the bags do not contain a rebate slip. The random variable x is the number of the 5 bags that do not contain a coupon.

1. To determine the probability that 1 bag will not contain a coupon, we need to calculate $P(1)$.

$$P(x) = C(n, x)p^x q^{n-x}$$
$$P(1) = C(5, 1)0.10^1 0.90^4$$
$$= \frac{5!}{1!4!}(0.10)(0.6561)$$
$$= 5(0.10)(0.6561)$$
$$= 0.3280$$

2. The probability that at least 1 bag will not contain a rebate slip can be found most easily by using the Complement rule for probabilities.

$$P(x \geq 1) = 1 - P(x < 1)$$
$$= 1 - P(0)$$
$$= 1 - C(5, 0)0.10^0 0.90^5$$
$$= 1 - \frac{5!}{0!5!}(1)(0.59049)$$
$$= 1 - (1)(1)(0.59049)$$
$$= 0.4095$$

Example 6.6   Safety tipped wooden matches pose less threat of an accidental fire but are more difficult to ignite than matches without this feature. Suppose a certain manufacturer has found that 25 percent of its safety tipped matches fail to ignite on the first strike. Find the probability that in a box of 20 matches,

1. 18 will light on the first strike,
2. more than 18 will ignite on the first strike.

Solution

This is a binomial experiment with $n = 20$ trials, where a trial is the striking of 1 match. Since we want probabilities concerning the number that light on the first strike, we will call this a success. Because 25 percent fail to ignite on the first strike, $p = 0.75$.

1. The probability that 18 will light on the first strike is given by $P(18)$.

$$P(18) = C(20, 18)0.75^{18}0.25^2$$
$$= \frac{20!}{18!2!}(0.00563771)(0.0625)$$
$$= (190)(0.00563771)(0.0625)$$
$$= 0.0669$$

2. The probability that more than 18 will ignite on the first strike is

$$P(x \geq 19) = P(19) + P(20)$$
$$= C(20, 19)0.75^{19}0.25^1 + C(20, 20)0.75^{20}0.25^0$$
$$= 0.0211 + 0.0032$$
$$= 0.0243.$$

The calculations in Example 6.6 were laborious. The probabilities needed for the solution could easily have been found using MINITAB. To obtain these, we will use the **PDF** command and the **BINOMIAL** subcommand. **PDF** denotes **probability density function** and is a term sometimes used in place of probability distribution. To obtain the desired probabilities, type

```
PDF;
BINOMIAL 20 0.75.
```

Recall that a command (**PDF** in this instance) must end with a semicolon whenever it is followed by a subcommand. The subcommand **BINOMIAL** must be followed by the values of n and p and end with a period. In response to these commands, MINITAB will produce the output given in Exhibit 6.1. We see that MINITAB has computed the probability for each value of x from 6 through 20. Normally the probabilities from 0 through 5 would also have been given, but MINITAB deletes them here because they are so small (0 when rounded to 4 decimal places).

Ⓜ **Example 6.7**   If a fair coin is tossed 100 times, the most probable outcome is an equal number of heads and tails. Use MINITAB to determine

1. the probability of obtaining 50 heads (and thus 50 tails);
2. the probability of obtaining "close to" 50 heads, say, between 45 and 55 inclusive.

Solution

This is a binomial experiment for which a success is tossing a head, $n = 100$ trials, and $p = 0.5$, since the coin is fair.

1. To find $P(50)$, we will use the **PDF** command with the **BINOMIAL** subcommand. In the previous example we saw that the **PDF** command resulted in the printing of the probabilities for all values of x, even though we only needed a few of these. If you don't want the unnecessary ones printed, then place the values of x being used in a column, say, C1, and follow the **PDF** command with C1. If you want the

Exhibit 6.1

```
MTB > PDF;
SUBC> BINOMIAL 20 0.75.
        BINOMIAL WITH N =  20  P = 0.750000
            K            P( X = K)
            6              0.0000
            7              0.0002
            8              0.0008
            9              0.0030
           10              0.0099
           11              0.0271
           12              0.0609
           13              0.1124
           14              0.1686
           15              0.2023
           16              0.1897
           17              0.1339
           18              0.0669
           19              0.0211
           20              0.0032
```

probability for just a single value of x, then just follow **PDF** with that number. Since we are interested in only $P(50)$, type

```
PDF 50;
BINOMIAL 100 0.5.
```

MINITAB will produce the output given in Exhibit 6.2. From this we see that the probability of exactly 50 heads in 100 tosses is 0.0796. Thus, even though 50 heads is the most likely outcome for 100 flips of a fair coin, it will only occur about 8 percent of the time.

Exhibit 6.2

```
MTB > PDF 50;
SUBC> BINOMIAL 100 0.5.
         K            P( X = K)
      50.00            0.0796
```

2. To find the probability of obtaining from 45 to 55 heads, we could use the **PDF** command to find each individual probability for the values 45 through 55, and then sum these. However, when we want probabilities for several consecutive values of x, it is easier to use the **CDF** command. **CDF** denotes **cumulative distribution function,** since it provides cumulative probabilities. For this problem, type the following.

```
CDF;
BINOMIAL 100 0.5.
```

This will produce the output given in Exhibit 6.3. To interpret the **CDF** output, consider the line

$$55 \qquad 0.8644.$$

The 0.8644 is the cumulative probability for $x \leq 55$. That is, 0.8644 is the sum of all the probabilities for values of x from 0 up to and including 55.

$$0.8644 = P(0) + P(1) + P(2) + \cdots + P(55)$$

Or stated another way,

$$0.8644 = P(x \leq 55).$$

Similarly, from Exhibit 6.3 we see that

$$P(x \leq 44) = 0.1356.$$

Since $\qquad P(x \leq 55) = P(0) + P(1) + P(2) + \cdots + P(55),$

and $\qquad P(x \leq 44) = P(0) + P(1) + P(2) + \cdots + P(44),$

we can obtain the probabilities $P(45) + P(46) + \cdots + P(55)$ by subtracting $P(x \leq 44)$ from $P(x \leq 55)$. Thus, from Exhibit 6.3, the probability of tossing from 45 to 55 heads is

$$P(45 \leq x \leq 55) = P(x \leq 55) - P(x \leq 44)$$
$$= 0.8644 - 0.1356$$
$$= 0.7288.$$

Table 1 in Appendix A contains cumulative probabilities for selected values of n and p. Use of the table is illustrated in the next example.

Example 6.8

For a binomial experiment with $n = 9$ and $p = 0.40$, determine the probability of obtaining:

1. fewer than 6 successes,
2. more than 2 successes,
3. 3 successes.

Solution

We first need to find the portion of Table 1 that is labeled $n = 9$ (See Appendix A). Next, locate the column that is headed by $p = .40$. The numbers in this column are cumulative probabilities for a binomial experiment with $n = 9$ and $p = .40$.

1. Since the table only gives cumulative probabilities, we must rewrite the desired probability P(fewer than 6 successes) as the cumulative probability $P(x \leq 5)$. This

Exhibit 6.3

```
MTB > CDF;
SUBC> BINOMIAL 100 0.5.
       K   P( X LESS OR = K)
      30          0.0000
      31          0.0001
      32          0.0002
      ..          ......
      ..          ......
      43          0.0967
      44          0.1356    ←── This is
      45          0.1841        P(0) + P(1) + . . . + P(43) + P(44)
      46          0.2421
      47          0.3086
      48          0.3822
      49          0.4602
      50          0.5398
      51          0.6178
      52          0.6914
      53          0.7579
      54          0.8159
      55          0.8644    ←── This is
      56          0.9033        P(0) + P(1) + . . . + P(54) + P(55)
      ..          ......
      ..          ......
      ..          ......
      67          0.9998
      68          0.9999
      69          1.0000
```

value is located in the column headed by $p = .40$ and the row that is labeled 5 in the left margin of the table. The table value is 0.901, and thus,

$$P(\text{fewer than 6 successes}) = P(x \le 5) = 0.901.$$

2. The probability $P(\text{more than 2 successes})$ can be expressed in terms of a cumulative probability by using the Complement rule.

$$P(\text{more than 2 successes}) = 1 - P(x \le 2)$$

$P(x \le 2)$ appears in the cell located in column $p = .40$ and the row labeled 2. This value is 0.232. Therefore,

$$P(\text{more than 2 successes}) = 1 - P(x \le 2)$$
$$= 1 - 0.232$$
$$= 0.768.$$

3. The probability of 3 successes, $P(3)$, can be expressed as the difference between 2 cumulative probabilities by noting that

$$P(3) = [P(0) + P(1) + P(2) + P(3)] - [P(0) + P(1) + P(2)]$$
$$= P(x \leq 3) - P(x \leq 2)$$
$$= 0.483 - 0.232$$
$$= 0.251.$$

Guidelines for Using Table 1

First: Use the value of n to locate the section of the table to be used.

Second: Use the column that is headed by the success probability p.

Third: The entry in a particular row, say k, gives the cumulative probability $P(x \leq k)$.

Fourth: Express the desired probability in terms of cumulative probabilities $P(x \leq k)$.

Example 6.9   For the binomial experiment in Example 6.8 ($n = 9$ and $p = 0.40$), use Table 1 to determine the following.

1. $P(x = 7)$
2. $P(x < 7)$
3. $P(x > 7)$
4. $P(x \geq 7)$
5. $P(2 \leq x \leq 7)$
6. $P(2 < x \leq 7)$
7. $P(2 \leq x < 7)$
8. $P(2 < x < 7)$

Solution

We again use the section of Appendix A Table 1 that is labeled $n = 9$. The column to be used is headed by $p = .40$.

1. $P(x = 7) = P(x \leq 7) - P(x \leq 6) = 0.996 - 0.975 = 0.021$
2. $P(x < 7) = P(x \leq 6) = 0.975$
3. $P(x > 7) = 1 - P(x \leq 7) = 1 - 0.996 = 0.004$
4. $P(x \geq 7) = 1 - P(x \leq 6) = 1 - 0.975 = 0.025$
5. $P(2 \leq x \leq 7) = P(x \leq 7) - P(x \leq 1) = 0.996 - 0.071 = 0.925$
6. $P(2 < x \leq 7) = P(x \leq 7) - P(x \leq 2) = 0.996 - 0.232 = 0.764$
7. $P(2 \leq x < 7) = P(x \leq 6) - P(x \leq 1) = 0.975 - 0.071 = 0.904$
8. $P(2 < x < 7) = P(x \leq 6) - P(x \leq 2) = 0.975 - 0.232 = 0.743$

> **TIP:**   To determine binomial probabilities, first try to use Table 1. This will be possible when the given values of n and p appear in the table. When Table 1 cannot be used, then use the binomial Formula 6.2 for simple calculations and MINITAB for more laborious calculations.

Section 6.2 Exercises
The Binomial Probability Distribution

For Exercises 6.12–6.20, evaluate the given expression.

6.12 $C(4, 2)0.2^2 0.8^2$

6.13 $C(6, 4)0.1^4 0.9^2$

6.14 $C(5, 0)(3/4)^0(1/4)^5$

6.15 $P(2)$ for a binomial experiment with $n = 5$ and $p = 2/3$.

6.16 $P(4)$ for a binomial experiment with $n = 10$ and $p = 0.1$.

6.17 $P(x \le 2)$, where x is a binomial random variable with $n = 5$ and $p = 1/4$.

6.18 $P(x \ge 1)$, where x is a binomial random variable with $n = 8$ and $p = 0.3$.

6.19 $P(4) + P(5)$ for a binomial experiment with $n = 6$ and $p = 0.2$.

6.20 $P(2 \le x \le 4)$, where x is a binomial random variable with $n = 7$ and $p = 0.3$.

For Exercises 6.21–6.29, use Table 1 to evaluate the following probabilities for a binomial random variable with $n = 10$ and $p = 0.6$.

6.21 $P(x \le 8)$

6.22 $P(x < 8)$

6.23 $P(x = 8)$

6.24 $P(x > 8)$

6.25 $P(x \ge 8)$

6.26 $P(4 \le x \le 8)$

6.27 $P(4 < x < 8)$

6.28 $P(4 \le x < 8)$

6.29 $P(4 < x \le 8)$

6.30   Use Table 1 to construct a probability histogram for a binomial random variable with $n = 5$ and each of the following success probabilities.
  a. $p = 0.1$ b. $p = 0.5$ c. $p = 0.9$
  (Observe how the shape of the graph is affected by p.)

6.31   *The Boston Globe* (December 31, 1989) reported that at the end of the 1980s, 65 percent of U.S. households had a video cassette recorder (VCR). If 5 U.S. households were randomly selected, what is the probability that 3 had a VCR?

6.32   In a smoking survey of 36 states, the Centers for Disease Control found that Maine had the highest percent of people who had smoked at some time in their lives (*The Ellsworth American,* December 21, 1989). This figure was 53 percent. If 4 Maine people were randomly selected, what is the probability that 2 had smoked at some time?

6.33 A lumberyard sells 2 × 4s in different types of wood, but it has found that 35 percent of its orders are for pine. Find the probability that 2 of its next 6 orders of 2 × 4s will be for pine.

6.34 Based on a recent survey of government workers, it is estimated that about 85 percent have had some college education. If 6 government workers are selected at random, what is the probability that at least 5 have had some college experience?

6.35 Fifty percent of all 1988 domestic cars were equipped with power windows *(Ward's Automotive Yearbook).* Suppose a driver encounters 5 1988 domestic cars on his next trip to work. Find the probability that at least 1 will have power windows. Solve this by using
a. the binomial probability distribution formula,
b. Table 1 in Appendix A.

6.36 Ninety percent of all 1988 domestic cars were equipped with air conditioning *(Ward's Automotive Yearbook).* Find the probability that at least 9 of 10 randomly selected 1988 domestic cars have air conditioning by using
a. the binomial probability distribution formula,
b. Table 1 in Appendix A.

6.37 *In Health* (January/February, 1990) reported on a study of the residents of Maryland's Montgomery County. It found that 5 percent are afflicted by full-fledged SAD (seasonal affective disorder, but usually referred to as the winter blues). Suppose 20 residents of this county are selected at random. Find the probability that
a. exactly 1 is afflicted by full-fledged SAD,
b. at least 1 is afflicted by full-fledged SAD.

6.38 Concerning the study cited in Exercise 6.37, it was also found that 13 percent are afflicted by a milder form of SAD. If 15 residents of Maryland's Montgomery County are selected at random, what is the probability that at least 2 are afflicted by this milder form?

6.39 Twenty percent of the tires manufactured by a company are snow tires. Suppose 15 tires made by this company are randomly selected. Find the probability that
a. 4 are snow tires,
b. at most 4 are snow tires,
c. at least 4 are snow tires,
d. at least 2 but not more than 4 are snow tires.

6.40 A chain of supermarkets is considering the purchase of a lot consisting of several thousand lobsters. It wants to buy the supply if at least 95 percent of the lobsters have both claws. To check on this, the chain will select a sample of 20 lobsters and buy the lot if no lobster is missing a claw.

a. Find the probability that the chain will commit the error of not buying the lot when in fact 95 percent of the lobsters have both claws;

b. Suppose only 90 percent of the lobsters have both claws. Find the probability that the chain will commit the error of buying the lot.

6.41 A NASA official estimated that there is a probability of 1/78 that any single space shuttle flight would result in a catastrophic failure (*Lancaster Sunday News,* April 9, 1989). Find the probability that for the next 10 flights, there will be

a. no shuttle disaster,

b. 1 shuttle disaster,

c. at least 1 shuttle disaster.

MINITAB Assignments

Ⓜ 6.42 Concerning Exercise 6.41, present plans call for more than 100 space shuttle missions for the decade of the 1990s. Assume that this number is exactly 100. Use MINITAB to find the probability that for these 100 flights, there will be

a. no shuttle disaster,

b. one shuttle disaster,

c. at least 1 shuttle disaster.

Ⓜ 6.43 Sixty-three percent of the voters in a large city favor a reduction in the school real estate tax. Use MINITAB to find the probability that more than 30 in a sample of 50 voters will favor a reduction.

Ⓜ 6.44 Ninety percent of all seeds that are sold by a certain company will germinate. Suppose 30 seeds are randomly selected from a large lot that are ready to be shipped. Use MINITAB to find the probability that fewer than 24 will germinate.

Ⓜ 6.45 Twelve percent of the bolts of fabric that are manufactured by a certain mill contain defects. Fifty bolts produced by this mill are randomly selected. Use MINITAB to find the probability that

a. seven bolts will contain defects,

b. fewer than seven will contain defects.

6.3 The Mean and the Standard Deviation of a Binomial Random Variable

General formulas were given in Chapter 5 for finding the mean and the standard deviation of any discrete random variable. Of course, these can be used for binomial random variables, but when the variable is binomial there is a much simpler and easier way to obtain μ and σ. We emphasize however that the formulas given below only apply when the random variable is binomial.

If the random variable x has a binomial probability distribution with parameters n and p, then

1. the **mean** of x is $\mu = np$,   (6.3)
2. the **variance** of x is $\sigma^2 = npq$,   (6.4)
3. the **standard deviation** of x is $\sigma = \sqrt{npq}$.   (6.5)

Example 6.10   A municipality offers a 2 percent discount on property taxes if they are paid 2 months in advance of the due date. Records indicate that 60 percent of taxpayers take advantage of the discount. Five tax bills from last year are selected at random. Find the mean and the standard deviation of the number in the sample of 5 that received the discount.

Solution

We have a binomial experiment with $n = 5$ trials, $p = 0.60$, and the random variable x is the number of the 5 tax bills that received the discount. The mean of x is

$$\mu = np = 5(0.60) = 3.$$

The standard deviation of x is

$$\sigma = \sqrt{npq} = \sqrt{5(0.60)(0.40)} = \sqrt{1.2} = 1.10.$$

The standard deviation also could have been obtained by using either of the general Formulas 5.5 (page 203) or 5.7 (page 204) given in the previous chapter. But whenever you are working with a binomial variable, $\sigma = \sqrt{npq}$ is the easiest way to find the standard deviation. To illustrate this point and the fact that we will obtain the same result, let's also find σ by using the general Formula 5.7. First, we need to use the binomial probability distribution to obtain $P(0)$, $P(1)$, $P(2)$, $P(3)$, $P(4)$, and $P(5)$.

$$P(0) = C(5, 0)0.60^0 0.40^5 = 0.01024$$
$$P(1) = C(5, 1)0.60^1 0.40^4 = 0.07680$$
$$P(2) = C(5, 2)0.60^2 0.40^3 = 0.23040$$
$$P(3) = C(5, 3)0.60^3 0.40^2 = 0.34560$$
$$P(4) = C(5, 4)0.60^4 0.40^1 = 0.25920$$
$$P(5) = C(5, 5)0.60^5 0.40^0 = 0.07776$$

x	$P(x)$	$xP(x)$	x^2	$x^2P(x)$
0	0.01024	0	0	0
1	0.07680	0.0768	1	0.0768
2	0.23040	0.4608	4	0.9216
3	0.34560	1.0368	9	3.1104
4	0.25920	1.0368	16	4.1472
5	0.07776	0.3888	25	1.9440
		$\mu = 3.0000$		10.2000

From Formula 5.7, the variance of x is

$$\sigma^2 = \Sigma x^2 P(x) - \mu^2$$
$$= 10.2 - 3^2 = 1.2.$$

Therefore, the standard deviation of x is

$$\sigma = \sqrt{1.2} = 1.10.$$

The moral of this example is to *always* use the special formulas $\mu = np$ and $\sigma = \sqrt{npq}$ to find the mean and standard deviation of a binomial random variable!

Example 6.11   The produce purchasing agent for a large chain of supermarkets is considering the purchase of several thousand pounds of oranges at a greatly reduced price. The reduction is because some of the oranges have major blemishes, but the agent has been told that only 20 percent are in unsatisfactory condition. If this is true, then the entire lot of oranges will be purchased. Before making the commitment, the agent samples 1,000 oranges and finds 240 with major blemishes. Should the agent make the purchase?

Solution

We can think of this situation as being a binomial experiment with $n = 1,000$ trials, a success is an orange with a major blemish, $p = 0.20$, and the random variable x is the number of unsatisfactory oranges in the sample of 1,000. By Formulas 6.3 and 6.5, the mean and standard deviation of x are

$$\mu = np = 1,000(0.20) = 200$$
$$\sigma = \sqrt{npq} = \sqrt{1,000(0.20)(0.80)} = 12.65.$$

We need to consider whether 240 oranges with major blemishes in a sample of 1,000 is unreasonably high if the claim of only 20 percent with serious flaws is true. Since $\mu = 200$ and $\sigma = 12.65$, an x-value of 240 has a z-value of

$$z = \frac{x - \text{mean}}{\text{std.dev.}} = \frac{x - \mu}{\sigma} = \frac{240 - 200}{12.65} = 3.16.$$

Thus, 240 is 3.16 standard deviations from the mean. Both the Empirical rule and Chebyshev's theorem tell us that it is highly improbable that a random variable will differ from its mean by more than 3 standard deviations. Thus, either an event with a

very small probability has occurred here, or the claim that only 20 percent have serious blemishes is false. Rather than believing that she has witnessed a rare event, the purchasing agent would likely reject the 20 percent claim and decline the purchase.

Section 6.3 Exercises
The Mean and Standard Deviation of a Binomial Random Variable

In Exercises 6.46–6.50, find the mean and the standard deviation of a binomial random variable with the specified values of n and p.

6.46 $n = 36, p = 0.5$

6.47 $n = 100, p = 0.9$

6.48 $n = 100, p = 0.1$

6.49 $n = 180, p = \frac{1}{6}$

6.50 $n = 10,000, p = \frac{4}{5}$

6.51 Find the mean and standard deviation of the number of heads observed in 400 tosses of a fair coin.

6.52 What are the mean and standard deviation of the number of sixes obtained in 720 tosses of a fair die?

6.53 Determine the mean and standard deviation of the number of hearts obtained in a sample of 48 cards if each card is selected singly, observed, and returned to the 52-card deck before the next card is selected (sampling with replacement).

6.54 Find the mean and standard deviation of the number of defective universal joints in a sample of 500 manufactured by a machine that produces 2 percent defectives.

6.55 Calculate the mean and standard deviation of the number of 60 true-false questions answered correctly by a student who has no knowledge of the material and guesses each answer.

6.56 What are the mean and standard deviation of the number of credit card users among the next 200 customers if 70 percent of all customers use a credit card?

6.57 For a sample of 80 homes, find the mean and standard deviation of the number that have a microwave oven, if three-quarters of all homes have a microwave.

6.58 For a sample of 500 adults, determine the mean and standard deviation of the number who are aware that a poor diet can increase their risk of cancer, if 82 percent of all adults are aware of this fact.

6.59 If a fair coin is tossed 4 times, the probability distribution of x, the number of heads, can be shown to be the following.

x	0	1	2	3	4
$P(x)$	$\frac{1}{16}$	$\frac{4}{16}$	$\frac{6}{16}$	$\frac{4}{16}$	$\frac{1}{16}$

Find the standard deviation of x by using
a. the general Formula 5.7 (page 204)
b. the special Formula 6.5 (page 234) for a binomial random variable.

6.60 A restaurant has found that the no-show rate for table reservations is 20 percent. Suppose the restaurant receives 6 reservations, and x denotes the number of parties that do not show.
a. Use Table 1 (Appendix A) to obtain the probabilities for x, and then use these with the general Formula 5.7 to obtain the standard deviation of x.
b. Find the standard deviation of x by using the special Formula 6.5 for a binomial random variable.

6.61 Identical twins occur when a single sperm fertilizes a single ovum, which ultimately splits during the developmental process. Identical twins constitute one-third of all twin births. If 128 twin births occur in a city during a given year, how many identical twins can be expected? Explain how it is possible for this value to be a noninteger.

6.62 *Newsweek* (March 27, 1989) reported that 60 percent of young children have blood lead levels that may impair their neurological development. If 300 young children are selected at random and tested, how many would you expect on the average to have a blood lead level that may impose this hazard?

6.63 A construction company claims that 25 percent of its subcontracting jobs are awarded to firms owned by minorities. However, a sample of 84 jobs revealed that only 9 went to minority firms.
a. What is the expected number of jobs that would be awarded to minority firms if the company's claim is correct?
b. Find the standard deviation of the number of jobs awarded to minority firms, if the company's claim is correct.

c. Do you think that the company's figure of 25 percent is correct? (Hint: Calculate a z-value and apply Chebyshev's theorem.)

6.64 An automobile dealer advertises that 90 percent of its customers would recommend them to a friend. A competitor contracts an independent firm to survey 89 of the dealer's customers. The survey found that 71 said that they would recommend the dealer to a friend. Do the survey results cast serious doubts on the veracity of the automobile dealer's claim? (Hint: Obtain a z-value and use Chebyshev's theorem.)

6.65 A student has a calculator with a random number generator. She generates 400 digits. Let x denote the number of odd digits obtained.
a. Find the mean of x.
b. Determine the standard deviation of x.
c. If this experiment were repeated many times, within what interval would you expect x to fall about 95 percent of the time? (Hint: Apply the Empirical rule.)

6.4 The Hypergeometric Probability Distribution (Optional)

In Example 6.4 of the first section, we considered a liquidator of surplus goods who has 20 fuel pumps obtained from a fire sale. These are being sold at 25% of their retail value, but on an as is basis. Unknown to anyone, 6 pumps are defective. A mechanic purchases at random 8 fuel pumps and is concerned about the number that will be in working order. We showed that this is not a binomial experiment because the trials (the selections of a pump) are dependent. It is, however, an example of a **hypergeometric experiment.**

Consider Figure 6.2. Suppose we have a population of N elements, where S of these have a particular characteristic of interest (success). In our example, the population consists of $N = 20$ pumps, and the characteristic of interest is that the pump is in good working order. Consequently, $S = 14$, since this is the number of good pumps. Suppose a sample of size n will be randomly selected from the population of size N. In the example, $n = 8$ pumps will be selected by the mechanic from the population of $N = 20$. If we define the random variable x to be the number of successes obtained in the sample of size n, then we have a **hypergeometric experiment.**

Figure 6.2
Hypergeometric experiment.

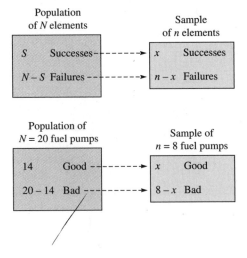

When **sampling without replacement** from a population of N elements that are considered successes and failures, the exact model to use is the **hypergeometric probability distribution.**

The **hypergeometric probability distribution** is given by Formula 6.6.

The hypergeometric probability distribution:

$$P(x) = \frac{C(S, x)C(N - S, n - x)}{C(N, n)}, \qquad \text{for } x = 0, 1, 2, \ldots n, \qquad \textbf{(6.6)}$$

where

N is the number of elements in the population,

S is the number of successes in the population,

n is the number of elements selected in the sample,

x is the number of successes obtained in the sample.

Example 6.12

To illustrate Formula 6.6, we will calculate the probability of the mechanic obtaining 5 good pumps in his purchase of 8 fuel pumps. The population of $N = 20$ pumps contains $S = 14$ successes (good pumps), and we want the probability of obtaining $x = 5$ successes in the sample of $n = 8$.

$$P(x) = \frac{C(S, x)C(N - S, n - x)}{C(N, n)}$$

$$P(5) = \frac{C(14, 5)C(6, 3)}{C(20, 8)} \qquad \textbf{(6.7)}$$

$$= \frac{2{,}002 \cdot 20}{125{,}970} = 0.318$$

To gain insight into how the hypergeometric probability distribution is obtained, consider Equation 6.7 above that gives the probability of obtaining 5 good pumps. The denominator, $C(20, 8)$, is the number of ways that a sample of 8 pumps can be selected from the population of 20. In the numerator, $C(14, 5)$ is the number of ways of selecting 5 good pumps from the 14, and $C(6, 3)$ is the number of ways of obtaining 3 defectives from the 6 in the population. Multiplying these to obtain $C(14, 5)C(6, 3)$ gives the number of possible ways that both 5 good and 3 defective pumps can be selected from the population. Finally, $C(14, 5)C(6, 3)$ is divided by $C(20, 8)$ because of the theoretical method of assigning probabilities. The definition says that the probability of obtaining 5 good and 3 defective pumps equals the number of ways that these can be selected, $C(14, 5)C(6, 3)$, divided by the number of possible samples of 8 pumps, $C(20, 8)$.

On April 26, 1989, Pennsylvania's Super 7 lottery paid a record $115.6 million dollar top prize to those who had selected 7 of the 11 winning numbers drawn that evening. The lottery can be modeled by the hypergeometric distribution, and the next example shows that the chances of obtaining a top prize ticket are only about 1 in 10 million.

Example 6.13

Thirty-three states and the District of Columbia now have lotteries. Ticket sales in 1990 were approximately $20 billion. About one-third of this amount was spent on the elderly, education, and other programs (*Wall Street Journal,* February 12, 1991).

In the Pennsylvania Super 7 lottery, for $1 a player selects 7 different numbers from 1 to 80. On the night of the drawing, lottery officials randomly choose 11 winning numbers. If a player's 7 numbers are among the 11 chosen numbers, then he or she wins the jackpot prize (if there is more than 1 winner, they equally share the prize). The order in which the numbers are selected is irrelevant. Suppose you decide to play 1 game. What is the probability that you will win the top prize?

Solution

We can consider this as a hypergeometric experiment by reasoning as follows (see Figure 6.3). Think of the population as consisting of the numbers from 1 to 80 ($N = 80$). Consider the successes to be those 11 winning numbers that will be drawn by lottery officials ($S = 11$). The failures are the remaining 69 losing numbers ($N - S = 69$). Your 7 picks are a sample ($n = 7$), and the random variable x is the number of winning numbers obtained in the sample. We want to find $P(7)$, the probability that your sample will contain $x = 7$ successes.

$$P(7) = \frac{C(11, 7)C(69, 0)}{C(80, 7)} = \frac{330 \cdot 1}{3,176,716,400}$$

$$= \frac{3}{28,879,240} \approx \frac{1}{9,626,413}$$

Thus, you have slightly better than 1 chance in 10 million of winning the jackpot prize!

Whenever one is sampling without replacement from a population whose elements are thought of as successes and failures, the exact model to use is the hypergeometric probability distribution. However, when the sample size is small relative to the population size (n is 10 percent of N or less), we usually prefer to use the binomial

Figure 6.3
Pennsylvania Super 7 lottery.

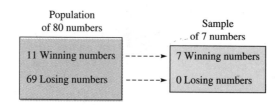

distribution as an approximate model because the calculations are easier. An illustration of this is given in the next example.

Example 6.14   An insurance company has purchased 200 laptop computers for its financial planners. Unknown to anyone, 8 of the machines are defective and require repair. If 5 laptops are sent to the Alaska office, find the probability that at least 3 of the machines are in good working order, by using

1. the hypergeometric distribution,
2. the binomial distribution as an approximation.

Solution

1. Since the 5 machines sent to Alaska represent a sample selected without replacement from the 200 computers, the exact model to use is the hypergeometric probability distribution. Let x denote the number of good machines sent to Alaska. Then $N = 200$, $S = 192$, and $n = 5$. We need to calculate the following.

$$P(x \geq 3) = P(3) + P(4) + P(5)$$

$$= \frac{C(192, 3)C(8, 2)}{C(200, 5)} + \frac{C(192, 4)C(8, 1)}{C(200, 5)} + \frac{C(192, 5)C(8, 0)}{C(200, 5)}$$

$$= \frac{(1,161,280)28 + (54,870,480)8 + (2,063,130,048)1}{2,535,650,040}$$

$$= \frac{2,534,609,728}{2,535,650,040} = 0.9996$$

2. We will now use a binomial approximation, where the random variable x is the number of good computers sent to Alaska, and $n = 5$ trials. The probability of a success on a trial is the population's success ratio S/N.

$$p = \frac{S}{N} = \frac{192}{200} = 0.96.$$

$$P(x \geq 3) = P(3) + P(4) + P(5)$$

$$= C(5, 3).96^3.04^2 + C(5, 4).96^4.04^1 + C(5, 5).96^5.04^0$$

$$= 0.014156 + 0.169869 + 0.815373$$

$$= 0.9994$$

There is little difference between this approximation and the exact answer obtained by using the hypergeometric probability distribution. The calculations, however, were considerably easier using the binomial distribution. Consequently, we usually prefer to use the binomial to approximate the hypergeometric whenever the sample size n is small relative to the population size N.

> **TIP:**   In a hypergeometric experiment for which the sample size n is no more than 10 percent of the population size N ($n \leq 0.1N$), use the binomial distribution as an approximate model.

As is the case for the binomial distribution, there are special formulas for finding the mean and the variance of a hypergeometric random variable. These formulas are given in Exercise 6.79.

Section 6.4 Exercises

The Hypergeometric Probability Distribution

For Exercises 6.66–6.70, evaluate the specified probability for the given hypergeometric distribution.

6.66 Find $P(2)$ if $N = 8$, $S = 5$, and $n = 3$.

6.67 Find $P(3)$ if $N = 10$, $S = 5$, and $n = 4$.

6.68 Find $P(7)$ if $N = 15$, $S = 9$, and $n = 10$.

6.69 Find $P(x \leq 2)$ if $N = 9$, $S = 4$, and $n = 5$.

6.70 Find $P(x > 1)$ if $N = 18$, $S = 10$, and $n = 7$.

6.71 A professional baseball player receives a shipment of 10 new bats that are identical in appearance, but 3 bats have internal flaws that will result in their breaking on impact with a baseball. The player selects 5 of the 10 bats to take with him on a road trip. Find the probability that he will choose the 3 defective bats.

6.72 A carton of 12 eggs contains 3 that are spoiled. Two eggs are selected to make an omelet. Find the probability that
a. one of the eggs is spoiled,
b. both eggs are spoiled.

6.73 To offer larger jackpots and appeal to a greater audience, Maine, New Hampshire, and Vermont joined forces to run Tri-State MEGABUCKS. For $1, a player selects 6 different numbers from 1 to 40. A drawing is held to choose the 6 winning numbers. If the player's 6 picks are the 6 chosen numbers, then he/she wins the jackpot prize (or an equal share if there is more than one winner). Determine the probability that a person purchasing 1 ticket will win the top prize.

6.74 A person is considering the purchase of a newly introduced camera. Hoping that he will be able to find an advertisement for the product, he buys two photography magazines from seven different ones available at a newsstand. If three of the magazines have an ad for the camera, what is the probability that at least one of the purchased magazines will contain such an ad?

6.75 A flea market stand has a box of 20 grab bags that can be purchased for $3 each. The owner claims that 2 of the bags contain a $10 bill. If this is true and a customer purchases 3 bags, what is the probability that

a. none of the purchased bags contains $10?

b. exactly one purchased bag contains $10?

6.76 In Exercise 6.75 suppose the owner of an adjacent stand has noticed that 13 of the 20 bags have been purchased, and that no one has yet obtained a $10 bill. Find the probability of this occurring, if in fact 2 of the 20 bags did contain $10.

6.77 If a player is dealt a 5-card poker hand, what is the probability of obtaining exactly 3 aces?

6.78 In bridge, the ultimate trump hand consists of all spades. If one is dealt a 13-card bridge hand, what is the probability of obtaining this dream hand?

6.79 The following are special formulas for finding the mean and variance of a hypergeometric random variable:

$$\mu = np \quad \text{and} \quad \sigma^2 = npq \frac{(N - n)}{(N - 1)},$$

where $p = \frac{S}{N}$, the population's success ratio, and $q = 1 - p$. For Exercise 6.71, use these formulas to find the mean, variance, and standard deviation of x, the number of defective bats selected for the road trip by the baseball player.

6.80 The Internal Revenue Service randomly selects a number of tax returns each year for its Taxpayer Compliance Measurement Program (TCMP). Returns selected under TCMP are subjected to a very detailed audit. Suppose from a group of 125 tax returns, the IRS is going to randomly select 6 for this program. If 5 of the 125 returns contain fraudulent information, what is the probability that the IRS's sample will include at least 1 of the fraudulent returns? Solve this by using

a. the hypergeometric distribution,

b. the binomial distribution as an approximation.

6.81 In 1976, some colleges participated in a bicentennial contest involving extemporaneous speaking. To determine a participant's topic, 13 slips of paper, each with a different topic, were prepared. Thirty minutes before the contest, a participant randomly selected 3 slips and was allowed to choose 1 of the 3 for his or her presentation. Several days before the debate, each participant knew the 13 topics from which a selection would have to be made. Suppose a participant had prepared in advance for 5 of the 13 topics. Find the probability that on the day of the contest, he/she would draw at least 1 topic for which the participant had prepared.

6.82 Concerning the extemporaneous speaking contest in Exercise 6.81, show that if a participant prepared for only 6 topics, the probability is nearly 90 percent that he/she would draw at least 1 topic for which the participant had prepared.

6.5   The Poisson Probability Distribution (Optional)

The Poisson probability distribution is often used to model the *number of random occurrences* of some phenomenon in a specified *unit of space or time*. The following are a few examples of counts for which the Poisson distribution might be used:

Simeon Denis Poisson (1781–1840) presented a derivation of the distribution that carries his name in his major probability work, *Researches on the Probability of Criminal and Civil Verdicts.* (Photo Courtesy of The Bettmann Archive.)

1. the number of chimney fires in Minnesota during January (the unit of time is the month of January),
2. the number of defects in a bolt of fabric (the unit of space is a bolt of fabric),
3. the number of people arriving at the emergency room of a hospital during an eight hour shift (the unit of time is an eight hour shift),
4. the number of typing errors per page made by a typist (the unit of space is one page),
5. the number of kernels of corn that fail to pop in a 3.5 ounce bag of microwave popcorn (the unit of space is a 3.5 ounce bag of popcorn),
6. the number of ticks picked up by a dog during a 30 minute walk in the woods (the unit of time is 30 minutes in the woods).

For both the binomial and the hypergeometric probability distributions, we specified the necessary conditions for these distributions to apply. While there are similar characteristics for a Poisson variable, they are more complex and difficult to recognize. We will be content here to give the distribution and to illustrate its use.

The **Poisson probability distribution** is given by the formula

$$P(x) = \frac{\mu^x e^{-\mu}}{x!}, \qquad \text{for } x = 0, 1, 2, \ldots, \qquad (6.8)$$

where x is the number of occurrences of the phenomenon in the specified interval of space or time,

μ is the mean of the random variable x and is the average number of occurrences in the specified interval,

e is the base of the natural logarithm and equals approximately 2.718.

Many calculators can compute the value of $e^{-\mu}$. If yours does not, you can use Table 2 in Appendix A.

Example 6.15   The owner of a tree farm specializes in growing blue spruce trees and selling them for Christmas trees. They are grown in rows of 300 trees, and the owner has found that, on average, about 6 trees per row will not be suitable to sell. Assuming that the Poisson distribution is applicable here, find the probability that

1. a row of trees selected for shipment will contain 2 unsalable trees,
2. a half row of trees will contain 2 unsalable trees.

Solution

1. Here the specified interval of space is 1 row of trees, the random variable x is the number of unsalable trees in 1 row, and $\mu = 6$, since the mean of x is 6 trees per row. We want to find $P(2)$.

$$P(x) = \frac{\mu^x e^{-\mu}}{x!}$$

$$P(2) = \frac{6^2 e^{-6}}{2!} = \frac{36(0.002479)}{2} = 0.045$$

2. Since we want a probability concerning a half row, the specified interval of space is a half row, and x is the number of unsalable trees in a half row. For this interval, the average value of x is $\mu = 3$ trees, because the average for a full row is 6.

$$P(x) = \frac{\mu^x e^{-\mu}}{x!}$$

$$P(2) = \frac{3^2 e^{-3}}{2!} = \frac{9(.049787)}{2} = 0.224$$

In MINITAB the **PDF** and **CDF** commands can be used to obtain Poisson probabilities in the same way that they were used in Section 6.2 with the binomial distribution. This is illustrated in Example 6.16.

ⓂExample 6.16   A large discount clothier has found that during the first day of its yearly summer clearance, customers will enter the store at an average rate of 4 per minute. Assuming that the Poisson distribution is applicable, use MINITAB to find the probability that during a 5 minute period

1. from 25 to 27 people will enter the store,
2. at most 30 will enter.

Solution

The specified interval of interest is a 5 minute time period. For this interval $\mu = 20$, since customers enter at the rate of 4 per minute. The random variable x is the number entering the store during a 5 minute period.

1. We need to find

$$P(25 \leq x \leq 27) = P(25) + P(26) + P(27).$$

We will obtain this using the **PDF** command by first placing the values 25, 26, and 27 in a column, say C1. Then apply the **PDF** command with the **POISSON** subcommand.

```
SET C1
25 26 27
END
PDF C1;
POISSON 20.
```

Note that in the **POISSON** subcommand we must specify the mean and end the subcommand with a period. MINITAB will produce the output presented in Exhibit 6.4.

Exhibit 6.4

```
MTB > SET C1
DATA> 25 26 27
DATA> END
MTB > PDF C1;
SUBC> POISSON 20.
        K            P( X = K)
      25.00            0.0446
      26.00            0.0343
      27.00            0.0254
```

From Exhibit 6.4, we have that

$$P(25) \leq x \leq 27) = P(25) + P(26) + P(27)$$
$$= 0.0446 + 0.0343 + 0.0254$$
$$= 0.1043.$$

2. To find the probability that at most 30 will enter, the **CDF** command has been used in Exhibit 6.5 to find $P(x \leq 30)$.

From Exhibit 6.5,

$$P(x \leq 30) = 0.9865.$$

Before closing this section, we wish to note that a Poisson random variable has the unusual property that its mean and variance are equal. (See Exercise 6.97).

Exhibit 6.5

```
MTB > CDF 30;
SUBC> POISSON 20.
      K  P( X LESS OR = K)
   30.00              0.9865
```

Section 6.5 Exercises
The Poisson Probability Distribution

For Exercises 6.83–6.87, assume that the random variable has a Poisson distribution with the given mean μ. Find the requested probability.

6.83 $P(3)$ if $\mu = 4$

6.84 $P(0)$ if $\mu = 5$

6.85 $P(4)$ if $\mu = 3.4$

6.86 $P(x > 2)$ if $\mu = 1.2$

6.87 $P(x \leq 1)$ if $\mu = 1.6$

In all the remaining exercises, assume that a Poisson distribution is applicable.

6.88 The number of false fire alarms in a certain city averages 2.4 per day. What is the probability that 4 false alarms will occur on a given day?

6.89 For a particular brand of popcorn cooked in an air popper, the average number of kernels of corn that fail to pop is 6.8 per quarter cup. Find the probability that all kernels will pop in a quarter cup batch.

6.90 During the preparation of a manuscript, a particular author averages 1.2 typing errors per page. Find the probability that a randomly selected page will contain 2 typing errors.

6.91 In Exercise 6.90, find the probability that 5 pages will contain a total of 2 typing errors.

6.92 A university in the Northeast has cancelled 90 days of classes because of inclement weather during the last 75 years. Find the probability that at least 1 day of classes will be cancelled next year at this university.

6.93 The Cleveland Browns' top draft choice in 1988 was Eric Metcalf. In a December game the following year, Metcalf sustained a foot injury. During an examination of the injury, doctors found that Metcalf had an extra bone in each foot and it was this extra bone that had been fractured. According to medical authorities, only about one person in 100,000 has this extra bone. In a city of 50,000 people, what is the probability that at least 1 person will have the extra bone? (*The Boston Globe,* December 31, 1989.)

6.94 A college athletic department claims that basketball ticket sales during the 1 hour preceding game time average 300. Find the probability of 4 sales during a particular 1 minute period within this time frame.

6.95 A radio call-in talk show has found that its switchboard receives an average of 72 calls during the 1 hour broadcast. Find the probability that there will be
a. fewer than 5 calls during a 10-minute period,
b. more than 2 calls during a 4-minute period.

6.96 In recent years there has been considerable concern about the health risks to people exposed to asbestos, particularly in schools. However, a recent article in *Forbes* (January 8, 1990) stated, "Scientists are convinced that asbestos in buildings and schools isn't so dangerous after all." The article cites that the number of expected deaths before age 65 from asbestos in school buildings is only 1 per 100,000 people. Assuming that these figures are correct, find the probability that for 5,000 randomly selected people, at least 1 person will die before age 65 from asbestos in school buildings.

6.97 It can be shown that the variance of a Poisson random variable is equal to the mean of the variable. That is,

$$\sigma^2 = \mu$$

Use this to find the variance and standard deviation of each Poisson random variable in Exercises 6.83–6.87.

MINITAB Assignments

Ⓜ 6.98   Use MINITAB to solve Exercise 6.95.

Ⓜ 6.99   Use MINITAB to solve Exercise 6.96.

Ⓜ 6.100 An auto body repair shop specializes in painting automobiles. It has found that an average of 1.6 spots per car need to be retouched after a paint job. Use MINITAB to find the probability that for the next 10 paint jobs, a total of 15 to 25 spots will need to be retouched.

Ⓜ 6.101 When n is large and p is small (we will take this to mean that $n \geq 30$ and $p \leq 0.1$), the Poisson distribution is sometimes used to approximate the binomial. The average used for the Poisson random variable is $\mu = np$. At a large university, 10 percent of the student body are members of a national honor society. Suppose 125 students are randomly selected. Use MINITAB to find the probability that more than 13 will be members of a national honor society, by using
a. the binomial distribution,
b. the Poisson distribution.

Ⓜ 6.102 The Royal Bank of Canada's computer center in Toronto processes more than 200,000 credit card transactions each day, yet it has an error rate of only 0.01% (*Financial Post,* Toronto, March 28, 1991). For a shift in which 100,000 transactions are processed, use MINITAB to find the probability that
a. fewer than 15 errors will occur,
b. the number of errors will exceed 18.
Hint: See Exercise 6.101.

Looking Back

In this chapter's title, **Discrete Probability Distributions, discrete** refers to the type of random variables studied in the chapter. For our purposes, discrete variables almost always consist of **count** variables that specify how many times some event occurs.

   Probability distributions are mathematical models for describing various physical phenomena. The particular one that applies in a given problem will depend on what assumptions are appropriate for that situation.

   We have seen that a random variable x can be modeled by the **binomial probability distribution** when the conditions of a **binomial experiment** are satisfied.

A **binomial experiment** is an experiment that satisfies the following properties:

1. The experiment consists of the repetition of n identical trials.
2. Each trial can result in one of two possibilities—success or failure.
3. The probability of a success on a trial is denoted by p, and p remains unchanged from trial to trial.
4. The trials are independent.
5. The random variable x of interest is the total number of successes observed on the n trials.

The **binomial probability distribution** is given by the formula

$$P(x) = C(n, x)p^x q^{n-x}, \qquad \text{for } x = 0, 1, 2, 3, \ldots n,$$

where

n is the number of trials,

x is the number of successes on the n trials,

p is the probability of a success on a trial,

q is the probability of a failure on a trial, and $q = 1 - p$.

The **mean** and **standard deviation** of a binomial variable are

$$\mu = np \quad \text{and} \quad \sigma = \sqrt{npq}.$$

The **hypergeometric probability distribution** is applicable when **sampling without replacement** from a population of N elements that are considered successes and failures. However, when the sample size is small relative to the population size (n is 10 percent of N or less), we usually prefer to use the binomial distribution as an approximate model because the calculations are easier.

The **hypergeometric probability distribution** is given by the formula

$$P(x) = \frac{C(S, x)C(N - S, n - x)}{C(N, n)}, \qquad \text{for } x = 0, 1, 2, \ldots n,$$

where

N is the number of elements in the population,

S is the number of successes in the population,

n is the number of elements selected in the sample,

x is the number of successes obtained in the sample.

The **Poisson probability distribution** is often used to model the number of random occurrences of some event in a specified unit of time or space. Although there are conditions that can be checked to see if the distribution is appropriate in a given situation, they are rather complex. You are not expected to recognize these conditions; rather, you should be able to use the Poisson distribution when told that it is applicable.

The **Poisson probability distribution** is given by the formula

$$P(x) = \frac{\mu^x e^{-\mu}}{x!}, \qquad \text{for } x = 0, 1, 2, \ldots,$$

where

x is the number of occurrences of the phenomenon in the specified interval of space or time,

μ is the mean of the random variable x and is the average number of occurrences in the specified interval,

e is the base of the natural logarithm and equals approximately 2.718.

Key Words

In reviewing this chapter, you should be able to define, explain, and illustrate each of the following.

probability distribution *(page 219)*

binomial experiment *(page 220)*

trial *(page 220)*

success *(page 220)*

failure *(page 220)*

binomial probability distribution *(page 224)*

mean of a binomial variable *(page 234)*

standard deviation of a binomial variable *(page 234)*

hypergeometric experiment *(page 238)*

hypergeometric probability distribution *(page 239)*

Poisson probability distribution *(page 244)*

ⓜ *MINITAB Commands*

PDF _; *(page 226)*
BINOMIAL _ _.

PDF _; *(page 246)*
POISSON _.

SET _ *(page 246)*

END *(page 246)*

CDF _; *(page 228)*
BINOMIAL _ _.

CDF _; *(page 246)*
POISSON _.

Review Exercises

For Exercises 6.103–6.108, determine the requested probability.

6.103 $P(3)$ for a binomial random variable with $n = 5$ and $p = \frac{1}{4}$.

6.104 $P(x > 1)$, if x is a binomial variable with $n = 9$, $p = \frac{1}{3}$.

6.105 $P(4)$ for a hypergeometric distribution with $N = 12$, $S = 7$, and $n = 6$.

6.106 $P(x \leq 1)$, where x is a hypergeometric random variable with $N = 13$, $S = 7$, and $n = 5$.

6.107 $P(1 \leq x \leq 3)$, where x is a Poisson variable with $\mu = 2$.

6.108 $P(5)$ for a Poisson distribution with $\mu = 4.8$.

6.109 Use Table 1 to evaluate $P(9 \leq x \leq 13)$, where x has a binomial distribution with $n = 20$ and $p = 0.7$.

6.110 A hardware store is having an end-of-summer clearance sale on lawn mower parts. It has eight blades that are being offered at the greatly reduced price of two dollars each. A customer, not knowing which blade will fit his mower, randomly selects four blades. Suppose three of the eight blades are suitable for his mower. What is the probability that his purchase will include at least one blade that will fit?

6.111 A high school stamp and coin club consists of six boys and seven girls. The club adviser has received four complementary passes to a coin show. The adviser randomly selects four club members to receive the passes. Find the probability that the selection consists of an equal number of boys and girls.

6.112 A spokeswoman for the Children's Defense Fund stated that one-fourth of all first grade pupils entering school in 1988 were poor (*Bangor Daily News,* September 29, 1988). What would be the expected number of poor students in a randomly selected sample of 3,000 first graders?

6.113 A survey conducted by Zenith Data Systems found that of those who plan to buy a lightweight laptop computer, 65 percent were women (*BusinessWeek,* January 15, 1990). Find the probability that in a sample of 10 people who plan to purchase such a computer, more than 7 will be women.

6.114 According to the National Coffee Association, 75 percent of American coffee drinkers use the drip-filtered method to make their coffee. For 5 randomly selected coffee drinkers, find the probability that
a. all use the drip-filtered method,
b. none use this method.

6.115 An automobile manufacturer has found that its new cars contain an average of 2.3 defects that require correction by the dealer. Use a Poisson distribution to find the probability that a newly purchased car of this type will be delivered with at least 1 defect.

6.116 Radon, the second largest cause of lung cancer after smoking, is a radioactive gas produced by the natural decay of radium in the ground. The number of expected deaths before age 65 from indoor radon is estimated to be 4 per 1,000 people (Harvard University symposium summary, August, 1989). Use a Poisson distribution to determine the probability that for a sample of 100 people, none will die before age 65 from indoor radon.

6.117 Based on a survey of several thousand students, ages 10 through 16, *USA WEEKEND* (August 18–20, 1989) found that 68 percent graded their schools an A or a B. Suppose 1,000 students in this age group are randomly selected, and the random variable x denotes the number of students who assign a grade of A or B to their school. Determine the mean and standard deviation of x.

6.118 A city inspector is going to select 10 vehicles from a trucking company's fleet of 180 trucks in order to test if the firm is complying with pollution emission standards. If the sample contains 2 or more vehicles that fail the test, then the company will be fined. Suppose 18 of the 180 trucks fail to meet the standards. Find the probability that the company will receive a fine, by using
a. the hypergeometric distribution,
b. the binomial distribution as an approximation.

6.119 Suppose 10,000 tosses of a coin resulted in 4,800 heads. Explain why it is very unlikely that the coin is fair.

6.120 The mean and the standard deviation of a binomial random variable are known to be 56 and 7, respectively. What must be the value of n and p?

6.121 We have seen that the variance of a binomial random variable is given by $\sigma^2 = npq$. In Chapter 9, we will need to know for what value of p this variance is a maximum (the solution can easily be determined using elementary calculus). To conduct a preliminary exploration of this, construct a table that shows the values of npq for $p = 0, 0.1, 0.2, 0.3, \ldots, 0.9, 1$. From an inspection of the table, make a guess concerning the p value that maximizes σ^2.

MINITAB Assignments

Ⓜ 6.122 A manufacturer of microprocessors has found that about 88 percent of its computer chips actually operate at speeds exceeding specifications. Use MINITAB to find the probability that a shipment of 30 microprocessors will contain more than 28 that exceed speed specifications.

Ⓜ 6.123 A blueberry canning plant has determined that about 4 percent of the labels fail to adhere to the cans. If a case contains 48 cans, use MINITAB to find the probability that more than 2 will have labels that fail to adhere properly.

Ⓜ 6.124 Assume that the number of calls received by a mail order company can be modeled with a Poisson probability function. The average number of calls per hour is 28. Use MINITAB to find the probability that fewer than 8 calls will be received during a 30-minute period.

(M) 6.125 An electronics company manufactures a component for a burglar alarm system. It has determined that about one-tenth of one percent of the components are defective. A lot of 10,000 components is ready for shipment. Use MINITAB and a Poisson approximation to determine the probability that at most 7 components contain defects.

(© *Goodwood Productions/The Image Bank.*)

7 Normal Probability Distributions

◀ *Companies Pursue Zero Defects Using the Six Sigma Strategy*
In 1987, Motorola Corporation devised a strategy to reduce the number of manufac-
turing defects to nearly zero. In just three years after its implementation, the program
resulted in a $500 million savings for the company. Since its inception, Digital
Equipment Corporation, Raytheon, Corning, IBM, and several other U.S. companies
have initiated similar programs to improve manufacturing quality.

Every manufacturing process has a certain amount of variability around its
target specification (the mean). Under the Six Sigma strategy, a company modifies
and improves its manufacturing process until the variation around the mean is
extremely small. Ideally, the variability should be so small that a six-standard devia-
tion (six-sigma) departure from the mean still falls within acceptable quality limits. If
this goal can be attained, then the number of manufacturing defects will be reduced to
practically zero.

The probability distributions discussed in this chapter are intrinsically involved
in Six Sigma programs. Many characteristics of manufactured items such as weight,
length, volume, strength, and so on, can be modeled by normal probability distribu-
tions.

Looking Ahead

In Chapter 5 we introduced the concept of a random variable and distinguished
between discrete and continuous variables. We saw that a discrete random variable
is limited in the number of possible values that it can assume. It can never have
more values than the counting numbers. On the other hand, a **continuous random
variable** can assume all the values in some interval on the real number line. These
are often measurement variables such as weight, length, time, distance, and speed.

Associated with each random variable is its **probability distribution.** These
serve as mathematical models for describing various physical phenomena, and
they can be used to calculate probabilities that the variable will equal various

values when an experiment is performed. In this chapter, we will consider **probability distributions of continuous random variables,** and we will focus on the most important of these—**normal probability distributions.** Several practical applications of normal distributions will be illustrated. We will also see how normal distributions can often be used to approximate closely binomial probabilities that would be laborious and time consuming to determine exactly.

7.1 Probability Distributions of Continuous Random Variables

In the last chapter, we saw that probability distributions are used in calculating probabilities of discrete random variables. This is also true for continuous variables, but their involvement differs considerably. For a discrete random variable x, one obtains the probability that x will equal a particular value such as 4 by substituting 4 into the probability distribution $P(x)$ and evaluating $P(4)$. For a continuous random variable, we always calculate the probability that x will assume a value within a specified interval, for example, $P(3.7 \leq x \leq 6.5)$. Moreover, this probability is obtained by determining the area under the graph of the probability distribution from $x = 3.7$ to $x = 6.5$. This is illustrated in Figure 7.1.

> **$P(a \leq x \leq b)$ for a Continuous Random Variable:**
> If x is a continuous random variable with probability distribution $f(x)$, then $P(a \leq x \leq b)$, the probability that x will assume a value within the interval from $x = a$ to $x = b$, equals the area under the graph of $f(x)$ from $x = a$ to b.

To determine probabilities of continuous random variables, we will utilize tables in which areas (probabilities) have been compiled for the probability distributions discussed in this book. It is important to keep in mind the point illustrated below.

> For continuous random variables we determine *probabilities of intervals,* and these are obtained by finding *areas* under the graph of the random variable's probability distribution.

Figure 7.1
$P(3.7 \leq x \leq 6.5)$ = Shaded area.

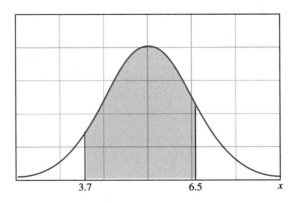

In Chapter 5, we saw that every discrete probability distribution $P(x)$ satisfies the following two basic properties.

1. $P(x) \geq 0$ for each value of x.
2. $\Sigma P(x) = 1$.

For probability distributions of continuous random variables, the analogous properties are given below.

Two Basic Properties of a Continuous Probability Distribution $f(x)$:
1. $f(x) \geq 0$ for each value of x.
2. The total area under the graph of $f(x)$ equals 1.

The present chapter will focus on continuous probability distributions of normal random variables. Probability distributions of other continuous random variables will be introduced in later chapters.

7.2 Normal Probability Distributions

Although we have not previously discussed normal probability distributions in a formal way, they are implicitly involved in the Empirical rule that was introduced in Chapter 2. This rule is applicable when a set of measurements has a relative frequency distribution that is approximately mound shaped. Three mound-shaped distributions were presented in Chapter 2 and are reproduced in Figure 7.2.

These mound-shaped distributions are also called bell-shaped distributions, and they occur frequently in nature and our daily surroundings. Normal probability distributions are the theoretical basis for the Empirical rule and its reference to mound-shaped distributions. Formally, a normal probability distribution is one that is of the form given in Formula 7.1 below.

The **normal probability distribution** with mean μ and standard deviation σ is given by the formula

$$f(x) = \frac{1}{\sigma\sqrt{2\pi}}\, e^{-(x-\mu)^2/2\sigma^2} \qquad \text{for} \qquad -\infty < x < \infty, \qquad \textbf{(7.1)}$$

where e and π are irrational numbers whose values are approximately 2.7183 and 3.1416, respectively.

Figure 7.2
Some mound-shaped distributions.

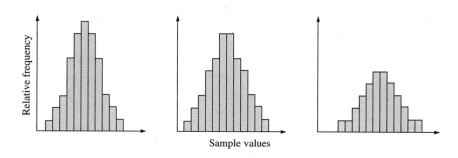

Figure 7.3
Graph of a normal probability distribution.

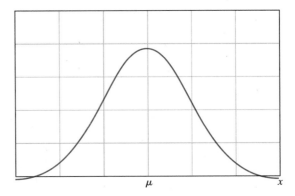

You do not need to memorize Formula 7.1, but you should be familiar with some characteristics of its graph, which appears in Figure 7.3.

The following points are worth noting concerning the graph of a normal probability distribution.

The first description in English of the normal distribution was given by Abraham De Moivre (1667–1754) in his probability book, *The Doctrine of Chances*. Major extensions of his work were later made by Pierre Simon LaPlace (1749–1827) and Carl Friedrich Gauss (1777–1855). (*Photo © G.M. Smith/ The Image Bank.*)

Some Characteristics of Normal Probability Curves:

1. A normal curve extends indefinitely far to the left and to the right, approaching more closely the *x*-axis as *x* increases in magnitude.
2. There are an infinite number of different normal probability curves. However, a particular curve is determined by only two parameters: its mean μ and its standard deviation σ.
3. The high point on the curve occurs at $x = \mu$, the mean of the normal random variable.
4. The standard deviation of the variable determines the spread of the curve about its mean. A curve with a large standard deviation will be more spread out, and consequently flatter, than one with a smaller standard deviation.
5. Since the curve is the graph of a continuous probability distribution, the total area under the curve equals 1.
6. The curve is symmetric about its mean, and, thus, the area to the left of the mean and the area to the right of the mean each equal 0.5.

To calculate probabilities concerning a normal random variable, we must be able to determine areas under the graph of the variable's probability distribution. As mentioned earlier, tables will be used to obtain areas for continuous random variables. However, since there are an infinite number of different normal distributions, it would seem that we are faced with an insurmountable task. Fortunately, only one table of areas is needed, namely, one for the normal distribution with $\mu = 0$ and $\sigma = 1$. With this single table, probabilities can be determined for all normal distributions.

The Standard Normal Distribution

> The normal random variable with mean $\mu = 0$ and standard deviation $\sigma = 1$ is called the **standard normal random variable,** and is denoted by z. Areas under the graph of its probability distribution are tabulated in Table 3 of Appendix A and on the inside of the front cover.

Since the standard normal random variable z has a mean of 0 and a standard deviation of 1, a **value of z** can be thought of as **the number of standard deviations from the mean.** For example, $z = 2.50$ is a value of z that is two and one-half standard deviations ($\sigma = 1$) from the mean ($\mu = 0$). Because of this interpretation of z, and because any normal random variable can be transformed to z, Table 3 in Appendix A is sufficient for finding areas concerning any normal probability distribution. This will become clear in the next section when we consider the transformation that changes a normal random variable x to z. At that time, we will present several illustrations and practical applications of this technique. The remainder of this section is concerned with illustrating the use of the standard normal table. A portion of Table 3 is reproduced in Table 7.1 on the following page.

Example 7.1   Determine $P(z \leq 0.98)$.

Solution

This probability equals the area under the standard normal curve that lies to the left of $z = 0.98$. Table 3 gives cumulative areas that lie to the left of specified values of z. To find the area to the left of $z = 0.98$, look in the first column of the table and find the row that contains 0.9 (the integer and tenths portion of 0.98). Next, look across the top of the table and find the column headed by .08 (the second decimal digit of 0.98). The intersection of the row and column gives the desired area of 0.8365. Thus,

$$P(z \leq 0.98) = 0.8365.$$

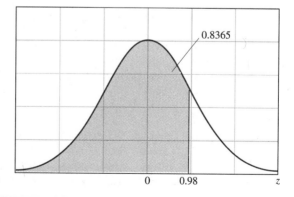

Table 7.1

A Portion of Table 3 from Appendix A

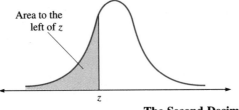

Area to the
left of z

Each table value is the cumulative area
to the left of the specified z-value

z	.00	.01	.02	.03	The Second Decimal Digit of z .04	.05	.06	.07	.08	.09
:	:	:	:	:	:	:	:	:	:	:
−1.4	.0808	.0793	.0778	.0764	.0749	.0735	.0721	.0708	.0694	.0681
−1.3	.0968	.0951	.0934	.0918	.0901	.0885	.0869	.0853	.0838	.0823
−1.2	.1151	.1131	.1112	.1093	.1075	.1056	.1038	.1020	.1003	.0985
−1.1	.1357	.1335	.1314	.1292	.1271	.1251	.1230	.1210	.1190	.1170
−1.0	.1587	.1562	.1539	.1515	.1492	.1469	.1446	.1423	.1401	.1379
−0.9	.1841	.1814	.1788	.1762	.1736	.1711	.1685	.1660	.1635	.1611
−0.8	.2119	.2090	.2061	.2033	.2005	.1977	.1949	.1922	.1894	.1867
−0.7	.2420	.2389	.2358	.2327	.2296	.2266	.2236	.2206	.2177	.2148
−0.6	.2743	.2709	.2676	.2643	.2611	.2578	.2546	.2514	.2483	.2451
−0.5	.3085	.3050	.3015	.2981	.2946	.2912	.2877	.2843	.2810	.2776
−0.4	.3446	.3409	.3372	.3336	.3300	.3264	.3228	.3192	.3156	.3121
−0.3	.3821	.3783	.3745	.3707	.3669	.3632	.3594	.3557	.3520	.3483
−0.2	.4207	.4168	.4129	.4090	.4052	.4013	.3974	.3936	.3897	.3859
−0.1	.4602	.4562	.4522	.4483	.4443	.4404	.4364	.4325	.4286	.4247
−0.0	.5000	.4960	.4920	.4880	.4840	.4801	.4761	.4721	.4681	.4641
0.0	.5000	.5040	.5080	.5120	.5160	.5199	.5239	.5279	.5319	.5359
0.1	.5398	.5438	.5478	.5517	.5557	.5596	.5636	.5675	.5714	.5753
0.2	.5793	.5832	.5871	.5910	.5948	.5987	.6026	.6064	.6103	.6141
0.3	.6179	.6217	.6255	.6293	.6331	.6368	.6406	.6443	.6480	.6517
0.4	.6554	.6591	.6628	.6664	.6700	.6736	.6772	.6808	.6844	.6879
0.5	.6915	.6950	.6985	.7019	.7054	.7088	.7123	.7157	.7190	.7224
0.6	.7257	.7291	.7324	.7357	.7389	.7422	.7454	.7486	.7517	.7549
0.7	.7580	.7611	.7642	.7673	.7704	.7734	.7764	.7794	.7823	.7852
0.8	.7881	.7910	.7939	.7967	.7995	.8023	.8051	.8078	.8106	.8133
0.9	.8159	.8186	.8212	.8238	.8264	.8289	.8315	.8340	.8365	.8389
1.0	.8413	.8438	.8461	.8485	.8508	.8531	.8554	.8577	.8599	.8621
1.1	.8643	.8665	.8686	.8708	.8729	.8749	.8770	.8790	.8810	.8830
1.2	.8849	.8869	.8888	.8907	.8925	.8944	.8962	.8980	.8997	.9015
1.3	.9032	.9049	.9066	.9082	.9099	.9115	.9131	.9147	.9162	.9177
:	:	:	:	:	:	:	:	:	:	:

You will find it helpful and will be less likely to make a mistake if you get in the habit of constructing a rough sketch of the area that is being sought.

Example 7.2   Find the area under the standard normal curve that lies to the right of $z = 1.34$.

Solution

To find the shaded area indicated in the figure, we first use Table 3 to find the area that lies to the left of $z = 1.34$ standard deviations. This value is 0.9099. Since the total area under the curve is 1, the shaded area is

$$1 - 0.9099 = 0.0901.$$

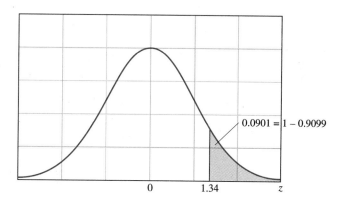

$0.0901 = 1 - 0.9099$

0 1.34 z

Example 7.3   Determine the following probabilities:

1. $P(z < -1.27)$,
2. $P(z \leq -1.27)$.

Solution

1. $P(z < -1.27)$ is equal to the area that lies to the left of -1.27. Using Table 3, the cumulative area under the z-curve to the left of -1.27 is 0.1020. Therefore,

$$P(z < -1.27) = 0.1020.$$

2. The difference between $P(z \leq -1.27)$ and $P(z < -1.27)$ in Part 1 is the probability that z equals -1.27, because

$$P(z \leq -1.27) = P(z < -1.27) + P(z = -1.27).$$

However, the probability that z equals a specified value such as -1.27 is 0, since $P(z = -1.27)$ equals the area above the interval consisting of the single point -1.27. Thus,

$$P(z \leq -1.27) = P(z < -1.27) = 0.1020.$$

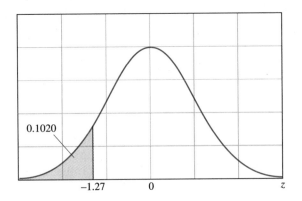

Example 7.4   Determine $P(-1.49 \le z \le 0)$.

Solution

The cumulative area to the left of $z = -1.49$ is 0.0681, and the cumulative area to the left of $z = 0$ is 0.5000. The difference between these areas gives the desired area shaded in the figure below. Thus,

$$P(-1.49 \le z \le 0) = 0.5000 - 0.0681 = 0.4319.$$

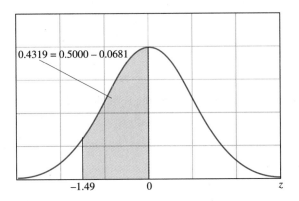

Example 7.5   Determine the proportion of the standard normal distribution that lies within 1 standard deviation of its mean; that is, find $P(-1 \le z \le 1)$.

Solution

The probability is equal to the area under the standard normal curve that lies between $z = -1$ and $z = 1$. From Table 3, the cumulative area to the left of $z = -1$ is 0.1587, and the area to the left of $z = 1$ is 0.8413. The difference between these areas gives the desired probability

$$P(-1 \le z \le 1) = 0.8413 - 0.1587 = 0.6826.$$

Notice that this result is consistent with the percentage given by the Empirical rule, namely, 68 percent.

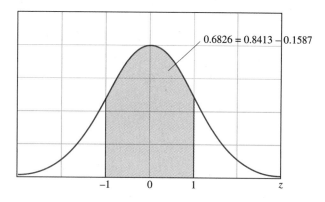

0.6826 = 0.8413 − 0.1587

Example 7.6   Find $P(-2.39 \leq z \leq -1.28)$.

Solution

The area under the standard normal curve that lies to the left of -2.39 standard deviations is 0.0084, and the area to the left of -1.28 standard deviations is 0.1003. The desired area is the difference between the two cumulative areas.

$$P(-2.39 \leq z \leq -1.28) = 0.1003 - 0.0084 = 0.0919$$

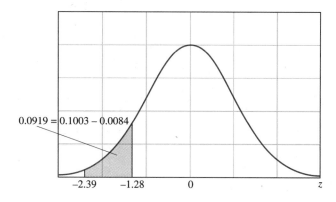

0.0919 = 0.1003 − 0.0084

The next three examples illustrate how Table 3 can be used to obtain values of z, rather than areas.

Example 7.7   Determine P_{93}, the ninety-third percentile of the standard normal distribution.

Solution

We need to find the value of z for which 93 percent of the distribution lies to the left of it. Using Table 3, we look for the z-value that corresponds to a cumulative area of 0.93. The nearest area to 0.9300 in Table 3 is 0.9306, and the associated z-value is 1.48. Thus, $P_{93} = 1.48$.

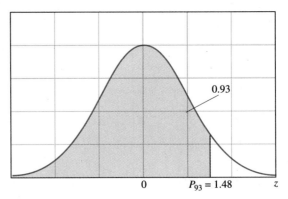

Example 7.8   Find the z-value for which the interval from $-z$ to z contains 95 percent of the total distribution.

Solution

Because of the symmetry of the standard normal distribution, the area to the left of $-z$ is

$$\frac{(1 - 0.95)}{2} = 0.025.$$

A cumulative area of 0.0250 appears exactly in Table 3, and the corresponding z-value is -1.96. Consequently, the interval from -1.96 to 1.96 contains 95 percent of the distribution.

This result is the basis for the statement in the Empirical rule that approximately 95 percent of a mound-shaped distribution falls within 2 standard deviations of its mean.

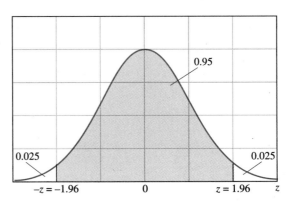

Example 7.9   For the standard normal distribution, find the number c such that $P(z > c) = 0.05$.

Solution

We want the z-value for which the area to the right of it is equal to 0.05. Consequently, the desired z-value will have a cumulative area to the left of it equal to

$$(1 - 0.05) = 0.95.$$

Two areas in Table 3 are equally close to 0.9500, and these are 0.9495 ($z = 1.64$) and 0.9505 ($z = 1.65$). Since 0.95 lies midway between the two areas, we will use a corresponding z-value of 1.645, which is halfway between 1.64 and 1.65. Thus, $c = 1.645$.

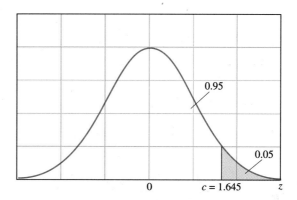

In future chapters, we will frequently have occasion to refer to the z-value that has an area to the right of it equal to some specified amount, such as a (see Figure 7.4). We will denote the z-value with this property by z_a. Thus, in the previous example we determined that $z_{.05} = 1.645$, since 1.645 is the z-value for which the area to the right of it equals 0.05.

z_a will denote the z-**value** that has an area to the right of it equal to the subscript a.

We will close this section by showing how MINITAB can be used to obtain a rough sketch of the graph of any normal probability distribution.

Figure 7.4
Z_a is the z-value with a right-tail area equal to a.

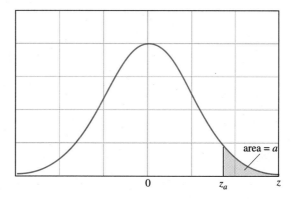

ⓜ**Example 7.10**   Use MINITAB to sketch the graph of the normal distribution with mean $\mu = 90$ and standard deviation $\sigma = 2$.

Solution

MINITAB sketches a probability distribution by plotting a set of (x, y) points. We begin by storing in column C1 the x-values of the points to be plotted. Since 3 standard deviations from the mean covers nearly all the distribution, we will start at 84 (3σ below the mean of 90) and stop at 96 (3σ above the mean). We arbitrarily select a spacing of 0.5 between each x. The x-values can be entered quickly in the following manner.

```
SET C1
84:96/0.5     # x-values 84 to 96 in increments of 0.5
END
```

The **PDF** command that was introduced in Chapter 5 is used to calculate the y-values for the points to be plotted. To obtain these for the x-values in C1, and then have them stored in C2, type

```
PDF C1 C2;    # puts y-values in C2 for x-values in C1.
NORMAL 90 2.  # 90 is the mean and 2 is the std. dev.
```

Next, the **PLOT** command is used to graph the points that have their x-values in C1 and y-values in C2.

```
PLOT C2 C1    # y-values are given first, then x-values
```

MINITAB will produce the plot given in Exhibit 7.1.

Exhibit 7.1

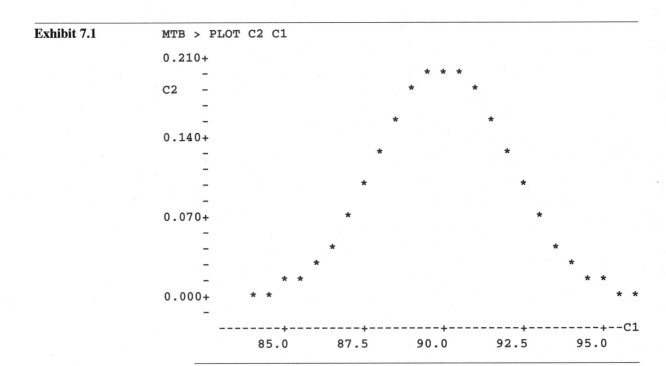

Section 7.2 Exercises
Normal Probability Distributions

Find the area under the standard normal curve between each of the following pairs of z-values.

7.1 $z = 0$ and $z = 1.23$

7.2 $z = 0$ and $z = 2.04$

7.3 $z = -2.98$ and $z = 0$

7.4 $z = -1.24$ and $z = 0$

7.5 $z = -0.51$ and $z = 1.12$

7.6 $z = -2.95$ and $z = 0.87$

7.7 $z = -2.31$ and $z = -1.47$

7.8 $z = -1.35$ and $z = -0.64$

7.9 $z = 0.20$ and $z = 1.31$

7.10 $z = 2.01$ and $z = 3.04$

Determine the area under the standard normal curve which lies to the left of each of the following z-values.

7.11 $z = 1.48$

7.12 $z = 2.37$

7.13 $z = -0.93$

7.14 $z = -1.44$

Determine the area under the standard normal curve that lies to the right of each of the following z-values.

7.15 $z = 2.18$

7.16 $z = 0.50$

7.17 $z = -1.66$

7.18 $z = -2.34$

Find the following probabilities for the standard normal random variable z.

7.19 $P(z \leq 2.37)$

7.20 $P(z \leq 0.99)$

7.21 $P(z \leq -1.32)$

7.22 $P(z \leq -2.22)$

7.23 $P(z \geq 1.17)$

7.24 $P(z \geq 0.59)$

7.25 $P(z \geq -2.85)$

7.26 $P(z \geq -1.53)$

7.27 $P(0 \leq z \leq 0.87)$

7.28 $P(0 \leq z \leq 1.25)$

7.29 $P(-0.55 \leq z \leq 0)$

7.30 $P(-2.21 \leq z \leq 0)$

7.31 $P(-2.15 \leq z \leq 1.96)$

7.32 $P(-2.59 \leq z \leq 1.31)$

7.33 $P(-0.64 \leq z \leq -0.12)$

7.34 $P(-2.93 \leq z \leq -2.03)$

7.35 $P(1.65 \leq z \leq 2.04)$

7.36 $P(2.13 \leq z \leq 2.98)$

Determine the following probabilities, and compare the results with those given by the Empirical rule and Chebyshev's theorem.

7.37 $P(-3 \leq z \leq 3)$

7.38 $P(-2 \leq z \leq 2)$

7.39 Find the z-value for which the area to the left of z is 0.9881.

7.40 Find the z-value for which the area to the left of z is 0.0125.

7.41 Find the z-value for which the area to the right of z is 0.3974.

7.42 Find the z-value for which the area to the right of z is 0.6217.

7.43 Find the z-value for which the area to the right of z is 0.9967.

7.44 Find the positive z-value for which the area between 0 and z is 0.2794.

7.45 Find the positive z-value for which the area between 0 and z is 0.4980.

7.46 Find the z-value for which the area between $-z$ and z is 0.7814.

7.47 Find the z-value for which the area between $-z$ and z is 0.4038.

Determine the following percentiles of the standard normal distribution.

7.48 P_{80} 7.49 P_{64} 7.50 P_{13}

7.51 P_{25} 7.52 P_{75} 7.53 P_{50}

Find the following values of z_a (the z-value that has an area to the right of it equal to the subscript a).

7.54 $z_{.10}$ 7.55 $z_{.05}$ 7.56 $z_{.025}$

7.57 $z_{.01}$ 7.58 $z_{.005}$ 7.59 $z_{.975}$

7.60 Explain why the following 4 probabilities have the same value.

$$P(-1.27 \leq z \leq 2.68)$$

$$P(-1.27 < z < 2.68)$$

$$P(-1.27 \leq z < 2.68)$$

$$P(-1.27 < z \leq 2.68)$$

MINITAB Assignments

Ⓜ 7.61 Use MINITAB to sketch the graph of the standard normal distribution.

Ⓜ 7.62 Using MINITAB, sketch the graph of the normal distribution with a mean of 500 and a standard deviation of 100.

Ⓜ 7.63 Use MINITAB to sketch the graph of the normal distribution with a mean of 100 and a standard deviation of (a) 20 and (b) 5.

7.3 Applications of Normal Distributions

The preceding section was primarily concerned with illustrating the use of Table 3. Now that we are familiar with obtaining probabilities concerning the standard normal random variable z, we can consider normal distributions in general.

Since z has a mean of 0 and a standard deviation of 1, a **value of z** can be thought of as **the number of standard deviations from the mean.** Because of this and the fact that any normal random variable can be transformed to z, Table 3 is the only table needed for finding areas concerning all normal probability distributions. The transformation that changes a normal random variable x to z was used in Chapter 2 to calculate z-scores, and it appears below in Formula 7.2.

If x is a normal random variable with mean μ and standard deviation σ, then

$$z = \frac{x - \mu}{\sigma} \qquad (7.2)$$

has the standard normal distribution with mean 0 and standard deviation 1.

To illustrate Formula 7.2, suppose we are working with a normal random variable that has a mean of 80 and a standard deviation of 10, and we need to determine the probability that x is greater than 105. Since

$$z = \frac{x - \mu}{\sigma} = \frac{105 - 80}{10} = 2.50,$$

the x-value of 105 has a z-value of 2.50, and x therefore lies two and one-half standard deviations above the mean. Thus,

$$P(x > 105) = P(z > 2.50).$$

Consequently, the probability $P(x > 105)$ can be determined by using the standard normal table (Table 3) to obtain the value of $P(z > 2.50)$.

We will now consider some examples of this technique and some applications of normal distributions.

Example 7.11 The popularity of notebook computers (lightweight laptops) experienced a sharp rise in 1991. With the introduction of less expensive and more powerful notebook computers, it is expected that their share of the personal computer market will continue to grow at a rapid rate. The manufacturer of a particular model claims in its manual that its battery will average 3 hours before needing a recharge. Assuming that the battery life has approximately a normal distribution with a standard deviation of 0.4 hour, what percentage of battery charges will power this brand for more than 3.6 hours?

Solution

We need to find $P(x > 3.6)$, where the random variable x is the number of hours that a battery charge will power this type of computer. The probability is equal to the shaded area in the following figure.

A survey conducted by Zenith Data Systems found that women comprised 65 percent of those planning to buy a notebook computer. (*Business Week,* January 15, 1990.) (*Photo © Goodwood Productions/The Image Bank.*)

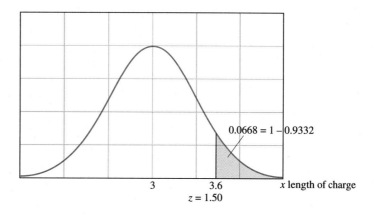

To determine the area, the z-value for $x = 3.6$ hours must be obtained.

$$z = \frac{x - \mu}{\sigma} = \frac{3.6 - 3}{0.4} = 1.50$$

A value of 3.6 hours is one and one-half standard deviations above the mean, and therefore, $P(x > 3.6) = P(z > 1.50)$. From Table 3, the cumulative area to the left of $z = 1.50$ standard deviations is 0.9332. Therefore, the shaded area to the right of $z = 1.50$ is $1 - 0.9332 = 0.0668$. Thus, approximately 7 percent of the time a battery charge will power this type of computer for more than 3.6 hours.

Example 7.12

The U.S. National Oceanic and Atmospheric Administration estimated that in 1988, U.S. fishermen received an average price of $2.99 per pound for lobsters. For what proportion of sales did they receive between $2.75 and $3.50 per pound? Assume that the distribution of prices is approximately normal with a standard deviation of 20 cents.

(© G.M. Smith/The Image Bank.)

Solution

The z-value for $2.75 is

$$z = \frac{x - \mu}{\sigma} = \frac{2.75 - 2.99}{0.20} = -1.20.$$

$3.50 has a z-value of

$$z = \frac{x - \mu}{\sigma} = \frac{3.50 - 2.99}{0.20} = 2.55.$$

From Table 3, the area to the left of -1.20 standard deviations is 0.1151, and the area to the left of 2.55 standard deviations is 0.9946. The difference between these areas gives the area of the shaded region in the following figure.

$$P(\$2.75 \le x \le \$3.50) = P(-1.20 \le z \le 2.55)$$
$$= 0.9946 - 0.1151$$
$$= 0.8795$$

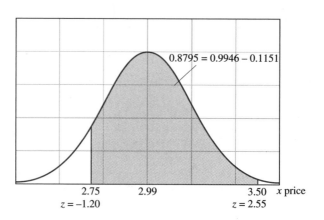

Example 7.13

By 1992, nearly all coffee manufacturers had discontinued the packaging of coffee in 1-pound containers. Most companies have reduced the contents by 3 ounces or more. Suppose a certain brand now has a label weight of 13 ounces, but in the interest of having few underweight cans, the average amount dispensed per can is 13.5 ounces. Further assume that the distribution of weights is approximately normal with a standard deviation of 0.2 ounce.

1. Determine the proportion of cans that are underweight.
2. What weight is exceeded by 67 percent of the cans?

Solution

1. We need to find the proportion of cans that contain less than the label weight of 13 ounces. This equals the probability $P(x < 13)$, which is given by the shaded area in the figure below. The z-value for 13 ounces is

$$z = \frac{x - \mu}{\sigma} = \frac{13 - 13.5}{0.2} = -2.50.$$

From Table 3, the area to the left of -2.50 standard deviations is 0.0062, and therefore,

$$P(x < 13) = P(z < -2.50) = 0.0062.$$

Thus, only about 0.6 percent of the cans will be underweight.

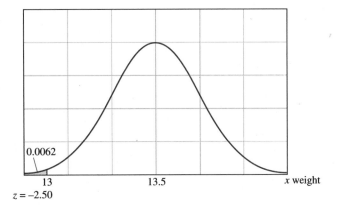

2. To find the weight that is exceeded by 67 percent of the cans, we first need to find the corresponding z-value. Since the area to the right of the desired z-value is 0.67, the area to the left of z is $1 - 0.67 = 0.33$. From Table 3, a cumulative area of 0.3300 has an associated z-value of -0.44. To find the weight x associated with $z = -0.44$, we use the relationship

$$z = \frac{x - \mu}{\sigma} \longrightarrow -0.44 = \frac{x - 13.5}{0.2}. \qquad \textbf{(7.3)}$$

Solving Equation 7.3, we have

$$x = 13.5 - 0.44(0.2) = 13.41.$$

Thus, 67 percent of the cans contain more than 13.41 ounces.

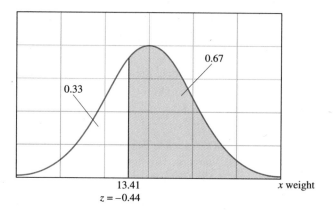

$$13.41$$
$$z = -0.44$$

ⓜ**Example 7.14**   For the coffee example in Example 7.13, use MINITAB to determine the proportion of underweight cans and also the weight that is exceeded by 67 percent of the cans.

Solution

We are given that the distribution of coffee weights is approximately normal with a mean of $\mu = 13.5$ ounces and a standard deviation of $\sigma = 0.2$ ounce.

1. To find the proportion of cans that contain less than 13 ounces, we need to use MINITAB's **CDF** (cumulative distribution function) command discussed in Chapter 6. Type the following.

```
CDF  13;
NORMAL 13.5 0.2.
```

   The main command, **CDF**, instructs MINITAB to obtain the cumulative area to the left of 13. Since a subcommand follows, the main command must end with a semicolon. The **NORMAL** subcommand specifies that the distribution is normal with mean 13.5 and standard deviation 0.2. This is terminated with a period since no additional subcommands follow. MINITAB will produce the output given in Exhibit 7.2, from which we have that

$$P(x < 13) = 0.0062.$$

2. We want to find the weight that 67 percent of the cans exceed. Note the fundamental difference between this type of problem and that in Part 1. There we wanted to find the area associated with the *x*-value 13, and we used the **CDF** command to obtain it. Now we are concerned with the inverse procedure, namely, finding the *x*-value that corresponds to a given area. For this we need to use the inverse of the **CDF** command. This is denoted by **INVCDF.** For the *x*-value that we are seeking, the cumulative area to the left of it is

$$1 - 0.67 = 0.33.$$

   To have MINITAB obtain the *x*-value corresponding to a left cumulative area (probability) of 0.33, type

```
INVCDF 0.33;
NORMAL 13.5 0.2.
```

Exhibit 7.2

```
MTB > CDF 13;                # 13 is the x-value
SUBC> NORMAL 13.5 0.2.       # 13.5 is mean, 0.2 is std dev.
     13.0000     0.0062
```

MINITAB's response to these commands appears in Exhibit 7.3, from which we see that 13.412 ounces is the weight that will be exceeded by 67 percent of the cans.

Exhibit 7.3

```
MTB > INVCDF 0.33;           # 0.33 is the cumulative prob.
SUBC> NORMAL 13.5 0.2.       # 13.5 is mean, 0.2 is std dev.
     0.3300     13.4120
```

Section 7.3 Exercises

Applications of Normal Distributions

The random variable x has a normal distribution with mean $\mu = 60$ and standard deviation $\sigma = 4$. For each of the following values of x, obtain the z-value in order to determine the number of standard deviations from the mean.

7.64 $x = 64$	7.65 $x = 68$	7.66 $x = 66$
7.67 $x = 56$	7.68 $x = 52$	7.69 $x = 54$
7.70 $x = 70$	7.71 $x = 47$	7.72 $x = 49$

x is a normal random variable with mean 100 and standard deviation 10. Evaluate the following probabilities.

7.73 $P(x \le 110)$	7.74 $P(x \le 90)$
7.75 $P(x \ge 95)$	7.76 $P(x \ge 125)$
7.77 $P(108 \le x \le 119)$	7.78 $P(113 \le x \le 129)$
7.79 $P(87 \le x \le 102)$	7.80 $P(72 \le x \le 122)$
7.81 $P(71 \le x \le 93)$	7.82 $P(84 \le x \le 99)$

The distribution of x is normal with $\mu = 40.8$ and $\sigma = 2$. Find the following probabilities.

7.83 $P(38.2 < x < 44.6)$	7.84 $P(36.4 < x < 41)$
7.85 $P(x < 45.1)$	7.86 $P(x < 35.7)$
7.87 $P(37.9 < x < 40.3)$	7.88 $P(41.7 < x < 43.9)$

Determine the following percentiles for a normal distribution with mean 280 and standard deviation 50.

7.89 P_{78}	7.90 P_{80}
7.91 P_{32}	7.92 P_9

7.93   A commuter airline has found that the time required for a flight between 2 cities has approximately a normal distribution with a mean of 54.8 minutes and a standard deviation of 1.2 minutes. Find the probability that a flight will take more than 56.6 minutes.

7.94   The amount of milk placed in cartons by a filling machine has approximately a normal distribution with a mean of 64.3 ounces and a standard deviation of 0.12 ounce. Find the probability that a carton will contain less than 64 ounces.

7.95   To get in step with the rest of the potato industry, the Maine legislature in 1989 passed a bill to eliminate the wooden barrel as the unit of measure for potato buyers and sellers. The barrels hold an average of 165 pounds. If the distribution of their contents is approximately normal with a standard deviation of 0.8 pound, what percentage of barrels contain at least 166 pounds of potatoes?

7.96   A random variable x has a normal distribution with $\mu = 164$ and $\sigma = 20$. What value of x is exceeded by 96 percent of the distribution?

7.97 x is a normal random variable with $\mu = 95$ and $\sigma = 12$. Fourteen percent of the distribution of x falls below what value?

7.98   According to the U.S. Department of Agriculture, a 3-ounce serving of trimmed sirloin beef contains an average of 7.4 grams of fat. Assume that the amount of fat for such servings has approximately a normal distribution with a standard deviation of 0.4 gram. Find the proportion of servings that contain
a. between 6.9 and 7.1 grams of fat,
b. more than 8.3 grams of fat.

7.99   With reference to Exercise 7.98, only one percent of servings will contain more than what amount of fat?

7.100   A forest products company claims that the amount of usable lumber in its harvested trees averages 172 cubic feet and has a standard deviation of 15.4 cubic feet. Assuming that these amounts have approximately a normal distribution, determine the proportion of trees that contain
a. less than 150 cubic feet,
b. more than 180 cubic feet,
c. anywhere from 175 to 190 cubic feet.

7.101   Concerning the harvested trees referred to in Exercise 7.100, 75 percent of the trees will contain more than what amount?

7.102   American homeowners have an average of 1,712 square feet of living space (*The Wall Street Journal,* April 4, 1989). Assume that the amount of living space is approximately a normal random variable with a standard deviation of 300 square feet. Find the percentage of homeowners that have
a. between 1,200 and 2,000 square feet of living space,
b. between 1,800 and 2,500 square feet of living space.

7.103   Former President Jimmy Carter has been actively involved in Habitat for Humanity, a charitable organization that assists the poor in building their own homes. Their average cost of supplies for houses built in the United States and Canada is $28,000. Assume that the distribution of costs is approximately

normal with a standard deviation of $2,500. Determine the percentage of homes for which the supplies cost
a. no more than $25,000,
b. from $20,000 to $30,000.

7.104 With Americans becoming more health conscious, many food companies are touting the benefits of oat bran in a wide variety of products from muffins to potato chips. One such company makes an oat bran doughnut that has an average of 5 grams of oat bran. Assume that the amount of oat bran per doughnut is approximately normally distributed with a standard deviation of 0.2 gram. Find the percentage of doughnuts that contain
a. more than 5.3 grams of oat bran,
b. at least 4.5 grams of oat bran.

7.105 Concerning the preceding exercise, 20 percent of the doughnuts will contain more than what amount of oat bran?

7.106 x is a normal random variable with a mean of 25. Determine its standard deviation, if 33 percent of the time x has a value below 23.9.

MINITAB Assignments

Ⓜ 7.107 A certain brand of electric range has a length of life that is approximately normal with a mean life of 15.7 years and a standard deviation of 4.2 years. Use MINITAB to find the proportion of ranges that last
a. fewer than 13.5 years,
b. more than 17 years,
c. between 10 and 20 years.

Ⓜ 7.108 The ranges referred to in the above problem are guaranteed for five years. Use MINITAB to determine the proportion of ranges that have to be replaced during the warranty period.

Ⓜ 7.109 For the ranges discussed in Problem 7.107, 95 percent of them will last at least how many years? Use MINITAB to solve.

Ⓜ 7.110 Use MINITAB to find the 80th percentile for a normal distribution with a mean of 517 and a standard deviation of 86.

7.4   Using Normal Distributions to Approximate Binomial Distributions

In Section 6.2 the binomial probability distribution was introduced to model binomial experiments. These consist of the repetition of n identical and independent trials, each of which results in either a success, with probability p, or a failure, with probability q. We saw that the binomial probability distribution is given by the formula

$$P(x) = C(n, x)p^x q^{n-x}, \qquad \text{for } x = 0, 1, 2, 3, \ldots n, \qquad \textbf{(7.4)}$$

where x is the number of successes obtained on the n trials.

Figure 7.5
Binomial distribution, $n = 10, p = 0.5$.*

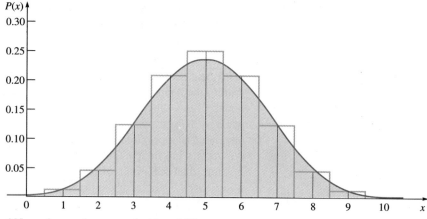

*Normal curve has mean 5, stdev. 1.58.

De Moivre spent several years working on the problem of estimating binomial probabilities. In 1733, at the age of 66, he published a complete solution to the use of normal curves in approximating binomial distributions.

When n is large, the evaluation of Formula 7.4 is laborious and time consuming. Fortunately, the evaluation of binomial probabilities can often be closely approximated by utilizing normal probability distributions. The viability of this is suggested by Figure 7.5, which shows a probability histogram for a binomial random variable with $n = 10, p = 0.5, \mu = np = 5$, and $\sigma = \sqrt{npq} = 1.58$. In the figure, we have also superimposed a normal distribution with the same mean and standard deviation.

The degree of accuracy in using a normal distribution to approximate a binomial distribution depends on the values of n and p. We will elaborate on this at the end of the section, but first we will introduce the technique employed. Let's reconsider Example 6.7 in which we were interested in determining the probability that 100 tosses of a fair coin will result in exactly 50 heads. Using the binomial probability function, the probability of 50 heads is given by

$$P(50) = C(100, 50)(0.5)^{50}(0.5)^{50}.$$

Because of the difficulty in evaluating the combination $C(100, 50)$, we used MINITAB to determine $P(50)$, and this was found to equal 0.0796. If MINITAB is not available, then $P(50)$ can be easily approximated by proceeding as follows.

1. *We need to select a particular normal curve to use for approximating the binomial distribution.*
   Since a binomial random variable has a mean $\mu = np$ and standard deviation $\sigma = \sqrt{npq}$, we will use the normal curve with

   $$\mu = 100(0.5) = 50 \quad \text{and} \quad \sigma = \sqrt{100(0.5)(0.5)} = 5.$$

2. *For continuous random variables, probabilities of only intervals can be computed. Consequently, to find P(50) an appropriate interval must be selected to represent $x = 50$.*
   As one might expect, the best choice is the interval that has 50 as its midpoint and is one unit in length. Thus, we make the following adjustment:

Binomial Value	Represented By	Normal Interval
$x = 50$	$\longleftrightarrow$	$49.5 \le x \le 50.5$

3. *Make a sketch of the normal curve selected in Step 1, and calculate the probability of the interval determined in Step 2.*

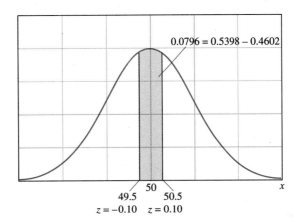

To find the area of the shaded region, we must determine the z-values for 49.5 and 50.5.

$$z = \frac{49.5 - 50}{5} = -0.10 \qquad \text{and} \qquad z = \frac{50.5 - 50}{5} = 0.10$$

From Table 3, the cumulative area to the left of $z = -0.10$ standard deviations is 0.4602, and the area to the left of $z = 0.10$ standard deviations is 0.5398. Therefore,

$$P(50) \approx P(49.5 \leq x \leq 50.5) = 0.5398 - 0.4602 = 0.0796.$$

Note that this is the same result as that obtained by using MINITAB in Chapter 6.

In Part 2 of Problem 6.7, MINITAB was used to obtain the probability of tossing between 45 and 55 heads, inclusive. To approximate this by using a normal distribution, each value of the binomial variable x is represented by a unit interval with the value of x as its midpoint.

Binomial Value	Represented By	Normal Interval
$x = 45$	← ⟶	$44.5 \leq x \leq 45.5$
$x = 46$	← ⟶	$45.5 \leq x \leq 46.5$
$x = 47$	← ⟶	$46.5 \leq x \leq 47.5$
$x = 48$	← ⟶	$47.5 \leq x \leq 48.5$
$x = 49$	← ⟶	$48.5 \leq x \leq 49.5$
$x = 50$	← ⟶	$49.5 \leq x \leq 50.5$
$x = 51$	← ⟶	$50.5 \leq x \leq 51.5$
$x = 52$	← ⟶	$51.5 \leq x \leq 52.5$
$x = 53$	← ⟶	$52.5 \leq x \leq 53.5$
$x = 54$	← ⟶	$53.5 \leq x \leq 54.5$
$x = 55$	← ⟶	$54.5 \leq x \leq 55.5$

When all the unit intervals are taken together, we obtain one interval starting at 44.5 and ending at 55.5. Therefore,

$$P(45) + P(46) + \cdots + P(55) \approx P(44.5 \leq x \leq 55.5).$$

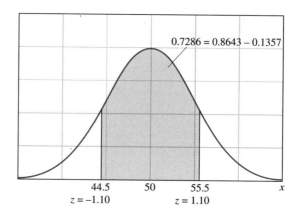

To find $P(44.5 \leq x \leq 55.5)$, we must first determine the z-values for 44.5 and 55.5.

$$z = \frac{44.5 - 50}{5} = -1.10 \quad \text{and} \quad z = \frac{55.5 - 50}{5} = 1.10$$

From Table 3, the cumulative area to the left of -1.10 standard deviations is 0.1357, and the area to the left of 1.10 standard deviations is 0.8643. The shaded area is thus

$$0.8643 - 0.1357 = 0.7286.$$

This approximation differs by only 0.0002 from MINITAB's value of 0.7288 that was obtained in Chapter 6.

The following summarizes the procedure for approximating a binomial distribution with a normal distribution.

Using a Normal Distribution to Approximate a Binomial:
1. For the approximating distribution, use the normal curve with mean $\mu = np$ and standard deviation $\sigma = \sqrt{npq}$.
2. Represent each value of x (the number of successes) by the interval from $(x - 0.5)$ to $(x + 0.5)$.
3. Use the normal curve selected in Step 1 to calculate the probability of the collection of unit intervals in Step 2.

Example 7.15   In 1988 the NCAA released the results of a year-long study on how the lives of male college athletes are affected by their participation in sports. Among the many results reported, it was found that business was the most popular intended major at time of enrollment for those playing either football or basketball, with 37 percent selecting

(© Melchior DiGiacomo/The Image Bank.)

this field. Assuming that this percentage is correct, find the probability that a random sample of 500 male college athletes in these 2 sports would contain more than 200 who intended to major in business at the time of enrollment.

Solution

This problem can be thought of as a binomial experiment for which a trial is the selection of 1 male college athlete, and a success is the selection of business as his intended major at time of enrollment. Then $n = 500$, $p = 0.37$, and the random variable x is the number in the sample who intended to major in business. We need to find the following.

$$P(x > 200) = P(201) + P(202) + P(203) + \cdots + P(500)$$

Since the calculation of these values using the binomial distribution would be overwhelming, we will use a normal approximation.

1. *A normal curve needs to be selected to use as the approximating distribution.*
   We use the normal curve with mean $\mu = np = 500(0.37) = 185$ and standard deviation $\sigma = \sqrt{npq} = \sqrt{500(0.37)(0.63)} = 10.80$.
2. *An interval must be selected to represent the whole numbers 201, 202, 203, . . . 500.*
   Since the first value, 201, is represented by the interval $200.5 \le x \le 201.5$, and the last value, 500, is represented by $499.5 \le x \le 500.5$, we will use the interval $200.5 \le x \le 500.5$.

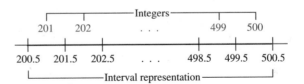

3. *Using the normal curve selected in Step 1, we now calculate the probability of the interval in Step 2.*

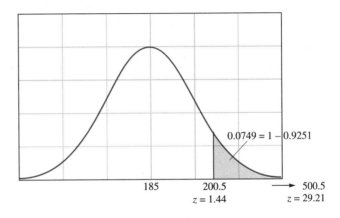

The z-values for 200.5 and 500.5 are

$$z = \frac{200.5 - 185}{10.80} = 1.44 \quad \text{and} \quad z = \frac{500.5 - 185}{10.80} = 29.21$$

From Table 3, the area to the left of 1.44 standard deviations is 0.9251. A z-value of 29.21 standard deviations is beyond the range of the table. When this occurs, for all practical purposes the corresponding area can be considered as 1. Thus, we have the following.

$$
\begin{aligned}
P(201) + P(202) + \cdots + P(500) &\approx P(200.5 \le x \le 500.5) \\
&= P(1.44 \le z \le 29.21) \\
&= 1 - 0.9251 \\
&= 0.0749
\end{aligned}
$$

Ⓜ Example 7.16   Use MINITAB to obtain the requested probability in Example 7.15.

Solution

We need to calculate the probability of obtaining more than 200 successes in a binomial experiment with $n = 500$ trials and $p = 0.37$. Even MINITAB cannot use the binomial probability distribution to solve this problem, since the value of n is so large. We must proceed as in Example 7.15 and use a normal approximation (with $\mu = 185$ and $\sigma = 10.80$) to evaluate $P(200.5 \le x \le 500.5)$. To use MINITAB, the command **CDF** needs to be applied to the values 200.5 and 500.5 to determine the cumulative probability to the left of each value. Instead of using the command **CDF** twice, it is easier to first store 200.5 and 500.5 in column C1 and then apply the command once to C1.

```
SET C1
200.5 500.5
END
CDF C1;
NORMAL 185 10.80.
```

In response to the above commands, MINITAB will produce the output shown in Exhibit 7.4. Using the output in Exhibit 7.4, we have the following.

Exhibit 7.4

```
MTB > SET C1
DATA> 200.5 500.5
DATA> END
MTB > CDF C1;
SUBC> NORMAL 185 10.80.
   200.5000      0.9244
   500.5000      1.0000
```

$$P(200.5 \le x \le 500.5) = P(x \le 500.5) - P(x < 200.5)$$
$$= 1 - 0.9244$$
$$= 0.0756$$

This answer differs slightly from the one obtained in the previous example because we rounded off the z-values to 2 decimal places, while MINITAB retained additional digits.

When a normal distribution is used to approximate a binomial distribution, the degree of accuracy is affected by the values of p and n. In general, the approximation improves as p gets closer to 0.5 and as n increases. When p is exactly 0.5, the graph of the binomial distribution is symmetric and similar in shape to a normal curve. This is true even for small values of n (see Figure 7.5, page 276). When the value of p is not equal to 0.5, the binomial distribution is not symmetric, and its graph becomes more skewed as p approaches its extreme values of 0 and 1. Figure 7.6 shows a probability histogram of the binomial distribution with $n = 10$ and $p = 0.8$. Notice that the distribution is negatively skewed, that is, skewed to the left.

Figure 7.6
Binomial distribution, $n = 10$, $p = 0.8$.

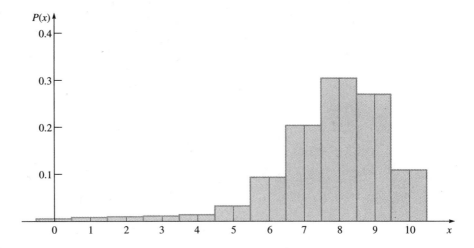

Even when p is not near 0.5, a normal distribution can still provide good approximations, provided that n is sufficiently large. There are various rules of thumb concerning appropriate combinations of n and p. Generally, good approximations can be expected when both $np \ge 5$ and $nq \ge 5$. Using this rule as a guide, n can be as small as 10 when $p = 0.5$. However, if $p = 0.8$, then n should be at least 25 in order to have $n(0.8) \ge 5$ and $n(0.2) \ge 5$.

Section 7.4 Exercises

Using Normal Distributions to Approximate Binomial Distributions

7.111 For the experiment of tossing a fair coin 20 times, find the probability of obtaining exactly 10 heads by using

    a. the binomial probability distribution formula,

    b. the table of binomial probabilities (Table 1, Appendix A),

    c. a normal curve approximation.

7.112 Find the probability of obtaining from 8 to 13 heads in 20 tosses of a fair coin. Solve this by using

    a. the table of binomial probabilities,

    b. a normal curve approximation.

7.113 x is a binomial random variable with $n = 15$ and $p = 0.6$. Use a normal curve approximation to find the probability of obtaining exactly 9 successes. Compare your answer with that given by using Table 1.

7.114 For the binomial random variable referred to in Exercise 7.113, evaluate $P(7 \leq x \leq 10)$ by using

    a. the table of binomial probabilities,

    b. a normal curve approximation.

7.115 According to a survey by the U.S. Department of Education, about 17 percent of America's public school teachers moonlight during the school year to supplement their salaries (*Lebanon Daily News,* February 3, 1989). If a public school district has 952 teachers, what is the probability that more than 175 moonlight during the school year?

7.116 During the 1980s, Caesarean deliveries increased significantly from 17 percent in 1980 to 25 percent in 1989 (*USA TODAY,* January 11, 1990). Suppose a community had 254 births in 1989. Determine the probability that fewer than 50 were by Caesarean delivery.

7.117 Twenty percent of consumer complaints against U.S. airlines in 1987 pertained to baggage problems *(Statistical Abstract of the United States).* What is the probability that for a random sample of 1,000 airline complaints, at least 220 are concerned with baggage?

7.118 Company cafeterias often subsidize the cost of their services to employees. The Society for Foodservice Management says that three of four firms follow this practice (The *Wall Street Journal,* January 9, 1990). In a particular geographical region there are 40 company cafeterias. Find the probability that

    a. exactly 30 offer subsidies

    b. more than 32 offer subsidies.

7.119 A restaurant has found that about 1 customer in 5 will order a dessert with the meal. If the restaurant served 580 meals on a certain day, find the probability that more than 125 were accompanied by dessert.

7.120 According to *Hippocrates Magazine* (March/April, 1989), 37 percent of all Americans take multiple vitamins regularly. If a company has 2,548 employees, what is the probability that fewer than 900 take a multiple vitamin regularly?

7.121 Approximately 90 percent of new automobiles now come equipped with all-season radial tires. A rental car agency has a fleet of 3,375 new cars of various makes and models. Find the probability that the number of cars equipped with this type of tire will be between 3,000 and 3,075, inclusive.

MINITAB Assignments

Ⓜ 7.122 In Exercise 6.119, the reader was asked to explain why it is very unlikely that a coin is fair if it showed 4,800 heads in 10,000 tosses. Use MINITAB to calculate the probability of obtaining as few as 4,800 heads in 10,000 tosses of a fair coin.

Ⓜ 7.123 A national chain of department stores has found that 12 percent of its charge account customers fail to make their monthly payment on time. Fifty accounts will be selected at random. Use MINITAB to find the probability that there will be fewer than 5 overdue accounts. Solve this by using the main command **CDF** with the following subcommands.

    a. **BINOMIAL**   b. **NORMAL**

Ⓜ 7.124 The Internal Revenue Service's annual report for fiscal 1988 revealed that taxpayers do not have great success in suits against the IRS. The IRS won about 85 percent of cases involving taxpayers' suits for refunds in the claims and district courts (The *Wall Street Journal,* April 5, 1989). Use MINITAB to determine the probability that for a sample of 80 such cases, the taxpayer was victorious in at least 19. Solve this by using the main command **CDF** with the following subcommands.

    a. **BINOMIAL**   b. **NORMAL**

Looking Back

Chapter 7 is concerned with **continuous random variables** and their **probability distributions.** While a discrete random variable can never have more values than the counting numbers, a continuous variable can assume all the values in some interval on the real number line. They are often measurement variables, such as time, distance, velocity, length, and weight. As with discrete variables, probability distributions are used in calculating probabilities of continuous variables. However, for a continuous random variable we only calculate the probability that the variable will assume a value within a specified interval. Moreover, this probability is obtained by determining the area under the graph of the random variable's probability distribution.

$P(a \leq x \leq b)$ **for a Continuous Random Variable:**
If x is a continuous random variable with probability distribution $f(x)$, then $P(a \leq x \leq b)$, the probability that x will assume a value within the interval from $x = a$ to $x = b$, equals the area under the graph of $f(x)$ from $x = a$ to b.

    Normal probability distributions are the most important of the continuous distributions. In addition to their vital role in the theory of statistics, normal distributions serve as mathematical models for many applications. There are an infinite number of different normal distributions, and a particular one is determined by two parameters: its mean μ, and its standard deviation σ. The normal random variable with $\mu = 0$ and $\sigma = 1$ is called the **standard normal random variable**

and is denoted by z. With only a table of areas for the z-distribution, we can determine probabilities for every normal random variable by using the following transformation.

> If x is a normal random variable with mean μ and standard deviation σ, then
>
> $$z = \frac{x - \mu}{\sigma}$$
>
> has the standard normal distribution with mean 0 and standard deviation 1.

Normal distributions can often be used to approximate probabilities concerning binomial random variables. When n is large, the evaluation of binomial probabilities is laborious and time consuming, while a normal approximation can be obtained with relative ease.

> **Using a Normal Distribution to Approximate a Binomial:**
> 1. For the approximating distribution, use the normal curve with mean $\mu = np$ and standard deviation $\sigma = \sqrt{npq}$.
> 2. Represent each value of x (the number of successes) by the interval from $(x - 0.5)$ to $(x + 0.5)$.
> 3. Use the normal curve selected in Step 1 to calculate the probability of the collection of unit intervals in Step 2.

The degree of accuracy is dependent on the values of n and p. In general, the approximation improves as p gets closer to 0.5 and as n increases.

Key Words

In reviewing this chapter, you should be able to define, explain, and illustrate each of the following.

continuous random variable *(page 255)*

probability distribution of a continuous random variable *(page 256)*

area under the graph of $f(x)$ *(page 256)*

basic properties of a continuous probability distribution *(page 257)*

normal probability distribution *(page 257)*

properties of a normal curve *(page 258)*

standard normal variable z *(page 259)*

standard normal probability distribution *(page 259)*

z-transformation *(page 268)*

normal approximation for a binomial distribution *(page 275)*

Ⓜ *MINITAB Commands*

```
CDF _; (page 272)
NORMAL _ _.
```

```
PDF _ _; (page 266)
NORMAL _ _.
```

PLOT _ _ *(page 266)*   SET _ *(page 266)*

INVCDF _; *(page 272)*   END *(page 266)*

NORMAL _ _ .

Review Exercises

7.125 Determine the area under the standard normal curve that lies to the left of $z = 2.17$.

7.126 Find the area under the standard normal curve that lies to the right of $z = 1.95$.

7.127 For the standard normal variable z, determine the following.
a. $P(-1.98 < z < -0.95)$ b. $P(-1.98 < z < 0.95)$
c. $P(0.95 < z < 1.98)$

7.128 Find the probability that the standard normal variable z will have a value that lies within 2.5 standard deviations of its mean. Compare this result with that given by Chebyshev's theorem.

7.129 Find the z-value for which the standard normal curve area to the right of z is 0.9726.

7.130 Determine the z-value such that the standard normal curve area between $-z$ and z is 0.7198.

7.131 Obtain the following percentiles for the standard normal distribution.
a. P_{70} b. P_{15}

7.132 Determine the following values of z_a.
a. $z_{.02}$ b. $z_{.08}$

7.133 Determine Q_3, the third quartile, for a normal distribution with mean $\mu = 150$ and standard deviation $\sigma = 10$.

7.134 The random variable x has a normal distribution with $\mu = 78.4$ and $\sigma = 3.6$. Find the following probabilities.
a. $P(x \geq 70.3)$ b. $P(73.9 \leq x \leq 89.2)$

7.135 The average U.S. per capita consumption of yogurt in 1987 was 4.6 pounds (U.S. Department of Agriculture). Assume that the per capita consumption can be approximated with a normal distribution having a standard deviation of 1.3 pounds. Determine the proportion of the population that consumed less than 1 pound of yogurt in 1987.

7.136 With reference to Exercise 7.135, 12 percent of the population consumed less than what amount of yogurt?

7.137 Use a normal approximation to the binomial distribution to find the probability of tossing more than 225 ones in 1,200 throws of a fair die.

7.138 The time required by a certain shop to serve a customer a pizza has approximately a normal distribution with a mean of 12.3 minutes and a standard deviation of 1.7 minutes. To attract customers, the shop does not charge for a pizza if a customer has to wait more than 15 minutes after placing an order. What proportion of pizzas are free because of this offer?

7.139 In its latest behavioral risk factor survey, the Centers for Disease Control found that only 15 percent of adults in Utah smoke (*The Bangor Daily News,* July 14, 1989). If 500 people in this state are surveyed, determine the probability that the number of smokers will be between 60 and 80, inclusive.

7.140 A tax return preparation firm has found that it takes an average of 57.8 minutes for its employees to complete a standard return. The distribution of times is approximately normal with a standard deviation of 9.6 minutes. Find the percentage of returns completed by this firm that require between 50 and 60 minutes to complete.

7.141 For the tax returns referred to in Exercise 7.140, 10 percent of the returns will require more than how many minutes to complete?

7.142 Real estate prices slumped in several parts of the Northeast in 1989. In spite of this, the average price of homes sold in Pennsylvania's Lancaster County increased from $87,141 in 1988 to $100,661 in 1989 (Lancaster County Association of Realtors). Assume that the distribution of home prices in this region during 1989 can be approximated with a normal distribution having a standard deviation of $24,000. Find the percentage of homes that cost between $50,000 and $150,000.

7.143 It is estimated that about 90 percent of the population in the United States will contract some form of gum disease during their lives. For a high school class of 682 students, what is the probability that more than 600 will eventually suffer some type of gum disease?

7.144 A normal pregnancy and hospital delivery now costs an average of $4,334 (*USA TODAY,* January 11, 1990). Assume that the distribution of costs is approximately normal with a standard deviation of $125. What proportion of the time will the cost exceed $4,500?

MINITAB Assignments

Ⓜ 7.145 *Newsweek* (March 27, 1989) reported that 60 percent of young children have blood lead levels that may impair their neurological development. In a class of 50 young children, what is the probability that at least 25 have a blood lead level that may impose this hazard? Solve using the MINITAB command **CDF** and the following subcommands.
   a. **BINOMIAL**   b. **NORMAL**

Ⓜ 7.146 x is a normal random variable with $\mu = 49.6$ and $\sigma = 4.8$. Use MINITAB to find
   a. $P(x < 43.3)$,   b. the 3rd quartile.

Ⓜ 7.147 Use MINITAB to determine
   a. $P(z \leq 3.57)$;
   b. P_{91}, the 91st percentile for the standard normal distribution.

Ⓜ 7.148 Using MINITAB, determine
   a. $P(z < -0.62)$,
   b. $P(z > 1.86)$,
   c. $P(-1.64 \leq z \leq 1.11)$.

Ⓜ 7.149 Use MINITAB to obtain the 83rd percentile for a normal distribution with mean 154 and standard deviation 12.3.

Ⓜ 7.150 Using MINITAB, find the value of $z_{.04}$.

Ⓜ 7.151 Use MINITAB to sketch the graph of the normal distribution with $\mu = 79$ and $\sigma = 8$.

Ⓜ 7.152 Using MINITAB, sketch a normal curve with a mean of 750 and a standard deviation of 32.

(© Andy Levin/Photo Researchers, Inc.)

8 Sampling Methods and Sampling Distributions

◀| *Unsafe Levels of Aflatoxin in Corn?*

Aflatoxin is a naturally occurring toxin that animal studies have shown to be a potent carcinogen. The dry, hot summer during 1988 resulted in environmental conditions that were conducive to the growth of the fungus that produces aflatoxin in corn crops. In the Letters to the Editor column of the Wall Street Journal *(March 20, 1989), the president of the Corn Refiners Association responded to an article that had questioned the safety of the country's food supply. Critical of the sampling methods that had been employed, he stated, "Regarding the charge, without attribution, that one-third of the official corn samples in Iowa and Illinois had 'dangerous levels of aflatoxin': Official samples are not selected randomly and are not representative of the corn crop as a whole. They are disproportionately made up of corn already suspected of containing aflatoxin as a result of initial screening. In fact, 80 percent of randomly sampled corn in areas susceptible to aflatoxin met the Food and Drug Administration standard as safe for humans."*

Looking Ahead

The main application of statistics is the making of inferences. In many fields of application, one is interested in obtaining information about a large collection of elements, called the **population.** Costs, time constraints, and other considerations usually dictate that data can be collected on only a relatively small portion of the population—a **sample.** Since the sample is the basis for making generalizations about the entire population, it is crucial that a proper sampling method is employed to select the sample. Inferences will have no validity if they are based on an improperly designed sampling procedure. This point was made at the beginning of the chapter by the president of the Corn Refiners Association in his letter to the *Wall Street Journal* concerning the safety of the nation's corn supply: "Official samples are not selected randomly and are not representative of the corn crop as a whole."

289

To produce valid inferences, the **sampling method** used must result in a **scientific sample** that has been selected according to known probabilities. The most frequently used sampling method is **random sampling,** and this method will be discussed in detail in Section 8.1. Although random sampling will nearly always be assumed in the chapters to follow, Section 8.2 will briefly consider three other scientific sampling methods: **systematic sampling, stratified random sampling,** and **cluster sampling.**

In the chapters to follow, many inferences will be made about a population **parameter**—a numerical characteristic of the population. For example, perhaps the current federal administration wants to estimate the **percentage** of registered voters who favor a tax increase to reduce the national budget deficit. A nutritionist might need to know the **mean** amount of fat contained in chocolate doughnuts of a particular brand. A cereal company may want to determine the **range** for the number of raisins placed in boxes of raisin bran cereal. A manufacturer might need to estimate the **standard deviation** of the amperage required to trip a circuit breaker.

Population parameters, such as percentage, mean, range, and standard deviation, are estimated by computing a similar characteristic of the sample, called a **statistic.** When a sample statistic is used to estimate a population parameter, it is not expected that its value will exactly equal the parameter. Consequently, a measure of the inference's reliability should be given. This will be in the form of some bound on the error of the estimate. To obtain this error bound, it is necessary to know the **probability distribution of the sample statistic.** These distributions of sample statistics are called **sampling distributions** and constitute a main topic of this chapter. In discussing sampling distributions, we will introduce one of the most important theorems in statistics—the **Central Limit theorem.**

8.1   Random Sampling

Newsweek reported on a study that suggested left-handers are more accident prone and may live an average nine years less than right-handers (April 15, 1991). The researchers "found that the typical right-hander lived to be 75. The average lefty died at 66." Their conclusions were based on a study of 987 death certificates from 2 southern California counties. Based on the sample used, do you think that the researchers' conclusions validly apply to all left-handers in general?

To make valid inferences about a population on the basis of information contained in a sample, it is necessary that careful attention be given to the method of selecting the sample. There are many different sampling methods, and some of these are discussed briefly in the next section. However, nearly all the methodology considered in an introductory statistics course assumes that a **random sample** has been selected. In addition to being the most frequently used sampling procedure, random sampling is the simplest to define.

Random Sample*

A sample of n elements selected from a population of N elements is a **random sample** if it is selected in such a way that each possible sample of size n has the same probability of being selected.

---

* The more precise term is a **simple random sample,** but the adjective "simple" is often omitted in reference to this type of sampling method.

The determining factor in whether a sample is random is not the composition of the sample, but *how* it was selected. The number of possible samples of size n that can be selected from a population of N elements is given by the combination $C(N, n)$. We will obtain a random sample of size n if the method of selection assures that each of the $C(N, n)$ possible samples has the same probability, namely, $1/C(N, n)$, of being selected. These concepts are illustrated in Example 8.1.

Example 8.1

A company specializing in the cleanup of hazardous waste materials structures its crews in teams of 5. Company policy dictates that after a crew completes an assignment, 2 members will be randomly selected and subjected to an extensive series of medical tests. Suppose we designate the members of a particular team as A, B, C, D, and E. The number of possible samples of size $n = 2$ that can be selected from this population of size $N = 5$ is

$$C(5, 2) = \frac{5!}{2!(5 - 2)!} = 10.$$

The 10 possible samples are as follows.

AB   AC   AD   AE   BC   BD   BE   CD   CE   DE

If the 2 selected members are to constitute a random sample, then the selection method must be such that each of these 10 possible samples has an equal probability (1/10) of being the chosen sample. One way of accomplishing this would be to list each possible sample on 10 slips of paper; place them in a container; thoroughly mix the slips; and, without looking, select one slip. While this method of selection is feasible for a small population of size $N = 5$, it is unrealistic when N is large. An easier and more practical alternative is illustrated in the next example.

(© Tom Kelly/Phototake, NYC.)

Example 8.2

With reference to the hazardous waste removal company, suppose a particular contract required the employment of 4 teams, each with 5 workers. At the conclusion of the job, management decides that 8 workers will be randomly selected from the 20 for submission to the series of medical tests.

The number of possible samples of size $n = 8$ that can be selected from a population of size $N = 20$ is

$$C(20, 8) = \frac{20!}{8!(20 - 8)!} = 125{,}970.$$

Clearly, it is no longer feasible to select the sample by first listing each of the 125,970 possible samples. A much easier procedure that can be shown to be equivalent involves listing only the 20 elements of the population. The required steps for then selecting a random sample of 8 workers from the 20 are outlined below.

1. *Construct a* **frame** *of the population, and number each member, as shown on the next page. A frame is just a listing of all the population's elements.*

1 Shubert, R.	11 Fitz, T.
2 Martinez, L.	12 Kim, A.
3 Kowalski, M.	13 Epstein, L.
4 Bernakel, A.	14 St. Lawrence, J.
5 O'Toole, W.	15 Baxter, A.
6 Durand, M.	16 Lehman, W.
7 Chao, L.	17 MacDonald, D.
8 Vavra, E.	18 Ferraro, F.
9 Schmidt, D.	19 Rao, D.
10 Hagopian, B.	20 Ellsworth, A.

2. *Using a randomizing device, select n = 8 of the numbers from 1 through 20.*
   There are many possibilities for the randomizing technique. It could consist of 20 slips of paper numbered from 1 through 20 that have been thoroughly mixed and placed in a container. Alternatively, a table of random numbers could be employed. Today, however, the wide availability of random number generators on pocket calculators and personal computers makes their use the preferred choice. We have used MINITAB to make the selection. For those interested, the commands required to accomplish this are detailed in Example 8.3. MINITAB selected the numbers 9, 4, 19, 2, 20, 13, 10, and 12.

3. *Identify the individuals associated with the 8 numbers selected in Step 2.*
   The following names are associated with the $n = 8$ randomly selected numbers and should be the 8 workers subjected to the battery of tests.

9 Schmidt, D.	20 Ellsworth, A.
4 Bernakel, A.	13 Epstein, L.
19 Rao, D.	10 Hagopian, B.
2 Martinez, L.	12 Kim, A.

ⓂExample 8.3   Use MINITAB to select a random sample of 8 numbers from the integers 1 through 20.

Solution

The first step is to create a column containing the numbers in the population. In this example, these are the integers from 1 to 20.

```
SET C1
1:20
END
```

The **SAMPLE** command is used to randomly select without replacement a number of elements from a specified column. To obtain a random sample of size $n = 8$ from the elements of C1, and then have the results placed in column C2, type

```
SAMPLE 8 C1 C2    # Places in C2 8 nos. from C1.
```

The contents of column C2 can be obtained by using the **PRINT** command. The output produced by MINITAB is displayed in Exhibit 8.1, from which we see that the random sample consists of the numbers 9, 4, 19, 2, 20, 13, 10, and 12.

Exhibit 8.1

```
MTB > SET C1
DATA> 1:20
DATA> END
MTB > SAMPLE 8 C1 C2
MTB > PRINT C2

C2
      9      4     19      2     20     13     10     12
```

Before closing this section, it should be noted that our discussion of a random sample has assumed that elements are selected without replacement from a population of finite size N. If **sampling with replacement** is used, or if the **population is infinite,** then the sample is random if it is composed of observations that have the same probability distribution and are **statistically independent.** For instance, in subsequent chapters, we will often assume that $X_1, X_2, X_3, \ldots, X_n$ is a random sample selected from a normal population. In such cases, the random measurements X_1, X_2, $X_3, \ldots, X_n$ are assumed to be independent random variables with the same normal distribution.

Section 8.1 Exercises

Random Sampling

In Exercises 8.1–8.6, a population size N and a sample size n are given. Determine the number of possible samples of the specified size.

8.1 $N = 8, n = 3$ 8.2 $N = 6, n = 2$

8.3 $N = 10, n = 2$ 8.4 $N = 15, n = 4$

8.5 $N = 100, n = 5$ 8.6 $N = 200, n = 2$

8.7   A random sample of size 4 will be selected from a population of 25 elements. What is the probability of each possible sample?

8.8   A random sample of 3 elements will be selected from a population of 16 elements. What is the probability of each possible sample?

8.9   A population consists of four elements labeled $W, X, Y,$ and Z. A random sample of two elements will be selected.
   a. Determine the number of different samples.
   b. List the possible samples.
   c. What is the probability of each possible sample?

8.10 A population consists of six elements that have been labeled $A, B, C, D, E,$ and F. A random sample of four elements will be selected.
   a. Determine the number of different samples.
   b. List the possible samples.
   c. What is the probability of each sample?

8.11 With reference to the population in Exercise 8.9, find the probability that a random sample of two elements will contain the element (a) *W*, (b) *X*, (c) *Y*, (d) *Z*.

8.12 For the population in Exercise 8.10, find the probability that a random sample of four elements will contain the element (a) *A*, (b) *B*, (c) *C*, (d) *D*, (e) *E*, (f) *F*.

8.13 A mathematics department has seven members, and its chairperson has received funding sufficient for sending only two faculty members to the annual meeting of the Mathematical Association of America. The chairperson decides to randomly select the two members who will go.
   a. Outline a procedure that the chairperson could use to assure that a random sample is selected.
   b. Two of the department members are married to each other. What is the probability that they will both be selected to go to the annual meeting?

8.14 According to the Maine Audubon Society, there are 109 bald eagle nest sites in the state. Suppose a sample of 25 sites will be randomly selected to observe nest activity.
   a. Would it be feasible to actually list all possible samples of 25 sites? Explain.
   b. Detail a practical procedure that could be used to assure that the 25 selected sites constitute a random sample.

MINITAB Assignments

Ⓜ 8.15 With reference to Exercise 8.14, use MINITAB to randomly select 25 of the integers from 1 through 109.

Ⓜ 8.16 A city has 87 food markets, and each Wednesday a newspaper randomly selects 10 stores to visit and determine the cost of a basket of goods. The results are then published in a special food section of the Thursday edition. Assume that a frame is available with the 87 stores numbered from 1 through 87. Use MINITAB to select a random sample of 10 stores.

Ⓜ 8.17 A computer manufacturer has developed a new personal computer model. From a production run of 240 machines, a random sample of 20 computers will be tested to see if FCC specifications for radio interference are being met. The 240 computers have consecutive serial numbers whose last 3 digits are from 510 through 749. Use MINITAB to randomly select 20 computers to be tested.

8.2 Other Scientific Sampling Methods

The concept of a random sample is simple to define and is the basis for nearly all the statistical methods considered in a first course. Its selection, however, is not always feasible, and some other scientific sampling method must be employed. In this section we will briefly discuss three sampling methods that are sometimes preferred to random sampling.

Systematic Sampling

Consider the problem of interviewing a sample of adults as they exit a shopping mall on a particular Saturday. It is not practical to obtain a listing (frame) of the population of shoppers for this time period, and consequently, a random sample is impossible. Similarly, a farmer who wants to sample his crop for the presence of a disease cannot realistically list and number the thousands of cabbages in his field. A similar problem is faced by a quality control inspector who needs to sample periodically from several thousand manufactured items. In these instances one often resorts to **systematic sampling** in which every kth item is selected for inclusion in the sample. Even when a sampling frame exists, a systematic sample is sometimes utilized because of its simplicity and ease of implementation. The method for selecting a systematic sample is illustrated in Example 8.4.

Example 8.4   An insurance agency has 8,000 accounts, each of which is represented by a file arranged alphabetically. An auditor is assigned the task of sampling 5 percent of the accounts. To obtain a 5 percent systematic sample, the following procedure is followed.

1. Since $\frac{100\%}{5\%} = 20$, one of every 20 files will need to be inspected.
2. So that each file will have an equal chance of being selected, a randomizing device is used to determine a starting position from the group consisting of the first 20 files. Let's assume that this yielded the number 7.
3. The first file selected will be the 7th, followed by every twentieth file thereafter. That is, the sampled files will be the 7th, 27th, 47th, . . . 7987th.

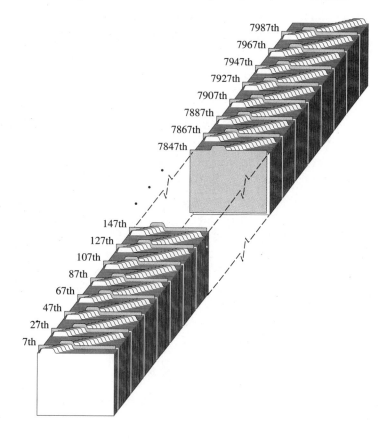

> **Systematic Sample:**
> To obtain a $(\frac{1}{k})100\%$ **systematic sample,** choose a starting element at random from the first k population elements, and thereafter select every kth element from the population.

A major risk of using a systematic sample is the possibility that the population may possess a cyclical characteristic that might introduce a bias not inherent in a random sample. For example, in sampling the daily receipts of a convenience store, it would not be appropriate to select every seventh day, since this would result in receipts for the same day of each week. Because the volume of sales is probably dependent on the particular week day, the resulting sample would not be representative of overall receipts.

Cluster Sampling

The cost of selecting a sample is often a primary consideration in choosing a sampling method. Sometimes a savings in both money and time can be realized by selecting elements in groups, rather than individually, from the population. Each group is called a **cluster,** and the sampling design is known as **cluster sampling.**

Example 8.5 A supermarket display contains fifty 1-pound packages of a particular brand of hot dogs, and each package contains 8 hot dogs. A food inspector plans to select 40 hot dogs for a fat analysis by first randomly selecting five packages (clusters) from the display, and then analyzing the entire contents of each package.

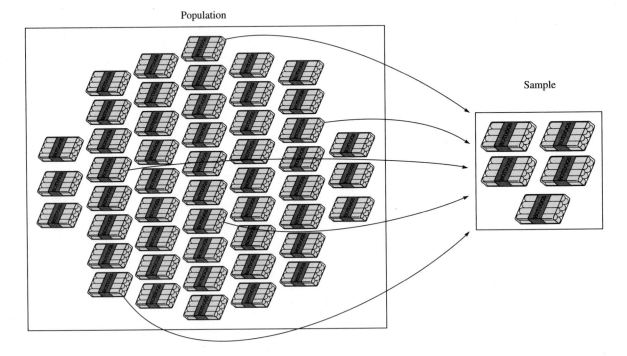

Population

Sample

> **Cluster Sample:**
> To obtain a **cluster sample,** the population elements are grouped in subsets, called **clusters.** A random sample of clusters is selected, and then the elements in each cluster are included in the sample.

Cluster sampling is prevalent in public opinion surveys and political polling in which a cluster may be a household, a building of occupants, or sometimes an entire group of neighboring housing units. By interviewing all adults in a cluster, considerable savings in time and money can be realized compared to randomly selecting people from a large geographical area. However, individuals within the same cluster are apt to be quite similar in many characteristics, such as consumer preferences, income, educational level, and political views. Consequently, cluster sampling is usually redundant and, thus, less informative about the population compared to a random sample of the same size.

Stratified Random Sampling

The focus in Chapter 9 is on estimating population parameters. We will see that the precision with which this can be done is dependent on the variability of the population. All other things being equal, more precise estimates can be realized for less variable populations. **Stratified random sampling** is a sampling design that uses this fact. In this sampling method, one attempts to divide a population into subpopulations that are relatively homogeneous in the variable of interest.

> **Stratified Random Sample:**
> To obtain a **stratified random sample,** the population is first divided into subpopulations, called **strata.** Then a random sample is selected from each stratum.

Example 8.6   A college official wants to estimate the average number of hours per week that students devote to homework. Because she believes that this figure will differ considerably among classes, stratified random sampling will be employed. The population

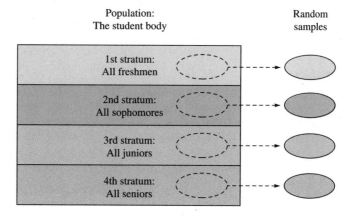

of students at this college will be grouped into four strata consisting of all freshmen, sophomores, juniors, and seniors. From each stratum, a random sample of students will then be selected. The resulting information can be combined to obtain an estimate that is expected to be more precise than that obtained from a random sample of the entire population.

Random sampling and the three sampling methods discussed in this section fall under the general heading of **scientific samples.** Such samples are selected on the basis of known probabilities, and thus they are also referred to as **probability samples.** The problem of choosing the most appropriate sampling method in a particular application involves many considerations that are outside the realm of a first course. In nearly all discussions of statistical methods that follow in this text, random sampling will be assumed.

Section 8.2 Exercises

Other Scientific Sampling Methods

For Exercises 8.18–8.23, state whether the sampling method employed is a systematic sample, cluster sample, or stratified random sample.

8.18 A supply of military ammunition consists of 1,000 boxes, with each box containing 50 rounds. To check if military specifications are being met, 10 boxes of ammunition are randomly selected, and each round from the 10 boxes is tested.

8.19 A large corporation wants to survey its employees to estimate the average amount that they would be willing to contribute to a company-assisted savings plan. The company will randomly select 50 employees from each of 3 groups consisting of management, skilled, and nonskilled workers.

8.20 A soft drink manufacturer has adopted a new bottling process. To check on the amount of carbonation, every twentieth bottle is removed from the bottling line, and the amount of carbon dioxide is measured.

8.21 The taxation board for a city wants to estimate the proportion of homes for which major improvements have been made without a building permit. To conduct a survey, officials will randomly select a sample of 100 city blocks, and then inspect each house in each selected block.

8.22 An auditing firm is going to check a company's accounts receivable by first grouping them into 2 categories consisting of all accounts with balances below $100, and those with balances of at least $100. The firm will then randomly select 50 accounts in the former category, and 200 accounts will be selected at random from those with balances of at least $100.

8.23 As part of its customer satisfaction program, a flooring installation company visits the site of every fifth sale and evaluates the quality of the installation.

8.24 Some television news and sports shows now utilize special 900 telephone numbers. By phoning 1 of 2 displayed numbers, a viewer can record a vote in favor of or against an issue. For each call made, the viewer is charged a certain amount on his/her phone bill. Do you think that this type of survey results in scientific samples that are representative of the entire population of television viewers? Explain.

8.3 Sampling Distributions

Earlier we mentioned that when a sample statistic is used to estimate a population parameter, it is not expected that a value will be obtained that is exactly equal to the parameter. Since a statistic is a random variable, it will fluctuate from sample to sample. Thus, a bound on the error of the estimate will need to be given. To obtain an error bound, it is necessary to know the **probability distribution of the sample statistic.** These distributions of sample statistics are referred to as **sampling distributions,** and they will be the focus of the remainder of the chapter. The basic concept of a sampling distribution is introduced in the following example.

Example 8.7

Consider the population consisting of the number of spots that show when a fair die is tossed. The population elements are the integers 1, 2, 3, 4, 5, and 6. The probability distribution of x, the number of spots tossed, is given by the probability histogram in Figure 8.1.

Figure 8.1
Proability distribution of x, the number of spots tossed.

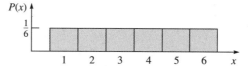

To illustrate the concept of a sampling distribution, we will find the **sampling distribution of the sample mean** for samples of size $n = 2$ tosses. The first step in obtaining this is to list all possible samples of 2 tosses and the corresponding sample means. Each possible sample is designated in Table 8.1 by an ordered pair, where the first and second numbers indicate the outcome on the first and second toss, respectively.

Each of the 36 samples has a probability of 1/36 of occurring when $n = 2$ tosses are made. Therefore, the probability of a particular value of the sample mean is

Table 8.1

Possible Samples and Their Means for $n = 2$ Tosses of a Die

Sample	Mean $\bar{x}$	Sample	Mean $\bar{x}$	Sample	Mean $\bar{x}$
(1, 1)	1.0	(3, 1)	2.0	(5, 1)	3.0
(1, 2)	1.5	(3, 2)	2.5	(5, 2)	3.5
(1, 3)	2.0	(3, 3)	3.0	(5, 3)	4.0
(1, 4)	2.5	(3, 4)	3.5	(5, 4)	4.5
(1, 5)	3.0	(3, 5)	4.0	(5, 5)	5.0
(1, 6)	3.5	(3, 6)	4.5	(5, 6)	5.5
(2, 1)	1.5	(4, 1)	2.5	(6, 1)	3.5
(2, 2)	2.0	(4, 2)	3.0	(6, 2)	4.0
(2, 3)	2.5	(4, 3)	3.5	(6, 3)	4.5
(2, 4)	3.0	(4, 4)	4.0	(6, 4)	5.0
(2, 5)	3.5	(4, 5)	4.5	(6, 5)	5.5
(2, 6)	4.0	(4, 6)	5.0	(6, 6)	6.0

obtained by counting the number of samples having this value and dividing the result by 36. Thus, the probability distribution of the sample mean, that is, the **sampling distribution of $\bar{x}$,** is as follows.

$\bar{x}$	1.0	1.5	2.0	2.5	3.0	3.5	4.0	4.5	5.0	5.5	6.0
$P(\bar{x})$	$\frac{1}{36}$	$\frac{2}{36}$	$\frac{3}{36}$	$\frac{4}{36}$	$\frac{5}{36}$	$\frac{6}{36}$	$\frac{5}{36}$	$\frac{4}{36}$	$\frac{3}{36}$	$\frac{2}{36}$	$\frac{1}{36}$

The graph of the sampling distribution of the sample mean appears in Figure 8.2. Notice that the sampling distribution of the sample mean has a shape that resembles a normal distribution, even though the distribution of x, given in Figure 8.1, is rectangular in shape. We will refer to this fact again in Example 8.10.

Figure 8.2
Sampling distribution of the sample mean $\bar{x}$.

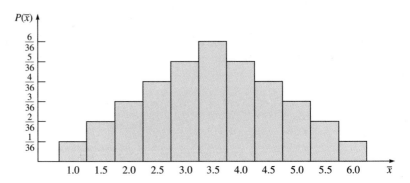

Example 8.8   In Example 8.7, a fair die was tossed two times. Find the sampling distribution of the sample range R.

Solution

A list of the possible samples of size $n = 2$ tosses is given in Table 8.2, together with the range for each sample. From Table 8.2, the possible values of the range R are 0, 1, 2, 3, 4, and 5. $P(0) = 6/36$, since there are 6 samples with a range of 0. Because 10 samples have a range of 1, $P(1) = 10/36$. Proceeding in a similar way, the sampling distribution of the sample range is found to be as follows.

R	0	1	2	3	4	5
$P(R)$	$\frac{6}{36}$	$\frac{10}{36}$	$\frac{8}{36}$	$\frac{6}{36}$	$\frac{4}{36}$	$\frac{2}{36}$

The graph of the sampling distribution appears in Figure 8.3.

Figure 8.3
Sampling distribution of the sample range R.

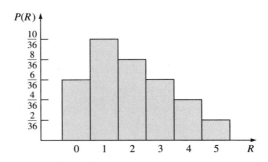

Table 8.2

Possible Samples and Their Ranges for $n = 2$ Tosses of a Die

Sample	Range R	Sample	Range R	Sample	Range R
(1, 1)	0	(3, 1)	2	(5, 1)	4
(1, 2)	1	(3, 2)	1	(5, 2)	3
(1, 3)	2	(3, 3)	0	(5, 3)	2
(1, 4)	3	(3, 4)	1	(5, 4)	1
(1, 5)	4	(3, 5)	2	(5, 5)	0
(1, 6)	5	(3, 6)	3	(5, 6)	1
(2, 1)	1	(4, 1)	3	(6, 1)	5
(2, 2)	0	(4, 2)	2	(6, 2)	4
(2, 3)	1	(4, 3)	1	(6, 3)	3
(2, 4)	2	(4, 4)	0	(6, 4)	2
(2, 5)	3	(4, 5)	1	(6, 5)	1
(2, 6)	4	(4, 6)	2	(6, 6)	0

The technique employed in Examples 8.7 and 8.8 for finding the sampling distribution of a sample statistic requires that one lists all possible samples of the specified size n. This is only feasible if n is very small. For instance, if the die is tossed $n = 10$ times, then, by the multiplication rule for counting, there are $6^{10} = 60,466,176$ different samples. In cases such as this, one sometimes approximates the sampling distribution by using simulation techniques. To illustrate this method, we had MINITAB generate some of the 60,466,176 possible samples of $n = 10$ tosses. We arbitrarily instructed MINITAB to produce 100 samples. (For those interested, the commands that produced the results are given in Example 8.9). The results appear in Table 8.3. A relative frequency distribution for the 100 sample means that appear in Table 8.3 is provided in Table 8.4. A relative frequency histogram follows in Figure 8.4.

Table 8.3

100 Random Samples and Their Means for $n = 10$ Tosses of a Die

Sample of 10 Tosses	Mean $\bar{x}$	Sample of 10 Tosses	Mean $\bar{x}$
6 3 4 2 2 6 1 2 6 4	3.6	2 5 6 1 1 1 2 4 1 2	2.5
1 4 5 2 2 3 3 3 5 6	3.4	4 4 3 5 5 2 5 3 6 2	3.9
6 5 3 5 2 3 3 2 1 6	3.6	4 1 2 6 4 2 6 5 1 5	3.6
2 2 2 6 5 5 6 3 1 5	3.7	6 2 5 1 3 6 1 4 4 2	3.4
2 6 5 5 2 6 1 5 5 5	4.2	1 3 4 6 5 2 6 3 5 2	3.7
2 6 1 2 4 6 1 1 5 2	3.0	6 5 1 4 1 2 5 3 6 3	3.6
1 2 1 4 5 2 6 1 5 1	2.8	5 4 4 3 2 1 1 1 3 5	2.9
6 1 5 5 6 3 2 4 3 2	3.7	5 5 6 4 6 4 1 4 5 3	4.3
1 4 4 6 2 3 2 5 6 1	3.4	2 3 5 4 2 6 1 1 1 6	3.1

(Continued on next page)

Table 8.3 *Continued*

100 Random Samples and Their Means for $n = 10$ Tosses of a Die

Sample of 10 Tosses	Mean $\bar{x}$	Sample of 10 Tosses	Mean $\bar{x}$
3 6 4 6 3 2 1 4 2 1	3.2	3 6 2 5 5 6 1 4 3 1	3.6
6 1 3 1 1 4 5 1 5 1	2.8	1 2 1 6 5 6 4 4 6 5	4.0
3 6 4 5 5 6 5 5 5 5	4.9	6 6 1 1 1 2 4 4 5 6	3.6
1 3 3 4 5 5 6 5 2 2	3.6	6 3 5 2 2 4 6 5 3 3	3.9
2 2 1 5 1 5 5 2 2 5	3.0	6 2 6 1 5 3 2 5 5 3	3.8
2 1 3 3 2 1 2 2 2 6	2.4	5 5 6 5 6 2 6 3 3 1	4.2
6 2 5 5 5 6 3 5 2	4.4	6 3 3 3 3 6 3 6 6 1	4.0
6 4 1 4 2 3 5 1 3 1	3.0	6 5 1 5 6 3 5 2 4 1	3.8
2 3 3 2 1 1 5 1 3 3	2.4	1 6 5 5 1 3 1 1 1 3	2.7
5 5 6 4 4 2 3 6 6 2	4.3	6 4 4 1 3 1 6 2 2 5	3.4
5 1 6 4 1 6 4 3 4 3	3.7	4 4 6 5 6 5 1 6 3 5	4.5
5 5 5 5 6 5 4 4 3 2	4.4	4 1 5 5 3 5 1 5 5 6	4.0
6 4 4 6 4 3 2 6 5 5	4.5	4 2 5 2 6 1 3 2 5 6	3.6
2 2 1 4 5 1 1 4 3 6	2.9	5 1 3 2 6 1 6 4 4 1	3.3
5 5 5 3 2 3 6 6 6 3	4.4	1 4 4 3 5 3 6 5 5 3	3.9
5 2 3 3 1 1 4 2 4 5	3.0	2 2 3 1 5 5 4 2 1 3	2.8
4 4 6 6 6 2 5 2 1 1	3.7	4 5 3 2 1 3 2 2 2 3	2.7
6 5 2 5 4 2 3 1 4 3	3.5	4 5 2 1 6 4 6 3 1 1	3.3
3 5 3 6 3 4 2 4 5 3	3.8	2 1 5 2 1 5 2 5 1 3	2.7
4 1 4 4 3 3 5 6 1 2	3.3	4 4 2 2 1 6 1 5 2 5	3.2
1 2 6 2 1 4 4 3 4 4	3.1	5 5 5 6 2 3 3 5 5 1	4.0
5 4 1 6 2 5 4 6 3 5	4.1	6 3 5 6 3 5 4 4 2 4	4.2
3 1 1 5 3 1 3 1 2 3	2.3	6 2 4 2 2 5 4 3 2 6	3.6
1 5 4 5 3 2 4 6 1 6	3.7	5 5 3 3 3 5 3 2 1 2	3.2
6 2 2 1 2 4 5 1 5 1	2.9	2 2 5 2 6 5 3 6 5 4	4.0
1 5 1 2 1 6 2 2 5 5	3.0	2 2 1 5 6 4 2 6 5 1	3.4
6 6 5 2 3 6 2 2 1 2	3.5	3 6 1 1 2 4 5 3 2 3	3.0
1 4 1 6 1 3 6 5 4 2	3.3	5 6 2 5 5 6 3 6 5 3	4.6
3 5 5 2 4 5 6 2 1 3	3.6	1 3 6 4 5 6 4 4 2 2	3.7
1 5 4 6 2 3 1 4 3 1	3.0	1 2 3 2 3 6 4 1 6 6	3.4
1 1 4 6 1 4 5 2 5 4	3.3	4 2 4 6 1 2 6 3 5 1	3.4
6 3 6 4 1 6 3 2 2 5	3.8	4 3 3 2 3 5 4 3 6 2	3.5
1 2 6 6 6 6 3 3 5 2	4.0	2 4 1 3 4 6 2 6 2 3	3.3
4 1 1 2 1 1 1 3 5 1	2.0	5 1 5 6 1 6 4 1 3 6	3.8
1 6 3 3 5 2 3 2 2 4	3.1	5 6 5 3 4 6 2 3 6 5	4.5
4 3 3 6 4 4 4 6 4 4	4.2	4 5 3 5 3 5 5 5 6 2	4.3
4 1 3 1 6 5 2 2 2 6	3.2	4 6 2 2 3 5 3 3 1 1	3.0
3 5 1 6 4 1 1 2 1 3	2.7	5 4 5 1 2 2 6 4 6 2	3.7
4 1 5 4 3 1 5 1 5 1	3.0	6 6 5 2 4 5 4 1 1 6	4.0
6 5 1 1 3 5 1 1 4 5	3.2	4 2 1 1 5 3 2 1 4 5	2.8
6 1 3 4 3 1 3 1 6 1	2.9	4 3 2 6 1 4 2 6 3 3	3.4

Table 8.4

Relative Frequency Distribution of the 100 Sample Means

Sample Mean	Relative Frequency
2.0–2.1	0.01
2.2–2.3	0.01
2.4–2.5	0.03
2.6–2.7	0.04
2.8–2.9	0.07
3.0–3.1	0.12
3.2–3.3	0.11
3.4–3.5	0.11
3.6–3.7	0.18
3.8–3.9	0.09
4.0–4.1	0.08
4.2–4.3	0.07
4.4–4.5	0.06
4.6–4.7	0.01
4.8–4.9	0.01

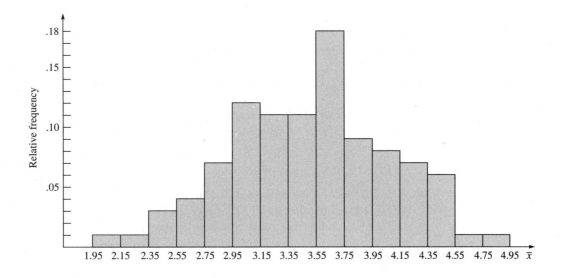

Figure 8.4
Relative frequency histogram of the 100 sample means.

The relative frequency histogram provides an approximation of the sampling distribution of the sample mean $\bar{x}$. The mean and the standard deviation of the histogram can be obtained by calculating these quantities for the 100 sample means in Table 8.3. They are equal to 3.493 and 0.573, respectively. The mound shape of the histogram suggests that a normal distribution might serve as an approximation for the sampling distribution of the sample mean. This feature is discussed in detail in Section 8.4.

Ⓜ **Example 8.9**   Many statistical experiments can be quickly and easily simulated by constructing a MINITAB **macro.** A macro is a set of commands that can be repeatedly executed a specified number of times. The commands **STORE** and **END** define the beginning and the end of a macro. The following commands utilize a MINITAB macro, and they were used to generate the 100 random samples of $n = 10$ die tosses and the corresponding sample means that appear in Table 8.3.

Commands	Comments
`LET K1 = 1`	`# Assigns the counter K1 an initial value 1`
`STORE`	`# Indicates the beginning of a macro`
`NOECHO`	`# Suppresses the printing of commands`
`RANDOM 10 C1;`	`# 10 random values are to be placed in C1`
`INTEGER 1 6.`	`# Random values selected from integers 1 to 6`
`LET C2(K1) = MEAN(C1)`	`# Puts sample mean in row K1 of column C2`
`LET K1 = K1 + 1`	`# Increases counter K1 by 1`
`END`	`# Indicates end of macro`
`EXECUTE 100`	`# Orders 100 executions of the macro`
`PRINT C2`	`# Prints the 100 sample means in C2`

Concerning the above macro, note the following points.

1. The statement (**LET K1 = 1**) gives the counter K1 an initial value of 1 at the beginning of the macro.

2. **STORE** informs MINITAB that a macro is about to be defined, in which case MINITAB will save for repeated execution everything that follows up to the **END** command.

3. **EXECUTE 100** instructs MINITAB to execute the macro 100 times.

4. Within the macro, the command (**RANDOM 10 C1;**) and its subcommand (**INTEGER 1 6.**) instruct MINITAB to generate 10 random values from the integers 1 through 6, and then store the results in column C1.

5. During the first pass through the macro, the statement (**LET K1 = K1 + 1**) increases the counter K1 by 1, so that K1 is 2 during the second execution. K1 increases by 1 each time the macro is executed.

6. **LET C2(K1) = MEAN(C1)** computes the mean of the random sample that is stored in C1, and this value is then placed in row K1 of column C2. Thus, during the first pass through the macro, the mean of the first sample is stored in row 1 of C2; during the second pass the mean of the second sample is stored in row 2 of C2. This continues until the mean of the 100th sample is stored in row 100 of C2.

7. The **NOECHO** command is used to suppress the printing of the macro commands during their 100 executions. Only output will be printed when **NOECHO** is in force. MINITAB can be instructed to resume the printing of commands by typing **ECHO.**

8. If you want MINITAB to print the 100 random samples, then insert the command **PRINT C1** just before the **END** statement.

Section 8.3 Exercises

Sampling Distributions

8.25 A random number generator is programmed so that it produces with equal frequency only the digits 1, 3, 5, and 7. A sample of 2 digits is generated.

    a. List all possible samples of $n = 2$ digits.
    b. For each possible sample, calculate the sample mean.
    c. Obtain the sampling distribution of the sample mean by constructing a table that shows the possible values of $\bar{x}$ and the associated probabilities.
    d. Use a histogram to graph the sampling distribution of the sample mean.

8.26 A random number generator is programmed so that it produces with equal frequency only the digits 3, 6, and 9. A sample of 3 digits is generated.

    a. List all possible samples of $n = 3$ digits.
    b. For each possible sample, calculate the sample mean.
    c. Obtain the sampling distribution of the sample mean by constructing a table that shows the possible values of $\bar{x}$ and the associated probabilities.
    d. Use a histogram to graph the sampling distribution of the sample mean.

8.27 A random variable x has the following probability distribution.

x	1	2	3	4	5
$P(x)$	$\frac{1}{5}$	$\frac{1}{5}$	$\frac{1}{5}$	$\frac{1}{5}$	$\frac{1}{5}$

Two independent observations of x are obtained.

    a. List all possible samples of $n = 2$ values.
    b. For each possible sample, calculate the sample mean.
    c. Obtain the sampling distribution of the sample mean by constructing a table that shows the possible values of $\bar{x}$ and the associated probabilities.
    d. Use a histogram to graph the sampling distribution of the sample mean.

8.28 A random variable x has the following probability distribution.

x	2	4
$P(x)$	0.5	0.5

Four independent observations of x are obtained.
a. List all possible samples of $n = 4$ values.
b. For each possible sample, calculate the sample mean.
c. Obtain the sampling distribution of the sample mean by constructing a table that shows the possible values of $\bar{x}$ and the associated probabilities.
d. Use a histogram to graph the sampling distribution of the sample mean.

8.29 For the random variable in Exercise 8.27, calculate the range for each of the possible samples of size $n = 2$. Use these values to obtain the sampling distribution of the sample range.

8.30 For the random variable in Exercise 8.28, calculate the range for each of the possible samples of size $n = 4$. Use these values to obtain the sampling distribution of the sample range.

MINITAB Assignments

Ⓜ 8.31 MINITAB can be used to simulate sampling from the probability distribution given in Exercise 8.27.
a. Generate 100 random samples of size $n = 2$ from that distribution.
b. Have MINITAB calculate the mean of each sample.
c. Use MINITAB to construct a histogram of the 100 sample means, and compare the results with that obtained in Exercise 8.27.

Ⓜ 8.32 Use MINITAB to simulate sampling from the probability distribution given in Exercise 8.28.
a. Generate 100 random samples of size $n = 4$ from that distribution.
b. Have MINITAB calculate the mean of each sample.
c. Have MINITAB construct a histogram of the 100 sample means, and compare the results with that obtained in Exercise 8.28.

8.4 The Central Limit Theorem

The previous section was concerned with the general problem of finding sampling distributions of sample statistics. The illustrations presented concerned the sample mean and range, but the procedures employed could have been applied to other statistics, such as the sample median, mode, variance, and so on. This section, however, is concerned only with investigating sampling distributions of the sample mean. Knowledge of these distributions is needed because a sample mean $\bar{x}$ is usually used to estimate the unknown mean μ of a population. Although it is unlikely that $\bar{x}$ will equal μ, we expect, with a high probability, that it will be close. The probability of this, and the closeness of the estimate, are both determined from the sampling distribution of the sample mean. Much of the information that the theory of statistics provides about the sampling distribution of the sample mean is summarized below. In the material to follow, the mean and the standard deviation of the sampling distribution of the sample mean are denoted by $\mu_{\bar{x}}$ and $\sigma_{\bar{x}}$, respectively. $\sigma_{\bar{x}}$ is often referred to as the **standard error of the mean.**

The Sampling Distribution of the Sample Mean $\bar{x}$:

For random samples of size n selected from a population with mean μ and standard deviation σ, the sampling distribution of the sample mean $\bar{x}$

1. has a mean of $\mu_{\bar{x}} = \mu$, the population mean;
2. has a standard deviation of $\sigma_{\bar{x}} = \sigma/\sqrt{n}$*;
3. has *exactly* a normal distribution when the population is normally distributed. For other types of populations, it will have *approximately* a normal distribution when n is large.

The Central Limit theorem was introduced to the academic world in 1810 by Pierre Simon Laplace (1749–1827). It was presented in a paper read by Laplace to the Academy of Sciences in Paris. (Photo courtesy of The Bettmann Archive.)

A major consequence of the above statement is that often a normal probability distribution can be used to find probabilities concerning a sample mean. The statement says that if the shape of the sampled population is normal, then the sampling distribution of the sample mean will also be normal. But more remarkably it says that even if the population is not normal, the sampling distribution of the sample mean will still be *approximately* normal, provided that the sample size n is large. This last result is one of the most important and frequently cited theorems in statistics. It is called the **Central Limit theorem** and is the basis for several inferential techniques in subsequent chapters.

The Central Limit Theorem:

For random samples selected from a nonnormal population, the sampling distribution of the sample mean $\bar{x}$ is approximately normal when the sample size n is large. The approximation improves as the sample size increases.

How large n must be to apply the Central Limit theorem depends on the actual distribution of the population, but a safe and conservative rule of thumb is 30 or more. We will adhere to this rule and define a **large sample** as $n \geq 30.$ However, for many populations the Central Limit theorem is still applicable for much smaller sample sizes. This point is illustrated in Example 8.10.

Example 8.10   Let the random variable x be the number of spots that show when a fair die is tossed.

1. Find the mean and the standard deviation of x.
2. For samples of $n = 2$ tosses, find $\mu_{\bar{x}}$, the mean of the sampling distribution of the sample mean.

---

* If the population is of finite size N, and the sampling is done without replacement, then $\sigma_{\bar{x}} = \dfrac{\sigma}{\sqrt{n}} \cdot \sqrt{\dfrac{N-n}{N-1}}$. The factor $\sqrt{\dfrac{N-n}{N-1}}$ is called the **finite population correction factor.** As is usually done in practice, we will ignore it since its value is close to 1 for most practical applications.

3. For samples of $n = 2$ tosses, find $\sigma_{\bar{x}}$, the standard deviation of the sampling distribution of the sample mean.
4. What can be said about the shape of the sampling distribution of the sample mean for samples of $n = 2$ tosses?

Solution

1. Since each of the 6 sides has an equal chance of showing when the die is tossed, the probability distribution of x is

x	1	2	3	4	5	6
$P(x)$	$\frac{1}{6}$	$\frac{1}{6}$	$\frac{1}{6}$	$\frac{1}{6}$	$\frac{1}{6}$	$\frac{1}{6}$

We will construct the following table to facilitate the calculations needed for finding the mean μ and standard deviation σ of the random variable x.

x	$P(x)$	$xP(x)$	$x^2P(x)$
1	$\frac{1}{6}$	$\frac{1}{6}$	$\frac{1}{6}$
2	$\frac{1}{6}$	$\frac{2}{6}$	$\frac{4}{6}$
3	$\frac{1}{6}$	$\frac{3}{6}$	$\frac{9}{6}$
4	$\frac{1}{6}$	$\frac{4}{6}$	$\frac{16}{6}$
5	$\frac{1}{6}$	$\frac{5}{6}$	$\frac{25}{6}$
6	$\frac{1}{6}$	$\frac{6}{6}$	$\frac{36}{6}$
		$\frac{21}{6}$	$\frac{91}{6}$

$$\mu = \Sigma xP(x) = \frac{21}{6} = 3.5$$

$$\sigma^2 = \Sigma x^2P(x) - \mu^2 = \frac{91}{6} - (3.5)^2 = 2.9167$$

$$\sigma = \sqrt{2.9167} = 1.71.$$

Thus, the mean and the standard deviation of the random variable x are $\mu = 3.5$ and $\sigma = 1.71$.

2. For samples of $n = 2$ tosses, the sampling distribution of $\bar{x}$ has the same mean as that of the population. Thus,

$$\mu_{\bar{x}} = \mu = 3.5.$$

3. The sampling distribution of $\bar{x}$ for samples of size $n = 2$ has a standard deviation equal to the population's standard deviation divided by the square root of the sample size. Therefore, the standard deviation of the sample mean (the standard error of the mean) is

$$\sigma_{\bar{x}} = \frac{\sigma}{\sqrt{n}} = \frac{1.71}{\sqrt{2}} = 1.21.$$

Note that since $\sigma_{\bar{x}}$ will always be smaller than σ for samples of size 2 or more, the distribution of sample means will always vary less than the population from which

Figure 8.5
Proability distribution of x,
the number of spots tossed.

the samples are selected. Moreover, $\sigma_{\bar{x}}$ will decrease if the sample size n is increased.

4. The graph of the probability distribution of x was given in Example 8.7 and is reproduced in Figure 8.5. The population's distribution has a uniform shape and is not normally distributed. Consequently, the sample size n must be large to use the Central Limit theorem to obtain information about the shape of the distribution of sample means. Since n is only 2, no conclusion in this regard can be drawn from the Central Limit theorem. However, in Example 8.7 we did obtain the exact sampling distribution of the mean $\bar{x}$, and its graph is reproduced in Figure 8.6.

Although n is only 2, the sampling distribution of $\bar{x}$ has a shape that resembles a normal distribution. As mentioned earlier, for many populations the sampling distribution of the sample mean will be approximately normal even when n is considerably less than 30.

The mean and standard deviation of $\bar{x}$ calculated in Parts 2 and 3 could also have been determined from the sampling distribution of the sample mean in Figure 8.6. By applying the mean and the standard deviation formulas used in Part 1, one will obtain $\mu_{\bar{x}} = 3.5$ and $\sigma_{\bar{x}} = 1.21$, the same values as those found in Parts 2 and 3.

Figure 8.6
Sampling distribution of the
sample mean $\bar{x}$ for $n = 2$.

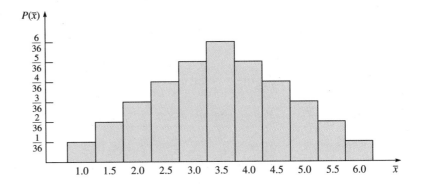

Example 8.11   In the previous section, the sampling distribution of the sample mean was simulated for $n = 10$ tosses of a die. The simulation produced 100 sample means from which estimates of $\mu_{\bar{x}}$ and $\sigma_{\bar{x}}$ were obtained, namely, 3.493 and 0.573, respectively. Use the properties of the sampling distribution of $\bar{x}$ to find the exact values of $\mu_{\bar{x}}$ and $\sigma_{\bar{x}}$.

Solution

In Part 1 of Example 8.10, the mean and standard deviation were obtained for x, the number of spots that show when a die is tossed. These values were found to be

$\mu = 3.5$ and $\sigma = 1.71$. For $n = 10$ tosses, the sampling distribution of the sample mean has a mean equal to

$$\mu_{\bar{x}} = \mu = 3.5.$$

The standard deviation of the sampling distribution of $\bar{x}$ (the standard error of the mean) is given by

$$\sigma_{\bar{x}} = \frac{\sigma}{\sqrt{n}} = \frac{1.71}{\sqrt{10}} = 0.54.$$

In Example 8.10, the mean and standard deviation were found for the sampling distribution of $\bar{x}$ for samples of $n = 2$ tosses, while in Example 8.11 these were found for samples of $n = 10$ tosses. In each case, $\mu_{\bar{x}}$, the mean of the sampling distribution equaled the population mean μ. The standard deviation, however, decreased from $\sigma_{\bar{x}} = 1.21$ for $n = 2$ to $\sigma_{\bar{x}} = 0.54$ for $n = 10$. The standard deviation of the mean decreases as the sample size n increases, and, consequently, the sample means will fluctuate less. The decrease in $\sigma_{\bar{x}}$ as n increases is illustrated in Table 8.5 for selected values of n.

The fact that sample means fluctuate less as the sample size n increases can be illustrated by applying the Empirical rule. From the Empirical rule we know that about 95 percent of the values of a random variable will lie within 2 standard deviations of its mean. Thus, from Table 8.5 for $n = 10$ tosses, about 95 percent of the time the sample mean will have a value within the interval $3.5 \pm 2(0.54)$, that is, between 2.42 and 4.58. However, if n is increased to 100 tosses, then about 95 percent of the time the sample mean will fall within the smaller interval $3.5 \pm 2(0.17)$, that is, from 3.16 to 3.84.

Table 8.5

$\sigma_{\bar{x}}$, the Standard Deviation of the Mean, for n Tosses of a Die

Sample Size n	2	5	10	50	100	1,000
Standard Deviation of the Mean, $\sigma_{\bar{x}} = 1.71/\sqrt{n}$	1.21	0.76	0.54	0.24	0.17	0.05

Section 8.4 Exercises
The Central Limit Theorem

8.33 Samples of size n were randomly selected from a population having a normal distribution. Which of the following statements best describes the sampling distribution of the sample mean?
   a. If n is large, then it will be approximately normal.
   b. It will be approximately normal, regardless of the size of n.
   c. It will be exactly normal, regardless of the size of n.

8.34 Random samples of size n were selected from a population that has a standard deviation $\sigma = 20$. Determine the standard deviation of the sampling distribution of the sample mean if
a. $n = 64$.
b. $n = 100$.

8.35 The standard deviation of a population is 18. For randomly selected samples of size n, determine the standard deviation of the sampling distribution of the sample mean if
a. $n = 144$.
b. $n = 81$.

8.36 Random samples of size n were selected from a population whose standard deviation is 48. How is the standard deviation of the sampling distribution of the sample mean affected if the sample size n is quadrupled from 36 to 144?

8.37 For the population referred to in Exercise 8.36, how large would the sample size n need to be for the standard deviation of the sampling distribution of the sample mean to have a value of 2?

8.38 A population has a mean of 100 and a standard deviation of 14. Sample means are computed for several thousand random samples of size $n = 49$, and a relative frequency histogram of the sample means is constructed.
a. What kind of shape would you expect the histogram of the sample means to have?
b. What value would you expect for the mean of the histogram of sample means?
c. What value would you expect for the standard deviation of the histogram of sample means?

8.39 A population has a mean of 79 and a standard deviation of 27. Sample means are computed for several thousand random samples of size $n = 81$, and a relative frequency histogram is constructed for the sample means.
a. What kind of shape would you expect the histogram of the sample means to have?
b. What value would you estimate for the mean of the histogram of sample means?
c. What value would you estimate for the standard deviation of the histogram of sample means?

8.40 In Exercise 8.27, the sampling distribution of the sample mean was obtained for samples of size $n = 2$ from a population with the following probability function.

x	1	2	3	4	5
$P(x)$	$\frac{1}{5}$	$\frac{1}{5}$	$\frac{1}{5}$	$\frac{1}{5}$	$\frac{1}{5}$

a. Show that the random variable x has mean $\mu = 3$ and standard deviation $\sigma = \sqrt{2}$.
b. Use the sampling distribution of $\bar{x}$ that was found in Exercise 8.27 to find $\mu_{\bar{x}}$ and $\sigma_{\bar{x}}$ for samples of size $n = 2$.
c. Check the results obtained in Part b by using the formulas in this section to obtain $\mu_{\bar{x}}$ and $\sigma_{\bar{x}}$.

8.41 In Exercise 8.28, the sampling distribution of the sample mean was obtained for samples of size $n = 4$ from a population with the following probability function.

x	2	4
$P(x)$	0.5	0.5

a. Show that the random variable x has mean $\mu = 3$ and standard deviation $\sigma = 1$.

b. Use the sampling distribution of $\bar{x}$ that was found in Exercise 8.28 to find $\mu_{\bar{x}}$ and $\sigma_{\bar{x}}$ for samples of size $n = 4$.

c. Check the results obtained in Part b by using the formulas in this section to obtain $\mu_{\bar{x}}$ and $\sigma_{\bar{x}}$.

MINITAB Assignments

Ⓜ 8.42 Use MINITAB to perform the following.

a. Generate 100 random samples of size $n = 9$ from a normal population with mean 50 and standard deviation 12.

b. For each sample, compute the sample mean and store them in column C2.

c. For the 100 sample means in C2, use MINITAB to construct a histogram and compute its mean and standard deviation.

d. Compare your empirical results with the properties given in this section for the theoretical sampling distribution of the sample mean.

Ⓜ 8.43 Use MINITAB to perform the following.

a. Generate 100 random samples of size $n = 25$ from a normal population with mean 50 and standard deviation 12.

b. For each sample, compute the sample mean and store them in column C2.

c. For the 100 sample means in C2, use MINITAB to construct a histogram and compute its mean and standard deviation.

d. Compare your empirical results with the properties given in this section for the theoretical sampling distribution of the sample mean.

Ⓜ 8.44 Consider a population with the probability distribution given in Exercise 8.40.

x	1	2	3	4	5
$P(x)$	$\frac{1}{5}$	$\frac{1}{5}$	$\frac{1}{5}$	$\frac{1}{5}$	$\frac{1}{5}$

a. Use MINITAB to generate 100 random samples of size $n = 30$ from this population with mean $\mu = 3$ and standard deviation $\sigma = \sqrt{2}$.

b. Have MINITAB calculate the mean of each sample and store the sample means in column C2.

c. For the 100 sample means, use MINITAB to construct a histogram and compute its mean and standard deviation.

d. Compare your empirical results with the properties given by the Central Limit theorem for the theoretical sampling distribution of the sample mean.

Ⓜ 8.45 Repeat Exercise 8.44 for 100 random samples of size $n = 60$.

8.5   Applications of the Central Limit Theorem

The Central Limit theorem and other properties of the sampling distribution of the sample mean were discussed in Section 8.4. We will now present some illustrations of how these results can be used to calculate probabilities concerning a sample mean $\bar{x}$.

Example 8.12   A population has a normal distribution with a mean of 98 and a standard deviation of 12. If 9 measurements are randomly selected from this population, what is the probability that the sample mean will exceed 104?

Solution

We are given that the population has a normal distribution with $\mu = 98$ and $\sigma = 12$. We are asked to find $P(\bar{x} > 104)$ for a sample of size $n = 9$. To obtain this probability concerning a sample mean, we need to know the probability distribution of all possible sample means for $n = 9$. This is just the sampling distribution of the sample mean $\bar{x}$, which was specified in the last section. Since we are told that the population has a normal distribution, the sampling distribution of $\bar{x}$ will also be normal, even though the sample size n is very small. Its distribution mean is $\mu_{\bar{x}} = \mu = 98$ and its standard deviation is $\sigma_{\bar{x}} = \sigma/\sqrt{n} = 12/\sqrt{9} = 4$. The sampling distribution of $\bar{x}$ is presented in Figure 8.7, and $P(\bar{x} > 104)$ is equal to the shaded area. To determine this probability, we need to find the z-value for $\bar{x} = 104$.

$$z = \frac{\bar{x} - \text{(its mean)}}{\text{(its std. dev.)}} = \frac{\bar{x} - \mu_{\bar{x}}}{\sigma_{\bar{x}}} = \frac{\bar{x} - \mu}{\sigma/\sqrt{n}}$$

$$z = \frac{104 - 98}{4} = 1.50$$

A value of $\bar{x} = 104$ is one-and-one-half standard deviations above $\mu_{\bar{x}}$. From Table 3, the cumulative area to the left of $z = 1.50$ standard deviations is 0.9332. Therefore, the shaded area is $1 - 0.9332 = 0.0668$, and thus,

$$P(\bar{x} > 104) = P(z > 1.50) = 0.0668.$$

Figure 8.7
Sampling distribution of $\bar{x}$.

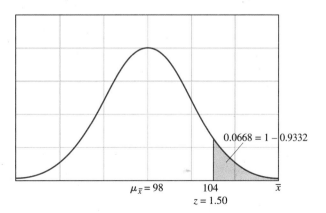

$0.0668 = 1 - 0.9332$

$\mu_{\bar{x}} = 98$ 104 $\bar{x}$

$z = 1.50$

Example 8.13

As mentioned in Example 7.13, several coffee manufacturers have discontinued the practice of packaging their product in 1-pound (454 grams) containers. Assume that the distribution of weights of a particular brand now has a mean of 383.0 grams and a standard deviation of 5.4 grams. Find the probability that a random sample of 36 containers will have an average weight of at least 382 grams.

Solution

The population of individual containers has a mean of $\mu = 383.0$ grams and a standard deviation of $\sigma = 5.4$ grams. Regardless of the shape of this population, the Central Limit theorem tells us that the sampling distribution of $\bar{x}$ is approximately normal because the sample size n is large. Furthermore, for samples of size $n = 36$ the distribution of $\bar{x}$ has a mean of $\mu_{\bar{x}} = \mu = 383.0$ grams and a standard deviation equal to $\sigma_{\bar{x}} = \sigma/\sqrt{n} = 5.4/\sqrt{36} = 0.9$ gram. The probability that a sample mean will be at least 382 is equal to the shaded area in Figure 8.8. To obtain this area, the z-value for 382 is needed.

$$z = \frac{\bar{x} - \mu_{\bar{x}}}{\sigma_{\bar{x}}} = \frac{382 - 383}{0.9} = -1.11$$

From Table 3, the cumulative area to the left of $z = -1.11$ is 0.1335. The shaded area is $1 - 0.1335 = 0.8665$. Therefore,

$$P(\bar{x} \geq 382) = P(z \geq -1.11) = 0.8665.$$

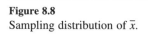

(© Bonnie Rauch/Photo Researchers, Inc.)

Figure 8.8
Sampling distribution of $\bar{x}$.

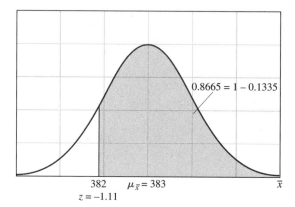

The next two examples illustrate how the Central Limit theorem can be used to find a probability concerning the error when a sample mean is used to estimate the population mean. To determine this probability, however, we need to know the population's standard deviation.

Example 8.14

To estimate the unknown mean of a population, the mean of a random sample of size $n = 50$ will be used. Find the probability that the estimate will be in error by less than 2, if the standard deviation of the population is 9.

Solution

Whenever we want to calculate a probability concerning a sample mean, we first need to obtain the distribution of all possible sample means for the specified sample size. Because $n = 50$ is large, the Central Limit theorem is applicable. It tells us that the sampling distribution of $\bar{x}$ is approximately normal. The mean of the distribution is $\mu_{\bar{x}} = \mu$ (the population mean), and the standard deviation is $\sigma_{\bar{x}} = \sigma/\sqrt{n} = 9/\sqrt{50} = 1.27$. The sampling distribution of $\bar{x}$ is presented in Figure 8.9. The probability that the sample mean $\bar{x}$ will differ from the population mean μ by less than 2 is equal to the shaded area. To determine this probability, we need to find the z-values for $(\mu - 2)$ and $(\mu + 2)$.

$$z = \frac{(\mu - 2) - \mu_{\bar{x}}}{\sigma_{\bar{x}}} = \frac{(\mu - 2) - \mu}{1.27} = \frac{-2}{1.27} = -1.57$$

$$z = \frac{(\mu + 2) - \mu_{\bar{x}}}{\sigma_{\bar{x}}} = \frac{(\mu + 2) - \mu}{1.27} = \frac{2}{1.27} = 1.57$$

From Table 3, the area to the left of $z = -1.57$ is 0.0582, and the area to the left of $z = 1.57$ is 0.9418. Therefore,

$$P(\mu - 2 < \bar{x} < \mu + 2) = P(-1.57 < z < 1.57)$$
$$= 0.9418 - 0.0582$$
$$= 0.8836.$$

Thus, there is about an 88 percent chance that the sample mean will be off from the population mean by less than 2.

Figure 8.9
Sampling distribution of $\bar{x}$.

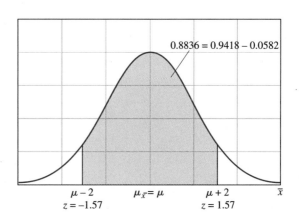

$0.8836 = 0.9418 - 0.0582$

Example 8.15   The National Center for Health Statistics conducted an extensive survey of house-holds in the United States and concluded that married people tend to be healthier than those who are single, divorced, or widowed. Part of the survey involved estimating the average amount of time that these groups were too sick to go to work.

Suppose that one is interested in estimating the mean number of work hours lost during the year for divorced men residing in a certain state. If the mean of a sample of 500 divorced males in this state will be used as an estimate, what is the probability that it will be in error by more than 4 hours? Assume that the standard deviation of the population is about 53 hours.

Solution

Since n is large, the Central Limit theorem can be applied to obtain the sampling distribution of all possible sample means for samples of size $n = 500$. The distribution is approximately normal with mean $\mu_{\bar{x}}$ equal to μ, the population's mean. The standard deviation of the sampling distribution is $\sigma_{\bar{x}} = \sigma/\sqrt{n} = 53/\sqrt{500} = 2.37$ hours. The sampling distribution of $\bar{x}$ is given in Figure 8.10. The probability that a sample mean will differ from the population mean μ by more than 4 hours is equal to the shaded area. To determine this, the z-values for $(\mu - 4)$ and $(\mu + 4)$ are needed.

$$z = \frac{(\mu - 4) - \mu_{\bar{x}}}{\sigma_{\bar{x}}} = \frac{(\mu - 4) - \mu}{2.37} = \frac{-4}{2.37} = -1.69$$

$$z = \frac{(\mu + 4) - \mu_{\bar{x}}}{\sigma_{\bar{x}}} = \frac{(\mu + 4) - \mu}{2.37} = \frac{4}{2.37} = 1.69$$

Using Table 3, the cumulative area to the left of $z = -1.69$ is 0.0455. Because of the symmetry of the z-distribution, the area to the right of $z = 1.69$ is also 0.0455. The requested probability is $2(0.0455) = 0.0910$, and we find that there is about a 9 percent chance that the sample mean will differ from the population mean by more than 4 hours.

Figure 8.10
Sampling distribution of $\bar{x}$.

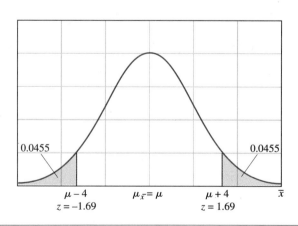

Section 8.5 Exercises
Applications of the Central Limit Theorem

8.46 A random sample of size $n = 4$ is selected from a normal population with mean $\mu = 63$ and standard deviation $\sigma = 3$. Find the following probabilities concerning the sample mean.
  a. $P(\bar{x} < 66)$
  b. $P(\bar{x} > 65)$
  c. $P(60 < \bar{x} < 66)$

8.47 A fruit-filled cereal is packaged in boxes that contain an average of 450 grams and for which the standard deviation is 12 grams. A sample of 36 boxes is randomly selected. Find the probability that the sample mean will be at least 454 grams.

8.48 A statistics department has found that the average grade on the final exam for its introductory course is 74.8, and the standard deviation is 8.2. Suppose a particular class of 32 students is considered typical of those who take this course. Determine the probability that their average score on the final examination will
  a. exceed 76,
  b. differ from the mean of 74.8 by less than 2 points, that is, will be between 72.8 and 76.8.

8.49 Lean, trimmed, 3-ounce tenderloin steaks contain an average of 174 calories (U.S. Department of Agriculture). Suppose the standard deviation of such steaks is 10 calories. If a person eats one of these steaks each week for a year, what is the probability that the average number of calories consumed per steak will be less than 175?

8.50 Off-price retailers use discounted prices to attract customers. Visitors to such stores are there an average of 40 minutes (*Forbes,* February 5, 1990). If the standard deviation is 12 minutes, what is the probability that a sample of 50 shoppers will spend an average of
  a. less than 38 minutes in the store,
  b. more than 45 minutes,
  c. between 38 and 45 minutes in the store?

8.51 The mean cost of a normal pregnancy and hospital delivery is $4,334 (*USA TODAY,* January 11, 1990). Assume that the standard deviation of costs is $125. If a health insurance company processes 35 claims for such services, what is the probability that the average cost exceeds $4,350?

8.52 A fast food restaurant claims that the average time required to fill an order is 5 minutes and that the standard deviation is 20 seconds. Over a 1 week period, a competitor timed a sample of 40 orders at this restaurant, and it obtained a sample mean of 310 seconds. Do you think that the restaurant's 5-minute claim is true? (Calculate the probability of obtaining a sample mean as large as 310 seconds if μ were actually 300 seconds.)

8.53 A catering service has found that people eat an average of 7.4 ounces of shrimp at affairs that it serves. The standard deviation is 2.4 ounces per person. The service is going to cater an event for 100 people, and it plans to bring 50 pounds of shrimp. Determine the probability that the caterer will run out of shrimp at the affair.

8.54 The mean of a random sample of size $n = 55$ will be used to estimate the mean of a population whose standard deviation is 23.
   a. Determine the probability that the sample mean will be off from the population mean by less than 5.
   b. What is the maximum amount by which you would expect the sample mean to differ from the population mean? (Use a probability of 0.95.)

8.55 An economist wants to estimate the average hourly wage of waitresses in a large metropolitan area. This will be done by randomly surveying 250 waitresses and calculating the sample mean. Find the probability that the estimate will be in error by more than 5 cents. Assume that the population's standard deviation is 45 cents.

8.56 A primary factor in the operating speed of a personal computer is the clock speed of its microprocessor chip, the system's central processing unit (CPU). The manufacturer's rating of a chip is conservative, and it will usually run at a faster speed. Suppose a computer manufacturer is considering the purchase of a large lot of several thousand chips. To estimate the true mean operating speed of all chips in the lot, 70 chips will be randomly selected and tested. The sample mean will then be used as an estimate. Determine the probability that the resulting value will be in error by less than 0.3 MHz (assume that $\sigma = 1.2$ MHz).

Looking Back

Making **inferences** is the main application of statistics. Frequently an inference will consist of estimating some numerical characteristic (a **parameter**) of a **population.** This is done by selecting a **sample** and calculating the value of a similar quantity (a **statistic**) for the sample. Because a sample usually is only a very small part of the population, valid inferences can be made only if a proper **sampling method** is employed. A **scientific (probability) sample** must be used, that is, the sample must be selected on the basis of known probabilities. The following four scientific samples were discussed in this chapter.

Scientific Sampling Methods:
1. A **random sample** is selected in such a way that each possible sample of the specified size has an equal probability of being selected.
2. In a **systematic sample,** a starting element is chosen at random from the first k population elements, and then every kth element is selected.

3. To obtain a **cluster sample,** the population elements are grouped in subsets, called clusters. Then a random sample of clusters is selected, and the elements in the chosen clusters are included in the sample.
4. In a **stratified random sample,** the population is first divided into subpopulations, called strata. Then a random sample is selected from each stratum.

For nearly all the statistical methods in this text, random sampling will be the assumed sampling method.

When a sample statistic is used to estimate a population parameter, it will usually be in error by some amount. After all, a statistic is a random variable that fluctuates from sample to sample. Consequently, in estimating a parameter, one should give a bound on the error of the estimate. To do this, one must know the **probability distribution of the sample statistic.** This is called the **sampling distribution of the statistic.** As illustrated in Section 8.3, sometimes sampling distributions can be determined exactly. They can also be approximated by using simulation techniques. Often the sampling distribution of interest pertains to the sample mean. From the theory of statistics, a great deal of information is known about **sampling distributions of the sample mean.**

The Sampling Distribution of the Sample Mean $\bar{x}$:
For random samples of size n selected from a population with mean μ and standard deviation σ, the sampling distribution of the sample mean $\bar{x}$

1. has a mean of $\mu_{\bar{x}} = \mu$, the population mean;
2. has a standard deviation of $\sigma_{\bar{x}} = \sigma/\sqrt{n}$;
3. has *exactly* a normal distribution when the population is normally distributed. For other types of populations, it will have *approximately* a normal distribution when n is large.

Contained in the above statement is one of the most important results in statistics —the **Central Limit theorem.**

The Central Limit Theorem:
For random samples selected from a nonnormal population, the sampling distribution of the sample mean $\bar{x}$ is approximately normal when the sample size n is large. The approximation improves as the sample size increases.

Key Words

In reviewing this chapter, you should be able to define, explain, and illustrate each of the following.

sampling method *(page 290)*

random sample *(page 290)*

frame *(page 291)*

systematic sample *(page 294)*

cluster sample *(page 296)*

stratified random sample *(page 297)*

scientific (probability) sample *(page 298)*

sampling distribution of a statistic *(page 299)*

Central Limit theorem *(page 307)*

standard error of the mean *(page 307)*

Ⓜ MINITAB Commands

SET _ *(page 292)*

END *(page 292)*

LET _ *(page 304)*

ECHO *(page 305)*

NOECHO *(page 304)*

RANDOM _ _; *(page 304)*

INTEGER _ _.

SAMPLE _ _ _ *(page 292)*

PRINT _ *(page 292)*

STORE *(page 304)*

EXECUTE _ *(page 304)*

MEAN _ *(page 304)*

Review Exercises

8.57 The average age of men at the time of their first marriage is 24.8 years *(Statistical Abstract of the United States)*. Suppose the standard deviation is 2.8 years, and 49 married males are selected at random and asked the age at which they were first married. Find the probability that the sample mean will be

a. more than 26,

b. less than 24,

c. between 24.2 and 25.5.

8.58 A population has a normal distribution with a mean of 76 and a standard deviation of 10. A random sample of size $n = 16$ is selected. Find the following probabilities concerning the sample mean.

a. $P(\bar{x} > 75)$

b. $P(\bar{x} < 72.5)$

c. $P(70 < \bar{x} < 80)$

8.59 A taxi company in a small city has found that its fares average $7.80 with a standard deviation of $1.40. On a particular day the company had 73 fares. What is the probability that the average fare for that day exceeded $8.00?

8.60 A population has a standard deviation of 22. For randomly selected samples of size n, determine the standard deviation of the sampling distribution of the sample mean if
a. $n = 121$,
b. $n = 400$.

8.61 If the size of a random sample is doubled, what effect will this have on the standard deviation of the sampling distribution of the sample mean?

8.62 The mean of a population is 88.5, and its standard deviation is 12. A very large number of random samples of size $n = 100$ is selected, and a relative frequency histogram of their sample means is constructed. What would be the expected shape, mean, and standard deviation of the histogram?

8.63 How many different samples of 5 elements can be selected from a population that contains 50 elements?

8.64 A fruit cannery is considering the purchase of several hundred baskets of peaches. To estimate the proportion that have serious blemishes, the company will randomly select 25 baskets and then inspect each peach in each basket. What type of sampling method is this?

8.65 To test the tensile strength of its fishing line, the manufacturer selects every 100th spool from a production run and submits the spool to a strength analysis. State the type of sampling method that is being utilized.

8.66 A city mayor is trying to decide if he should run for the state governorship. To assess his popularity, a pollster will randomly select a sample of 1,000 voters from each of the western, central, and eastern parts of the state. This is an example of what type of sampling method?

8.67 An answering service employs 10 people. To decide which 3 employees will work New Year's day, the owner lists each worker's name on a slip of paper, places the 10 slips in a hat, and has someone select 3 slips. What type of sampling method is the owner using?

8.68 The mean weight of oranges in an orange grove is 6.4 ounces, and the standard deviation is 0.6 ounce. The owner of the grove has a roadside stand at which boxes of 36 oranges are sold. What proportion of boxes contain at least 223.2 ounces of oranges?

8.69 To estimate the mean cost of automobile insurance policies in a certain state, the mean of a random sample of 125 policies will be calculated. If the population standard deviation is $35, what is the probability that the estimate will be in error by less than $7?

8.70 A random sample of 39 measurements will be selected from a population that has a standard deviation of 25. The sample mean will be used to estimate the population mean. Determine the probability that the estimate will be in error by more than 5.

8.71 An optical laboratory advertises that it only charges an average of $55.00 for a pair of prescription glasses. In response to numerous consumer complaints, the state department of consumer affairs sampled 33 prescriptions filled by this company. For the sample, the mean price charged was $65.48. Do you believe there is justification to doubt the laboratory's claim? (Calculate the probability of obtaining a sample mean as large as $65.48, if the population mean were really $55.00. Assume that $\sigma = \$19.00$.)

8.72 A pet store has the following puppies for sale: an Airedale, Boston Terrier, Collie, Dalmatian, Elkhound, Fox Terrier, and Great Dane. The owner has agreed to let the activities director of a nursing home borrow any two puppies for use in a pet therapy program.
   a. In how many different ways can the activities director select the two puppies?
   b. List the possible samples of two dogs.
   c. Suppose the director decides to randomly select the two puppies. Give a procedure that could be used that would assure the selection of a random sample.
   d. If the dogs are randomly selected, what is the probability that the Airedale and the Fox Terrier will be chosen?

8.73 A population has the following probability distribution.

x	3	4	5	6	7	8
$P(x)$	$\frac{1}{6}$	$\frac{1}{6}$	$\frac{1}{6}$	$\frac{1}{6}$	$\frac{1}{6}$	$\frac{1}{6}$

   a. Find the mean and the standard deviation of x.
   b. List all possible samples of two independent observations from this population.
   c. Calculate the sample mean of each possible sample.
   d. Obtain the sampling distribution of the sample mean by constructing a table that gives the possible values of $\bar{x}$ and the associated probabilities.
   e. Use the sampling distribution in (d) to find the mean and the standard deviation of the sampling distribution of the sample mean.
   f. Use the formulas $\mu_{\bar{x}} = \mu$ and $\sigma_{\bar{x}} = \sigma/\sqrt{n}$ to find the mean and the standard deviation of the sampling distribution of the sample mean.

MINITAB Assignments

Ⓜ 8.74 Simulate sampling from the population in Exercise 8.73 by using MINITAB to performing the following:
   a. generate 100 random samples of size $n = 2$ from the population;
   b. calculate the mean of each sample;
   c. construct a histogram of the 100 sample means;
   d. calculate the mean and the standard deviation of the 100 sample means.

Ⓜ 8.75 Repeat Exercise 8.74 for 100 random samples of size $n = 32$.

Ⓜ 8.76 Use MINITAB to generate 100 random samples of size $n = 20$ from a normal population with $\mu = 500$ and $\sigma = 75$. For each sample, compute the sample mean and store them in column C2. For the 100 sample means, use MINITAB to construct a histogram and compute its mean and standard deviation. Compare the results with the theoretical distribution of sample means.

Ⓜ 8.77 A clinical trial will be conducted on 100 diabetics to test the effectiveness of a special diet on controlling blood sugar levels. Fifty patients will be randomly selected and placed on the diet. The remaining 50 participants will be given a standard diet for diabetics. Assume that the 100 patients are numbered from 1 through 100. Use MINITAB to select the 50 patients who will receive the special diet.

Ⓜ 8.78 Three hundred and eight employees will attend their company's annual Christmas party. Thirty door prizes will be given away by a random drawing, and no one can receive more than one prize. Use MINITAB to simulate the results of the drawing.

(© Grant Heilman/Grant Heilman Photography, Inc.)

9 Estimating Means, Proportions, and Variances: Single Sample

◀ *Chilean Grapes*
*On March 2, 1989, the United States Embassy in Santiago received an anonymous phone call claiming that some of Chile's exported grapes contained the poison cyanide. After a second similar call, the U.S. Food and Drug Administration ordered an inspection of Chilean grapes at the off-loading facility in Philadelphia. Over 2,000 crates were examined, and only two of the millions of sampled grapes were found to contain a small amount of cyanide. Several thousand additional crates were subsequently sampled, with no additional contaminated grapes found. Although the **estimated proportion** of tainted Chilean grapes was practically zero, the negative economic ramifications from the two tainted grapes were enormous. Millions of tons of Chilean grapes on store shelves and in warehouses were destroyed, and the jobs of thousands of Chilean workers were jeopardized.*

Looking Ahead

Making statistical inferences is the primary concern of this and the chapters to follow. Many practical problems involve the process of generalizing from a sample to a population. Sometimes the inference pertains to a **problem of estimation,** such as in the grape scare of 1989. The U.S. Food and Drug Administration sampled crates of Chilean grapes to estimate the proportion of this imported fruit that had been contaminated with cyanide. In this situation, the inferential process was concerned with estimating the value of a parameter, the topic of this chapter. Some inferential problems deal with **tests of hypotheses,** the subject of Chapter 10. While each involves inferences concerning population parameters, their foci of interest differ. An estimation problem is concerned with approximating the value of a parameter, while a test of hypotheses problem checks whether a hypothesized value of the parameter is consistent with the results of a sample.

In Chapters 9 and 10 we will consider inferences for parameters of a single population. In Chapter 11 we will investigate inferences concerning two populations. This chapter will be concerned with estimating the following types of population parameters.

1. A **population mean μ:** A manufacturer of laptop computers needs to know the average amount of time that a battery charge will power the system, since many buyers consider this information an important factor in the selection of a particular brand.
2. A **population proportion *p*:** In planning the amount of food that must be prepared for a weekend, the manager of a college cafeteria needs to estimate the proportion of the student body who will vacate the campus during that period.
3. A **population variance σ²:** A supplier of engine pistons for an automobile manufacturer must control the variability in the diameters of its pistons. Knowledge of the diameters' variance is essential to assure that quality control specifications are being satisfied.

We will see that the preceding parameters μ, *p*, and σ² can be estimated with a single value **(point estimate)** or with an interval estimate **(confidence interval).**

9.1   Basic Concepts of Estimating Population Parameters

A parameter can be estimated with either a single number or an interval. For example, one might calculate a sample mean $\bar{x}$ and use its value of $527 to estimate μ, the mean cost of an uncontested divorce in the United States. Since the estimate consists of a single value, it is called a **point estimate.** On the other hand, the mean cost μ could be estimated by saying that it is some amount between $517 and $537. This is an **interval estimate** and will always have some amount of confidence (such as 95 percent) associated with it. Consequently, it is referred to as a **confidence interval estimate.**

> **Estimating a Parameter:**
> A **point estimate** of a parameter is a single value of some statistic.
>
> **A confidence interval estimate** consists of an interval of values determined from sample information. Associated with the interval is a percentage that measures one's confidence that the parameter lies within the interval.

A point estimate is less informative than a confidence interval estimate because a point estimate does not reveal by how much one can expect it to be off from the parameter that it estimates. For instance, if $527 is used to estimate the true mean cost μ of an uncontested divorce, it is highly improbable that μ is exactly equal to this figure. We have no idea whether $527 is likely to be off by a small or a large amount. For this reason, one usually prefers to estimate a parameter by using a confidence interval. We will soon see, however, that the first step in constructing a confidence interval is to obtain a point estimate of the parameter.

For large-sample situations, the Central Limit theorem is the basis for deriving many confidence interval formulas. As discussed in Chapter 8, a sample size will be considered large if it consists of 30 or more values. To illustrate how a large-sample

Figure 9.1
Sampling distribution of the sample mean $\bar{x}$.

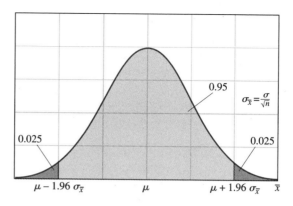

(© Ulrike Welsch/Photo Researchers, Inc.)

confidence interval can be derived, consider the following example. The purchasing agent for a national chain of seafood restaurants wants to estimate the mean weight of shrimp in a large catch. To accomplish this, a random sample of $n = 50$ shrimp will be selected from the population of all shrimp in the catch. The sample mean $\bar{x}$ will be used to estimate μ, the mean weight of the entire catch. Since n is large, the Central Limit theorem is applicable. It tells us that the sampling distribution of $\bar{x}$ is approximately normal. Moreover, the mean of the sampling distribution is $\mu_{\bar{x}} = \mu$, the mean of the population, and the standard deviation is $\sigma_{\bar{x}} = \sigma/\sqrt{n}$, the population standard deviation divided by the square root of the sample size. From the Empirical rule, we know that if random samples of 50 shrimp were repeatedly drawn from the catch, then about 95 percent of the resulting sample means would fall within two standard deviations of μ. In Example 7.8, it was shown that a more precise number of standard deviations is 1.96, and consequently, we will use this value in place of 2 (see Figure 9.1).

Thus, for each sample of size n, there is a probability of approximately 0.95 that the sample mean will satisfy the following inequality.

$$\mu - 1.96 \frac{\sigma}{\sqrt{n}} < \bar{x} < \mu + 1.96 \frac{\sigma}{\sqrt{n}}$$

This inequality can easily be rewritten to obtain the following interval that is centered around μ.

$$\bar{x} - 1.96 \frac{\sigma}{\sqrt{n}} < \mu < \bar{x} + 1.96 \frac{\sigma}{\sqrt{n}} \tag{9.1}$$

Formula 9.1 is called a **95 percent confidence interval for μ,** because the probability is approximately 0.95 that a randomly selected sample of n shrimp will produce a sample mean that satisfies the inequality. In other words, for a random sample and its associated sample mean $\bar{x}$, the random interval

$$\text{from } \bar{x} - 1.96 \frac{\sigma}{\sqrt{n}} \qquad \text{to } \bar{x} + 1.96 \frac{\sigma}{\sqrt{n}}$$

has about a 95 percent probability of containing the actual value of μ. The left endpoint of the interval is called the **lower confidence limit,** and the right endpoint is

Figure 9.2
95 percent confidence interval for μ.

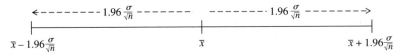

the **upper confidence limit.** (see Figure 9.2). A simpler way to denote the interval is by writing

$$\bar{x} \pm 1.96 \, \frac{\sigma}{\sqrt{n}}. \tag{9.2}$$

To illustrate Formula 9.2, suppose the population of shrimp in the catch has a standard deviation of $\sigma = 4.3$ grams, and a sample of $n = 50$ shrimp produces a mean weight of $\bar{x} = 38.6$ grams. Then a 95 percent confidence interval for μ is given by the following.

$$\bar{x} \pm 1.96 \, \frac{\sigma}{\sqrt{n}}$$

$$38.6 \pm 1.96 \, \frac{4.3}{\sqrt{50}}$$

$$38.6 \pm 1.2$$

Thus, the purchasing agent for the chain of restaurants can be 95 percent confident that the mean weight of shrimp in the entire catch is some value between 37.4 and 39.8 grams.

When a 95 percent confidence interval such as the above is constructed to estimate a population mean, the interval obtained will or will not contain the actual value of μ. In the above illustration, the 95 percent confidence interval from 37.4 to 39.8 either does or does not contain the real value of μ. The 95 percent confidence means that if the procedure used to construct the interval is repeatedly applied over the long run, then about 95 percent of the time it will produce an interval that contains μ. To illustrate this, we had MINITAB construct one hundred 95 percent confidence intervals for a normal population for which the mean was known to be 50. The resulting 95 percent confidence intervals appear in Table 9.1. Six of the 100 intervals that appear in Table 9.1 fail to enclose the population mean of 50. Thus, 94 of the one hundred 95 percent confidence intervals do contain μ. If we had generated several thousand intervals, then the percentage of intervals that enclose 50 would be expected to be even closer to 95 percent.

Confidence Level:
The amount of confidence associated with an interval is called the **confidence level,** and this may be expressed as a percentage or a decimal.

Although the choice of the confidence level is at the discretion of the experimenter, the most commonly used values are 90 percent, 95 percent, and 99 percent.

Table 9.1

100 Confidence Intervals for a Population with μ = 50

(48.774, 52.545)	(47.959, 52.058)	(46.680, 51.180)	(47.732, 51.546)
(45.920, 50.450)	(47.510, 52.630)	(46.662, 50.606)	(46.850, 50.970)
(47.600, 52.230)	(49.951, 53.506)	(47.172, 50.130)	(48.366, 52.222)
(47.360, 51.650)	(46.450, 51.980)	(48.580, 52.880)	(47.900, 52.170)
*(50.210, 53.968)	(46.130, 50.440)	(48.760, 53.000)	(47.240, 52.680)
(46.750, 50.930)	(48.690, 53.560)	(47.187, 50.431)	(47.820, 52.460)
(49.520, 54.050)	*(50.484, 53.900)	(47.760, 52.280)	(47.550, 52.450)
(47.240, 51.860)	(48.469, 52.225)	(47.365, 51.425)	(49.009, 52.670)
(48.316, 51.601)	(47.884, 52.006)	(46.470, 52.320)	(46.980, 51.700)
(48.177, 51.869)	(48.500, 52.850)	(48.650, 53.140)	(47.600, 52.080)
(49.510, 54.070)	(47.330, 51.680)	(47.830, 51.586)	(47.020, 51.340)
(48.375, 52.401)	(48.140, 52.380)	(47.422, 51.222)	(47.950, 53.310)
(46.320, 51.080)	(47.637, 51.360)	(47.620, 51.650)	(48.025, 52.121)
(48.934, 51.505)	(47.530, 52.860)	(47.691, 51.373)	(48.920, 53.200)
(48.480, 52.650)	(47.846, 51.792)	(48.813, 52.681)	(47.930, 52.380)
(46.856, 50.828)	(47.820, 53.030)	*(50.417, 54.117)	(48.590, 52.780)
(48.690, 53.060)	(46.400, 50.670)	*(50.526, 53.747)	(48.030, 51.371)
(47.374, 51.165)	(49.238, 53.129)	(48.410, 52.980)	(47.580, 52.450)
(47.500, 52.470)	(47.670, 51.597)	*(45.025, 49.058)	(48.640, 52.900)
(46.960, 51.530)	(47.384, 51.061)	(48.248, 51.085)	(48.578, 52.318)
(47.874, 51.618)	(46.110, 50.420)	(48.780, 52.662)	(49.023, 52.705)
*(50.687, 54.476)	(49.514, 53.174)	(48.690, 53.000)	(48.540, 52.970)
(47.320, 51.129)	(48.072, 52.070)	(46.704, 50.603)	(46.710, 51.370)
(48.230, 52.900)	(46.340, 51.480)	(48.103, 51.442)	(47.149, 50.919)
(48.081, 51.854)	(49.240, 54.240)	(48.438, 51.433)	(47.820, 52.320)

* Intervals that fail to enclose the population mean of 50.

For the illustration involving the shrimp catch, how would the confidence interval given in Formula 9.2 be affected if a confidence level of 99 percent had been chosen? This can be answered by referring back to Figure 9.1, which was the starting point in deriving the 95 percent confidence interval. For a confidence level of 99 percent, the value of 1.96 is replaced by the number of standard deviations (z-value) for which the interval around μ will contain 99 percent of the sample means. From Figure 9.3, we see that this number is $z_{.005}$, the z-value that has an area to the right of it equal to 0.005.

Table 3 in the Appendix reveals that $z_{.005}$ is a value between 2.57 and 2.58. Its value correct to 3 decimal places is 2.576, and thus a 99 percent confidence interval for μ is given by

$$\bar{x} \pm 2.576 \frac{\sigma}{\sqrt{n}}. \tag{9.3}$$

Formula 9.3 differs from Formula 9.2 only in regard to the z-value that is used.

Figure 9.3
Sampling distribution of the
sample mean $\bar{x}$.

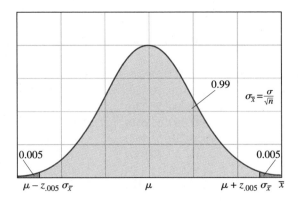

Comparing the 2 formulas, we see that, when the confidence level is increased from 95 percent to 99 percent, the z-value also increases from 1.960 to 2.576, thus resulting in a wider interval. As one might expect, the confidence interval estimate of a parameter becomes less precise (wider interval, and thus less specific) as the confidence level is increased. Formula 9.3 reveals that the precision of the confidence interval is also affected by the choice of the sample size n. The precision can be increased (interval width decreased) by using a larger sample size.

We can use the same procedure as employed above to obtain a confidence interval for μ with any specified confidence level. It is customary to denote a general confidence level by $(1 - \alpha)$. If in Figure 9.3 the probability of 0.99 is replaced by $(1 - \alpha)$, then each tail area of 0.005 is replaced by $\alpha/2$, and $z_{.005}$ changes to $z_{\alpha/2}$ (see Figure 9.4). Then the 0.99 confidence interval given in Formula 9.3 becomes the following $(1 - \alpha)$ confidence interval.

$$\bar{x} \pm z_{\alpha/2} \frac{\sigma}{\sqrt{n}} \qquad (9.4)$$

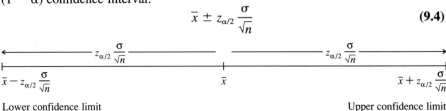

Figure 9.4
Sampling distribution of the
sample mean $\bar{x}$.

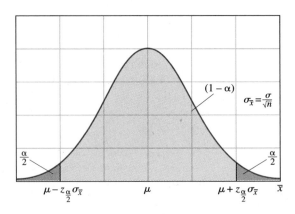

Table 9.2

z-Values for Selected Confidence Levels

Confidence Level $(1 - \alpha)$	α	Subscript for z $\alpha/2$	z-Value $z_{\alpha/2}$
0.90	0.10	0.05	1.645
0.95	0.05	0.025	1.960
0.98	0.02	0.01	2.326
0.99	0.01	0.005	2.576

If, for example, the desired confidence level is 0.90, then $(1 - \alpha) = 0.90$, which implies that $\alpha = 0.10$ and $\alpha/2 = 0.05$. Consequently, the z-value (from Table 3) used in Formula 9.4 is

$$z_{\alpha/2} = z_{.05} = 1.645.$$

Table 9.2 shows the z-values for a few selected choices of the confidence level $(1 - \alpha)$.

TIP: The subscript for z can be remembered by noting that it satisfies the following relationship:

$$\text{Subscript} = \frac{(1 - \text{confidence level})}{2}$$

For instance, for a 90 percent confidence interval, the subscript for the z-value is

$$\text{Subscript} = \frac{(1 - 0.90)}{2} = 0.05.$$

Thus, the confidence interval formula would use $z_{.05} = 1.645$.

We have shown how the Central Limit theorem can be used to obtain a large-sample confidence interval for μ, the mean of a population. The method employed is also applicable for nearly all the large-sample estimation problems that are considered in a first course. Instead of using μ, suppose we let θ denote a general parameter that needs to be estimated. A point estimator of a parameter is denoted by placing a "hat" over the parameter. Thus, $\hat{\theta}$ (theta hat) denotes a point estimator of θ. $\hat{\theta}$ will be some statistic computed from the sample measurements (when θ is μ, then $\hat{\theta}$ is $\bar{x}$). Since $\hat{\theta}$ is a statistic, it will have a sampling distribution with a mean denoted by $\mu_{\hat{\theta}}$ and a standard deviation denoted by $\sigma_{\hat{\theta}}$. If $\mu_{\hat{\theta}} = \theta$ and if the distribution of $\hat{\theta}$ is approximately normal for large n, then the following large-sample confidence interval for θ can be obtained.

Large-Sample $(1 - \alpha)$ Confidence Interval for θ:

If $\hat{\theta}$ is a point estimator whose sampling distribution has a mean equal to θ and is approximately normal for large n, then a large-sample confidence interval for the parameter θ is

$$\hat{\theta} \pm z_{\alpha/2}\sigma_{\hat{\theta}}, \tag{9.5}$$

where $\sigma_{\hat{\theta}}$ is the standard deviation of the estimator $\hat{\theta}$.

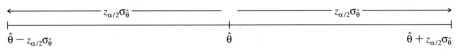

$\hat{\theta} - z_{\alpha/2}\sigma_{\hat{\theta}}$ $\hat{\theta}$ $\hat{\theta} + z_{\alpha/2}\sigma_{\hat{\theta}}$

Lower confidence limit   Upper confidence limit

Formula 9.5 is the basis for several large-sample confidence interval formulas that will be used in this text.

9.2 Confidence Interval for a Mean: Large Sample

In the previous section, an approximate large-sample $(1 - \alpha)$ confidence interval was derived for μ, the mean of a population. This was given in Formula 9.4 and is

$$\bar{x} \pm z_{\alpha/2} \frac{\sigma}{\sqrt{n}}.$$

At first glance it appears that the formula is impractical since it involves the population standard deviation σ, and this value would seldom be known. However, since the formula is based on the assumption that n is large, the sample standard deviation s will usually provide a good approximation for σ.

Large-Sample $(1 - \alpha)$ Confidence Interval for μ:

$$\bar{x} \pm z_{\alpha/2} \frac{\sigma}{\sqrt{n}} \tag{9.6}$$

Assumptions:
1. The sample is random.
2. The sample size n is large ($n \geq 30$).

Note:
1. $z_{\alpha/2}$ is the z-value whose right-tail area is $\alpha/2$.
2. σ, the population standard deviation, is usually unknown and is approximated with the sample standard deviation s.
3. If σ *is known* and if *the population has a normal distribution,* then the confidence interval is *exact* for any sample size n, large or small.

Example 9.1

The Tax Reform Act of 1986 resulted in substantial tax law changes and had a major impact on the preparation of the federal income tax return. Prior to its implementation, a certain tax preparation firm had charged a fixed fee for completing a ''standard'' return. To determine what price should be charged under the new laws, the firm needed to estimate the mean time now required by its employees to complete the new return.

The company randomly sampled 45 standard returns from its recent work, and it was found that the returns required an average of 57.8 minutes. The standard deviation of the sample was 9.6 minutes. Estimate with 99 percent confidence the actual mean time now required by this firm to complete a standard return.

Solution

The general form of the confidence interval for μ is

$$\bar{x} \pm z_{\alpha/2} \frac{\sigma}{\sqrt{n}},$$

where $\bar{x} = 57.8$ minutes and $n = 45$. Since σ, the population standard deviation, is unknown we must approximate it with the sample standard deviation $s = 9.6$ minutes. To determine the z-value, its subscript, $\alpha/2$, must first be found. This is

$$\frac{\alpha}{2} = \frac{(1 - \text{confidence level})}{2} = \frac{(1 - 0.99)}{2} = 0.005.$$

Thus, from Table 9.2 (see page 331), the z-value is

$$z_{\alpha/2} = z_{.005} = 2.576$$

Therefore, a 99 percent confidence interval for μ, the population mean, is given by

$$\bar{x} \pm z_{.005} \frac{s}{\sqrt{n}}$$

$$57.8 \pm (2.576)\frac{9.6}{\sqrt{45}}$$

$$57.8 \pm 3.7.$$

Thus, the firm can be 99 percent confident that the mean time required to complete a standard return is between 54.1 and 61.5 minutes. This information can now be used to establish a new fixed fee for preparing a standard return.

Ⓜ**Example 9.2**

To estimate the average cost of dining at an expensive restaurant, a random sample of 40 checks from this month's receipts produced the following dollar expenses per individual.

38.00	36.25	37.75	38.62	38.50	40.50	41.25	43.75	44.50	42.50
43.00	42.25	42.75	42.88	44.25	45.75	45.50	35.75	35.12	39.95
35.88	35.50	36.75	35.50	36.25	36.62	37.25	37.75	37.62	36.80
38.50	38.25	37.88	38.88	38.25	46.00	46.50	45.25	34.60	39.50

Use MINITAB to estimate with a 99 percent confidence interval the current mean charge per person at this restaurant.

Solution

The command **ZINTERVAL** can be used to obtain a large-sample confidence interval for μ. The command requires that a value be specified for the population standard deviation σ. Since this is unknown, we will approximate it by using the standard deviation of the sample. To obtain s, type the following.

```
SET C1
38.00  36.25  37.75  38.62  38.50  40.50  41.25  43.75  44.50  42.50
43.00  42.25  42.75  42.88  44.25  45.75  45.50  35.75  35.12  39.95
35.88  35.50  36.75  35.50  36.25  36.62  37.25  37.75  37.62  36.80
38.50  38.25  37.88  38.88  38.25  46.00  46.50  45.25  34.60  39.50
END
STDEV  C1
```

MINITAB will respond with the following.

```
MTB > STDEV C1
   ST.DEV. =      3.5476
```

Using $s = \$3.55$ to approximate σ, we can obtain a 99 percent confidence interval for μ with the following command.

```
ZINTERVAL 99 3.55 C1
```

The required information above must appear in the exact order specified, that is, the command **ZINTERVAL** is followed by the confidence level; then the population standard deviation; and, lastly, the column that contains the sample values. In response, MINITAB will produce the output in Exhibit 9.1. Thus, with 99 percent confidence we estimate that the mean cost per person for dining at this restaurant is between \$38.26 and \$41.16.

Exhibit 9.1

```
MTB > ZINTERVAL 99 3.55 C1

THE ASSUMED SIGMA =3.55

            N      MEAN     STDEV   SE MEAN    99.0 PERCENT C.I.
C1         40     39.709    3.548    0.561   ( 38.260,  41.158)
```

Sections 9.1 & 9.2 Exercises

Basic Concepts of Estimating Population Parameters
Confidence Interval for a Mean: Large Sample

9.1 For each of the following confidence levels, use Table 3 to determine the z-value that would be used to construct a large-sample confidence interval for a population mean.

a. 91 percent   b. 94 percent   c. 96 percent   d. 97 percent

9.2   Use Table 3 to determine the z-value that would be used to construct a large-sample confidence interval for μ with the following confidence levels.
   a. 86 percent   b. 85 percent   c. 87 percent   d. 84 percent

9.3   What is the difference between a point estimate and a confidence interval estimate for a parameter?

9.4   A researcher estimated a population mean by constructing a 99 percent confidence interval for μ. She then stated that, "I am 99 percent confident that μ lies within this interval." Explain what is meant by the statement.

9.5   Obtain a 90 percent confidence interval for μ, if $\bar{x} = 75.9$, $s = 8$, and $n = 100$.

9.6   Estimate μ with a 99 percent confidence interval, if $\bar{x} = 135.6$, $s = 24$, and $n = 64$.

9.7   A population has a standard deviation of 12, but its mean is unknown. To estimate μ, a random sample of size $n = 36$ was selected, and the resulting sample mean was 25.4.
   a. Construct a 95 percent confidence interval for μ.
   b. Construct a 99 percent confidence interval for μ.
   c. How is the width of the confidence interval affected by increasing the confidence from 95 percent to 99 percent?

9.8   A random sample of n measurements was selected from a population whose standard deviation is 32. The sample mean equaled 223.8.
   a. Assuming that $n = 64$, construct a 95 percent confidence interval for the population mean.
   b. Assuming that $n = 256$, construct a 95 percent confidence interval for the population mean.
   c. How is the width of the confidence interval affected by increasing the sample size from 64 to 256?

9.9   A random sample of size $n = 100$ was selected from a population with standard deviation σ. The sample mean was 98.84.
   a. Assume that $\sigma = 10$ and obtain a 95 percent confidence interval for the population mean.
   b. Assume that $\sigma = 20$ and obtain a 95 percent confidence interval for the population mean.
   c. How does the variability of the population affect the width of the confidence interval?

9.10 For a sample of 97 sales at a newspaper stand, the average purchase was 88 cents and the standard deviation was 16 cents. Estimate the mean of all purchases at this stand by constructing a 95 percent confidence interval.

9.11 A coffee manufacturer has decided to use the vacuum brick method in place of cans for packaging its coffee. To check if its newly installed packaging equipment is filling appropriately, it samples 31 packages and obtains a sample mean of 13.36 ounces and a standard deviation of 0.22 ounce. Construct a 99 percent confidence interval to estimate the actual mean weight of packages filled by this machine.

9.12 The *Wall Street Journal* (January 30, 1990) reported that a sample of 1,943 medical plans had an average yearly cost to the employer of $2,748 per worker.

Use a 95 percent confidence interval to estimate the mean employer cost of all employee medical plans. Assume that $\sigma = \$500$.

9.13   Lactose is a carbohydrate that is only present in milk, and it is a primary source of calories for infants. A nutritionist, investigating the eating habits of a particular ethnic group, sampled 52 one year olds and found that they consumed a daily average of 850 calories of lactose. The standard deviation was 80 calories. Construct a 90 percent confidence interval to estimate the mean daily consumption of calories from lactose for all one year olds of this ethnic group.

9.14   A psychologist, studying cognitive deficits in level 5 head trauma victims, administered a neuropsychological test to 35 patients. The mean score was 92.6 and the standard deviation was 9.4. Construct a 95 percent confidence interval for the mean score of level 5 head trauma patients on this test.

9.15   A social worker for a state's department of human services conducted a study of social security benefits of retirees. For a sample of 924, the mean benefit was $577 and the standard deviation equaled $173. Estimate with 99 percent confidence the mean amount of social security benefits received by all retirees in this state.

9.16   A sample of 30 former patients at a treatment center for alcohol rehabilitation revealed that the average length of stay was 27.8 days, with a standard deviation of 3.2 days. Estimate with a 95 percent confidence interval the mean length of stay for all patients at this center.

9.17   An economist surveyed 48 heating oil dealers and obtained a mean cost per gallon of 97.8 cents, with a standard deviation of 3.1 cents. Estimate with 90 percent confidence the mean cost of a gallon of heating oil in the region surveyed.

9.18   The following are the percentage rates charged for a 30-year home mortgage by a sample of 41 financial institutions in eastern Pennsylvania (*Lancaster Sunday News,* February 4, 1990).

9.625	10	9.875	9.875	9.9
9.75	10.1	10.125	9.5	10.125
9.875	9.5	9.875	9.625	9.75
10	9.375	9.5	9.875	9.5
10.1	9.95	9.75	9.875	11.5
9.875	9.95	9.875	9.875	9.375
10	9.625	9.75	9.75	9.375
10.5	9.5	10	9.875	9.75
9.875				

For the sample, $\Sigma x = 403.875$ and $\Sigma x^2 = 3,983.456768$. Assume that the survey can be considered representative of all financial institutions in this region. Estimate with 99 percent confidence the true mean rate for a 30-year home mortgage in eastern Pennsylvania at the time of the survey.

MINITAB Assignments

Ⓜ 9.19 Use MINITAB to do Exercise 9.18.

Ⓜ 9.20 The fat content was determined for 40 hamburgers sold by a fast food restaurant. The results in grams appear below.

20.0	18.7	21.6	20.9	21.5	21.8	20.2	19.7	18.9	19.5
19.3	21.2	18.4	21.0	21.6	20.6	20.7	21.9	20.1	17.1
18.1	21.1	19.3	21.5	20.1	16.5	18.9	17.4	20.8	18.5
21.6	23.1	20.5	22.0	20.6	17.5	16.1	20.1	21.8	19.4

Use MINITAB to obtain a 90 percent confidence interval for estimating the mean fat content of all hamburgers sold by this restaurant.

Ⓜ 9.21 Redo Exercise 9.20 under the assumption that the standard deviation of the population is known and equals 1.6 grams.

9.3 Student's *t* Probability Distributions

In Section 9.4 we will consider the problem of obtaining a confidence interval for a population mean when the sample size *n* is small. To do this, we first need to discuss a probability distribution that in many ways resembles the *z*-distribution. It is called the **Student's *t*-distribution,** although it is usually referred to more simply as the *t*-distribution. The formula that defines the *t*-distribution is complex, and it would serve no useful purpose to give it here. You should, however, be familiar with some important characteristics of a *t*-distribution's graph.

The *t*-distribution was introduced in 1908 in a paper published by William S. Gosset (1876–1936). Gosset, who was employed by an Irish brewery, chose to publish his findings using the pen name, Student. In his honor, the distribution is known today as Student's *t*-distribution. *(Photo courtesy The Granger Collection, New York.)*

Some Characteristics of *t* Probability Curves:
1. The graph of a *t*-distribution extends indefinitely far to the left and to the right, approaching more closely the *x*-axis as *x* increases in magnitude.
2. There are an infinite number of different *t*-curves. Each particular curve is determined by a single parameter, called **degrees of freedom*** (denoted by *df*). The possible values for *df* are the positive integers, and each integer defines a different *t*-distribution.
3. The high point on the graph of a *t*-distribution occurs at its mean, which is always equal to 0.
4. A *t*-curve is symmetric about its mean of 0, and, therefore, each of the areas to the left of 0 and to the right of 0 is equal to 0.5.

A *t*-distribution has many characteristics similar to those of the standard normal distribution. In fact, it can be shown that the *t*-distribution becomes more like the *z*-distribution as the degrees of freedom *df* increase.

* For an introductory statistics course, degrees of freedom can simply be thought of as an index number that identifies a particular *t*-distribution. The origin of the term can be traced to the probability distribution of the sample variance s^2. For *n* measurements, s^2 involves *n* squared deviations but it only has $(n - 1)$ degrees of freedom since only $(n - 1)$ squared deviations can vary freely. This follows from the fact that the *n* deviations must always sum to 0.

> If the degrees of freedom df are allowed to increase beyond bounds (to infinity), then the t-curve becomes the z-curve.
>
> For this reason, the t-distribution with $df = \infty$ is defined to be the standard normal distribution.

In Figure 9.5, the z-distribution and the t-distribution with $df = 4$ are sketched. Even for this small value of degrees of freedom, the t-curve is very similar to the z-curve. The main difference is that the t-curve has a larger amount of area in its tails and less area concentrated near its mean of 0.

Tables are used to provide frequently used t-values for various t-distributions. Table 4 in Appendix A is such a table, and it also appears on the inside of the back cover. Thirty t-distributions are represented, those having $df = 1, 2, 3, \ldots 29$, and also $df = \infty$ (the z-distribution). A portion of Table 4 is reproduced below in Table 9.3.

Table 9.3

A Portion of Table 4 from Appendix A

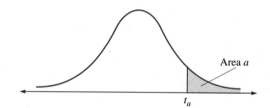

Degrees of	Amount of Area in One Tail				
Freedom (df)	.100	.050	.025	.010	.005
1	3.078	6.314	12.706	31.821	63.657
2	1.886	2.920	4.303	6.965	9.925
3	1.638	2.353	3.182	4.541	5.841
4	1.533	2.132	2.776	3.747	4.604
5	1.476	2.015	2.571	3.365	4.032
6	1.440	1.943	2.447	3.143	3.707
7	1.415	1.895	2.365	2.998	3.499
8	1.397	1.860	2.306	2.896	3.355
9	1.383	1.833	2.262	2.821	3.250
10	1.372	1.812	2.228	2.764	3.169
11	1.363	1.796	2.201	2.718	3.106
⋮	⋮	⋮	⋮	⋮	⋮

Figure 9.5
Standard normal and t-distributions.*

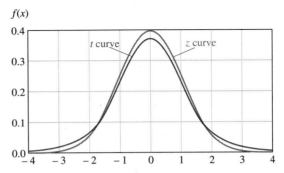

*t - distribution has 4 degrees of freedom.

Notice that the t-table is constructed much differently from the z-table. Only five areas are given in the table, and these appear as the headings of columns two through six. Moreover, these are right-tail areas, not left cumulative areas as appear in the z-table. Column one contains the degrees of freedom for the particular t-distribution under consideration. The contents of each remaining column are the t-values that have an area to the right equal to that at the top of the column. In other words, each column contains values of t_a, where a is the specified area that heads the column. The next two examples illustrate how the table is used.

Example 9.3   For the t-distribution with 10 degrees of freedom, find the number c such that $P(t \geq c) = 0.05$.

Solution

We want $t_{.05}$, the t-value for which the area to the right of it is equal to 0.05. Consequently, we use the column headed by this area. The applicable row is that for which $df = 10$, since this is the t-distribution being considered. The value that appears at the intersection of the identified row and column is 1.812. Therefore,

$$P(t \geq 1.812) = 0.05$$

and thus,

$$t_{.05} = 1.812.$$

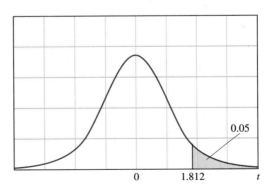

Example 9.4   Find P_{90}, the 90th percentile, for the t-distribution with 7 degrees of freedom.

Solution

The 90th percentile is the value of t for which the cumulative area to the left is 0.90. Therefore, the area to the right of t is $1 - 0.90 = 0.10$. From Table 4, the t-value with 7 degrees of freedom that has a right-tail area of 0.10 is 1.415. Thus,

$$P_{90} = t_{.10} = 1.415.$$

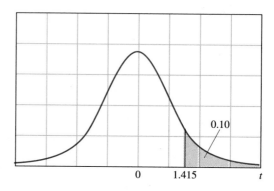

Although Table 4 only gives t-values with tail areas of 0.10, 0.05, 0.025, 0.01, and 0.005, MINITAB can be used to obtain t-values for any area. It can also be employed to determine probabilities concerning any value of t.

Ⓜ **Example 9.5**   For the t-distribution with 23 degrees of freedom, use MINITAB to find the following.

1. $P(t > 2.65)$
2. P_{85}, the 85th percentile

Solution

1. We have seen that the **CDF** command can be used to obtain cumulative probabilities. Since

$$P(t > 2.65) = 1 - P(t \le 2.65),$$

we need to find the cumulative probability to the left of 2.65 and then subtract this value from 1. Type the following commands.

```
CDF 2.65;
T 23.
```

In the subcommand that appears on the second line, **T** indicates that we want a cumulative probability for a t-distribution, and 23 specifies its degrees of freedom. MINITAB will produce the results appearing in Exhibit 9.2, from which we obtain

$$P(t > 2.65) = 1 - 0.9928 = 0.0072.$$

Exhibit 9.2

```
MTB > CDF 2.65;          # 2.65 is the t-value
SUBC> T 23.              # 23 indicates value of df
    2.6500    0.9928
```

2. To find the 85th percentile, the **INVCDF** command is used to obtain the *t*-value that has a cumulative probability of 0.85. Type the following.

```
INVCDF 0.85;
T 23.
```

From the results in Exhibit 9.3, we have that $P_{85} = 1.0603$.

Exhibit 9.3

```
MTB > INVCDF 0.85;
SUBC> T 23.
    0.8500    1.0603
```

Section 9.3 Exercises

Student's t *Probability Distributions*

For each of the Exercises 9.22–9.31, find the *t*-value that has a right-tail area equal to the specified subscript.

9.22 $t_{.05}$ with $df = 15$

9.23 $t_{.10}$ with $df = 23$

9.24 $t_{.10}$ with $df = 6$

9.25 $t_{.005}$ with $df = 14$

9.26 $t_{.025}$ with $df = 28$

9.27 $t_{.025}$ with $df = 11$

9.28 $t_{.01}$ with $df = \infty$

9.29 $t_{.01}$ with $df = 29$

9.30 $t_{.005}$ with $df = 21$

9.31 $t_{.05}$ with $df = \infty$

Determine the following percentiles.

9.32 P_{99} with $df = 13$

9.33 P_{95} with $df = 26$

9.34 P_{90} with $df = \infty$

9.35 P_{99} with $df = 15$

9.36 P_{95} with $df = 18$

9.37 P_{90} with $df = 3$

9.38 P_{10} with $df = 9$

9.39 P_5 with $df = 12$

9.40 P_1 with $df = 22$

9.41 P_1 with $df = \infty$

9.42 P_5 with $df = 11$

9.43 P_{10} with $df = 17$

Use Table 4 in Appendix A to determine the following probabilities.

9.44 $P(t > 2.086)$ with $df = 20$

9.45 $P(t > 2.650)$ with $df = 13$

9.46 $P(t < 2.787)$ with $df = 25$

9.47 $P(t < 1.476)$ with $df = 5$

9.48 $P(t > -1.701)$ with $df = 28$

9.49 $P(t < -2.093)$ with $df = 19$

9.50 $P(-1.397 < t < 2.896)$ with $df = 8$

9.51 $P(-1.708 < t < 2.787)$ with $df = 25$

For each of the Exercises 9.52 − 9.57, find the value of c.

9.52 $P(t > c) = 0.05$ with $df = 19$

9.53 $P(t > c) = 0.01$ with $df = 11$

9.54 $P(t < c) = 0.005$ with $df = 27$

9.55 $P(t > c) = 0.90$ with $df = 3$

9.56 $P(-c < t < c) = 0.95$ with $df = 10$

9.57 $P(-c < t < c) = 0.99$ with $df = 26$

MINITAB Assignments

Ⓜ 9.58 For the t-distribution with $df = 37$, use MINITAB to find
   a. the 75th percentile,
   b. $P(t < 1.96)$.

Ⓜ 9.59 For the t-distribution with $df = 48$, use MINITAB to find
   a. $P(t > 2.22)$,
   b. the 68th percentile.

9.4 Confidence Interval for a Mean: Small Sample

In Section 9.2 we saw that for a large sample the construction of a confidence interval for μ is quite straightforward, even when the population's standard deviation σ is unknown. When n is small, the derivation of a confidence interval for μ is only slightly more complicated. In this case it is necessary to assume that the sampled population has a normal distribution. Then a small-sample confidence for μ can be derived by using the fact that the sampling distribution of the statistic

$$t = \frac{\bar{x} - \mu}{\dfrac{s}{\sqrt{n}}}$$

is a t-distribution with degrees of freedom $df = n - 1$. The resulting confidence interval is given in Formula 9.7.

Small-Sample $(1 - \alpha)$ Confidence Interval for μ:

$$\bar{x} \pm t_{\alpha/2}\, \frac{s}{\sqrt{n}} \tag{9.7}$$

Assumptions:
1. The sample is random.
2. The sampled population has a normal distribution.

> *Note:*
> $t_{\alpha/2}$ is the *t*-value whose right-tail area is $\alpha/2$, and it's value is obtained from the *t*-distribution (Table 4) with $df = n - 1$.

Statisticians generally try to base their inferences on large samples. However, it is not always feasible to have *n* large. Sometimes we are using data that have been obtained by someone else, and thus, we are restricted to the data at hand; or we may be involved in **destructive sampling** in which each sampled item must be destroyed to obtain the desired measurement. For example, suppose a laboratory is gathering safety data on an expensive automobile model, and each vehicle has to be demolished to obtain the desired information. In this case, it would not be feasible to take a large sample. Thus, there is a need for small-sample confidence interval formulas.

In obtaining a confidence interval for μ, the basic difference between the large- and small-sample formulas is that a *z*-value is used in the large-sample situation, while this is replaced by a *t*-value when *n* is small. Moreover, when *n* is small, the population should have approximately a normal distribution. At first this may seem to be a very restrictive assumption. In reality it often does not present a problem since many things in life have mound-shaped distributions.

Before illustrating Formula 9.7, we should point out that, although it is usually referred to as the small-sample confidence interval for μ, it actually applies for any sample size *n*, as long as the sampled population is normally distributed. However, when *n* is large ($n \geq 30$), there is little difference between the values of $t_{\alpha/2}$ and $z_{\alpha/2}$. Consequently, we will adhere to the usual practice of using a *z*-value for $n \geq 30$ and a *t*-value for $n < 30$.

Example 9.6

A primary factor in the operating speed of a personal computer is the clock speed of its microprocessor chip, the system's central processing unit (CPU). The manufacturer's

In 1978, Intel developed the 8086 CPU that was used in the first IBM personal computers. Today, microprocessors smaller than a postage stamp are available with over a million transistors and more computing power than many older mainframe computers. (Photo reprinted by permission of Intel Corporation, Copyright 1993.)

rating of a chip is conservative, and it will usually run at a higher speed. A computer manufacturer is considering the purchase of a large lot of chips that are rated at a speed of 33 MHz (megahertz). It randomly selects and tests eight chips, obtaining the following speeds.

34.3   33.6   32.9   35.2   34.0   32.8   33.8   33.0

Construct a 90 percent confidence interval to estimate the mean operating speed of all chips in the lot. Assume that operating speeds are approximately normally distributed.

Solution

Since n is small, the confidence interval for μ is

$$\bar{x} \pm t_{\alpha/2} \frac{s}{\sqrt{n}}.$$

To use this, we first must calculate the sample mean and standard deviation.

$$\bar{x} = \frac{\Sigma x}{n} = \frac{269.6}{8} = 33.7$$

$$SS(x) = \Sigma x^2 - \frac{(\Sigma x)^2}{n} = 9{,}090.18 - \frac{(269.6)^2}{8} = 4.66$$

$$s^2 = \frac{SS(x)}{n-1} = \frac{4.66}{7} = 0.6657$$

$$s = \sqrt{0.6657} = 0.816$$

For a confidence level of 90 percent, the subscript for the t-value is

$$\frac{\alpha}{2} = \frac{(1 - 0.90)}{2} = 0.05.$$

We now need to use Table 4 to look up $t_{.05}$ based on degrees of freedom $df = (n - 1) = 7$. This value is 1.895. Therefore, a 90 percent confidence interval for μ is

$$\bar{x} \pm t_{.05} \frac{s}{\sqrt{n}}$$

$$33.7 \pm (1.895) \frac{0.816}{\sqrt{8}}$$

$$33.7 \pm 0.55.$$

Thus, we are 90 percent confident that the mean speed of the entire lot of microprocessors is between 33.15 MHz and 34.25 MHz.

Ⓜ**Example 9.7**   Use MINITAB to find the confidence interval in Example 9.6.

Solution

First, we enter the sample values into MINITAB with the **SET** command.

```
SET C1
34.3  33.6  32.9  35.2  34.0  32.8  33.8  33.0
END
```

To obtain a confidence interval for μ, the command **TINTERVAL** is used, followed by the confidence level (percentage or decimal) and the column that contains the sample. Thus, to obtain a 90 percent confidence interval based on the sample contained in column C1, type the following.

```
TINTERVAL 90 C1
```

MINITAB will produce the output presented in Exhibit 9.4.

Note that unlike the **ZINTERVAL** command, **TINTERVAL** does not require that the population standard deviation be specified. This is because the small-sample formula always uses the sample standard deviation, and MINITAB automatically computes s and uses it with the **TINTERVAL** command.

Exhibit 9.4

```
MTB > SET C1
DATA> 34.3 33.6 32.9 35.2 34.0 32.8 33.8 33.0
DATA> END
MTB > TINTERVAL 90 C1

             N      MEAN     STDEV    SE MEAN    90.0 PERCENT C.I.
    C1       8     33.700    0.816     0.288    ( 33.153,  34.247)
```

Normal Probability Plot (Optional)

Ⓜ **Example 9.8**

The small-sample confidence interval for μ is based on the assumption that the sampled population is normally distributed. Gross violations of this assumption can be checked by having MINITAB construct a **normal probability plot.** The procedure consists of first determining the so-called **normal scores** of the sample. Normal scores are basically the z-scores that you would expect to obtain for the largest value, the next-to-largest value, and so on, in a sample of n-values randomly selected from the z-distribution. Once obtained, each sample value is plotted with its associated normal score. If the sample is from a normal population, then the points are expected to lie approximately on a straight line. The "closeness" of the points to a straight line can be measured by calculating the correlation coefficient (a formal test for this is given in Chapter 14).

We will illustrate the procedure for the sample of eight microprocessor speeds that were stored in column C1 in the preceding example. The command **NSCORES** calculates the normal scores for the sample in C1 and places the results in column C2. The **PLOT** command graphs the normal scores against the corresponding sample values. The **CORRELATION** command is then applied to the points.

```
NSCORES C1 C2 # puts in C2 normal scores of sample C1
PLOT C2 C1
CORRELATION C1 C2
```

The resulting output appears in Exhibit 9.5.

The normal scores plot appears reasonably straight, and this judgment is supported by the fact that the correlation coefficient of 0.967 is close to 1. Thus, the normality assumption concerning the population of microprocessor speeds seems to be reasonable.

Exhibit 9.5

```
MTB > SET C1
DATA> 34.3   33.6   32.9   35.2   34.0   32.8   33.8   33.0
DATA> END
MTB > NSCORES C1 C2     # puts in C2 normal scores of sample C1
MTB > PLOT C2 C1

C2   -
     -
     -                                                        *
 1.0+
     -                                          *
     -                                   *
     -                              *
 0.0+
     -                        *
     -             *
     -       *
-1.0+
     -
     -    *
     -
     +-------+---------+---------+---------+---------+--------C1
      32.50   33.00     33.50     34.00     34.50     35.00

MTB > CORRELATION C1 C2

Correlation of C1 and C2 = 0.967
```

Section 9.4 Exercises
Confidence Interval for a Mean: Small Sample

In the following exercises, assume that the sampled populations have approximately normal distributions.

9.60 For each of the following confidence levels and sample sizes, use Table 4 to determine the t-value that would be used to construct a small-sample confidence interval for a population mean.
   a. 95 percent, $n = 21$ b. 99 percent, $n = 28$
   c. 90 percent, $n = 13$ d. 98 percent, $n = 7$

9.61 Use Table 4 to determine the *t*-value that would be used to construct a small-sample confidence interval for μ with the following confidence levels and sample sizes.

 a. 80 percent, $n = 15$ b. 90 percent, $n = 22$
 c. 99 percent, $n = 29$ d. 95 percent, $n = 9$

9.62 Obtain a 95 percent confidence interval for μ if $\bar{x} = 85.0$, $s = 6$, and $n = 16$.

9.63 Estimate μ with a 99 percent confidence interval if $\bar{x} = 575$, $s = 95$, and $n = 25$.

9.64 A random sample of n measurements had a mean $\bar{x} = 150$ and a standard deviation $s = 20$.

 a. Assume that $n = 16$ and construct a 95 percent confidence interval for the population mean.
 b. Assume that $n = 25$ and construct a 95 percent confidence interval for the population mean.
 c. Determine the width of each confidence interval in Parts a and b.

9.65 Use the following random sample to obtain a 90 percent confidence interval for the mean of the sampled population.

$$x_1 = 28, \quad x_2 = 20, \quad x_3 = 25, \quad x_4 = 27, \quad x_5 = 25$$

9.66 A sample of 24 students in the college of science at a large university revealed that they had used the institution's computing facilities an average of 39.8 hours last semester. The sample standard deviation was 6.3 hours. Construct a 95 percent confidence interval to estimate the mean number of hours of usage last semester by all science students at this university.

9.67 To see if enough fruit is being placed in cups of fruit-flavored yogurt, a quality inspector measured the amount in 27 containers. She found an average of 28.9 grams of fruit and a sample standard deviation of 2.8 grams. Estimate with 99 percent confidence the mean amount of fruit for all cups of this type of yogurt.

9.68 A normal pregnancy and hospital delivery now costs an average of $4,334 (*USA TODAY*, January 11, 1990). To see how charges in a particular state compare with this figure, a sample of 29 recent births was investigated. For the sample, the mean cost was $4,725 and the standard deviation was $197. Construct a 90 percent confidence interval to estimate the true mean cost of a normal delivery in this state.

9.69 A sociologist, studying the use that older people make of their environmental space, surveyed 21 residents of a large retirement community. For the sample, she found that the residents averaged 34.3 destinations per month (a destination is a trip outside the residence). The standard deviation was 3.9. For residents of this community, estimate with 95 percent confidence the mean number of destinations per month.

9.70 The U.S. Department of Education claims that the average teacher (elementary and high school) puts in a 50-hour work week (*Wall Street Journal*, January 9, 1990). Suppose a superintendent of schools wants to estimate the average work week of teachers in his district. Twenty-five teachers are randomly selected, and it is found that they work an average of 52.7 hours a week. The standard deviation is 4.2 hours. Estimate with 95 percent confidence the mean work week for all teachers in this district.

9.71 The amounts, in ounces, of beverage dispensed by a vending machine are given below for a sample of 10 cups.

8.7   8.5   9.1   9.2   8.5   8.9   8.7   9.1   8.9   9.2

Estimate with a 90 percent confidence interval the true mean amount dispensed by the machine.

9.72 Background radiation is the normal radiation that is present in the environment from natural sources such as bricks, concrete, rocks, the ground, cosmic rays, and so on. To determine the average background level of a laboratory, an engineer measured radiation levels with a Geiger counter, and obtained the following values in microroentgens/hour.

5.5   6.1   5.7   5.9   5.5   6.0   5.8   5.9

Estimate with 90 percent confidence the mean background level of the laboratory.

9.73 For a random sample of 15 withdrawals from an automated teller machine, the average amount withdrawn was $115 and the standard deviation was $25. Obtain a 95 percent confidence interval to estimate the true mean amount withdrawn per transaction from this machine.

MINITAB Assignments

Ⓜ 9.74 The balances of a random sample of 13 charge accounts at a large department store appear below. The amounts are in dollars.

1300.43   795.78   1028.94   376.86   982.22   1002.23   254.33
1729.64   103.45   1112.87   202.32   875.68   1212.21

Use MINITAB to construct a 90 percent confidence interval for the mean balance of all accounts at this store.

Ⓜ 9.75 A survey of 15 plumbers in a midwestern state revealed the following hourly rates. The figures are in dollars.

18.50   17.50   17.50   16.00   17.00   15.00   16.00   15.50
15.00   18.00   18.50   15.50   17.00   17.50   19.00

Use MINITAB to estimate with 95 percent confidence the average hourly rate charged by all plumbers in this state.

Ⓜ 9.76 For the account balances in Exercise 9.74, use MINITAB to construct a normal scores plot in order to check on the reasonableness of the normality assumption.

Ⓜ 9.77 Using MINITAB, check if the normality assumption appears to be violated for the sample of hourly rates in Exercise 9.75.

9.5 Confidence Interval for a Proportion: Large Sample

In our daily exposures to television, radio, and the written media, we are frequently presented with estimates of population proportions or, equivalently, percentages. We might be told that 8 percent of all companies surveyed have a policy against hiring smokers; that 23 percent of the automobiles driven in a given city fail to meet emissions control standards; that, based on a recent poll, only 44 percent of a state's eligible voters will participate in an upcoming referendum.

To estimate a population proportion, we need to consider the problem of estimating the parameter p (probability of a success) in a binomial experiment. For example, automobile manufacturers often survey recent purchasers of its cars to estimate the proportion of satisfied owners. Suppose 200,000 people have bought a recently introduced model, and the company wants to estimate p, the proportion of this population of owners who are satisfied with their purchase. To accomplish this, 1,000 buyers were surveyed and 850 expressed satisfaction with their car. A point estimate of the population proportion p is obtained by computing the sample proportion of satisfied owners. The sample proportion is denoted by $\hat{p}$ and equals x/n, where x is the number of successes (satisfied owners) in the sample and n is the sample size. Thus, the proportion of the 200,000 owners who are satisfied is estimated by

$$\hat{p} = \frac{850}{1,000} = 0.85$$

A large-sample confidence interval for p can be obtained from Formula 9.5 by utilizing the following properties of $\hat{p}$.

The Sampling Distribution of $\hat{p}$, the Sample Proportion:
1. The mean of $\hat{p}$ is $\mu_{\hat{p}} = p$, the population proportion.
2. The standard deviation of $\hat{p}$ is $\sigma_{\hat{p}} = \sqrt{pq/n}$.
3. The shape of the sampling distribution is approximately normal for large samples.

In Section 9.1, Formula 9.5 gave the following large-sample confidence interval for a parameter θ.

$$\hat{\theta} \pm z_{\alpha/2}\sigma_{\hat{\theta}}$$

To obtain a confidence interval for p, we replace $\hat{\theta}$ by $\hat{p}$ and $\sigma_{\hat{\theta}}$ by $\sigma_{\hat{p}}$. However, for $\sigma_{\hat{p}}$ we must use $\sqrt{\hat{p}\hat{q}/n}$ in place of $\sqrt{pq/n}$, since p is unknown. These substitutions lead to the following approximate large-sample confidence interval for p.

Large-Sample $(1 - \alpha)$ Confidence Interval for p:

$$\hat{p} \pm z_{\alpha/2}\sqrt{\frac{\hat{p}\hat{q}}{n}} \qquad\qquad (9.8)$$

Assumptions:
1. The sample is random.
2. The sample size n is large ($n \geq 30$, $n\hat{p} \geq 5$, and $n\hat{q} \geq 5$).

Note:
1. $z_{\alpha/2}$ is the z-value whose right-tail area is $\alpha/2$.
2. $\hat{p}$ is the sample proportion and equals x/n, where x is the number of successes in the sample.
3. $\hat{q} = 1 - \hat{p}$.

Previously, we stated that a sample is considered large if $n \geq 30$. This rule of thumb will usually suffice as a guide for applying Formula 9.8. However, exceptions to this rule can occur if p or q is small. Consequently, in addition to being at least 30, n should also be large enough so that the guidelines discussed in Section 7.4 are also satisfied, namely, $n\hat{p} \geq 5$ and $n\hat{q} \geq 5$.

Example 9.9   Because of recent expansions in new housing, a city's present sewage treatment facility needs to be expanded. In planning for this project, the city wants to estimate the proportion of houses that have an in-sink garbage disposal. A survey of 880 randomly selected houses was taken, and 308 were found to have disposals. Estimate with 95 percent confidence the proportion of all houses in the city that have an in-sink garbage disposal.

Solution

The point estimate of p, the population proportion, is the sample proportion

$$\hat{p} = \frac{x}{n} = \frac{308}{880} = 0.35.$$

A 95 percent confidence interval for p is given by Formula 9.8.

$$\hat{p} \pm z_{\alpha/2} \sqrt{\frac{\hat{p}\hat{q}}{n}}$$

$$0.35 \pm z_{.025} \sqrt{\frac{(0.35)(1 - 0.35)}{880}}$$

$$0.35 \pm 1.96 \sqrt{\frac{(0.35)(0.65)}{880}}$$

$$0.35 \pm 0.032$$

Thus, with 95 percent confidence, we estimate that the percentage of houses in the city with in-sink garbage disposals is some value between 31.8 percent and 38.2 percent.

Example 9.10   One hundred college football players eligible for the NFL draft were surveyed by the Campos Market Research Company (*The Tampa Tribune,* January 16, 1992). Eighty-two percent of those sampled believe that the NFL should have mandatory testing for the HIV virus. Assume that the sample can be regarded as random and estimate with 90 percent confidence the percentage of all draft-eligible college players who believe in mandatory testing for the virus.

Solution

The sample size n equals 100 and the sample proportion is

$$\hat{p} = 0.82.$$

Applying Formula 9.8, a 90 percent confidence interval for the population proportion p is

$$\hat{p} \pm z_{\alpha/2} \sqrt{\frac{\hat{p}\hat{q}}{n}}$$

$$0.82 \pm z_{.05} \sqrt{\frac{(0.82)(1 - 0.82)}{100}}$$

$$0.82 \pm 1.645 \sqrt{\frac{(0.82)(0.18)}{100}}$$

$$0.82 \pm 0.063.$$

We estimate with 90 percent confidence that between 75.7 percent and 88.3 percent of draft-eligible college football players believe that the NFL should have mandatory testing for the HIV virus.

Formula 9.8 is applicable for estimating p when n is large. This is the case in nearly all practical applications pertaining to proportions. Professionally conducted public opinion surveys are usually based on a minimum of 1,000 people. For samples based on a small value of n, one can utilize extensive tables* that have been constructed for obtaining confidence intervals for p. For these reasons, we will not consider small-sample inferences for a proportion.

Section 9.5 Exercises
Confidence Interval for a Proportion: Large Sample

9.78 A random sample of size $n = 500$ resulted in $x = 325$ successes. Obtain a 99 percent confidence interval for p, the proportion of successes in the population.

9.79 A random sample of $n = 400$ observations yielded $x = 120$ successes. Estimate the proportion of successes in the population by constructing a 90 percent confidence interval.

9.80 To determine if a coin is fair, it was tossed 1,000 times, and 535 heads were observed. Estimate with 95 percent confidence the probability of tossing a head with this coin.

9.81 A die was tossed 600 times, and an ace (one spot) occurred on 90 tosses. Construct a 90 percent confidence interval to estimate the probability of tossing an ace with this die.

* See, for example, National Bureau of Standards Handbook 91, *Experimental Statistics.* U.S. Government Printing Office, Washington, D.C., 1963.

9.82   The National Transportation Safety Board conducted a study of truck drivers killed in highway accidents. They found that 24 of 185 drivers tested positive for alcohol (The *Wall Street Journal,* February 7, 1990). Obtain a 90 percent confidence interval for the true percentage of truck driver deaths in which the truck driver had a positive level of alcohol.

9.83   An insurance company wants to estimate what proportion of its policyholders would be interested in a $500 deductible on their collision coverage. The company randomly sampled 1,200 policyholders, and 780 said they would adopt this deductible if it were available. Estimate with a 95 percent confidence interval the proportion of the company's policyholders who would add the $500 deductible.

9.84   Capitalizing on America's love affair with oat bran, food companies are promoting its benefits in a wide variety of products. To illustrate the extreme to which this has been carried, the *Lempert Report,* a food industry newsletter, asked 160 shoppers if they would buy bottles of Coke and packages of Life Savers with oat bran added (*Newsweek,* January 29, 1990). Fifty-one shoppers responded ''yes.'' Assume that the sample can be regarded as random and estimate with 95 percent confidence the proportion of all shoppers who would be willing to purchase such products with oat bran added.

9.85   One health insurance company now offers a discount on group policies to companies with at least 90 percent nonsmoking employees. Suppose a company with several thousand workers surveys 200 workers and finds that 184 are nonsmokers. Estimate with 95 percent confidence the percentage of all employees for this company who do not smoke.

9.86   Radial keratotomy is a surgical procedure used to correct myopia and other vision problems. The cornea is reshaped by making several incisions in a spoke-like pattern. In a clinical study of its effectiveness, 331 out of 435 patients achieved visual improvement to the 20/40 level (*Business Week,* January 29, 1990). Estimate with a 90 percent confidence interval the probability that this procedure will improve a patient's vision to the 20/40 level.

9.87   A computer software company has purchased a new cellophane wrapping machine for packaging its products. An inspector randomly selects 250 boxes wrapped by this machine and finds that 35 have excessive wrinkles in the wrapping. Construct a 99 percent confidence interval for the probability that a package wrapped by this machine will contain excessive wrinkles.

9.88   A major pharmaceutical firm recently received approval by the U.S. Food and Drug Administration to market a prescription drug for hair loss. Its potential for success, however, is uncertain. In a test of 2,300 men, only 39 percent showed dense or moderate new hair growth. The other 61 percent showed little or no new growth. Estimate with 95 percent confidence the success rate for this product.

9.89   The *Wall Street Journal* (January 15, 1990) reported that a recent survey showed that 16 percent of employers give their employees the day off on their birthdays. Assume that 1,000 employers were surveyed (the actual figure was not given). Estimate with 90 percent confidence the percentage of all employers that let their workers stay home on their birthdays.

9.90 Of 860 houses burglarized in an affluent New York community, only 10 percent had alarm systems (*Temple Review,* Winter 1990).
   a. Obtain a 90 percent confidence interval to estimate the percentage of all houses in this community with alarm systems.
   b. Explain why one might question the validity of the confidence interval obtained in Part a.

9.91 The Rand Corporation coordinated a study of adult outpatients who were treated in a variety of settings such as HMOs, group medical practices, and solo practices. Researchers surveyed 11,242 subjects and found that 2,467 experienced depressive symptoms such as feeling "down in the dumps" or chronically tired (*Bangor Daily News,* August 18, 1989). Estimate the percentage of all outpatients at these health centers who suffer from depressive symptoms. Use a confidence level of 95 percent.

9.6 Determining the Required Sample Size

Determining the sample size is an important consideration when designing an experiment to estimate a parameter such as a population mean. The problem of deciding how large n should be is usually solved by referring back to the large-sample confidence interval formula for the parameter of interest. We will first illustrate the general technique for μ, a population mean, and then for a population proportion p.

Finding n for Estimating a Mean

From Formula 9.6, a $(1 - \alpha)$ confidence interval for μ is

$$\bar{x} \pm z_{\alpha/2} \frac{\sigma}{\sqrt{n}}.$$

As Figure 9.6 illustrates, the center of the confidence interval is $\bar{x}$, and the confidence limits differ in magnitude from $\bar{x}$ by the quantity

$$z_{\alpha/2} \frac{\sigma}{\sqrt{n}}.$$

We have seen that, prior to selecting the sample, the probability is $(1 - \alpha)$ that the random interval in Figure 9.6 will contain the population mean μ. Consequently, if we were to use only the value of the sample mean $\bar{x}$ to estimate μ, we would expect, with probability $(1 - \alpha)$, that this point estimate will differ from μ by at most $z_{\alpha/2} \dfrac{\sigma}{\sqrt{n}}$.

This quantity is called the **maximum error of the estimate,** and is denoted by E.

Figure 9.6
$(1 - \alpha)$ confidence interval for μ.

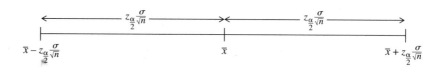

Maximum Error of the Estimate:
When the sample mean is used to estimate the population mean, the probability is $(1 - \alpha)$ that $\bar{x}$ will be off from μ by at most E. E is called the **maximum error of the estimate** and equals one-half the width of the confidence interval.

$$E = z_{\alpha/2} \frac{\sigma}{\sqrt{n}} \qquad (9.9)$$

Formula 9.9 can be used to determine how large the sample size n must be. Using algebra, the equation can be rewritten so that n is explicitly expressed in terms of the other quantities. The result is given in Formula 9.10.

Sample Size Formula When $\bar{x}$ is Used to Estimate μ:

$$n = \left[\frac{\sigma z_{\alpha/2}}{E} \right]^2 \qquad (9.10)$$

where
σ is the standard deviation of the population,
E is the maximum error of the estimate,
$(1 - \alpha)$ is the associated confidence level.

To use Formula 9.10 to find n, we must have a value for σ, the standard deviation of the population. In practice, σ is seldom known and it must be approximated. We may be able to approximate it with the standard deviation s of a previously selected sample. If this is not feasible, then knowledge of the range of the data can be used to obtain the following rough approximation for σ (see Exercise 2.98, page 79).

$$\sigma \approx \frac{\text{range}}{4}$$

Example 9.11   Earlier, we considered a purchasing agent for a national chain of seafood restaurants who wanted to estimate the mean weight of shrimp in a large catch. How many shrimp should be sampled from the catch to be 95 percent confident that the sample mean will be off from the actual mean weight by no more than 1.4 grams? Assume that the range of the weights is expected to be about 17 grams.

Solution

The population standard deviation can be approximated by

$$\sigma \approx \frac{\text{range}}{4} = \frac{17}{4} = 4.25 \text{ grams.}$$

We want the maximum error of the estimate to be $E = 1.4$ grams. For 95 percent confidence, the z-value is $z_{\alpha/2} = z_{.025} = 1.96$. The required sample size is

$$n = \left[\frac{\sigma z_{\alpha/2}}{E} \right]^2 = \left[\frac{(4.25)(1.96)}{1.4} \right]^2 = 35.4025 \approx 36.$$

Notice that instead of rounding n off to 35, we have rounded it up to 36. This is because n must be a whole number that is at least as large as 35.4025. For sample size problems, we will *always round* n *up to the nearest integer.*

Example 9.12   Suppose in the previous example the purchasing agent needs to estimate the mean weight of the catch with greater precision than 1.4 grams. Assume that the agent wants the sample mean to be off by at most 1 gram. How many shrimp in the catch will need to be sampled? Again assume that $\sigma = 4.25$ grams and that a confidence level of 95 percent is desired.

Solution

Now the maximum error is to be $E = 1$ with probability 0.95.

$$n = \left[\frac{\sigma z_{\alpha/2}}{E}\right]^2 = \left[\frac{(4.25)(1.96)}{1}\right]^2 = 69.3889$$

Rounding 69.3889 up to the nearest integer, we have that 70 shrimp need to be sampled. Thus, consistent with our intuition, a larger sample size is needed to estimate μ with more precision (a smaller maximum error) and the same amount of confidence.

The previous examples illustrate that the size of n depends on the specified maximum error. In addition, an examination of Formula 9.10 reveals that n is also affected by the confidence level and the standard deviation of the population. Increasing the confidence will increase the value of $z_{\alpha/2}$, and, thus, the size of n. For instance, if in Example 9.12 the confidence is increased from 95 percent to 99 percent, then $z_{\alpha/2}$ changes from 1.96 to 2.576, and n would need to be 120. As one would likely expect, the population's variability adversely affects the required sample size. If, for example, the standard deviation of the catch were 5.3 instead of 4.25 grams, then $n = 187$ shrimp would have to be sampled to have the maximum error be 1 gram with 99 percent confidence.

Finding n for Estimating a Proportion

We can obtain a sample size formula for estimating a population proportion p by following the same general method that was used for a mean. We started with the fact that when $\bar{x}$ is used to estimate μ, the maximum error of the estimate is as follows.

$$E = z_{\alpha/2}\frac{\sigma}{\sqrt{n}}$$
$$= z_{\alpha/2} \cdot \text{(the standard deviation of the estimate)}$$

Similarly, when the sample proportion $\hat{p}$ is used to estimate the population proportion p, the **maximum error of the estimate** is

$$E = z_{\alpha/2} \cdot \text{(the standard deviation of the estimate)}$$
$$= z_{\alpha/2}\sqrt{pq/n}.$$

Table 9.4

Values of pq for Different Values of p

p	0	0.1	0.2	0.3	0.4	**0.5**	0.6	0.7	0.8	0.9	1
q	1	0.9	0.8	0.7	0.6	**0.5**	0.4	0.3	0.2	0.1	0
pq	0	0.09	0.16	0.21	0.24	**0.25**	0.24	0.21	0.16	0.09	0

By rewriting this equation so that it is solved for n, we have

$$n = pq \left[\frac{z_{\alpha/2}}{E} \right]^2.$$

A problem with using this formula is that it depends on p, which is unknown. However, this can be circumvented by using 0.5 for p, since this is the value that maximizes pq (see Table 9.4). By using 0.5 for p, the formula will give a value of n that may be larger than actually needed, and the associated confidence level will be at least $(1 - \alpha)$. On some occasions, the experimenter will have some prior information about p, such as a range of possible values. In this case we use the value for p that is closest to 0.5.

Sample Size Formula when $\hat{p}$ is Used to Estimate p:

$$n = pq \left[\frac{z_{\alpha/2}}{E} \right]^2 \qquad (9.11)$$

where E is the maximum error of the estimate,
$(1 - \alpha)$ is the associated confidence level.
Note:
When prior information indicates that p will fall within some range, use for p the value nearest 0.5; with no prior knowledge about p, use 0.5.

Example 9.13 Concerned with the rising cost of health care insurance, the National Association of Manufacturers surveyed its members about health care benefits to employees (The *Wall Street Journal,* May 24, 1989). One of several objectives was to estimate the proportion of companies that pay the entire cost of worker health care coverage. Determine the number of companies that would have to be sampled to be 99 percent confident that the sample proportion would be off from the true value by no more than 0.03.

Solution

The maximum error of the estimate is $E = 0.03$, and the z-value for 99 percent confidence is $z_{\alpha/2} = z_{.005} = 2.576$. Since no information is given concerning p, the population proportion, we will use 0.5.

$$n = pq \left[\frac{z_{\alpha/2}}{E} \right]^2 = (0.5)(1 - 0.5) \left[\frac{2.576}{0.03} \right]^2 = 1,843.27 \approx 1,844$$

A sample size of $n = 1{,}844$ is probably larger than necessary since 0.5 was used as an approximation for p. If the actual value of p differs from 0.5, then a smaller sample size would satisfy the stated requirements.

Example 9.14   How many companies in Example 9.13 would have to be surveyed if it is expected that the percentage of companies that pay the entire cost is somewhere between 55 percent and 65 percent? Use the same maximum error of 0.03 with a 99 percent confidence level.

Solution

Since p is expected to lie somewhere in the interval from 0.55 to 0.65, we will use a value of 0.55 for p in Formula 9.11 (this is the value nearest to 0.5).

$$n = pq \left[\frac{z_{\alpha/2}}{E} \right]^2 = (0.55)(1 - 0.55) \left[\frac{2.576}{0.03} \right]^2 = 1{,}824.84$$

Rounding this up to the nearest integer, we have that 1,825 companies should be sampled. Notice that the prior information concerning p has resulted in a value of n that is smaller than that indicated in Example 9.13.

Section 9.6 Exercises
Determining the Required Sample Size

9.92   A researcher wants to estimate the mean of a population by taking a random sample and computing the sample mean. He wants to be 95 percent confident that the maximum error of the estimate is 2. If the population standard deviation is 20, how many observations need to be included in the sample?

9.93   An experimenter is investigating a population of measurements whose range is about 50. She will use the mean of a random sample to estimate the population mean. How large must the sample size be if she wants to be 99 percent confident that the error will be at most 2?

9.94   A politician wants to estimate the proportion of homeowners who favor restrictions on overnight parking on public streets. Determine how many people will have to be sampled to be 95 percent confident that the sample proportion will be off from the true proportion by at most 0.04.

9.95   With reference to Exercise 9.94, determine the required sample size if it is believed that the true proportion is somewhere between 0.65 and 0.80.

9.96   A state's fish and game commission wants to estimate the mean number of fish caught this year during the fishing season. From previous surveys, it estimates that the standard deviation will be about 10. How many license holders will need to be surveyed if the commission wants the sample mean to be off from the true mean by at most 2 with 95 percent confidence?

9.97   A university safety official plans to estimate the percentage of dormitory rooms with a microwave oven. He believes that the figure will be about 30 percent. Determine the number of rooms that need to be surveyed, if it is desired to have the estimate accurate to within 5 percent with probability 0.90.

9.98 In a previous exercise, reference was made to the U.S. Department of Education's claim that the average elementary and high school teacher puts in a 50-hour work week. Suppose a superintendent of schools wants to estimate the average work week of teachers in his district and he wants to be 95 percent sure that the estimate is off by at most 1.5 hours. It is expected that the number of hours will vary from 40 to 60. How many teachers should be sampled?

9.99 After a severe frost, a produce grower needs to estimate the proportion of tomato plants that have been damaged. Determine the number of plants that would have to be examined in order to be 90 percent confident that the sample proportion will be in error by not more than 0.05.

9.100 A newspaper reporter is preparing a story on automobile inspection stations, and she wants to estimate the average cost of a state inspection (excluding the cost of repairs). She believes that costs will vary from $10 to $25. How many inspection stations should be surveyed if she wants the error of the estimate to be at most $1 with 95 percent confidence?

9.101 With reference to Exercise 9.99, determine the required sample size if it is believed that about 25 percent of the plants have sustained damage.

9.102 Each year the Internal Revenue Service randomly selects a number of tax returns for its Taxpayer Compliance Measurement Program (TCMP). Returns selected under TCMP are subjected to a very detailed audit. Suppose the IRS is interested in estimating the percentage of returns that contain fraudulent information. How many returns would have to be audited in order to be 99 percent confident that the error of the estimate is at most 2 percent?

9.103 Consistent with the trend toward fewer children and more single-parent families, the U.S. Census Bureau reported that the average size of households declined to a record low in 1989. In order for the estimate of the average size to be in error by at most 0.1 with confidence 95 percent, how many families would have to be surveyed? Assume that $\sigma = 1.5$.

9.104 A computer manufacturer recently introduced a lightweight laptop computer, and it wants to estimate the mean price charged by retailers. It expects that the selling price will have a range of $400. Determine the number of dealers that need to be sampled in order to be 95 percent confident that the sample mean will be accurate to within $25.

9.105 The *Wall Street Journal* estimated that 25 percent of 13-year-old U.S. students use calculators in their math class (February 9, 1990). Suppose we want to estimate this percentage for high school students. How many students would have to be sampled if it is desired to have the sample estimate correct to within 2 percent with probability 0.95? Assume we have no prior information as to what the actual percentage will be.

9.106 A food company is planning to market a new frozen dinner, and it needs to determine the mean number of calories per serving. Find the number of servings that will have to be analyzed in order to have the sample mean accurate to within 5 calories with probability 0.99. The company believes that the range in the number of calories is about 45.

9.7 Chi-Square Probability Distributions

We have discussed how to estimate the mean μ of a population, as well as p, the proportion of the population that possesses some characteristic of interest. Sometimes of equal importance is the problem of estimating a population's variance σ^2. This can be accomplished by using the procedure in Section 9.8 to construct a confidence interval for σ^2. The confidence interval formula is based on the **chi-square probability distribution** that we will now introduce.

We have seen that the normal and t probability distributions are positive for all values on the real number line. In contrast, the chi-square (χ^2) probability distribution is positive only for positive values of the random variable. Furthermore, the distribution is not symmetric, but is skewed to the right (positively skewed). Like the t-distribution, there are an infinite number of different chi-square distributions, and a particular distribution is determined by a parameter called **degrees of freedom (df)**. The possible values of df are the positive integers, and each integer defines a different chi-square distribution. In Figure 9.7, the chi-square distribution having 5 degrees of freedom is illustrated.

Figure 9.7
Chi-square distribution.*

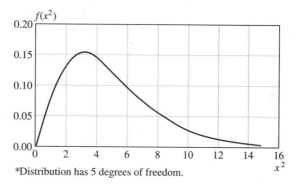

*Distribution has 5 degrees of freedom.

As with the other continuous variables in this text, tables are used when working with probabilities concerning chi-square distributions. Table 5 in Appendix A contains selected probabilities for various degrees of freedom. A portion of the table is duplicated in Table 9.5.

The construction of the table is similar to that for the t-distribution. The first column contains the degrees of freedom that identify the particular chi-square distribution being considered. The other columns contain chi-square values that have an area to their right equal to the amount appearing at the top of the column. Thus, each column contains values of χ^2_a, where a is the specified area that heads the column.

Table 9.5

A Portion of Table 5 from Appendix A

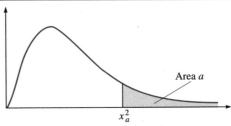

Area a

x_a^2

df	.995	.990	.975	.950	.900	.100	.050	.025	.010	.005
					Amount of Area to the Right of the Table Value					
1	0.00	0.00	0.00	0.00	0.02	2.71	3.84	5.02	6.63	7.88
2	0.01	0.02	0.05	0.10	0.21	4.61	5.99	7.38	9.21	10.60
3	0.07	0.11	0.22	0.35	0.58	6.25	7.81	9.35	11.34	12.84
4	0.21	0.30	0.48	0.71	1.06	7.78	9.49	11.14	13.28	14.86
5	0.41	0.55	0.83	1.15	1.61	9.24	11.07	12.83	15.09	16.75
6	0.68	0.87	1.24	1.64	2.20	10.64	12.59	14.45	16.81	18.55
7	0.99	1.24	1.69	2.17	2.83	12.02	14.07	16.01	18.48	20.28
8	1.34	1.65	2.18	2.73	3.49	13.36	15.51	17.53	20.09	21.95
9	1.73	2.09	2.70	3.33	4.17	14.68	16.92	19.02	21.67	23.59
10	2.16	2.56	3.25	3.94	4.87	15.99	18.31	20.48	23.21	25.19
11	2.60	3.05	3.82	4.57	5.58	17.28	19.68	21.92	24.72	26.76
⋮	⋮	⋮	⋮	⋮	⋮	⋮	⋮	⋮	⋮	⋮

Example 9.15   For the chi-square distribution with $df = 8$,

1. find P_{99}, the ninety-ninth percentile;
2. determine the number c such that $P(\chi^2 \leq c) = 0.10$.

Solution

1. P_{99} is the chi-square value that has a cumulative probability to the left of it equal to 0.99. Since Table 5 lists right-hand areas, we must first determine this area for P_{99} before using the table. The area to the right of P_{99} is $1 - 0.99 = 0.01$. Using the column headed by .010 and the row for $df = 8$, we obtain a value of 20.09. Thus,
$$P_{99} = \chi_{.01}^2 = 20.09.$$

$f(x^2)$

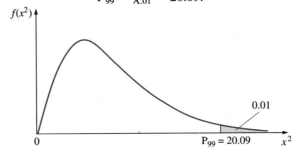

0.01

$P_{99} = 20.09$

x^2

0

2. We are seeking the chi-square value c that has an area to the left of it equal to 0.10. Consequently, the area to the right of c must be $1 - 0.10 = 0.90$. From Table 5, the intersection of the row with $df = 8$ and the column headed by a right-hand area of 0.90 is 3.49. Therefore,

$$c = \chi^2_{.90} = 3.49.$$

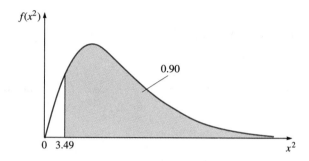

ⓂExample 9.16   For the chi-square distribution with $df = 33$, use MINITAB to find

1. $P(\chi^2 \le 21.18)$,
2. $\chi^2_{.25}$.

Solution

1. Since we want the cumulative probability up to 21.18, the **CDF** command is applied to 21.18. The subcommand specifying the applicable distribution is **CHISQUARE,** and this must be followed by its degrees of freedom.

```
CDF 21.18;
CHISQUARE 33. # You can just type CHIS 33.
```

From Exhibit 9.6, we have that $P(\chi^2 \le 21.18) = 0.0556$.

2. Since $\chi^2_{.25}$ is the chi-square value with an area of 0.25 to the right of it, the cumulative area to the left is 0.75. $\chi^2_{.25}$ can be obtained by using the **INVCDF** command to find the chi-square value that has a cumulative area of 0.75. Type the following commands.

```
INVCDF 0.75;
CHISQ 33.
```

From Exhibit 9.6, $\chi^2_{.25} = P_{75} = 38.0575$.

Exhibit 9.6

```
MTB > CDF 21.18;
SUBC> CHISQUARE 33.
      21.1800     0.0556
MTB > INVCDF 0.75;
SUBC> CHISQ 33.
      0.7500     38.0575
```

MINITAB can be used to obtain a rough sketch of the graph of any probability distribution discussed in this text. We will demonstrate the technique by sketching the chi-square distribution that was illustrated in Figure 9.7 at the beginning of this section.

Ⓜ**Example 9.17**   Use MINITAB to sketch the graph of the chi-square probability distribution having 5 degrees of freedom.

Solution

MINITAB sketches a probability distribution by plotting a set of (x, y) points. We begin by storing in C1 the x-values of the points to be plotted. These should start at 0 and extend sufficiently far until the y-values are nearly 0. We arbitrarily will use the values from 0 to 14, with a spacing of 0.5. If these points are not adequate, more can be added later. The x-values are entered quickly in the following manner.

```
SET C1
0:14/0.5
END
```

The **PDF** command that was introduced in Chapter 5 can be used to calculate the y-values for the points to be plotted. To obtain these for the x-values in C1, and then have them stored in column C2, type the following.

```
PDF C1 C2;
CHISQ 5.
```

Next, the **PLOT** command is used to graph the points that have their x-values in C1 and their y-values in C2.

```
PLOT C2 C1
```

MINITAB will produce the plot given in Exhibit 9.7.

Exhibit 9.7

```
MTB > SET C1
DATA> 0:14/0.5
DATA> END
MTB > PDF C1 C2;
SUBC> CHISQ 5.
MTB > PLOT C2 C1
```

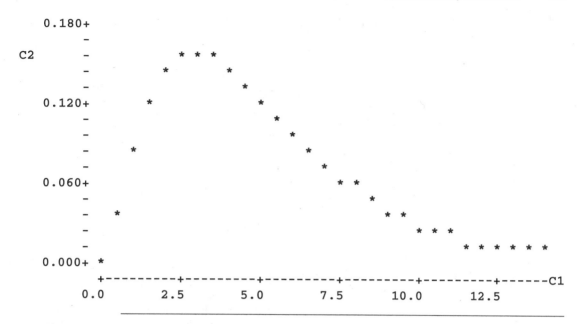

Section 9.7 Exercises

Chi-Square Probability Distributions

For each of the Exercises 9.107–9.116, find the chi-square value with a right-hand area equal to the specified subscript.

9.107 $\chi^2_{.01}$ with $df = 17$

9.108 $\chi^2_{.01}$ with $df = 19$

9.109 $\chi^2_{.005}$ with $df = 13$

9.110 $\chi^2_{.025}$ with $df = 24$

9.111 $\chi^2_{.025}$ with $df = 25$

9.112 $\chi^2_{.005}$ with $df = 21$

9.113 $\chi^2_{.05}$ with $df = 50$

9.114 $\chi^2_{.10}$ with $df = 60$

9.115 $\chi^2_{.995}$ with $df = 90$

9.116 $\chi^2_{.95}$ with $df = 50$

Determine the following percentiles.

9.117 P_{10} with $df = 11$

9.118 P_{90} with $df = 18$

9.119 P_1 with $df = 80$

9.120 P_{99} with $df = 12$

9.121 P_5 with $df = 14$

9.122 P_5 with $df = 6$

9.123 P_{90} with $df = 9$

9.124 P_{95} with $df = 11$

9.125 P_{99} with $df = 16$

9.126 P_1 with $df = 90$

9.127 P_{95} with $df = 21$

9.128 P_{10} with $df = 13$

Use Table 5 to determine the following probabilities.

9.129 $P(\chi^2 > 8.23)$ with $df = 18$

9.132 $P(\chi^2 < 2.16)$ with $df = 10$

9.130 $P(\chi^2 > 23.68)$ with $df = 14$

9.133 $P(57.15 < \chi^2 < 96.58)$ with $df = 80$

9.131 $P(\chi^2 < 40.29)$ with $df = 22$

9.134 $P(40.26 < \chi^2 < 53.67)$ with $df = 30$

For each of the Exercises 9.135–9.138, find the value of c.

9.135 $P(\chi^2 > c) = 0.01$ with $df = 9$

9.137 $P(\chi^2 < c) = 0.005$ with $df = 30$

9.136 $P(\chi^2 > c) = 0.95$ with $df = 13$

9.138 $P(\chi^2 < c) = 0.90$ with $df = 27$

MINITAB Assignments

Ⓜ 9.139 Use MINITAB to determine $\chi^2_{.15}$ and $\chi^2_{.85}$ for the chi-square distribution with 42 degrees of freedom.

Ⓜ 9.140 For the chi-square distribution with $df = 37$, use MINITAB to find (a) $P(\chi^2 < 20)$, (b) the median.

Ⓜ 9.141 For the chi-square distribution with $df = 19$, use MINITAB to find the value of c for which $P(\chi^2 < c) = 0.35$.

Ⓜ 9.142 Use MINITAB to sketch the graph of the chi-square distribution with 8 degrees of freedom.

9.8 Confidence Interval for a Variance

In some situations, an experimenter must be concerned with estimating the variance σ^2 of a population. Earlier we saw that a value for σ is needed to determine the required sample size when estimating μ. Knowledge of the variance is also important because it measures the variability in a population, which in turn may affect the precision of some process. In practical applications, it is often necessary to monitor the variance to assure that it does not exceed a certain amount. For example:

1. It is important that measuring instruments provide readings that do not deviate from actual values by more than a prescribed precision.
2. Critical measurements of machined parts must fall within certain tolerances of specified values.
3. Equipment used by manufacturers to fill containers should not dispense amounts that deviate from the label figure by an excessive amount.

As one might expect, a confidence interval for the population variance σ^2 involves the sample variance s^2. Its derivation is based on the fact that the quantity

$$\frac{(n-1)s^2}{\sigma^2}$$

has a sampling distribution that is chi-square when the sampled population is normally distributed. Using this result, one can derive a confidence interval for σ^2 that is applicable for both small and large samples. The only requirements are that a random sample is selected from a population whose distribution is approximately normal.

Large- or Small-Sample $(1 - \alpha)$ Confidence Interval for σ^2:

$$\frac{(n-1)s^2}{\chi^2_{\alpha/2}} < \sigma^2 < \frac{(n-1)s^2}{\chi^2_{(1-\alpha/2)}} \qquad (9.12)$$

Assumptions:

1. The sample is random.
2. The sampled population has a normal distribution.

Note:

1. s^2 is the sample variance.
2. $\chi^2_{\alpha/2}$ and $\chi^2_{(1-\alpha/2)}$ are chi-square values whose right-hand areas are $\alpha/2$ and $(1-\alpha/2)$, and these are based on the chi-square distribution (Table 5) with $df = n - 1$.

Example 9.18   An elevated level of blood cholesterol is believed to increase a person's risk of heart disease. To measure one's serum cholesterol, a chemical analyzer is used, but its accuracy varies. Suppose a certain laboratory checks on the variability of its analyzer by performing a sample of 10 analyses of blood drawn from the same source. If a sample variance of 26.70 was obtained for such a procedure, estimate with 95 percent confidence the variance of all measurements by this analyzer.

Solution

The confidence interval for σ^2 is given by

$$\frac{(n-1)s^2}{\chi^2_{\alpha/2}} < \sigma^2 < \frac{(n-1)s^2}{\chi^2_{(1-\alpha/2)}}$$

We are given that $s^2 = 26.70$ and $n = 10$. For 95 percent confidence, $1 - \alpha = 0.95$, and therefore $\alpha = 0.05$. Thus, the chi-square table values that need to be looked up are $\chi^2_{\alpha/2} = \chi^2_{.025}$ and $\chi^2_{(1-\alpha/2)} = \chi^2_{.975}$. These are illustrated in Figure 9.8 for the chi-square distribution with degrees of freedom $df = n - 1 = 9$. From Table 5, $\chi^2_{.025} = 19.02$ and $\chi^2_{.975} = 2.70$. Substituting the table values into the confidence interval formula, we have the following.

$$\frac{(10-1)(26.70)}{19.02} < \sigma^2 < \frac{(10-1)(26.70)}{2.70}$$

$$12.63 < \sigma^2 < 89.00$$

Thus, with 95 percent confidence we can assert that the variance of all measurements with this analyzer is between 12.63 and 89.

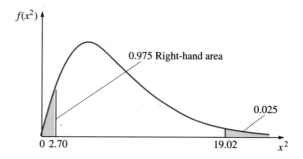

Figure 9.8
Chi-square distribution with $df = 9$.

Example 9.19 For the chemical analyzer referred to in Example 9.18, estimate with 95 percent confidence the standard deviation of the analyzer's measurements.

Solution

Since standard deviation is the square root of variance, a confidence interval for σ can be obtained by taking the square root of the confidence limits for σ^2 that were obtained in Example 9.18. This will give the following 95 percent confidence interval for the analyzer's standard deviation.

$$\sqrt{12.63} < \sqrt{\sigma^2} < \sqrt{89.00}$$

$$3.6 < \sigma < 9.4$$

Section 9.8 Exercises

Confidence Interval for a Variance

In the following exercises, assume that the sampled populations have approximately normal distributions.

9.143 A random sample of $n = 20$ measurements produced a variance of $s^2 = 15.5$. Construct a 90 percent confidence interval for the population variance σ^2.

9.144 Twenty-four measurements were randomly selected from a population, and the sample variance was $s^2 = 48.8$. Estimate the population variance σ^2 with a 95 percent confidence interval.

9.145 Obtain a 90 percent confidence interval for the population standard deviation σ in Exercise 9.143.

9.146 Estimate the population standard deviation σ in Exercise 9.144 by constructing a 95 percent confidence interval.

9.147 A random sample produced the following values.

$$x_1 = 15, \quad x_2 = 12, \quad x_3 = 12, \quad x_4 = 13, \quad x_5 = 18$$

Find a 99 percent confidence interval for the population variance.

9.148 Use the following random sample of measurements to estimate the population variance σ^2 with a 90 percent confidence interval.

$$x_1 = 2, \quad x_2 = 5, \quad x_3 = 7, \quad x_4 = 7, \quad x_5 = 3, \quad x_6 = 6, \quad x_7 = 7, \quad x_8 = 3$$

9.149 A 1-ounce serving of a certain breakfast cereal is suppose to contain 9 grams of psyllium, a high fiber food product that may be beneficial in lowering cholesterol levels. Twenty-four servings were analyzed for the amount of psyllium, and the sample variance was $s^2 = 0.52$. Estimate with 95 percent confidence the true variance of the amount of psyllium per serving of this cereal.

9.150 For Exercise 9.149, construct a 95 percent confidence interval for the standard deviation of the amount of psyllium per cereal serving.

9.151 To assure the proper functioning of an electric heat pump, the diameter of a critical component must fall within strict tolerances. The manufacturer randomly selected and measured the diameters of 20 components from a production run. The sample mean and standard deviation were (in millimeters)

$$\bar{x} = 25.3 \text{ mm} \quad \text{and} \quad s = 0.24 \text{ mm}.$$

Estimate the standard deviation of the entire production run by constructing a 99 percent confidence interval.

9.152 A supersensitive test for measuring extremely low levels of the thyroid stimulating hormone (TSH) has recently been employed in Canada. To assure the consistency and accuracy of independent laboratories, the Department of Health periodically sends identical samples to several labs for analysis of TSH. Suppose identical blood specimens were sent to 10 labs and the following results were obtained, where the measurements are in micromoles/liter.

0.5   0.5   0.4   0.2   0.4   0.8   0.4   0.5   0.6   0.5

Construct a 95 percent confidence interval for σ^2, the true variance of the laboratory readings of TSH measurements.

9.153 In Exercise 9.152, obtain a 95 percent confidence interval for σ, the true standard deviation of the laboratory readings of TSH measurements.

9.154 A technician checked the precision of a newly purchased tachometer by taking five readings of an electric motor running at a constant speed of 900 rpm (revolutions per minute). The obtained values were 893, 901, 889, 899, and 902. Use a 90 percent confidence interval to estimate the variance of the tachometer's readings.

9.155 Concentrations of airborne asbestos levels are measured by counting the number of fibers identified through a microscope in 100-size view fields. To test the precision of the monitoring protocol, 8 air samples were collected simultaneously from the same room. An analyst obtained the following fiber counts per 100-size view fields.

8   6   12   10   10   8   9   7

Construct a 95 percent confidence interval to estimate the true variance of counts for the monitoring protocol.

9.156 For Exercise 9.155, construct a 95 percent confidence interval for the standard deviation of counts for the monitoring protocol.

Looking Back

A primary application of statistics is using information contained in a sample to make an inference about the population from which the sample was drawn. The inferences will usually take one of two forms: **estimating parameters** or **testing hypotheses** concerning the parameters. This chapter is concerned with estimation, and the next chapter introduces the concept of hypotheses testing. Both chapters pertain to inferences for a single population. Inferences pertaining to two populations will be discussed in Chapter 11.

Many estimation problems are concerned with approximating the value of a population's **mean** μ, its **proportion** p of successes, or its **variance** σ^2. While these parameters can be estimated with a single value, called a **point estimate,** we usually prefer a **confidence interval** since it provides information about the maximum error of the estimate. This chapter discussed the following four $(1 - \alpha)$ confidence intervals.

1. **Large-sample ($n \geq 30$) confidence interval for μ**

$$\bar{x} \pm z_{\alpha/2} \frac{\sigma}{\sqrt{n}},$$

where in practice one must nearly always use the sample standard deviation s in place of the unknown population value σ.

2. **Small-sample ($n < 30$) confidence interval for μ**

$$\bar{x} \pm t_{\alpha/2} \frac{s}{\sqrt{n}},$$

where the degrees of freedom of the t-distribution are $df = n - 1$.

3. **Large-sample confidence interval for p**

$$\hat{p} \pm z_{\alpha/2} \sqrt{\frac{\hat{p}\hat{q}}{n}},$$

where $\hat{p}$ is the sample proportion of successes, and $\hat{q} = 1 - \hat{p}$.

4. **Large- or small-sample confidence interval for σ^2**

$$\frac{(n-1)s^2}{\chi^2_{\alpha/2}} < \sigma^2 < \frac{(n-1)s^2}{\chi^2_{(1-\alpha/2)}},$$

where the degrees of freedom of the chi-square distribution are $df = n - 1$.

In designing an experiment to estimate a parameter, one of the first things that must be decided is how large n should be. In Section 9.6, the following formulas for determining the required sample size are given when estimating means or proportions. E is the maximum error of the estimate, and $(1 - \alpha)$ is the associated confidence coefficient.

1. **Sample size n when $\bar{x}$ is used to estimate μ**

$$n = \left[\frac{\sigma z_{\alpha/2}}{E}\right]^2.$$

2. **Sample size n when $\hat{p}$ is used to estimate p**

$$n = pq \left[\frac{z_{\alpha/2}}{E}\right]^2.$$

When prior information indicates that p will fall within some range, we use in the above formula the value of p that is closest to 0.5. If no information is available about p, then 0.5 is used.

Key Words

In reviewing this chapter, you should be able to define, explain, and illustrate each of the following.

point estimate *(page 326)*

confidence interval estimate *(page 326)*

population mean μ *(page 327)*

lower confidence limit *(page 327)*

upper confidence limit *(page 328)*

confidence level *(page 328)*

Student's t probability distribution *(page 337)*

normal probability plot *(page 345)*

normal scores *(page 345)*

population proportion p *(page 349)*

required sample size *(page 353)*

maximum error of the estimate *(page 353)*

chi-square probability distribution *(page 359)*

population variance σ^2 *(page 364)*

Ⓜ MINITAB Commands

SET _ *(page 334)*

END *(page 334)*

STDEV _ *(page 334)*

ZINTERVAL _ _ _ *(page 334)*

CDF _; *(page 340)*
T _.

INVCDF _; *(page 341)*
T _.

TINTERVAL _ _ *(page 345)*

CORRELATION _ _ *(page 345)*

NSCORES _ _ *(page 345)*

PLOT _ _ *(page 345)*

CDF _; *(page 361)*
CHISQUARE _.

INVCDF _; *(page 361)*
CHISQUARE _.

PDF _ _; *(page 362)*
CHISQUARE _.

Review Exercises

9.157 A random sample of $n = 17$ measurements was selected from a normal population, and the sample mean was 478.7 and the standard deviation was 29.5. Estimate the mean of the population by constructing a 90 percent confidence interval.

9.158 For Exercise 9.157, obtain a 90 percent confidence interval for estimating the variance of the population.

9.159 A shop manual gives 6.5 hours as the average time required to perform a 30,000 mile major maintenance service on a Porsche 911. Last month a mechanic performed 10 such services, and his required times were the following.

6.3 6.6 6.7 5.9 6.3 6.0 6.5 6.1 6.2 6.4

Estimate with 95 percent confidence the mean time required by this mechanic to perform the maintenance service.

9.160 For Exercise 9.159, estimate with 90 percent confidence the standard deviation of the times required by the mechanic to perform the maintenance service.

9.161 A vitamin company manufactures 1,000 mg oyster shell calcium tablets. Seventy-one tablets were randomly selected from a production run, and each tablet was analyzed for its calcium content. The sample mean was 1,002.8 mg and the standard deviation was 3.4 mg. Estimate with 99 percent confidence the mean calcium content of all tablets in the production run.

9.162 Use the results in Exercise 9.161 to estimate with 95 percent confidence the standard deviation of the calcium content.

9.163 For several years there has been a shortage of nurses. Suppose a research sociologist wants to estimate the proportion of hospitals in the United States that have openings for nurses. Determine the number of hospitals that need to be sampled, if the researcher wants to be 90 percent confident that the sample proportion will be in error by less than 0.04.

9.164 A furniture manufacturer wants to estimate the average drying time of stained chairs. It is desired to have the sample mean differ from the true mean drying time by less than 2 minutes with probability 0.95. How many chairs will have to be sampled? Assume that σ is about 12 minutes.

9.165 A highway safety engineer wants to determine the percentage of trucks that exceed the speed limit in a certain residential area. The speeds of how many trucks will have to be checked, if the engineer wants to be 99 percent confident that his estimate will be off by not more than 4 percent?

9.166 In Exercise 9.165, suppose from an earlier study the engineer believes that the percentage of trucks exceeding the speed limit will be about 35 percent. Use this information to determine the number of trucks that will have to be sampled.

9.167 Exercise 9.18 gave the percentage rates charged for a 30-year home mortgage by a sample of 41 financial institutions in eastern Pennsylvania. The figures are repeated below, for which $\Sigma x = 403.875$ and $\Sigma x^2 = 3,983.456768$.

9.625	10	9.875	9.875	9.9
9.75	10.1	10.125	9.5	10.125
9.875	9.5	9.875	9.625	9.75
10	9.375	9.5	9.875	9.5
10.1	9.95	9.75	9.875	11.5
9.875	9.95	9.875	9.875	9.375
10	9.625	9.75	9.75	9.375
10.5	9.5	10	9.875	9.75
				9.875

Estimate the variability in rates for a 30-year mortgage in eastern Pennsylvania at the time of the survey by constructing a 95 percent confidence interval for σ^2, the variance of the rates.

9.168 For Exercise 9.167, obtain a 95 percent confidence interval for σ, the standard deviation of the rates.

9.169 For the 1987–88 academic year, The *Wall Street Journal* estimated that 58 percent of U.S. colleges required foreign language training for the bachelor of arts degree (February 9, 1990). Suppose a researcher wants to estimate this percentage for the present academic year. How many colleges would have to be surveyed in order for the sample estimate to be off by no more than 4 percent with 90 percent confidence? Assume that the researcher expects the figure to be between 55 percent and 60 percent.

9.170 To increase productivity, a manufacturing company made a major capital investment in new equipment. A study was then conducted to estimate the mean amount of time for a production assembler to complete one finished product. For a sample of 374 items, the mean number of minutes required was 52.8 and the standard deviation was 4.8 minutes. Estimate the mean time required with a 99 percent confidence interval.

9.171 An economist investigated 300 recent home sales in her state, and found that 81 were financed with an adjustable-rate mortgage. Estimate with 90 percent confidence the proportion of all recent home sales in this state that were financed with this type of mortgage.

9.172 Recent surveys have suggested that many U.S. students and adults are seriously deficient in a knowledge of geography. The *Wall Street Journal* (February 8, 1990) reported that a poll conducted by the National Geographical Society revealed that only 55 percent of adults could locate the state of New York on a map of the United States. Assume that the poll sampled 1,200 adults (the actual size of n was not given). Estimate with 95 percent confidence the true percentage of all U.S. adults that cannot locate New York on a U.S. map.

9.173 A marketing agency wants to conduct a survey to estimate the mean number of radios owned per household. The agency expects that the number will vary from 0 to 10. Determine the number of households that need to be surveyed if the agency wants to be 99 percent confident that the sample mean will be off by less than 1.

9.174 A student plans to estimate the probability p that a particular coin lands heads by performing a number of tosses and using the sample proportion as an estimate of p. How many tosses will be needed if the student wants to be 95 percent confident that the maximum error of the estimate is 0.02?

9.175 Use Table 4 to determine the following t-values.
   a. $t_{.025}$ with $df = 23$ b. $t_{.01}$ with $df = 10$

9.176 Use Table 5 to find the following chi-square values.
   a. $\chi^2_{.005}$ with $df = 28$ b. $\chi^2_{.05}$ with $df = 14$

9.177 Determine the 10th percentile for the chi-square distribution with $df = 50$.

9.178 Obtain P_{95}, the 95th percentile, for the t-distribution with $df = 29$.

9.179 To raise funds, a civic organization sells an energy saving light bulb that uses less energy and lasts longer than conventional incandescent bulbs. A sample of 8 light bulbs were tested, and their lives in hours appear below.

3,675   3,597   3,412   3,976   2,831   3,597   3,211   3,725

Estimate the mean life of this type of light bulb by constructing a 99 percent confidence interval.

MINITAB Assignments

Ⓜ **9.180** Use MINITAB to do Exercise 9.179.

Ⓜ **9.181** Using MINITAB, construct a normal scores plot to check on the reasonableness of the normality assumption for the data in Exercise 9.179.

Ⓜ **9.182** Use MINITAB to sketch the graph of the chi-square distribution with $df = 11$.

Ⓜ **9.183** Use MINITAB to sketch the graph of the t-distribution with $df = 34$.

Ⓜ **9.184** For a chi-square random variable with $df = 23$, use MINITAB to find the following.
   a. $P(\chi^2 > 15)$ b. the 77th percentile

Ⓜ **9.185** For the t-distribution with $df = 43$, use MINITAB to find the following.
   a. $P(t > 1.96)$ b. the 59th percentile

Ⓜ **9.186** A fish wholesaler has a catch of several thousand lobsters. A prospective buyer selected 50 at random and obtained the following weights in ounces.

21.3	21.1	21.4	18.9	20.2	19.3	19.1	18.3	19.9	22.0
20.6	20.7	21.9	20.1	17.1	19.3	21.2	18.4	21.0	21.6
16.5	18.9	17.4	20.8	18.5	18.1	21.1	19.3	21.5	20.1
21.8	20.2	19.7	18.9	19.5	20.0	18.7	21.6	20.9	21.5
17.5	16.1	20.1	21.8	19.4	21.6	23.1	20.5	22.0	20.6

Use MINITAB to obtain a 99 percent confidence interval to estimate the mean weight of lobsters in the catch.

Ⓜ **9.187** The ages of the 72 members of a senior citizens club are given below. Assume that this group is representative of all members of such clubs. Use MINITAB to construct a 90 percent confidence interval for estimating the mean age of all people who belong to a senior citizens club.

69	82	68	82	76	74	69	76	59	62
72	58	68	90	58	67	59	56	62	71
63	82	64	59	56	65	67	69	64	55
74	78	75	71	67	74	81	59	67	63
62	64	63	59	68	67	62	64	59	57
67	68	62	64	59	56	57	62	63	64
91	75	87	56	65	64	69	79	57	75
98	58								

Ⓜ **9.188** Use MINITAB to redo Exercise 9.187 under the assumption that the population standard deviation is known and equals 9 years.

Ⓜ 9.189 A random sample of invoices for 20 home heating oil deliveries by a particular dealer revealed the following numbers of gallons purchased.

165 217 200 134 175 153 108 174 219 183
123 198 261 247 176 121 202 100 207 200

Use MINITAB to construct a 90 percent confidence interval for the mean number of gallons per delivery by this dealer.

Ⓜ 9.190 For the gallons purchased in Exercise 9.189, use MINITAB to construct a normal scores plot to check if the normality assumption appears to be violated.

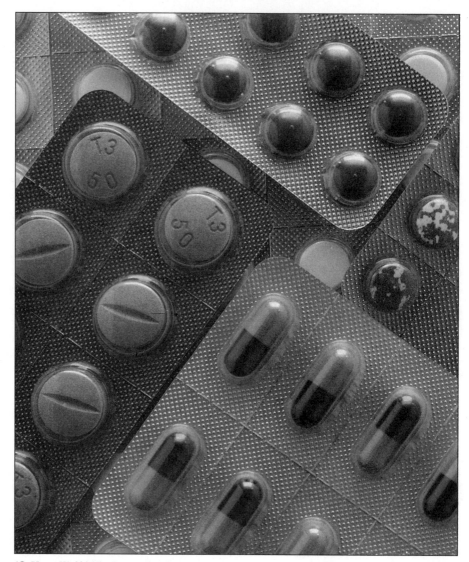

(© *Hans Wolf / The Image Bank*)

10 Tests of Hypotheses: Single Sample

Should You Believe that Claim?

*Each day, television, radio, and the print media bombard us with a barrage of product claims. We are told that brand A toothpaste has maximum effectiveness in controlling tartar, batteries of brand A outlast those of its closest competitor, deodorant A provides over 48 hours of odor protection, a majority of people prefer the taste of cola A to that of the other leading brand, more doctors recommend product A for headache relief than any other pain reliever, or repellent A provides more than 8 hours of protection against mosquitos and flies. Sample results are used as the basis for justifying superiority claims such as these. In this chapter you will learn how to **test** if the collected data actually do provide compelling evidence to warrant the acceptance of a manufacturer's **hypothesis** about its product.*

Looking Ahead

In this chapter, we will investigate statistical inference problems that involve **testing hypotheses.** In an estimation problem, interest is focused on approximating the value of a parameter. In a problem involving a test of hypothesis, we formulate a hypothesis concerning the parameter, sample the population pertaining to it, and then check whether the sample results substantiate the hypothesis of interest. For example, a researcher who has developed a natural food preservative may hypothesize that it will increase the mean shelf life of bread to more than eight days. This hypothesis could be tested by baking a sample of loaves with the new preservative and then determining the number of days that each loaf remains fresh. Supporting evidence for the researcher's hypothesis would be provided by a mean shelf life for the sample that is significantly greater than eight days.

This chapter is concerned with testing hypotheses for a single population. Specifically, we will discuss tests of hypotheses pertaining to the population parameters μ, a **mean;** p, a **proportion;** and σ^2, a **variance.** In Chapter 11, attention will be directed to inferences concerning two populations.

10.1   Basic Concepts of Testing Statistical Hypotheses

In Chapter 9 we saw that a confidence interval can be used to make an inference concerning the value of a population's parameter. There are many occasions when one must make a decision that is based on the parameter's value. In such instances a test of hypothesis about the parameter may be more appropriate.

Statistical Hypothesis:
A **statistical hypothesis** is an assertion about one or more characteristics of a population. It is often a claim about the value of a parameter.

For instance, a major logging company in the Pacific Northwest supplies its loggers with an insect repellent that provides an average protection of eight hours against mosquitos and flies. Suppose the company is considering a change to a new product that its manufacturer claims will repel these insects for an average of more than eight hours. Because this product is more expensive than the currently used brand, the company will buy the new product only if there is strong evidence that it does provide more than eight hours of protection. To decide whether the repellent lasts longer than eight hours, the logging company will formulate and test two hypotheses.

Null Hypothesis H_0:
The first hypothesis is called the **null hypothesis** because it usually represents a state of no change or no difference from the experimenter's point of view. The null hypothesis is denoted by H_0 and is assumed to be true until sufficient evidence is obtained to warrant its rejection.

For this example, the null hypothesis is that the new product does not offer a longer protection time than that of the currently used repellent. If we let μ denote the

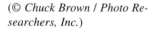

(© Chuck Brown / Photo Researchers, Inc.)

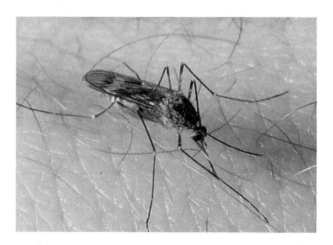

mean number of hours of protection for the new product, then the null hypothesis can be stated mathematically as follows:

$H_0: \mu = 8$ (The effectiveness of the new product is the same as that of the product in use.)

> **Alternative Hypothesis H_a:**
> The second hypothesis is called the **alternative hypothesis** and is denoted by H_a. It is the hypothesis on which the researcher places the burden of proof. Substantial supporting evidence is required before it will be accepted.

Since the logging company will purchase the new product only if it is shown to last longer than eight hours, the alternative hypothesis is as follows.

$H_a: \mu > 8$ (The effectiveness of the new product is superior to that of the product in use.)

In an attempt to obtain sufficient evidence to justify the acceptance of the alternative hypothesis H_a, the logging company had a sample of 36 loggers use the product and record the number of hours that the repellent was effective. For this sample of size $n = 36$, the mean was $\bar{x} = 8.9$ hours, and the sample standard deviation was $s = 1.8$ hours. What kind of results must the sample yield to justify the rejection of the null hypothesis ($\mu = 8$) and, thus, acceptance of the alternative hypothesis ($\mu > 8$)? Since we are attempting to show that the population mean exceeds 8 hours, it seems reasonable to base the decision on the value of the sample mean. If $\bar{x}$ is "much larger" than 8 hours, this would provide compelling evidence that $\mu > 8$. To determine if $\bar{x} = 8.9$ exceeds 8 by an amount that would be considered unlikely to occur by chance, we will calculate the z-value for $\bar{x}$. The z-value is called the **test statistic.**

> **Test Statistic:**
> The **test statistic** is a random variable whose value is calculated from the sample information. Its value determines if the sample contains sufficient evidence to warrant the rejection of the null hypothesis H_0 and, thus, the acceptance of the alternative hypothesis H_a.

The test statistic (z-value for $\bar{x}$) and its value can be obtained as follows. By the Central Limit theorem, for large n the sampling distribution of $\bar{x}$ is approximately normal, and its mean and standard deviation are $\mu_{\bar{x}} = \mu$ and $\sigma_{\bar{x}} = \sigma/\sqrt{n}$. Since the null hypothesis is assumed to be true until sufficient evidence is obtained to warrant its rejection, we assume that $\mu_{\bar{x}} = 8$. The population standard deviation σ is unknown (which is nearly always the case), but we can approximate it with s since n is large. Therefore,

$$\sigma_{\bar{x}} \approx \frac{s}{\sqrt{n}} = \frac{1.8}{\sqrt{36}},$$

and we have that the z-value for $\bar{x}$ is approximately

$$z = \frac{\bar{x} - \mu}{\dfrac{\sigma}{\sqrt{n}}} \approx \frac{8.9 - 8}{\dfrac{1.8}{\sqrt{36}}} = 3.00.$$

Thus, the value of $\bar{x}$ is 3 standard deviations above its mean of 8. Is this sufficiently large (and thus sufficiently improbable) to warrant the rejection of the null hypothesis and the acceptance of the alternative hypothesis? We will definitively answer this question shortly. Right now we will just venture a tentative opinion. We know from the Empirical rule that the probability of obtaining a z-value as great as 3 is practically zero. Therefore, either the null hypothesis is true and a very rare event has occurred, or a large z-value has been obtained because the null hypothesis is false. We will soon see that the large z-value will be interpreted as compelling evidence against the null hypothesis that $\mu = 8$ and as support in favor of the alternative hypothesis that $\mu > 8$. Our tentative conclusion, which will be justified shortly, is that sufficient evidence exists to reject the null hypothesis and accept the alternative hypothesis that the average effectiveness of the new repellent exceeds 8 hours.

Whenever incomplete information such as a sample is used to make an inference about a population, there is a risk of making a mistake. In a problem involving a test of hypothesis, there are two types of erroneous conclusions that could be made.

The Two Types of Erroneous Decisions and Their Probabilities:
A **Type I error** is falsely rejecting H_0. That is, a Type I error occurs when we reject a true null hypothesis.

The *probability of committing a Type I error* is denoted by α (the Greek letter alpha) and is called the **significance level** of the test.

A **Type II error** is falsely accepting H_0. That is, a Type II error occurs when we accept a false null hypothesis.

The *probability of committing a Type II error* is denoted by β (the Greek letter beta).

Table 10.1 summarizes the possible decisions and their consequences when a test of hypothesis is performed.

In the insect repellent example, a Type I error would occur if the logging company concluded that the new product lasts longer (H_0 is rejected), but, in fact, its average repelling time is the same as the one presently used (H_0 is true). A Type II error would occur if the company decides that the new repellent only lasts eight hours (H_0 is accepted), when it actually has a longer lasting time (H_0 is false).

Before conducting a hypothesis test, the experimenter decides on an acceptable level of risk for committing a Type I error. This is accomplished by specifying what the value of α (significance level) should be. The choice for α depends on the seriousness of a Type I mistake. The most commonly used value is $\alpha = 0.05$, except in situations when a Type I error is a very serious mistake, in which case $\alpha = 0.01$ is usually selected.

Table 10.1

Possible Decisions and Their Consequences

		The Actual State of Nature	
		H_0 **Is True**	H_0 **Is False**
The Decision	Reject H_0	Type I error	Correct decision
	Do not reject H_0	Correct decision	Type II error

For our example, suppose the logging company selects $\alpha = 0.05$. Setting the value of α will automatically determine how large the z-value for $\bar{x}$ must be to reject the null hypothesis. Since the probability of rejecting H_0 is to be 0.05 when H_0 is true, H_0 should be rejected if the z-value for $\bar{x}$ exceeds $z_{.05} = 1.645$. This is illustrated in Figure 10.1.

> **Rejection Region of H_0:**
> The **rejection region** consists of the values of the test statistic for which the null hypothesis H_0 is rejected.
>
> The **nonrejection region** consists of the values of the test statistic for which the null hypothesis is not rejected.*

The rejection region is shown in Figure 10.1, and it consists of all values of z that are to the right of $z_{.05} = 1.645$.

If the logging company had chosen a significance level of $\alpha = 0.01$, then the rejection region would consist of the values of z that are greater than $z_{.01} = 2.326$.

Since the experimenter has the liberty of selecting the significance level α, you might be wondering why α is not always chosen to be an extremely small value such

Figure 10.1
Values of the test statistic for which H_0 is rejected.

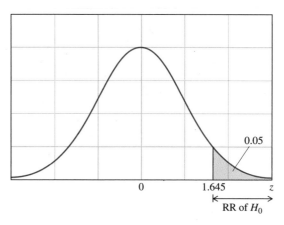

* The nonrejection region of H_0 is sometimes called the acceptance region, but we will not use this phrase.

as 0.01. The reason is that there is an inverse relationship between α, the probability of a Type I error, and β, the probability of a Type II error. For a fixed sample size n, decreasing α will increase β. Consequently, the experimenter often prefers to use $\alpha = 0.05$ instead of a smaller value, since the value of β will be reduced.

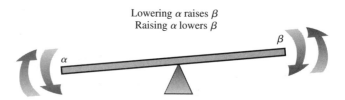

By appropriately choosing the sample size, it is often possible to specify the value of both α and β. Because of its complexity, however, we will follow the usual approach in a first course and only specify α.

Statisticians usually recommend that an inferential decision be made only when the risk of making a mistake is known. Whenever the null hypothesis is rejected, the probability of committing an error (Type I) is α, which is specified by the experimenter. However, when the null hypothesis is accepted, the probability of making an error (Type II) is β, which is unknown. Consequently, accepting the null hypothesis will only mean that there is insufficient evidence to warrant its rejection.* We will reserve judgment as to whether the null hypothesis is true. The only hypothesis that we will be able to "prove" (in the statistical sense) is H_a, the alternative hypothesis.

Because of the many concepts and terms introduced in this section, testing hypotheses may appear to be a complex process. Conducting a test can be facilitated by following the five basic steps outlined in the procedure below.†

Procedure for Conducting a Test of Hypothesis:
Step 1: Formulate the null and the alternative hypotheses.
The null hypothesis (H_0) usually represents the status quo, and the alternative hypothesis (H_a) is chosen to be the hypothesis on which the burden of proof is placed.

Step 2: Select the significance level α at which the test is to be conducted.
Usually α is chosen to be 0.05, except when a Type I error is very serious, in which case 0.01 is used. (In the exercises the value of α will be specified.)

Step 3: Using the sample data, calculate the value of the test statistic.
The test statistic is the random variable whose value determines the conclusion.

Step 4: From the tables in Appendix A, determine the rejection region of H_0.
The rejection region consists of the values of the test statistic for which the null hypothesis H_0 is rejected.

---

* We will use the phrase "fail to reject the null hypothesis" in place of "accept the null hypothesis."
† Some instructors prefer to reverse the order of steps 3 and 4.

Step 5: State the conclusion of the test.
If the value of the test statistic is in the rejection region, then reject H_0 and accept H_a; if it is not in the rejection region, then fail to reject H_0. Conclude that insufficient evidence exists to support the alternative hypothesis.

To illustrate the application of the five basic steps involved in a test of hypothesis problem, we will apply them to the insect repellent example.

Step 1: Formulate the null and the alternative hypotheses.

$$H_0: \mu = 8$$

$$H_a: \mu > 8$$

Step 2: Select the significance level for the test.
We will accept a 5 percent risk of committing a Type I error. Therefore, we will let $\alpha = 0.05$.

Step 3: Calculate the value of the test statistic.
The test statistic is z, and its value is

$$z = \frac{\bar{x} - \mu}{\dfrac{\sigma}{\sqrt{n}}} \approx \frac{8.9 - 8}{\dfrac{1.8}{\sqrt{36}}} = \frac{0.9}{0.3} = 3.00.$$

Step 4: Determine the rejection region of H_0 from the tables.
We reject H_0 for values of z that exceed $z_{.05} = 1.645$. The rejection region is illustrated in Figure 10.2.

Step 5: State the conclusion of the test.
Since the value of the test statistic is $z = 3.00$, and this value lies in the rejection region, we reject the null hypothesis H_0. Thus, we accept the alternative hypothesis H_a that the new product is effective in repelling insects for more than 8 hours.

Note that although we cannot be certain, we are quite confident that our conclusion is correct. The test was conducted at the 5 percent significance level, and, thus, the probability of rejecting the null hypothesis if it were actually true is only 0.05.

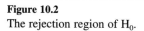

Figure 10.2
The rejection region of H_0.

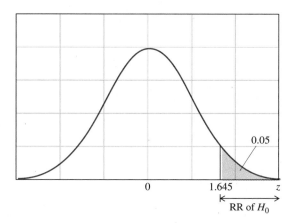

Before closing this section, we will consider three examples that illustrate how one distinguishes between the null and the alternative hypotheses. Earlier we emphasized that the alternative hypothesis is formulated so that it reflects what the experimenter is attempting to demonstrate (''prove'' in the statistical sense). Its form will be one of the following when testing a population mean μ.

$$H_a: \mu < \text{(some value)}$$

$$H_a: \mu > \text{(some value)}$$

$$H_a: \mu \neq \text{(some value)}$$

How the alternative hypothesis is stated will depend on where the experimenter wants to place the burden of proof. We will consider some examples to illustrate these points.

Example 10.1

Nutrition experts recommend that one's daily diet contain a minimum of 20 grams of fiber. The director of a summer camp for teenagers wants to show that the camp provides meals that exceed this amount. What null and alternative hypotheses should be tested?

Solution

Let μ denote the mean daily amount of fiber per person that is provided by the camp. Since the director wants to show that this is more than 20, the alternative hypothesis is $\mu > 20$. The burden of proof is put on this hypothesis, and until sufficient evidence is obtained to support it, the null hypothesis that $\mu = 20$ will be assumed. Thus, the hypotheses to be tested are the following.

$$H_0: \mu = 20*$$

$$H_a: \mu > 20$$

Example 10.2

For the summer camp mentioned in Example 10.1, suppose a competitor believes that the camp's claim is an exaggeration, and that the camp is not meeting the nutritional needs of its participants with regard to fiber content. In particular, the competitor suspects that teenagers are provided a daily average of fewer than 20 grams of fiber. What set of hypotheses would the competitor be interested in testing?

Solution

The competitor believes that μ is less than 20 and would be interested in demonstrating this belief. Consequently, this would be formulated as the alternative hypothesis, and the competitor would test the following.

$$H_0: \mu = 20$$

$$H_a: \mu < 20$$

---

* The null hypothesis could be written as $H_0: \mu \leq 20$ since the opposite of *more than 20* is *less than or equal to 20*. However, when H_0 involves more than 1 value, we use the one that is nearest those in H_a. The $=$ symbol will always be in H_0.

In Examples 10.1 and 10.2, the alternative hypothesis was $\mu > 20$ and $\mu < 20$, respectively. Each of these is an example of a **one-sided hypothesis,** since each is a statement about values of μ on "one side" of 20. To see this, think of the points on the real number line. The hypothesis $\mu > 20$ is concerned with points that lie to the right of (above) 20; $\mu < 20$ involves the points that lie to the left of (below) 20. In the next example, H_a is $\mu \neq 20$. This is called a **two-sided hypothesis** since it pertains to values on both sides of 20 ($\neq$ could be written as the double inequality "$<$ or $>$").

Example 10.3   Suppose an unbiased third party is only interested in determining if the camp's mean daily amount of fiber differs from 20 grams. It has no preconceived notion as to whether the actual mean is more or less than this figure. What set of hypotheses would this independent group want to test?

Solution

The group is only concerned with determining if sufficient evidence exists to conclude that μ is a value that is different from 20. Consequently, the alternative hypothesis should be that μ does not equal 20. Thus, this third party would be interested in testing the following.

$$H_0: \mu = 20$$

$$H_a: \mu \neq 20$$

Examples in the following sections will illustrate that how the alternative hypothesis is formulated (whether it contains $<$ or $\neq$ or $>$) affects the type of rejection region for a test.

Section 10.1 Exercises
Basic Concepts of Testing Statistical Hypotheses

10.1   State the type of error committed in each of the following situations.
   a. The null hypothesis is rejected when it is really true.
   b. The null hypothesis is accepted when it is actually false.
   c. The alternative hypothesis is accepted when in fact it is false.
   d. The alternative hypothesis is not accepted when it is actually true.

10.2   Explain each of the following.

a. Null hypothesis	b. Alternative hypothesis
c. Type I error	d. Type II error
e. α	f. β
g. Test statistic	h. Rejection region
i. Significance level	

10.3   Since an experimenter is free to choose the significance level α to be any value, why not always select α to be 0 or very close to this?

10.4 A researcher tests the following hypotheses.

$$H_0: \mu = 50$$

$$H_a: \mu > 50$$

a. What type of error would be committed if the mean is actually 50, but the researcher concludes that the mean exceeds 50?

b. Suppose the population mean is actually 52. What type of error would be committed if the researcher concludes that the mean is 50?

10.5 The following hypotheses are tested.

$$H_0: \mu = 100$$

$$H_a: \mu \neq 100$$

a. What type of error would be committed if one concludes that the mean equals 100 when it really does not?

b. If one concludes that the mean differs from 100 when it actually equals 100, what type of error has been committed?

10.6 Under the jury trial system employed in the United States, a person is assumed innocent until proven guilty. Consequently, the null and alternative hypotheses are as follows.

$$H_0: \text{the person being tried is innocent}$$

$$H_a: \text{the person being tried is guilty}$$

a. What type of error is committed when a guilty person is found innocent?

b. What type of error is committed when an innocent person is judged guilty?

c. Which of the Type I and Type II errors do you believe our justice system considers to be the more serious mistake?

10.7 An automobile manufacturer is evaluating a new plastic for use in the manufacture of its bumpers. As part of the evaluation process, it tests the following hypotheses.

$$H_0: \text{the strength of the plastic bumper does not exceed that of the presently used metal bumper}$$

$$H_a: \text{the plastic bumper is stronger}$$

a. If the plastic bumper is weaker, and the manufacturer decides that it is stronger, which type of error has been committed?

b. If it is decided that the plastic bumper is not stronger, but in reality it is, which type of error has been made?

10.8 A pharmaceutical company conducts an investigation of two pain relievers, A and B, for the treatment of migraine headaches. It tests the following hypotheses.

$$H_0: \text{A and B are equally effective}$$

$$H_a: \text{A is more effective than B}$$

a. What decision would result in a Type I error?

b. What decision would result in a Type II error?

10.9   A government agency has received numerous complaints that a particular restaurant has been selling underweight hamburgers. As part of its investigation, it tests the following hypotheses.

H_0: the restaurant is not selling underweight hamburgers

H_a: the restaurant is selling underweight hamburgers

a. A Type I error will occur if what conclusion is drawn?

b. What conclusion will result in a Type II error?

c. Which error would the restaurant consider to be more serious?

10.10   A manufacturer of a rechargeable flashlight currently uses a battery with a mean charge life of 5 hours. It is considering a change to a new type of battery.

a. What set of hypotheses would the manufacturer test if the new battery will be adopted only if it is shown to have a longer charge life?

b. Suppose the new battery costs less, and the manufacturer will adopt it unless evidence suggests that its average charge life is inferior. What set of hypotheses would the manufacturer test?

10.11   The speed of a personal computer is measured by its clock speed, which indicates how fast its central processing unit (CPU) can process data. A computer manufacturer now sells a desktop computer whose nominal operating speed is a blazing 66 MHz (megahertz, which is one million cycles per second).

a. What set of hypotheses should be tested if the manufacturer wants to demonstrate that the mean speed actually exceeds the nominal speed of 66 MHz?

b. What set of hypotheses about this computer's speed would a competing manufacturer likely be interested in testing?

10.12   A doctor has devised a new surgical procedure that he believes will reduce the mean length of hospital stay from its present value of 3 days. What set of hypotheses would be tested if the surgeon wants to demonstrate that the new procedure does decrease the mean length of stay?

10.13   An aviation design engineer has proposed a modification in the wing span of a plane that she believes will decrease the coefficient of drag and thus increase the mean top speed of this model from the present value of 950 miles-per-hour. What set of hypotheses would the engineer formulate if she wants to demonstrate this belief?

10.2   Hypothesis Test for a Mean: Large Sample

This and the next section are concerned with testing hypotheses that involve a single population mean. Recall that in the previous chapter we considered two situations for constructing a confidence interval for μ: a large-sample case and a small-sample case.

Tests of hypotheses involving μ will also be categorized into large-sample ($n \geq 30$) and small-sample ($n < 30$) cases. As was true with confidence intervals, we will see that the methodologies for the large and small situations are very similar, basically differing in the use of a z-value or a t-value as the test statistic. When the sample size n is large, the only assumption that we need to make is that the sample is random.

All problems regarding tests of hypotheses involve the 5 basic steps that were outlined in the last section. In Step 3, the value of a test statistic is computed. In the previous section, the following test statistic was obtained for testing hypotheses concerning a population mean when n is large.

Large-Sample Hypothesis Test for μ:
For testing the null hypothesis $H_0: \mu = \mu_0$ (μ_0 is a constant), the test statistic is

$$z = \frac{\bar{x} - \mu_0}{\dfrac{\sigma}{\sqrt{n}}}. \tag{10.1}$$

Assumptions:
1. The sample is random.
2. The sample size n is large ($n \geq 30$).

Note:
1. σ, the population standard deviation, is usually unknown and is approximated with the sample standard deviation s.
2. If σ *is known* and if *the population has a normal distribution,* then the test is *exact* for any sample size n, large or small.

We should emphasize that if the population is normally distributed and its standard deviation σ is known, then the test statistic in Formula 10.1 can be used for any sample size n. Since the value of σ is seldom known in practice, the test statistic is usually used only for large samples. A test statistic for small-sample situations will be given in the next section.

Example 10.4   An engineer for a tire manufacturer has developed a new tread design that, she believes, will decrease the braking time required to bring a vehicle to a complete stop. Thirty-four tires with the new design were subjected to a standard test of braking. For the sample of 34 measurements of the required braking time, the mean was 2.5 seconds and the standard deviation was 0.3 second. If tires that are currently being produced require an average stopping time of 2.7 seconds for this test, is there sufficient evidence at the 1 percent significance level to support the engineer's belief?

Solution

Step 1: Formulate the null and the alternative hypotheses.
The engineer is trying to show that the new design requires less braking time than the present design, which has a mean stopping time of 2.7 seconds. Therefore, the

alternative hypothesis is that μ is less than 2.7 seconds, where μ is the average braking time required for tires with the new design.

$$H_0: \mu = 2.7$$

$$H_a: \mu < 2.7$$

Step 2: Select the significance level for the test.
We are given that $\alpha = 0.01$.

Step 3: Calculate the value of the test statistic.
For the sample, $n = 34$, $\bar{x} = 2.5$, and $s = 0.3$ is used to approximate σ, since it is unknown.

$$z = \frac{\bar{x} - \mu_0}{\dfrac{\sigma}{\sqrt{n}}} = \frac{2.5 - 2.7}{\dfrac{0.3}{\sqrt{34}}} = -3.89$$

Step 4: Determine the rejection region of H_0 from the tables.
H_0 should be rejected if $\bar{x}$ is significantly smaller than 2.7, since this would be supportive of the alternative hypothesis. This will correspond to a value of the test statistic z that is significantly smaller than 0. Therefore, the rejection region of H_0 consists of z-values that are less than $-z_{.01}$. From Table 3 in Appendix A, $z_{.01} = 2.326$. Thus, H_0 is rejected for $z < -2.326$. The rejection region (RR) is shown in Figure 10.3.

Figure 10.3
Rejection region for Example 10.4.

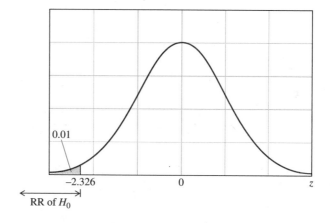

Step 5: State the conclusion of the test.
The value of the test statistic in Step 3 is $z = -3.89$. Since this lies to the left of -2.326, it is in the rejection region, and we reject H_0. Thus, there is sufficient evidence at the 1 percent significance level to conclude that the new tread design requires a smaller mean stopping time.

Example 10.5   Some stock market investors rely on advisory newsletters for guidance in the purchase and sale of stocks. The author of one such newsletter monitors the interest rates paid by banks in 5 key states for 90-day certificates of deposit (CDs). To check whether the mean rate has changed from last week's value of 9.28 percent, he surveys 50 banks in these states and calculates a mean rate of 9.33 percent and a standard deviation of 0.37 percent. Is there sufficient evidence at the 5 percent level to conclude that the mean rate has changed?

Solution

Step 1: Hypotheses.
The author of the newsletter wants to determine if the present mean rate μ differs from (does not equal) 9.28 percent. Consequently, H_a is the two-sided hypothesis $\mu \neq$ 9.28.

$$H_0: \mu = 9.28$$

$$H_a: \mu \neq 9.28$$

Step 2: Significance level.

$$\alpha = 0.05.$$

Step 3: Calculations.
For the sample, $n = 50$, $\bar{x} = 9.33$, and $s = 0.37$, which is used as an approximation for σ.

$$z = \frac{\bar{x} - \mu_0}{\dfrac{\sigma}{\sqrt{n}}} = \frac{9.33 - 9.28}{\dfrac{0.37}{\sqrt{50}}} = 0.96$$

Step 4: Rejection region of H_0.
Evidence against H_0 will occur when the value of $\bar{x}$ is either significantly less than or more than 9.28. These cases will result in a test statistic value that is either significantly smaller than or more than 0. Consequently, the rejection region consists of the extreme values in the lower and upper tails of the z-distribution. Since the total probability of the rejection region equals the α value of 0.05, half of 0.05 is placed in each of the two tails. Thus, we reject H_0 for values of z below $-z_{.025} = -1.96$, and for values of z above $z_{.025} = 1.96$. The rejection region is two-tailed and is shown in Figure 10.4.

Figure 10.4
Rejection region for Example 10.5.

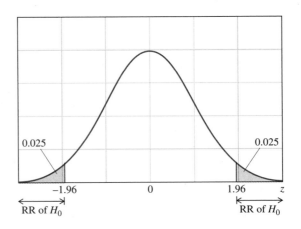

Step 5: Conclusion.

Since the value of the test statistic is $z = 0.96$, which does not fall in the rejection region, there is not enough evidence at the 5 percent significance level to reject the null hypothesis. Thus, there is insufficient evidence to conclude that the current mean rate has changed from last week's value of 9.28 percent.

In stating the conclusion, we have been careful not to imply that the null hypothesis has been "proven." Recall that when we fail to reject H_0, we will reserve judgment about its truthfulness. This is because we want to avoid the possibility of committing a Type II error, for which the probability β is unknown.

Ⓜ**Example 10.6** This example shows how MINITAB can be used to perform a large-sample test of hypothesis concerning a population mean. In the last chapter, Example 9.2 was concerned with the current mean cost μ of dining in a particular restaurant. The sample of 40 charges (in dollars) per person is repeated below.

38.00	36.25	37.75	38.62	38.50	40.50	41.25	43.75	44.50	42.50
43.00	42.25	42.75	42.88	44.25	45.75	45.50	35.75	35.12	39.95
35.88	35.50	36.75	35.50	36.25	36.62	37.25	37.75	37.62	36.80
38.50	38.25	37.88	38.88	38.25	46.00	46.50	45.25	34.60	39.50

Is there sufficient evidence at the 1 percent level to conclude that the mean cost per person exceeds $38.00?

Solution

As was done for obtaining a large-sample confidence interval, we must first approximate the population standard deviation σ by determining the sample standard deviation s. To do this, type the following.

```
SET C1
38.00 36.25 37.75 38.62 38.50 40.50 41.25 43.75 44.50 42.50
43.00 42.25 42.75 42.88 44.25 45.75 45.50 35.75 35.12 39.95
35.88 35.50 36.75 35.50 36.25 36.62 37.25 37.75 37.62 36.80
38.50 38.25 37.88 38.88 38.25 46.00 46.50 45.25 34.60 39.50
END
STDEV C1
```

MINITAB will respond with the following.

```
MTB > STDEV C1
   ST.DEV.  =        3.5476
```

Since $s = 3.5476$, we will use $3.55 to approximate σ. Because we want to determine if the mean cost exceeds $38.00, the alternative hypothesis is $\mu > 38.00$, and we will test the following hypotheses.

$$H_0: \mu = 38.00$$

$$H_a: \mu > 38.00$$

The following code numbers are used to inform MINITAB as to which type of relationship is involved in the alternative hypothesis.

If H_a Is	Then Use
<	-1
$\neq$	0
>	1

In this example the relationship in H_a is >, so we will use a code of 1. To have MINITAB test the hypotheses for the sample that has been stored in column C1, type the following.

```
ZTEST 38.00 3.55 C1;
ALTERNATIVE 1.
```

Before continuing, note the following important points:

1. The main command is **ZTEST,** and the subcommand is **ALTERNATIVE.**
2. With the main command **ZTEST** we specify:
   a. the value of μ given in H_0 (38.00 in this case),
   b. the population standard deviation σ (3.55 here),
   c. the column that contains the sample data (C1),
   d. a semicolon to indicate that the next line will be a subcommand.
3. In the subcommand **ALTERNATIVE** we specify:
   a. the code number that indicates the type of relationship in H_a (1 for >),
   b. a period to indicate the end of the instructions.

In response to the preceding commands, MINITAB will produce the output contained in Exhibit 10.1.

From Exhibit 10.1, the value of the test statistic is $z = 3.04$. H_0 should be rejected if $\bar{x}$ is significantly greater than \$38.00, which would result in z being significantly greater than 0. Using Table 3 in Appendix A with $\alpha = 0.01$, the rejection region is $z > z_{.01} = 2.326$. Since the test statistic equals 3.04 and this is in the rejection region, we reject H_0 and conclude that the mean cost does exceed \$38.00 per person.

A p-value is provided as part of the output in Exhibit 10.1. P-values will be discussed in Section 10.6.

Exhibit 10.1

```
MTB > ZTEST 38.00 3.55 C1;
SUBC> ALTERNATIVE 1.

TEST OF MU = 38.000 VS MU G.T. 38.000
THE ASSUMED SIGMA = 3.55
```

	N	MEAN	STDEV	SE MEAN	Z	P VALUE
C1	40	39.709	3.548	0.561	3.04	0.0012

A review of the examples in this and the preceding section will reveal that when H_a was a one-sided hypothesis, the rejection region was one tailed, and when H_a was a

Table 10.2

Relationship in H_a and the Type of Rejection Region

If the Relation in H_a Is	Then the Rejection Region Is
One sided of the form $<$	One tailed (on the left side)
One sided of the form $>$	One tailed (on the right side)
Two sided of the form $\neq$	Two tailed

two-sided hypothesis, the rejection region was two tailed. More specifically, if the relationship in H_a was $<$, then the rejection region was a lower left-hand tail; when the relationship was $>$, then the rejection region was an upper right-hand tail; and when H_a contained $\neq$, the rejection region consisted of both a left and a right tail. These conditions apply for all tests of hypotheses in this and the following chapter, and they are summarized in Table 10.2.

> **TIP:**   The results of Table 10.2 might be remembered more easily by noting that the point of the inequality in H_a is in the direction of the rejection region.
>
> Note that $<$ "points" to the left, while $>$ "points" to the right. Since $\neq$ is equivalent to the double inequality $<$ or $>$, it can be thought of as pointing to both the left and the right.

Section 10.2 Exercises
Hypothesis Test for a Mean: Large Sample

10.14 For each of the following hypotheses, determine the rejection region that would be used to perform a large-sample test. Use a significance level $\alpha = 0.05$.

a. $H_0: \mu = 150$
   $H_a: \mu > 150$

b. $H_0: \mu = 150$
   $H_a: \mu < 150$

c. $H_0: \mu = 150$
   $H_a: \mu \neq 150$

10.15 For each of the following hypotheses, determine the rejection region that would be used to perform a large-sample test. Use a significance level $\alpha = 0.01$.

a. $H_0: \mu = 87.6$
   $H_a: \mu < 87.6$

b. $H_0: \mu = 87.6$
   $H_a: \mu \neq 87.6$

c. $H_0: \mu = 87.6$
   $H_a: \mu > 87.6$

10.16 A researcher believes that a population mean exceeds 250. To test this belief, a random sample of 100 measurements was selected. The sample mean was 259 and the standard deviation was 40. The test will be conducted with $\alpha = 0.05$.

a. Formulate the null and alternative hypotheses.
b. Give the significance level of the test, and explain its meaning.
c. Calculate the value of the test statistic.
d. Determine the rejection region.
e. State the conclusion of the test.

10.17   In the past, the mean of a population has been 95, but a researcher theorizes that it has changed. To test this theory, a random sample of size $n = 64$ was selected. For the sample, the mean was 90 and the standard deviation was 16. A value of $\alpha = 0.01$ was chosen.

   a. Formulate the null and alternative hypotheses.
   b. Give the significance level of the test, and explain its meaning.
   c. Calculate the value of the test statistic.
   d. Determine the rejection region.
   e. State the conclusion of the test.

10.18   Management for a national chain of department stores claims that the mean hourly rate of its part-time employees is at least $6.75. A random sample of 75 parttime workers had a mean rate of $6.65 and a standard deviation of $0.55. Is there sufficient evidence at the 1 percent level to refute management's claim?

10.19   A regional magazine states that the mean household income of its subscribers exceeds $80,000. To show this, a random sample of 539 subscribers was surveyed. For the sample, the mean income was $80,634 and the standard deviation was $7,874. Is there sufficient evidence at the 0.05 significance level to support the magazine's statement?

10.20   A laboratory analyzed the potency of one hundred forty-five 5-grain aspirin tablets of a particular brand. For the sample, the average strength was 5.03 grains and the standard deviation was 0.24 grain. Do the sample results indicate that the mean potency of this brand is not 5 grains? Test at the 5 percent significance level.

10.21   A company manufactures an electronic flea collar that repels pests from pets by generating a high frequency sound that is repulsive to pests but beyond the range of human and pet hearing. The device is powered by a battery whose mean life is 150 days. The company is testing a new battery that it believes will extend the mean life. Thirty-five units were tested with the new battery, and the mean life was 159 days. The sample standard deviation was 27 days. Do these results provide sufficient evidence at the 5 percent level to conclude that the new battery has a longer mean life?

10.22   Tree stands are elevated platforms that are sometimes used by hunters. The *Journal of the American Medical Association* (November 10, 1989) reported on a 10-year study of tree stand-related injuries among deer hunters in Georgia. During this period there were 594 injuries, and the average age of injured hunters was 38. The ages ranged from 8 to 72. Suppose a researcher has theorized that tree stand-related injuries in deer hunting occur for younger hunters with a mean age below 40. Do the results of this study support the researcher's belief? Test at the 5 percent significance level.

10.23   In recent years, applicants for admission to a particular college have had a mean SAT score of 573. A random sample of 235 applications for next fall's class had a mean score of 579 and a standard deviation of 39. Test at the 0.05 level to see whether these results indicate a significant change in the mean SAT score for this year's applicants.

10.24   The engines of many U.S. automobiles use hydraulic valve lifters that do not require periodic valve adjustments. To squeeze more horsepower and performance from smaller engines, many foreign cars utilize a valve train configuration that requires regular valve adjustments. One foreign manufacturer recommends an adjustment every 15,000 miles. Suppose the company's research division is experimenting with a new alloy for valves that it believes will extend the mean number of miles between adjustments to more than 30,000. Tests were conducted on 33 engines with the new valves. The average miles between required adjustments was found to be 30,650 and the standard deviation was 1,425 miles. Formulate a suitable set of hypotheses and test at the 1 percent level.

10.25 During the preparation of a canned soup, a machine is supposed to dispense an average of 1,600 mg of sodium per can. The amount dispensed was determined for a random sample of 30 cans. The mean amount of sodium was 1,594 mg and the standard deviation was 19 mg. Is there sufficient evidence to conclude that the machine is not dispensing an average of 1,600 mg? Test at the 0.05 significance level.

10.26 To determine the order in which customers are served, the meat department of a large supermarket utilizes an electronic device that issues a number to each person waiting to be served. To collect data on customer waiting times, the clerk enters the customer's number in the machine when he or she is served. The store will add additional clerks if it finds that the mean wait exceeds 90 seconds. A sample of 38 customers revealed an average waiting time of 91.5 seconds and a standard deviation of 7.8 seconds. Does this constitute sufficient evidence at the 0.05 level to indicate that additional help is needed?

10.27 Asbestos, a group of fibrous minerals that have been shown to cause cancer, was widely used after World War II in the construction of schools. Today, school systems are mandated by federal law to eliminate unsafe exposures to asbestos. The EPA has estimated that secondary schools will have to spend more than $5,000 a year to remain in compliance with the new regulations (The *Lancaster Sunday News,* March 20, 1988). Suppose a random sample of 47 secondary schools last year revealed that the mean expenditure for compliance was $5,153 and the standard deviation was $598. Would these results support the EPA's contention at the 0.05 significance level?

MINITAB Assignments

Ⓜ 10.28 A carnival ride is supposed to last at least five minutes. Thirty-six operations of the ride are timed, and the following data (in seconds) were obtained.

283	274	296	301	294	288	302	275	297
291	306	316	285	296	289	295	300	291
305	289	298	287	281	295	291	290	316
275	296	284	303	295	303	290	278	299

Use MINITAB to determine if there is sufficient evidence at the 0.05 level to conclude that the mean duration of the ride is less than 5 minutes.

Ⓜ 10.29 A fish wholesaler has a catch of several thousand lobsters. A prospective buyer selected 50 at random and obtained the following weights in ounces.

21.3	21.1	21.4	18.9	20.2	19.3	19.1	18.3	19.9	22.0
20.6	20.7	21.9	20.1	17.1	19.3	21.2	18.4	21.0	21.6
16.5	18.9	17.4	20.8	18.5	18.1	21.1	19.3	21.5	20.1
21.8	20.2	19.7	18.9	19.5	20.0	18.7	21.6	20.9	21.5
17.5	16.1	20.1	21.8	19.4	21.6	23.1	20.5	22.0	20.6

The prospective buyer will purchase the entire catch if it can be shown that the mean weight exceeds 19.9 ounces. Formulate a suitable set of hypotheses, and use MINITAB to conduct the test at the 1 percent significance level.

10.3   Hypothesis Test for a Mean: Small Sample

In the previous section we saw that a hypothesis test concerning a population's mean μ involves the population's standard deviation σ. When the sample size n is large, σ can be estimated with s, the standard deviation of the sample. If n is small, s cannot be relied upon to provide an adequate estimate of σ. Consequently, the large-sample test statistic z in Section 10.2 should not be used for testing hypotheses concerning μ. We can, however, use the following t-statistic as long as the sampled population has approximately a normal distribution.

Small-Sample Hypothesis Test for μ:
For testing the null hypothesis H_0: $\mu = \mu_0$ (μ_0 is a constant), the test statistic is

$$t = \frac{\bar{x} - \mu_0}{\dfrac{s}{\sqrt{n}}}. \tag{10.2}$$

The rejection region is given by the following.

For H_a: $\mu < \mu_0$, H_0 is rejected when $t < -t_\alpha$.

For H_a: $\mu > \mu_0$, H_0 is rejected when $t > t_\alpha$.

For H_a: $\mu \neq \mu_0$, H_0 is rejected when $t < -t_{\alpha/2}$ or $t > t_{\alpha/2}$.

Assumptions:
1. The sample is random.
2. The sampled population has a normal distribution.

Note:
The t-distribution has degrees of freedom $df = n - 1$.

Although the above is usually referred to as the small-sample test for μ, it is applicable for any value of n as long as the sampled population is normally distributed. However, when n is large there is little difference between table values from the z-distribution and a t-distribution. Consequently, statisticians usually use the z-statistic when $n \geq 30$, and the t-statistic is used for $n < 30$.

Example 10.7   An optical laboratory advertises that it only charges an average of $55.00 for a pair of prescription glasses. In response to many consumer complaints about excessive charges, the State Department of Consumer Affairs sampled 23 prescriptions filled by this company. For the sample, the mean price charged was $65.25, and the standard deviation was $18.88. Is there sufficient evidence to accuse the laboratory of false advertising? Assume that the distribution of prescription costs is approximately normal and test at the 1 percent level of significance.

Solution

Step 1: Hypotheses.
The department of consumer affairs is interested in determining if the mean prescription cost μ is more than the advertised figure of $55.00. Therefore, the alternative hypothesis is $\mu > 55.00$.

$$H_0: \mu = 55.00$$
$$H_a: \mu > 55.00$$

Step 2: Significance level.

$$\alpha = 0.01.$$

Step 3: Calculations.
Since the sample size is less than 30, a t-test is used, where $n = 23$, $\bar{x} = \$65.25$, and $s = \$18.88$.

$$t = \frac{\bar{x} - \mu_0}{\dfrac{s}{\sqrt{n}}} = \frac{65.25 - 55.00}{\dfrac{18.88}{\sqrt{23}}} = 2.60$$

Step 4: Rejection region of H_0.
Since H_a involves the relation $>$, the rejection region is in the right-hand tail (upper tail) of a t-distribution. It consists of values for which $t > t_{.01}$. To determine $t_{.01}$, Table 4 of Appendix A is used with degrees of freedom $df = n - 1 = 22$. Thus, the rejection region is $t > 2.508$ and is illustrated in Figure 10.5.

Figure 10.5
Rejection region for Example 10.7.

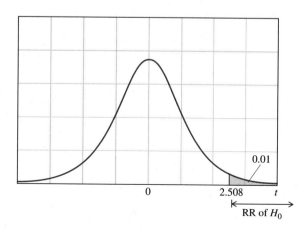

Step 5: Conclusion.
The value of the test statistic is $t = 2.60$, which does lie in the rejection region. Therefore, H_0 is rejected at the 1 percent level, and we conclude that the mean price charged by the optical company is greater than the advertised amount of $55.00. Since the significance level was set at only 1 percent, we can be quite confident that the correct conclusion has been drawn.

Ⓜ**Example 10.8**   We will illustrate how MINITAB can be used to perform a small-sample test of hypothesis concerning a mean. Consider a vending machine that is supposed to dispense 8 ounces of a soft drink. A sample of 20 cups taken over a one-week period contained the following amounts in ounces.

$$8.1 \quad 7.7 \quad 7.9 \quad 8.0 \quad 7.7 \quad 7.8 \quad 7.9 \quad 8.0 \quad 7.6 \quad 7.9$$
$$8.0 \quad 7.9 \quad 7.6 \quad 7.5 \quad 8.1 \quad 7.8 \quad 7.8 \quad 7.9 \quad 8.2 \quad 7.5$$

Is there sufficient evidence at the 5 percent significance level to conclude that the mean amount dispensed by the machine is less than 8 ounces?

Solution

First, the **SET** command is used to enter the sample into MINITAB.

```
SET C1
8.1   7.7   7.9   8.0   7.7   7.8   7.9   8.0   7.6   7.9
8.0   7.9   7.6   7.5   8.1   7.8   7.8   7.9   8.2   7.5
END
```

We want to test the following hypotheses.

$$H_0: \mu = 8$$
$$H_a: \mu < 8$$

We saw in Example 10.6 that a code number is used to inform MINITAB as to which type of relationship is involved in the alternative hypothesis. Since the relationship in H_a is $<$, a code of -1 is used. To have MINITAB test the hypotheses for sample values located in column C1, type the following commands.

```
TTEST 8 C1;
ALTERNATIVE -1.
```

Note the following concerning these commands.

1. The main command is **TTEST** and the subcommand is **ALTERNATIVE.**
2. With the main command we specify:
   a. the value of μ contained in H_0 (8),
   b. the column that contains the sample values (C1),
   c. a semicolon to indicate that a subcommand follows.
3. The subcommand specifies:
   a. the code number indicating the relationship in H_a (-1 for $<$),
   b. a period to indicate the end of the instructions.

The preceding commands will produce the output displayed in Exhibit 10.2. From Exhibit 10.2, the value of the test statistic is $t = -3.49$. H_0 should be rejected if $\bar{x}$ is significantly less than 8, which corresponds to a t-value significantly less than 0. Using Table 4 from Appendix A with $\alpha = 0.05$ and $df = n - 1 = 19$, the rejection region is $t < -t_{.05} = -1.729$. Since the computed t of -3.49 is less than -1.729, H_0 is rejected at the 5 percent significance level. We conclude that the mean amount dispensed by the machine is less than 8 ounces.

Exhibit 10.2	MTB > SET C1									

```
MTB > SET C1
DATA> 8.1   7.7   7.9   8.0   7.7   7.8   7.9   8.0   7.6   7.9
DATA> 8.0   7.9   7.6   7.5   8.1   7.8   7.8   7.9   8.2   7.5
DATA> END
MTB > TTEST 8 C1;
SUBC> ALTERNATIVE -1.

TEST OF MU = 8.0000 VS MU L.T. 8.0000
```

	N	MEAN	STDEV	SE MEAN	T	P VALUE
C1	20	7.8450	0.1986	0.0444	-3.49	0.0012

The small-sample t-test of this section assumes that the sampled population is normally distributed. We showed in Chapter 9 (see Example 9.8, page 345) that serious violations of the normality assumption can be checked by having MINITAB construct a normal probability plot. If the plot reveals that the assumption is questionable, then one can use a procedure from an area of statistical methodology referred to as nonparametric statistics. These techniques usually require few assumptions about the population. Nonparametric tests are discussed in Chapter 15, and a nonparametric alternative to the t-test of this section is considered there.

Section 10.3 Exercises
Hypothesis Test for a Mean: Small Sample

In the following exercises, assume that the sampled populations have approximately normal distributions.

For Exercises 10.30 through 10.35, determine the rejection region for testing the given hypotheses with the specified significance level α and sample size n.

10.30 $H_0: \mu = 45$ $\alpha = 0.05$ $n = 23$
$\qquad H_a: \mu < 45$

10.31 $H_0: \mu = 76$ $\alpha = 0.01$ $n = 16$
$\qquad H_a: \mu > 76$

10.32 $H_0: \mu = 197$ $\alpha = 0.01$ $n = 19$
$\qquad H_a: \mu \neq 197$

10.33 $H_0: \mu = 16.2$ $\alpha = 0.10$ $n = 11$
$\qquad H_a: \mu \neq 16.2$

10.34 $H_0: \mu = 2,000$ $\alpha = 0.05$ $n = 28$
$H_a: \mu > 2,000$

10.35 $H_0: \mu = 172$ $\alpha = 0.05$ $n = 9$
$H_a: \mu < 172$

10.36 A random sample of size $n = 20$ had a mean $\bar{x} = 32.6$ and a standard deviation $s = 4.7$. Test the following hypotheses at the 0.05 significance level.

$$H_0: \mu = 30$$

$$H_a: \mu > 30$$

10.37 For a random sample of 15 measurements, $\bar{x} = 576$ and $s = 44$. With $\alpha = 0.01$, test the following hypotheses.

$$H_0: \mu = 590$$

$$H_a: \mu \neq 590$$

10.38 Based on the following random sample, is there sufficient evidence at the 0.05 level to conclude that the mean of the sampled population exceeds 12?

$$x_1 = 15, \quad x_2 = 17, \quad x_3 = 16, \quad x_4 = 10, \quad x_5 = 18$$

10.39 Use the following random sample to determine if the mean of the sampled population differs from 25. Test at the 1 percent significance level.

$$x_1 = 21, \quad x_2 = 20, \quad x_3 = 22, \quad x_4 = 19, \quad x_5 = 24, \quad x_6 = 20$$

10.40 Last year the number of false fire alarms in a large city averaged 10.4 a day. In an effort to reduce this number, the fire department conducted a safety program in the city's schools. Six months after completion of the program, a sample of 21 days had a mean of 8.1 false alarms and a standard deviation of 3.4 days. Does it appear that the department's program was successful? Test at the 0.05 level.

10.41 A credit card company has found that account holders use its card an average of 8.4 times per month. To increase usage, the company conducted a promotion for one month during which prizes could be won by using the card. During the promotion, a random sample of 28 account holders used the card an average of 9.7 times. The sample standard deviation was 2.6. Do these results provide sufficient evidence at the 0.05 level to conclude that card usage increased during the promotion period?

10.42 A fast food restaurant claims that the average time required to fill an order is 5 minutes. A competing restaurant doesn't believe it and wants to prove that the actual time is more. Over a period of several days, it measured the waiting times for a sample of 20 orders, and it obtained a mean of 309.5 seconds and a standard deviation of 18.5 seconds. Formulate a suitable set of hypotheses and test at the 1 percent level.

10.43 In the United States, the average age of men at the time of their first marriage is 24.8 years *(Statistical Abstract of the United States)*. To determine if this differs for males in a certain county, a sociologist randomly surveyed 24 male residents who were or had been married. She obtained a sample mean and standard deviation of 23.5 and 3.2 years, respectively. Formulate a suitable set of hypotheses and test at the 0.05 significance level.

10.44 As part of its fund-raising program, a civic organization sells an energy-saving light bulb that it claims has an average life of more than 3,200 hours. A sample of 8 lights was tested, and their lives in hours appear below.

3,675   3,597   3,412   3,976   2,831   3,597   3,211   3,725

Is there sufficient evidence at the 0.05 level to support the organization's claim?

10.45 A shop manual gives 6.5 hours as the average time required to perform a 30,000 mile major maintenance service on a Porsche 911. Last month a mechanic performed 10 such services, and his required times were as follows.

6.3   6.6   6.7   5.9   6.3   6.0   6.5   6.1   6.2   6.4

Is there sufficient evidence at the 0.05 significance level to conclude that the mechanic can perform this service in less time than specified by the service manual?

10.46 A sociologist conducted a study of the use that older people make of their environmental space. She surveyed 21 residents of a large retirement community and found that the residents averaged 34.3 destinations per month (a destination is a trip outside the residence). The standard deviation was 3.9 trips. Do the results support her hypothesis that the mean number of destinations per month for all residents of the retirement community exceeds 30? Test at the 0.05 level.

10.47 According to the U.S. Department of Education, elementary and high school teachers work an average of 50 hours per week during the school year (*The Wall Street Journal,* January 9, 1990). To see if this differs for his school district, the superintendent of schools surveyed 25 randomly selected teachers and found that they work an average of 52.7 hours a week. The standard deviation was 4.2 hours. Is the sample mean significantly different from 50 hours at the 0.05 level?

MINITAB Assignments

Ⓜ 10.48 Quart cartons of milk should contain at least 32 ounces. A sample of 22 cartons contained the following amounts in ounces.

31.5   32.2   31.9   31.8   31.7   32.1   31.5   31.6
32.4   31.6   31.8   32.2   32.1   31.8   31.6   32.0
31.6   31.7   32.0   31.5   31.9   31.8

Use MINITAB to determine if sufficient evidence exists at the 0.05 level to conclude that the mean amount of milk in all cartons of this brand is less than 32 ounces.

Ⓜ 10.49 Last year the tutoring center at a college found that the average length of tutoring sessions was 38 minutes. A sample of 23 sessions this year revealed the following lengths in minutes.

$$29 \quad 37 \quad 21 \quad 17 \quad 32 \quad 22 \quad 19 \quad 43 \quad 23 \quad 35 \quad 48 \quad 37$$
$$14 \quad 19 \quad 29 \quad 38 \quad 30 \quad 28 \quad 26 \quad 25 \quad 21 \quad 31 \quad 15$$

Is there sufficient evidence that there has been a change in the mean length of tutoring sessions? Use MINITAB to test at the 0.05 significance level.

10.4 Hypothesis Test for a Proportion: Large Sample

In Chapter 9 we discussed inferences concerning population proportions (or percentages). To estimate a population proportion, we had to consider the problem of estimating the parameter p, the probability of a success in a binomial experiment. We saw that a point estimate of p is the sample porportion, $\hat{p} = x/n$, where x is the number of successes in a sample of size n. The following properties of $\hat{p}$ were considered in Chapter 9. They are repeated here because they form the basis for obtaining a large-sample hypothesis test concerning p.

The Sampling Distribution of $\hat{p}$, the Sample Proportion:
1. The mean of $\hat{p}$ is $\mu_{\hat{p}} = p$, the population proportion.
2. The standard deviation of $\hat{p}$ is $\sigma_{\hat{p}} = \sqrt{pq/n}$.
3. The shape of the sampling distribution is approximately normal for large samples.

By utilizing the above properties, the z-value of $\hat{p}$ can be obtained, and it can be used as the test statistic for testing the null hypothesis.

$$H_0: p = p_0 \quad (p_0 \text{ is a constant})$$

Under the assumption that H_0 is true, the z-value of $\hat{p}$ is

$$z = \frac{\hat{p} - \mu_{\hat{p}}}{\sigma_{\hat{p}}} = \frac{\hat{p} - p_0}{\sqrt{\dfrac{p_0(1 - p_0)}{n}}}.$$

These results are summarized below.

Large-Sample Hypothesis Test for p:
For testing the null hypothesis $H_0: p = p_0$ (p_0 is a constant), the test statistic is

$$z = \frac{\hat{p} - p_0}{\sqrt{\dfrac{p_0(1 - p_0)}{n}}}. \tag{10.3}$$

The rejection region is given by the following.

For $H_a: p < p_0$, H_0 is rejected when $z < -z_\alpha$.

For $H_a: p > p_0$, H_0 is rejected when $z > z_\alpha$.

For $H_a: p \neq p_0$, H_0 is rejected when $z < -z_{\alpha/2}$ or $z > z_{\alpha/2}$.

Assumptions:

1. The sample is random.
2. The sample size n is large ($n \geq 30$, $np_0 \geq 5$, and $n(1 - p_0) \geq 5$).

Note: $\hat{p}$ is the sample proportion and equals x/n, where x is the number of successes in the sample.

With regard to the size of n, it should be sufficiently large so that the sampling distribution of $\hat{p}$ is approximately normal. As discussed in Section 9.5, this will generally be true when $n \geq 30$ and n also satisfies the inequalities $np_0 \geq 5$ and $n(1 - p_0) \geq 5$.

Example 10.9   The manufacturer of a particular brand of microwave popcorn claims that only 2 percent of its kernels of corn fail to pop. A competitor, believing that the actual percentage is larger, tests 2,000 kernels and finds that 44 failed to pop. Do these results provide sufficient evidence to support the competitor's belief? Test at the 0.01 significance level.

(© *Larry Lefever / Grant Heilman Photography, Inc.*)

Solution

Step 1: Hypotheses.
The competitor wants to show that this brand's proportion of kernels that fail to pop, p, is more than 0.02. Thus, the alternative hypothesis is $p > 0.02$.

$$H_0: p = 0.02$$
$$H_a: p > 0.02$$

Step 2: Significance level.

$$\alpha = 0.01.$$

Step 3: Calculations.
The point estimate of p is the sample proportion

$$\hat{p} = \frac{x}{n} = \frac{44}{2,000} = 0.022.$$

The value of the test statistic is

$$z = \frac{\hat{p} - p_0}{\sqrt{\dfrac{p_0(1 - p_0)}{n}}} = \frac{0.022 - 0.02}{\sqrt{\dfrac{(0.02)(0.98)}{2,000}}} = 0.64.$$

Note that the value of p_0, and not $\hat{p}$, is used in calculating the denominator.

Step 4: Rejection region of H_0.
Since H_a involves the relation $>$, the rejection region is in the right-hand tail of the z-distribution. We reject H_0 for $z > z_{.01} = 2.326$. The rejection region is shown in Figure 10.6.

Step 5: Conclusion.
Since the value of the test statistic, $z = 0.64$, is not in the rejection region, do not reject H_0. Thus, there is not sufficient evidence to conclude that more than 2 percent of the kernels fail to pop.

Figure 10.6
Rejection region for Example 10.9.

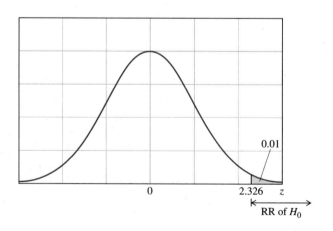

Example 10.10   *Hippocrates* magazine (May/June, 1989) reported that 30 percent of American children skip breakfast. Suppose a foreign social scientist wants to determine if this figure differs for children in her country. What can one conclude in this regard if a sample of 500 children revealed that 172 skip breakfast? Test at the 0.05 significance level.

Solution

Step 1: Hypotheses.
Let p denote the proportion of children in the scientist's country who skip breakfast. Since she wants to determine if p differs from (is not equal to) 0.30, the alternative hypothesis is $p \neq 0.30$.

$$H_0: p = 0.30$$

$$H_a: p \neq 0.30$$

Step 2: Significance level.

$$\alpha = 0.05.$$

Step 3: Calculations.
For the sample, $n = 500$, $x = 172$, and the sample proportion is $\hat{p} = x/n = 172/500 = 0.344$.

$$z = \frac{\hat{p} - p_0}{\sqrt{\dfrac{p_0(1 - p_0)}{n}}} = \frac{0.344 - 0.30}{\sqrt{\dfrac{(0.30)(0.70)}{500}}} = 2.15$$

Step 4: Rejection region of H_0.
Since the alternative hypothesis involves the relation $\neq$, we have a two-tailed rejection region. H_0 is rejected for values of $z < -z_{\alpha/2}$ or $z > z_{\alpha/2}$. That is, H_0 is rejected if $z < -1.96$ or $z > 1.96$. The rejection region is illustrated in Figure 10.7.

Step 5: Conclusion.
Since $z = 2.15$ is in the rejection region, H_0 is rejected, and the alternative hypothesis is accepted. Thus, there is sufficient evidence at the 0.05 significance level to conclude

Figure 10.7
Rejection region for Example 10.10.

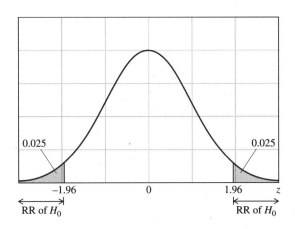

0.025   0.025

−1.96   0   1.96   z

RR of H_0 RR of H_0

that the percentage of children in the scientist's country who skip breakfast is different from 30 percent. The sample proportion of 0.344 suggests that the actual figure is higher.

Section 10.4 Exercises

Hypothesis Test for a Proportion: Large Sample

10.50 A random sample of $n = 800$ observations resulted in 496 successes. Test the following hypotheses at the $\alpha = 0.05$ significance level.

$$H_0: p = 0.60$$

$$H_a: p \neq 0.60$$

10.51 For a random sample of 500 observations, 182 successes were obtained. With $\alpha = 0.01$, test the following hypotheses.

$$H_0: p = 0.40$$

$$H_a: p < 0.40$$

10.52 A random sample of size $n = 200$ resulted in $x = 174$ successes. Test the following at the 0.01 significance level.

$$H_0: p = 0.80$$

$$H_a: p > 0.80$$

10.53 A random sample of $n = 100$ observations produced a sample proportion of 0.78. Would this result provide sufficient evidence at the 5 percent level to conclude that the population proportion differs from 0.70?

10.54 Suppose in Exercise 10.53, the sample size were $n = 1,000$, and the sample proportion equaled 0.73. Would this result provide sufficient evidence at the 5 percent level to conclude that the population proportion differs from 0.70?

10.55 The *American Scientist* (September/October, 1988) reported that a survey indicated that nearly half of American adults do not know that the sun is a star. Suppose 1,000 adults are sampled, and 52.5 percent know that the sun is a star. Would this constitute sufficient evidence at the 0.05 level to conclude that more than 50 percent of American adults are aware of this fact?

10.56 Exercise 10.22 referred to an article in the *Journal of the American Medical Association* (November 10, 1989) that reported on a ten-year study of tree stand-related injuries among deer hunters in Georgia. There were 594 deer hunting-related injuries reported during the period studied, and 214 were tree stand-related. Do the results provide sufficient evidence to conclude that more than one-third of all deer hunting-related injuries in Georgia are tree stand-related? Test at the 0.05 significance level.

10.57 An automobile dealer advertises that 90 percent of its customers would recommend the dealership to a friend. A competitor believes the percentage is less and contracts an independent firm to survey 89 of the dealer's customers. The survey found that 71 said that they would recommend the dealer to a friend. Do the survey results cast serious doubts on the veracity of the automobile dealer's claim? Test at the 0.01 significance level.

10.58   A fruit cannery will purchase several thousand pounds of peaches if it is convinced that less than 10 percent have serious blemishes. Inspectors for the company examined 1,500 peaches and found that 138 had serious blemishes. Is there sufficient evidence at the 0.05 level to conclude that the peaches should be purchased?

10.59   *Hippocrates* magazine (March/April, 1989) states that 37 percent of all Americans take multiple vitamins regularly. Suppose a researcher surveyed 750 people to test this claim, and he found that 290 regularly take a multiple vitamin. Is this sufficient evidence at the 0.01 level to conclude that the actual percentage is different from 37 percent?

10.60   A state official, running for the U.S. Senate, claims that more than 50 percent of voters favor him. To assess his popularity, a polling organization surveyed 1,200 voters and found 618 who said they prefer this candidate. Do the results of the poll support the official's claim at the 0.05 level?

10.61   A student, investigating the fairness of a particular coin, tossed it 1,000 times and obtained 535 heads. Do these results provide sufficient at the 0.05 significance level to conclude that the coin is biased?

10.62   The *Wall Street Journal* estimated that 25 percent of 13-year-old U.S. students use calculators in their math class (February 9, 1990). Suppose a state board of education wants to determine if this percentage differs for 13 year olds in its state. A random sample of 688 students in this age group were surveyed, and the sample revealed that 206 use calculators in their math courses. Formulate a set of hypotheses and test at the 0.05 level.

10.5   Hypothesis Test for a Variance

Although inferences frequently concern population means, there are many practical situations where interest in a population's variance is equally important. In a manufacturing process, knowledge of variance is essential because it measures variability, which in turn indicates the precision of the process. Quality control engineers must monitor the variance of the process to assure that it remains within specified tolerance limits.

A test of hypothesis concerning a population variance σ^2 is based on the same statistic that was used in the preceding chapter to obtain a confidence interval. There, we used the fact that the quantity

$$\frac{(n - 1)s^2}{\sigma^2}$$

has a sampling distribution that is chi-square for samples selected from a normal population.

The hypothesis test is summarized below and is applicable for both large and small samples, as long as the sampled population has approximately a normal distribution.

Large- or Small-Sample Hypothesis Test for σ^2:

To test the null hypothesis H_0: $\sigma^2 = \sigma_0^2$ (σ_0^2 is a constant), the test statistic is

$$\chi^2 = \frac{(n-1)s^2}{\sigma_0^2}. \tag{10.4}$$

The rejection region is given by the following.

For H_a: $\sigma^2 < \sigma_0^2$, H_0 is rejected when $\chi^2 < \chi^2_{(1-\alpha)}$.

For H_a: $\sigma^2 > \sigma_0^2$, H_0 is rejected when $\chi^2 > \chi^2_\alpha$.

For H_a: $\sigma^2 \neq \sigma_0^2$, H_0 is rejected when $\chi^2 < \chi^2_{(1-\alpha/2)}$ or $\chi^2 > \chi^2_{\alpha/2}$.

Assumptions:

1. The sample is random.
2. The sampled population has a normal distribution.

Note:

1. s^2 is the sample variance.
2. χ^2_a denotes a chi-square value whose right-hand area is the subscript a. This value is based on the chi-square distribution (Table 5, Appendix A) with $df = n - 1$.

Example 10.11 A supplier of cam shafts for an engine manufacturer guarantees that their lengths will vary with a maximum standard deviation of 0.003 inch. The engine manufacturer suspects that a large shipment recently received fails to meet this specification. To verify this, 10 cam shafts are randomly selected from the shipment and measured. If the standard deviation of the sample is 0.005 inch, does this provide sufficient evidence that the standard deviation for the entire lot exceeds the 0.003 specification? Assume that the cam shaft lengths are approximately normally distributed, and test at the 5 percent significance level.

Solution

Step 1: Hypotheses.
The engine manufacturer suspects that the standard deviation σ of the cam shaft lengths is greater than 0.003, and it is interested in detecting this. Thus, the alternative hypothesis is that the variance σ^2 is more than $(0.003)^2$.

$$H_0: \sigma^2 = (0.003)^2$$

$$H_a: \sigma^2 > (0.003)^2$$

Step 2: Significance level.

$$\alpha = 0.05.$$

Step 3: Calculations.
The sample size is $n = 10$, and the sample standard deviation is $s = 0.005$. The test statistic is a chi-square variable whose value is given by

$$\chi^2 = \frac{(n-1)s^2}{\sigma_0^2} = \frac{(10-1)(0.005)^2}{(0.003)^2} = 25.00.$$

Step 4: Rejection region of H_0.

The weight of evidence in support of the alternative hypothesis H_a increases with larger values of s^2, which in turn increase the value of the test statistic χ^2. Therefore, we reject H_0 for large values of χ^2. Also note that H_a involves the $>$ relation, which "points" to the right, thus indicating a right-tailed rejection region. The rejection region consists of values for which $\chi^2 > \chi_{.05}^2 = 16.92$. The value of 16.92 is obtained from Table 5, Appendix A, where the applicable degrees of freedom are $df = n - 1 = 9$. The rejection region is shown in Figure 10.8.

Figure 10.8
Rejection region for Example 10.11.

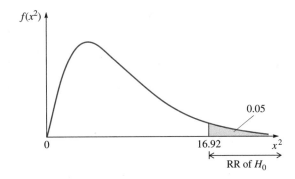

Step 5: Conclusion.

The value of the test statistic is $\chi^2 = 25.00$, and this lies in the rejection region. Therefore, H_0 is rejected at the 5 percent significance level, and we conclude that the standard deviation of the lot exceeds the specification of 0.003 inch.

Example 10.12   A food processing company strictly controls the amount of salt that its cans of soup contain, as this is critical to the soup's flavor. With the present canning process, the amount of sodium per can has a mean of 2,440 mg and varies with a standard deviation of 13 mg. An alteration to the process is being considered that the company believes will decrease the variation. To test this, 25 cans produced under the new process were analyzed, and the standard deviation of the sodium amounts was found to be 9 mg. Is there sufficient evidence at the 5 percent level to conclude that the new process does decrease the standard deviation?

Solution

Step 1: Hypotheses.

The company is concerned with showing that the new process results in a standard deviation σ that is less than that of the present process, which is 13. Consequently, the alternative hypothesis is that the variance satisfies the inequality $\sigma^2 < (13)^2$.

$$H_0: \sigma^2 = 169$$

$$H_a: \sigma^2 < 169$$

Step 2: Significance level.

$$\alpha = 0.05.$$

Step 3: Calculations.
For the sample, we are given that $n = 25$ and $s = 9$. The value of the chi-square test statistic is

$$\chi^2 = \frac{(n-1)s^2}{\sigma_0^2} = \frac{(25-1)(9)^2}{169} = 11.50$$

Step 4: Rejection region of H_0.
Small values of s^2, and thus of χ^2, support H_a. Therefore, the rejection region is in the left (lower) tail of a chi-square distribution, and the left-tail area is $\alpha = 0.05$ (note that the relation in H_a is $<$, which "points" to the left). To find this left-tailed rejection region, Table 5 from Appendix A is used. Recall, however, that the table only gives right-hand areas, and therefore we must look up the chi-square value for which the area to the right of it equals $1 - 0.05 = 0.95$. In Table 5 the column used is headed by a right-hand area of 0.95, and we use the row that has degrees of freedom $df = n - 1$ $= 24$. The rejection region is illustrated in Figure 10.9, and it consists of values for which $\chi^2 < \chi^2_{.95} = 13.85$.

Figure 10.9
Rejection region for Example 10.12.

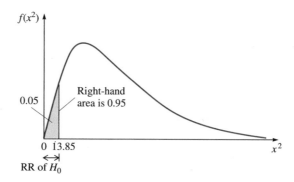

Step 5: Conclusion.
Since the value of the test statistic in Step 3 is 11.50, which is less than the table value of 13.85, we reject H_0. Thus, there is sufficient evidence at the 5 percent significance level to conclude that the new process does decrease the standard deviation.

Example 10.13   The precision of a tachometer is checked by taking 5 readings of an electric motor running at a constant speed of 900 rpm. The obtained values are 893, 901, 889, 899, and 902. Does this sample of measurements provide sufficient evidence at the 0.05 significance level to conclude that the variance in the tachometer's readings differs from 25?

Solution

Step 1: Hypotheses.
We want to determine if σ^2 differs from (does not equal) 25. Consequently, H_a is $\sigma^2 \neq 25$.

$$H_0: \sigma^2 = 25$$
$$H_a: \sigma^2 \neq 25$$

Step 2: Significance level.

$$\alpha = 0.05.$$

Step 3: Calculations.

To find the value of the test statistic χ^2, we first need to calculate the sample variance s^2.

$$SS(x) = \Sigma x^2 - \frac{(\Sigma x)^2}{n} = 4{,}021{,}376 - \frac{(4{,}484)^2}{5} = 124.8$$

$$s^2 = \frac{SS(x)}{n-1} = \frac{124.8}{4} = 31.2$$

$$\chi^2 = \frac{(n-1)s^2}{\sigma_0^2} = \frac{(5-1)(31.2)}{25} = 4.99$$

Step 4: Rejection region of H_0.

Values of χ^2 that are significantly small or large will support H_a. The rejection region is two tailed (note that $\neq$ is equivalent to the double inequality $<$ or $>$, which "point" to the left and to the right, thus indicating both a left- and right-tailed rejection region). Since the total probability of the rejection region equals the α value of 0.05, half of 0.05 is placed in each of the two tails. Thus, H_0 is rejected for values of χ^2 below $\chi^2_{.975} = 0.48$, and for values of χ^2 above $\chi^2_{.025} = 11.14$. The values of 0.48 and 11.14 are obtained from Table 5 using $df = n - 1 = 4$. The rejection region is shown in Figure 10.10.

Figure 10.10
Rejection region for Example 10.13.

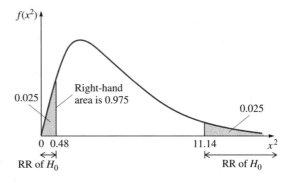

Step 5: Conclusion.

Since the test statistic $\chi^2 = 4.99$ is not in the rejection region, do not reject the null hypothesis H_0. Thus, with a sample of only $n = 5$ measurements, the difference between 25 and the sample variance of 31.2 is not statistically significant at the 0.05 level.

Section 10.5 Exercises

Hypothesis Test for a Variance

In the following exercises, assume that the sampled populations have approximately normal distributions.

10.63 Determine the rejection region that would be used to test each of the following sets of hypotheses at the $\alpha = 0.05$ significance level. Assume a sample size of $n = 20$.

a. $H_0: \sigma^2 = 200$
   $H_a: \sigma^2 < 200$

b. $H_0: \sigma^2 = 14.9$
   $H_a: \sigma^2 > 14.9$

c. $H_0: \sigma^2 = 76$
   $H_a: \sigma^2 \neq 76$

10.64 For each of the following, obtain the rejection region for performing a test of hypotheses at the $\alpha = 0.01$ significance level. Assume a sample size of $n = 22$.

a. $H_0: \sigma^2 = 8$
   $H_a: \sigma^2 \neq 8$

b. $H_0: \sigma^2 = 57$
   $H_a: \sigma^2 < 57$

c. $H_0: \sigma^2 = 9.8$
   $H_a: \sigma^2 > 9.8$

10.65 A random sample of size $n = 21$ had a mean $\bar{x} = 96.6$ and a standard deviation $s = 4.8$. Test the following hypotheses at the 0.05 significance level.

$$H_0: \sigma^2 = 28$$

$$H_a: \sigma^2 < 28$$

10.66 For a random sample of 12 measurements, $\bar{x} = 277$ and $s = 24$. With $\alpha = 0.01$, test the following hypotheses.

$$H_0: \sigma^2 = 500$$

$$H_a: \sigma^2 > 500$$

10.67 The variance equaled 59.69 for a random sample of 20 measurements. Test the following hypotheses at the 10 percent significance level.

$$H_0: \sigma = 6$$

$$H_a: \sigma \neq 6$$

10.68 Based on the following random sample, is there sufficient evidence at the 0.10 level to conclude that the variance of the sampled population is more than two?

$$x_1 = 10, \quad x_2 = 11, \quad x_3 = 8, \quad x_4 = 9, \quad x_5 = 12$$

10.69 Use the following random sample to determine if the standard deviation of the sampled population differs from 2.4. Test at the 5 percent significance level.

$$x_1 = 25, \quad x_2 = 21, \quad x_3 = 20, \quad x_4 = 17, \quad x_5 = 24, \quad x_6 = 28$$

10.70 Reference was made in Chapter 9 to a chemical analyzer used to measure serum cholesterol levels. A laboratory checked the variability of the analyzer by performing a sample of 10 analyses of blood drawn from the same source. The sample variance was 26.7. Does this constitute sufficient evidence that the variance of measurements by this analyzer is less than 30? Test at the 0.05 significance level.

10.71 A beverage machine is adjusted so that it dispenses an average of 8 ounces of soda. If the variation in the amounts dispensed, as measured by the standard deviation, exceeds 0.4 ounce, then the machine is overhauled. A technician measured the contents of a sample of 10 cups filled by the machine and obtained the following amounts in ounces. Is there sufficient evidence at the 0.05 level to conclude that the machine is in need of an overhaul?

10.2 10.1 10.5 9.7 10.8 9.1 9.3 10.9 9.1 10.0

10.72 In order for a computer hard disk drive to function properly, the length of a critical component must vary with a standard deviation of less than 0.15 mm. Thirty of these components are examined and found to have a standard deviation of 0.13 mm. Does this provide sufficient evidence at the 0.05 level to conclude that the specification concerning variation of lengths is being met?

10.73 A vitamin company manufactures 1,000-mg oyster shell calcium tablets. Seventy-one tablets were randomly selected from a production run, and each tablet was analyzed for its calcium content. The sample mean was 1,002.8 mg and the standard deviation was 3.4 mg. Do the data provide sufficient evidence to indicate that the standard deviation of the calcium content of all tablets in the production run differs from 2 mg? Test at the 10 percent significance level.

10.74 A college mathematics department gives an examination each year to high school seniors who compete for scholarship awards. The department attempts to design the examination so that over the life of the test, the range of scores will be about 40 points. The examination was recently given to 28 seniors, and the mean and standard deviation were 78.3 and 8.9, respectively. Is there sufficient evidence at the 10 percent level to conclude that the test does not have the desired variation in scores?

10.75 The following are the percentage rates charged for a 30-year home mortgage by a sample of 41 financial institutions in eastern Pennsylvania (*Lancaster Sunday News,* February 4, 1990).

9.625	10	9.875	9.875	9.9
9.75	10.1	10.125	9.5	10.125
9.875	9.5	9.875	9.625	9.75
10	9.375	9.5	9.875	9.5
10.1	9.95	9.75	9.875	11.5
9.875	9.95	9.875	9.875	9.375
10	9.625	9.75	9.75	9.375
10.5	9.5	10	9.875	9.75
				9.875

For the sample, $\Sigma x = 403.875$ and $\Sigma x^2 = 3,983.456768$. Assume that the survey can be considered representative of all financial institutions in this region. Is there sufficient evidence to conclude that the standard deviation in rates at the time of the survey exceeds one-quarter of a percentage point? Test at the 0.05 level.

10.6 Using p-Values to Report Test Results

We have seen that the size of the test statistic is an indication of the amount of evidence against the null hypothesis H_0, and, thus, in support of the alternative hypothesis H_a. For example, suppose we were testing at the 5 percent level of significance the following hypotheses.

$$H_0: \mu = 25$$

$$H_a: \mu > 25$$

A test statistic value of $z = 2.75$ is indicative of more evidence against H_0 than would be a value of $z = 2.00$, even though each z-value would result in sufficient evidence to reject H_0. A method of measuring the strength of the evidence against H_0 (and in favor of H_a) is to determine the **p-value** for the obtained value of the test statistic.

For this illustration, the p-value for $z = 2.75$ is the probability of obtaining a value of z *as large as* 2.75 (≥ 2.75) under the tentative assumption that H_0 were true (see Figure 10.11). Using Table 3 from Appendix A, this probability is

$$p\text{-value} = P(z \geq 2.75) = 1 - 0.9970 = 0.0030.$$

If the value of the test statistic had been $z = 2.00$, then its p-value would be the probability of observing a value of z as large as 2.00 (≥ 2.00) when H_0 is true (see Figure 10.11). Again, from Table 3, this probability is

$$p\text{-value} = P(z \geq 2.00) = 1 - 0.9772 = 0.0228.$$

Figure 10.11
P-value for testing
$H_0: \mu = 25$ versus
$H_a: \mu > 25$.

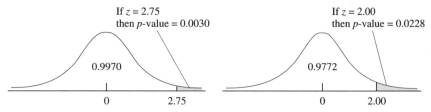

The situation on the left provides more evidence against H_0 because the p-value is smaller.

Since $z = 2.75$ has the smaller p-value, it is less probable and therefore more contradictory to H_0 than is $z = 2.00$. In general, *the smaller the p-value, the more evidence against the null hypothesis.* This is because a p-value is the probability of obtaining a value of the test statistic that is at least as extreme as the actual sample value, under the tentative assumption that H_0 were true.

The p-value is also referred to as the **observed significance level.** When testing a set of hypotheses, if the observed significance level is as small as the specified significance level α, then the null hypothesis is rejected. In the examples above, a test statistic value of either $z = 2.75$ (p-value of 0.0030) or $z = 2.00$ (p-value of 0.0228) would result in the rejection of H_0, since each p-value is less than the specified significance level $\alpha = 0.05$.

The **p-value** is the probability (assuming H_0 is true) of obtaining a test statistic value that is at least as extreme (contradictory) to H_0 as the sample value that was actually obtained.

The *p*-value is also called the **observed significance level,** and if the *p*-value satisfies

(a): *p*-value $\leq \alpha$, then reject H_0

(b): *p*-value $> \alpha$, then fail to reject H_0

where α is the specified significance level.

The *p*-value can also be thought of as the smallest possible choice of α for which H_0 can be rejected.

Example 10.14   Find the *p*-value for the large-sample test in Example 10.9 that involved a population proportion.

Solution

Example 10.9 tested the following hypotheses with a specified significance level of $\alpha = 0.01$.

$$H_0: p = 0.02$$

$$H_a: p > 0.02$$

The value of the test statistic computed in Step 3 was $z = 0.64$. The *p*-value is the probability of obtaining a value of z that contradicts H_0 at least as much as 0.64. Such values of z are those that are at least as extreme as 0.64. These are the values for which $z \geq 0.64$. Therefore, the *p*-value is

$$p\text{-value} = P(z \geq 0.64) = 1 - 0.7389 = 0.2611.$$

The *p*-value is illustrated in Figure 10.12 and is obtained by using Table 3 from Appendix A.

Figure 10.12
Determining the *p*-value for Example 10.9.

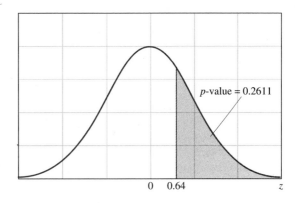

p-value $= 0.2611$

0   0.64   *z*

Note that the large p-value indicates that the null hypothesis should not be rejected since the p-value is more than the specified α level of 0.01. That is,

$$p\text{-value} > \alpha \Rightarrow \text{fail to reject } H_0$$

The next example illustrates how the p-value is determined for a two-tailed test.

Example 10.15   Find the p-value for the large-sample test concerning the population mean μ in Example 10.5.

Solution

The following hypotheses were tested with $\alpha = 0.05$.

$$H_0\!: \mu = 9.28$$
$$H_a\!: \mu \neq 9.28$$

The value of the test statistic was found to be $z = 0.96$. Since H_a is a two-sided hypothesis, the rejection region was two tailed. Consequently, the values of z that are more extreme (contradictory) to H_0 than 0.96 are those that lie either to the left of -0.96 or to the right of 0.96 (the values that exceed 0.96 in absolute value). The p-value is found by using Table 3, and its value equals (see Figure 10.13)

$$
\begin{aligned}
p\text{-value} &= P(z \leq -0.96 \text{ or } z \geq 0.96) \\
&= P(z \leq -0.96) + P(z \geq 0.96) \\
&= 2(0.1685) \\
&= 0.3370.
\end{aligned}
$$

Figure 10.13
Determining the p-value for Example 10.5.

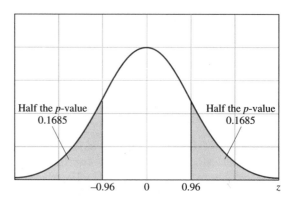

Since the p-value equals 0.337 and is more than $\alpha = 0.05$, the null hypothesis would not be rejected. That is,

$$p\text{-value} > \alpha \Rightarrow \text{fail to reject } H_0$$

> **NOTE:**   When the alternative hypothesis H_a involves the relation $\neq$ and the rejection region is two tailed, the *p*-value is obtained by doubling the extreme tail area for the test statistic value.
>
> $$p\text{-value} = 2(\text{tail area})$$

Because the chi-square and the *t*-tables only list a few selected tail areas, *p*-values must be approximated when these tables are used. MINITAB, however, can be utilized in these instances to obtain *p*-values that are correct to several decimal places. This is illustrated in the next example.

Ⓜ**Example 10.16**   For the small-sample *t*-test concerning μ in Example 10.7,

1. use Table 4 from Appendix A to approximate the *p*-value,
2. use MINITAB to obtain the exact *p*-value (to 4 decimals).

Solution

Example 10.7 involved testing the following hypotheses.

$$H_0: \mu = \$55.00$$

$$H_a: \mu > \$55.00$$

The significance level was 1 percent, the sample size was $n = 23$, and the value of the test statistic was $t = 2.60$. The values of *t* that are more contradictory to H_0 are those that lie to the right of 2.60. Therefore, the *p*-value is given by

$$p\text{-value} = P(t \geq 2.60).$$

1. Using Table 4 to find the *p*-value, we use the row for which the degrees of freedom are $df = n - 1 = 22$. The value of the test statistic, 2.60, lies between $t_{.01} = 2.508$ and $t_{.005} = 2.819$ (see Figure 10.14). Therefore, the area to the right of 2.60 is a value between 0.005 and 0.01. Thus,

$$0.005 < p\text{-value} < 0.01.$$

Figure 10.14
Determining the *p*-value for Example 10.7.

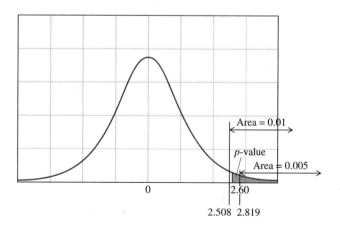

Since $\alpha = 0.01$ and the p-value is smaller than 0.01, the null hypothesis H_0 would be rejected. That is,

$$p\text{-value} < \alpha \Rightarrow \text{reject } H_0$$

Note that if α had been chosen as 0.005, then H_0 would not be rejected because the p-value exceeds 0.005.

2. To have MINITAB determine the p-value, we use the **CDF** (cumulative distribution function) command that was employed in Example 9.5. Type the following.

```
CDF 2.60;
T 22.
```

The first line of the above commands instructs MINITAB to find the cumulative probability up to (probability to the left of) 2.60. The second line indicates that this is to be done for the t-distribution with $df = 22$. In response to these commands, MINITAB will give the output in Exhibit 10.3. From Exhibit 10.3, the probability to the left of 2.60 is 0.9918. The p-value is the area to the right of 2.60, and this is given by the following.

$$
\begin{aligned}
p\text{-value} &= P(t \geq 2.60) \\
&= 1 - P(t < 2.60) \\
&= 1 - 0.9918 \\
&= 0.0082
\end{aligned}
$$

Exhibit 10.3

```
MTB > CDF 2.60;
SUBC> T 22.
        2.6000      0.9918
```

In recent years the use of p-values has become common practice in many scientific journals. By reporting the p-value for a statistical test, the reader is informed of the strength of the evidence against H_0. In this way, each person can weigh the evidence in relation to their own standards of proof and decide if the rejection of H_0 and, thus, the acceptance of the alternative hypothesis is warranted. Additional illustrations of p-values will be given in subsequent chapters.

Section 10.6 Exercises
Using p-*Values to Report Test Results*

10.76 For each of the following, state whether the null hypothesis should be rejected for the given p-value and significance level α.
  a. p-value $= 0.0431$, $\alpha = 0.05$
  b. p-value $= 0.0226$, $\alpha = 0.01$
  c. p-value $= 0.0828$, $\alpha = 0.10$
  d. p-value $= 0.0531$, $\alpha = 0.05$

10.77 For each of the following, should the null hypothesis be rejected for the given *p*-value and significance level α?
  a. $0.05 < p\text{-value} < 0.10$, $\alpha = 0.05$
  b. $0.05 < p\text{-value} < 0.10$, $\alpha = 0.10$
  c. $0.025 < p\text{-value} < 0.05$, $\alpha = 0.01$
  d. $0.005 < p\text{-value} < 0.01$, $\alpha = 0.01$
  e. $p\text{-value} > 0.10$, $\alpha = 0.05$
  f. $p\text{-value} < 0.005$, $\alpha = 0.01$

10.78 Find the *p*-value for testing the following hypotheses, if the test statistic $z = 1.51$.

$$H_0: \mu = 100$$
$$H_a: \mu > 100$$

10.79 Find the *p*-value for testing the following hypotheses, if the test statistic $z = -1.66$.

$$H_0: \mu = 230$$
$$H_a: \mu < 230$$

10.80 Find the *p*-value for testing the following hypotheses, if the test statistic $z = 2.23$.

$$H_0: \mu = 28$$
$$H_a: \mu \neq 28$$

10.81 Find the *p*-value for testing the following hypotheses, if the test statistic $z = 0.98$.

$$H_0: \mu = 89.5$$
$$H_a: \mu \neq 89.5$$

10.82 Find the *p*-value for testing the following hypotheses, if the test statistic $t = 1.90$ and the sample size $n = 16$.

$$H_0: \mu = 12.8$$
$$H_a: \mu > 12.8$$

10.83 Find the *p*-value for testing the following hypotheses, if the test statistic $t = -2.83$ and the sample size $n = 25$.

$$H_0: \mu = 347$$
$$H_a: \mu < 347$$

10.84 Find the *p*-value for testing the following hypotheses, if the test statistic $t = 2.61$ and the sample size $n = 20$.

$$H_0: \mu = 37$$
$$H_a: \mu \neq 37$$

10.85 Find the p-value for testing the following hypotheses, if the test statistic $\chi^2 = 23.52$ and the sample size $n = 16$.

$$H_0: \sigma^2 = 28$$

$$H_a: \sigma^2 > 28$$

10.86 Find the p-value for testing the following hypotheses, if the test statistic $\chi^2 = 14.89$ and the sample size $n = 30$.

$$H_0: \sigma^2 = 65$$

$$H_a: \sigma^2 < 65$$

10.87 Calculate the p-value and use it to draw the conclusion for the hypothesis test in Exercise 10.21 repeated here. A company manufactures an electronic flea collar that repels pests from pets by generating a high frequency sound that is repulsive to pests, but beyond the range of human and pet hearing. The device is powered by a battery whose mean life is 150 days. The company is testing a new battery that it believes will extend the mean life. Thirty-five units were tested with the new battery, and the mean life was 159 days. The sample standard deviation was 27 days. Do these results provide sufficient evidence at the 5 percent level to conclude that the new battery has a longer mean life?

10.88 Calculate the p-value and use it to draw the conclusion for the hypothesis test in Exercise 10.20 repeated here. A laboratory analyzed the potency of one hundred forty-five 5-grain aspirin tablets of a particular brand. For the sample, the average strength was 5.03 grains and the standard deviation was 0.24 grain. Do the sample results indicate that the mean potency of this brand is not 5 grains? Test at the 5 percent significance level.

10.89 Calculate the p-value and use it to draw the conclusion for the hypothesis test in Exercise 10.45 repeated here. A shop manual gives 6.5 hours as the average time required to perform a 30,000 mile major maintenance service on a Porsche 911. Last month a mechanic performed 10 such services, and his required times were as follows.

6.3   6.6   6.7   5.9   6.3   6.0   6.5   6.1   6.2   6.4

Is there sufficient evidence at the 0.05 significance level to conclude that the mechanic can perform this service in less time than specified by the service manual?

10.90 Calculate the p-value and use it to draw the conclusion for the hypothesis test in Exercise 10.47 repeated here. According to the U.S. Department of Education, elementary and high school teachers work an average of 50 hours per week during the school year (The *Wall Street Journal,* January 9, 1990). To see if this differs for his school district, the superintendent of schools surveyed 25 randomly selected teachers and found that they work an average of 52.7 hours a week. The standard deviation was 4.2 hours. Is the sample mean significantly different from 50 hours at the 0.05 level?

10.91 Calculate the p-value and use it to draw the conclusion for the hypothesis test in Exercise 10.58 repeated here. A fruit cannery will purchase several thousand pounds of peaches if it is convinced that less than 10 percent have serious blemishes. Inspectors for the company examined 1,500 peaches and found that 138 had serious blemishes. Is there sufficient evidence at the 0.05 level to conclude that the peaches should be purchased?

10.92 Calculate the p-value and use it to draw the conclusion for the hypothesis test in Exercise 10.62 repeated here. The *Wall Street Journal* estimated that 25 percent of 13-year-old U.S. students use calculators in their math class (February 9, 1990). Suppose a state board of education wants to determine if this percentage differs for 13 year olds in its state. A random sample of 688 students in this age group was surveyed, and the sample revealed that 206 use calculators in their math courses. Formulate a set of hypotheses and test at the 0.05 level.

10.93 Calculate the p-value and use it to draw the conclusion for the hypothesis test in Exercise 10.72 repeated here. For a computer hard disk drive to function properly, the length of a critical component must vary with a standard deviation of less than 0.15 mm. Thirty of these components are examined and found to have a standard deviation of 0.13 mm. Does this provide sufficient evidence at the 0.05 level to conclude that the specification concerning variation of lengths is being met?

10.94 Calculate the p-value and use it to draw the conclusion for the hypothesis test in Exercise 10.71 repeated here. A beverage machine is adjusted so that it dispenses an average of 8 ounces of soda. If the variation in the amounts dispensed, as measured by the standard deviation, exceeds 0.4 ounce, then the machine is overhauled. A technician measured the contents of a sample of 10 cups filled by the machine and obtained the following amounts in ounces. Is there sufficient evidence at the 0.05 level to conclude that the machine is in need of an overhaul?

$$10.2 \quad 10.1 \quad 10.5 \quad 9.7 \quad 10.8 \quad 9.1 \quad 9.3 \quad 10.9 \quad 9.1 \quad 10.0$$

MINITAB Assignments

Ⓜ 10.95 Use MINITAB to find the p-value for testing the following hypotheses, if the test statistic $z = -1.27$.

$$H_0: \mu = 13.9$$

$$H_a: \mu < 13.9$$

Ⓜ 10.96 Use MINITAB to find the p-value for testing the following hypotheses, if the test statistic $t = 1.39$ and the sample size $n = 18$.

$$H_0: \mu = 100$$

$$H_a: \mu > 100$$

Ⓜ 10.97 Use MINITAB to find the p-value for testing the following hypotheses, if the test statistic $\chi^2 = 22.14$ and the sample size $n = 15$.

$$H_0: \sigma^2 = 28$$
$$H_a: \sigma^2 > 28$$

Looking Back

This chapter is concerned with **testing statistical hypotheses** that involve a single population **mean μ, proportion p,** or **variance σ².** In a test of hypothesis problem, a researcher might make a claim (hypothesis) concerning the value of one of these parameters. To demonstrate its validity, the claim is tested by conducting the following test of hypothesis procedure.

Procedure for Conducting a Test of Hypothesis:

Step 1: Formulate the null and the alternative hypotheses.
The null hypothesis (H_0) usually represents the status quo, and the alternative hypothesis (H_a) is chosen as the hypothesis on which the burden of proof is placed.

Step 2: Select the significance level α at which the test is to be conducted.
Usually α is chosen to be 0.05, except when a Type I error is very serious, in which case 0.01 is used.

Step 3: Using the sample data, calculate the value of the test statistic.
The test statistic is the random variable whose value determines the conclusion.

Step 4: From the tables in the Appendix, determine the rejection region of H_0.
The rejection region consists of the values of the test statistic for which the null hypothesis H_0 is rejected.

Step 5: State the conclusion of the test.
If the value of the test statistic is in the rejection region, then reject H_0 and accept H_a; if it is not in the rejection region, then fail to reject H_0. Conclude that insufficient evidence exists to support the alternative hypothesis.

In this chapter, the following four tests of hypotheses are discussed.

1. **Large-sample ($n \geq 30$) test of H_0: $\mu = \mu_0$.**

The test statistic is

$$z = \frac{\bar{x} - \mu_0}{\dfrac{\sigma}{\sqrt{n}}},$$

where nearly always the population standard deviation σ must be approximated by the sample standard deviation s.

---

2. **Small-sample ($n < 30$) test of H_0: $\mu = \mu_0$.**

The test statistic is

$$t = \frac{\bar{x} - \mu_0}{\dfrac{s}{\sqrt{n}}},$$

where the degrees of freedom of the t-distribution are $df = n - 1$, and the sampled population is assumed to have approximately a normal distribution.

3. **Large-sample test of H_0: $p = p_0$.**

The test statistic is

$$z = \frac{\hat{p} - p_0}{\sqrt{\dfrac{p_0(1 - p_0)}{n}}},$$

where $\hat{p}$ is the sample proportion of successes.

4. **Large- or small-sample test of H_0: $\sigma^2 = \sigma_0^2$.**

The test statistic is

$$\chi^2 = \frac{(n - 1)s^2}{\sigma_0^2},$$

where the chi-square distribution has $df = n - 1$, and the sampled population is assumed to have approximately a normal distribution.

Key Words

In reviewing this chapter, you should be able to define, explain, and illustrate each of the following.

statistical hypothesis *(page 376)*   significance level *(page 378)*

null hypothesis *(page 376)*   rejection region *(page 379)*

alternative hypothesis *(page 377)*   nonrejection region *(page 379)*

test statistic *(page 377)*   one-tailed rejection region *(page 391)*

Type I error *(page 378)*   two-tailed rejection region *(page 391)*

Type II error *(page 378)* p-value *(page 412)*

α (alpha) *(page 378)*   observed significance level *(page 412)*

β (beta) *(page 378)*

Ⓜ   *MINITAB Commands*

SET _ *(page 389)*   TTEST _ _; *(page 396)*
END *(page 389)*   ALTERNATIVE _.
STDEV _ *(page 389)*   CDF _; *(page 416)*
ZTEST _ _ _; *(page 390)*   T _.
ALTERNATIVE _.

Review Exercises

For Exercises 10.98 through 10.106, determine the rejection region for testing the given hypotheses with the specified significance level α and sample size n.

10.98 $H_0: \mu = 163$ $\alpha = 0.01$ $n = 89$
       $H_a: \mu > 163$

10.99 $H_0: \mu = 23.6$ $\alpha = 0.05$ $n = 76$
       $H_a: \mu < 23.6$

10.100 $H_0: \mu = 1.87$ $\alpha = 0.05$ $n = 11$
       $H_a: \mu \neq 1.87$

10.101 $H_0: \mu = 16.2$ $\alpha = 0.01$ $n = 23$
       $H_a: \mu < 16.2$

10.102 $H_0: \mu = 5$ $\alpha = 0.10$ $n = 100$
       $H_a: \mu \neq 5$

10.103 $H_0: p = \frac{1}{3}$ $\alpha = 0.10$ $n = 1,200$
       $H_a: p < \frac{1}{3}$

10.104 $H_0: p = 0.59$ $\alpha = 0.05$ $n = 750$
       $H_a: p \neq 0.59$

10.105 $H_0: \sigma^2 = 0.81$ $\alpha = 0.05$ $n = 28$
       $H_a: \sigma^2 > 0.81$

10.106 $H_0: \sigma^2 = 120$ $\alpha = 0.10$ $n = 25$
       $H_a: \sigma^2 \neq 120$

10.107 A researcher wants to test the following hypotheses.

$$H_0: \mu = 60$$

$$H_a: \mu < 60$$

a. What type of error would be committed if the population mean is actually 58, but the researcher concludes that the mean is 60?
b. What type of error would be committed if the population mean is 60, but it is concluded that the mean is smaller than 60?
c. To test the above hypotheses, a random sample of 32 measurements was selected, and the sample mean was $\bar{x} = 59.2$ and the standard deviation was $s = 3.9$. Conduct the test at the 0.05 significance level.
d. Determine the p-value for the test.

10.108 According to the American Society of Plastic and Reconstructive Surgeons, there were more than 48,000 facelifts performed in the United States in 1988, and 39 percent were performed on people who were 50 or younger. Suppose a particular plastic surgeon believes that less than one-third of the facelifts performed in his state are on people 50 or younger. To prove his claim, the surgeon sampled 200 cases and found that 59 were for people in this age group. Do the data support the surgeon's belief? Test at the 0.05 level.

10.109 A manufacturer of pressure-treated lumber believes that by adding a new chemical to the treatment process, the mean life of telephone poles can be increased to more than 50 years. To prove this, a simulation test was performed on 35 posts treated under the new process. For the sample, the mean simulated life was 52.1 years and the standard deviation was 5.7 years. Is there sufficient evidence at the 0.05 level to accept the manufacturer's belief?

10.110 Obtain the p-value for the test in Exercise 10.109, and use its value to determine if the null hypothesis should be rejected.

10.111 An administrator of a large community hospital claims that the average age of blood donors during the last 12 months has fallen below the previous value of 40.3 years. To check on this belief, 20 records were randomly selected from those of all donors during the last 12 months. The ages of the 20 donors are given below ($\Sigma x = 689$, $\Sigma x^2 = 25,837$).

$$32 \quad 43 \quad 51 \quad 43 \quad 21 \quad 42 \quad 23 \quad 29 \quad 35 \quad 37$$
$$39 \quad 28 \quad 27 \quad 64 \quad 27 \quad 31 \quad 36 \quad 28 \quad 31 \quad 22$$

a. Test at the 0.05 level if sufficient evidence exists to support the administrator's claim.
b. In order for the test performed in Part a to be valid, what assumptions must be made?

10.112 Obtain the p-value for the test in Exercise 10.111.

10.113 During the growth period of a laboratory culture, the ambient temperature of the incubation unit must be maintained at a constant temperature that varies with a standard deviation of at most 0.6 degree. A sample of 15 temperature readings had a standard deviation of 0.74 degree. Does this provide sufficient evidence to conclude that the true standard deviation of the temperature exceeds 0.6 degree? Test at the 0.05 significance level.

10.114 Obtain the p-value for the test in Exercise 10.113.

10.115 The average percentage of saturated fat in butter is 66 percent (*In Health*, March/April, 1990). To determine if its brand differs in this regard, a food company analyzed ninety-six 1-pound packages. The mean percentage of saturated fat was found to be 65.6, and the standard deviation was 1.4. Formulate a suitable set of hypotheses and test at the 0.01 level.

10.116 Obtain the p-value for the test in Exercise 10.115.

10.117 A stock market advisory service bases its weekly recommendations on the percentage of financial analysts who are bullish on stocks. For its next newsletter, the advisory service surveyed 150 analysts, and 60 said that they were bullish. Do these results provide sufficient evidence to conclude that the percentage of bullish analysts has changed from the previous week's value of 35 percent? Test at the 0.05 significance level.

10.118 Obtain the p-value for the test in Exercise 10.117.

10.119 In the United States, the average yield of honey per bee colony is about 53 pounds (*American Scientist,* September/October, 1988). A researcher believes that a new strain of bees will produce more honey. Eighteen colonies of this strain were investigated, and the mean honey yield was 59.7 pounds. The standard deviation was 3.6 pounds. Do the data suggest that the mean honey yield for this type of bee is greater than 53 pounds? Test at the 0.01 level.

10.120 A random sample of 5 measurements from a normal population resulted in the following values.

$$x_1 = 1.6, \quad x_2 = 1.7, \quad x_3 = 0.9, \quad x_4 = 1.4, \quad x_5 = 1.9$$

Is there sufficient evidence at the 0.01 significance level to conclude that the population variance differs from 0.04?

10.121 Obtain the p-value for the test in Exercise 10.120.

MINITAB Assignments

Ⓜ 10.122 A consumer prefers a particular brand of potato chips packaged with a label weight of 15 ounces. However, he believes that the bags tend to be underweight. In an attempt to prove this belief, 15 bags are purchased over a period of several weeks. The resulting weights are given below.

14.1   14.0   14.3   15.1   14.2   14.7   15.3   14.9
15.1   14.2   13.7   14.1   14.2   14.3   14.2

Use MINITAB to determine if there is sufficient evidence at the 0.05 significance level to support the consumer's belief.

Ⓜ 10.123 The balances of a random sample of 13 regular savings accounts at a large financial institution appear below (the amounts are in dollars).

1300.43   795.78   1028.94   376.86   982.22   1002.23   254.33
1729.64   103.45   1112.87   202.32   875.68   1212.21

Using MINITAB, test at the 0.01 level if sufficient evidence exists to conclude that the average balance of all regular savings accounts at this institution exceeds $800.00.

Ⓜ 10.124 A certain candy bar is suppose to weigh 75 grams. Fifty bars were randomly selected from a production run, and the following weights in grams were recorded.

75.4	68.8	71.3	74.4	77.0	73.0	70.8	75.2	70.4	73.6
73.4	75.2	73.7	77.2	74.9	73.1	68.8	73.5	77.9	76.2
75.2	74.1	73.9	75.8	74.1	74.3	70.5	74.0	73.5	71.6
75.6	76.1	71.9	71.1	75.4	74.0	77.4	76.4	78.8	74.2
73.9	74.5	72.5	71.6	73.0	75.2	71.0	75.9	73.5	74.1

Use MINITAB to test at the 0.05 level if there is sufficient evidence to conclude that the mean weight of all bars in this production run differs from 75 grams.

Ⓜ 10.125 Use MINITAB to find the *p*-value for testing the following hypotheses, if the test statistic $z = 1.78$.

$$H_0: \mu = 92.3$$

$$H_a: \mu > 92.3$$

Ⓜ 10.126 Use MINITAB to find the *p*-value for testing the following hypotheses, if the test statistic $t = 2.01$ and the sample size $n = 21$.

$$H_0: \mu = 184$$

$$H_a: \mu \neq 184$$

Ⓜ 10.127 Use MINITAB to find the *p*-value for testing the following hypotheses, if the test statistic $\chi^2 = 18.53$ and the sample size $n = 23$.

$$H_0: \sigma^2 = 37$$

$$H_a: \sigma^2 > 37$$

Ⓜ 10.128 The following are the heights in inches of 36 blue spruce trees randomly selected from a 10-acre tree farm.

84	79	67	86	75	89	76	91	83
74	87	78	86	90	84	79	73	88
87	79	73	76	87	86	80	82	74
85	87	69	84	83	74	81	85	79

Use MINITAB to test if sufficient evidence exists at the 0.05 level to conclude that the mean height of all trees on the farm exceeds 80 inches.

(© *Michael Salas/The Image Bank*)

11

Tests of Hypotheses and Estimation: Two Samples

◀ *Are VDTs a Health Risk?*

Recently, several concerns have been raised about possible health risks from expo-sures to low level sources of radiation such as electrical lines, home appliances, and video display terminals (VDTs). With the increasing use of personal computers in the workplace, much controversy has been generated as to what constitutes a safe level of monitor radiation. Some researchers believe that exposures over a long period of time may increase the risk of health problems such as cancer and miscarriages. The Wall Street Journal *(March 14, 1991) reported on a recent study conducted by the govern-ment's National Institute of Occupational Safety and Health. The project investigated 882 pregnancies among directory assistance telephone operators in several south-eastern states. One group of operators used VDTs, and a similar age group used non-VDT equipment that emitted much lower levels of radiation. The investigators found no significant difference in miscarriage rates for the two groups of operators during the first trimester of their pregnancies.*

This chapter is concerned with problems such as the above that involve compari-sons between two groups. In the following sections, you will learn how to compare two populations with regard to their means, variances, and proportions.

Looking Ahead

In Chapter 9, we considered inferences that involve estimating a parameter with a confidence interval. Chapter 10 was concerned with inferences in the form of test-ing a set of hypotheses concerning the value of a parameter. In both chapters, the inferences pertained to a single population parameter and they were based on a single sample.

This chapter is concerned with *inferences about two populations*. The meth-odologies of hypothesis testing and constructing confidence intervals for one pop-ulation will now be extended to two. Specifically, we will be concerned with com-paring two populations with regard to their **means,** μ_1 and μ_2; their **variances,** σ_1^2 and σ_2^2; and their **proportions,** p_1 and p_2.

Problems concerning the comparison of two population parameters frequently arise in practical applications.

- A university's director of career development may want to know if a difference exists in the mean starting salary of ultrasound technicians and radiologic technologists.
- A stove manufacturer may wish to compare the mean burning time for two models of wood stoves.
- A company might want to determine if the proportion of workers who are satisfied with their jobs differs between those on fixed and those on flexitime schedules.
- A pharmaceutical company that produces a quit-smoking drug would want to compare the proportion of its users who are successful with the proportion of smokers who can quit on their own.
- A cereal manufacturer may be interested in comparing two machines with regard to the variability in the amounts of cereal placed in boxes.

11.1 Large-Sample Inferences for Two Means: Independent Samples

To compare the means of two populations, we first need to introduce some notation. We will arbitrarily label the populations as one and two and denote their means and standard deviations by μ_1, μ_2, σ_1, and σ_2. A random sample will be selected from each population, and the sample sizes, sample means, and sample standard deviations will be denoted by n_1, n_2, $\bar{x}_1$, $\bar{x}_2$, s_1, and s_2. This notation is illustrated in Figure 11.1. The random samples are assumed to be **independent.** This means that the selection of one sample is not affected by nor related to the selection of the other sample. (Samples that are not independent will be considered in Section 11.4.)

Figure 11.1
Notation for two populations and their samples.

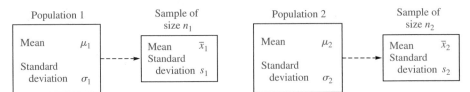

We will compare the means of the two populations by considering their difference, $(\mu_1 - \mu_2)$. The usual **point estimate** of this difference between the population means is $(\bar{x}_1 - \bar{x}_2)$, the difference between the corresponding sample means. The following properties of $(\bar{x}_1 - \bar{x}_2)$ form the basis for obtaining a large-sample confidence interval and a test of hypothesis concerning $(\mu_1 - \mu_2)$.

The Sampling Distribution of $(\bar{x}_1 - \bar{x}_2)$ for Independent Samples:

1. The mean of $(\bar{x}_1 - \bar{x}_2)$ is
$$\mu_{(\bar{x}_1 - \bar{x}_2)} = (\mu_1 - \mu_2).$$

2. The standard deviation of $(\bar{x}_1 - \bar{x}_2)$ is
$$\sigma_{(\bar{x}_1 - \bar{x}_2)} = \sqrt{\frac{\sigma_1^2}{n_1} + \frac{\sigma_2^2}{n_2}}.$$

3. For large samples, the shape of the distribution is approximately normal.

Sampling distribution of $(\bar{x}_1 - \bar{x}_2)$ for large independent samples.

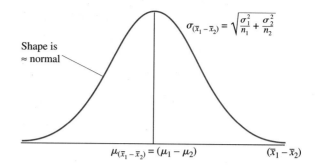

In Chapter 9, Formula 9.5 gave the following large-sample confidence interval for a parameter θ.

$$\hat{\theta} \pm z_{\alpha/2}\sigma_{\hat{\theta}}$$

Lower Confidence Limit   Upper Confidence Limit

A confidence interval for $(\mu_1 - \mu_2)$ can be obtained from this by replacing $\hat{\theta}$ with $(\bar{x}_1 - \bar{x}_2)$ and $\sigma_{\hat{\theta}}$ with $\sigma_{(\bar{x}_1 - \bar{x}_2)}$, which equals $\sqrt{\sigma_1^2/n_1 + \sigma_2^2/n_2}$. These substitutions lead to the following large-sample confidence interval for $(\mu_1 - \mu_2)$.

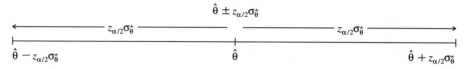

Lower Confidence Limit   Upper Confidence Limit

Large-Sample $(1 - \alpha)$ Confidence Interval for $(\mu_1 - \mu_2)$: Independent Samples

$$(\bar{x}_1 - \bar{x}_2) \pm z_{\alpha/2}\sqrt{\frac{\sigma_1^2}{n_1} + \frac{\sigma_2^2}{n_2}} \qquad \textbf{(11.1)}$$

Assumptions:
1. Independent random samples.
2. Each sample size is large (30 or more).

Note: In nearly all practical situations, σ_1 and σ_2 are unknown and must be replaced with the approximations s_1 and s_2, respectively.

Example 11.1   A company will choose one of two cities in which to build a new manufacturing plant. As part of the decision process, it wants to compare the mean number of school years completed by the labor force in each city. Samples of 125 workers were randomly selected from each city's labor force, and the sample statistics are given below.

	City 1	City 2
Sample size	$n_1 = 125$	$n_2 = 125$
Sample mean	$\bar{x}_1 = 12.9$ years	$\bar{x}_2 = 12.3$ years
Sample std. dev.	$s_1 = 0.8$ year	$s_2 = 1.1$ years

Use a 95 percent confidence interval to estimate the difference in the mean number of school years completed by the two labor forces.

Solution

The general form of the confidence interval for $(\mu_1 - \mu_2)$, the difference between the population means, is

$$(\bar{x}_1 - \bar{x}_2) \pm z_{\alpha/2} \sqrt{\frac{\sigma_1^2}{n_1} + \frac{\sigma_2^2}{n_2}}.$$

Since the population standard deviations, σ_1 and σ_2, are unknown, they must be approximated using the sample standard deviations, s_1 and s_2. To determine the z-value, its subscript must first be found, and this is given by

$$\alpha/2 = \frac{(1 - \text{confidence level})}{2} = \frac{(1 - 0.95)}{2} = 0.025.$$

The z-value is $z_{\alpha/2} = z_{.025} = 1.96$. Therefore, a 95 percent confidence interval for $(\mu_1 - \mu_2)$ is as follows.

$$(12.9 - 12.3) \pm 1.96 \sqrt{\frac{0.8^2}{125} + \frac{1.1^2}{125}}$$

$$0.6 \pm 0.24$$

The lower and upper confidence limits are 0.36 and 0.84, and the confidence interval can also be written as

$$0.36 < (\mu_1 - \mu_2) < 0.84.$$

Note that since the interval contains only positive values, we are 95 percent confident that μ_1 is more than μ_2, and this difference is some amount between 0.36 and 0.84. Thus, with 95 percent confidence, we estimate that the mean number of school years for the first city's labor force is between 0.36 and 0.84 year more than that for the second city.

Testing hypotheses about $(\mu_1 - \mu_2)$ can be accomplished by converting $(\bar{x}_1 - \bar{x}_2)$ to standard units and using the resulting z-value as the test statistic. The z-value for $(\bar{x}_1 - \bar{x}_2)$ is

$$z = \frac{(\bar{x}_1 - \bar{x}_2) - \mu_{(\bar{x}_1 - \bar{x}_2)}}{\sigma_{(\bar{x}_1 - \bar{x}_2)}} = \frac{(\bar{x}_1 - \bar{x}_2) - (\mu_1 - \mu_2)}{\sqrt{\dfrac{\sigma_1^2}{n_1} + \dfrac{\sigma_2^2}{n_2}}} \qquad (11.2)$$

For testing the null hypothesis

$$H_0: (\mu_1 - \mu_2) = d_0 \ (d_0 \text{ is a constant}),$$

$(\mu_1 - \mu_2)$ in Equation 11.2 is replaced by d_0, thus giving the following as the test statistic.

$$z = \frac{(\bar{x}_1 - \bar{x}_2) - d_0}{\sqrt{\dfrac{\sigma_1^2}{n_1} + \dfrac{\sigma_2^2}{n_2}}}$$

The large-sample test of hypothesis concerning $(\mu_1 - \mu_2)$ is summarized below.

Large-Sample Hypothesis Test for $(\mu_1 - \mu_2)$: Independent Samples
For testing the null hypothesis $H_0: (\mu_1 - \mu_2) = d_0$ (d_0 is a constant), the test statistic is

$$z = \frac{(\bar{x}_1 - \bar{x}_2) - d_0}{\sqrt{\dfrac{\sigma_1^2}{n_1} + \dfrac{\sigma_2^2}{n_2}}}. \qquad (11.3)$$

The rejection region is given by the following.

For $H_a: (\mu_1 - \mu_2) < d_0$, H_0 is rejected when $z < -z_\alpha$.

For $H_a: (\mu_1 - \mu_2) > d_0$, H_0 is rejected when $z > z_\alpha$.

For $H_a: (\mu_1 - \mu_2) \neq d_0$, H_0 is rejected when $z < -z_{\alpha/2}$ or $z > z_{\alpha/2}$.

Assumptions:
1. Independent random samples.
2. Each sample size is large ($n_1 \geq 30$ and $n_2 \geq 30$).

Note:
1. σ_1 and σ_2 are usually unknown and must be approximated by the sample standard deviations s_1 and s_2.
2. Usually the null hypothesis is that μ_1 and μ_2 are the same, in which case the constant d_0 is 0.

The above test of hypothesis is sometimes referred to as the **two-sample z-test for means.** In the unlikely event that both populations are normally distributed and their standard deviations are known, the test is applicable for any values of n_1 and n_2. Since in practice these conditions would seldom be satisfied, the two-sample z-test is usually only used for large samples. A small-sample test is given in the next section.

Example 11.2

The manager of a department store believes that camera film sales can be increased by relocating the film display from the middle to the front of the store. To investigate this, the manager recorded the number of rolls of film sold for a random sample of 35 days at the old location and for 30 days at the new location. Do the results given below provide sufficient evidence to support the manager's belief? Test at the 5 percent level of significance.

New Location 7	Old Location
$n_1 = 30$	$n_2 = 35$
$\bar{x}_1 = 98.8$	$\bar{x}_2 = 95.3$
$s_1 = 8.9$	$s_2 = 7.4$

(Courtesy of Fuji Photo Film U.S.A., Inc.)

Solution

Step 1: Hypotheses.

The manager is attempting to show that the mean film sales at the new location (μ_1) are greater than the mean sales at the previous location (μ_2). Thus, the alternative hypothesis is that μ_1 differs from μ_2 by a positive amount, that is ($\mu_1 - \mu_2$) > 0.

$$H_0: (\mu_1 - \mu_2) = 0 \quad \text{(the constant } d_0 \text{ is 0)}$$

$$H_a: (\mu_1 - \mu_2) > 0$$

Step 2: Significance level.

$$\alpha = 0.05.$$

Step 3: Calculations.

The test statistic is

$$z = \frac{(\bar{x}_1 - \bar{x}_2) - d_0}{\sqrt{\dfrac{\sigma_1^2}{n_1} + \dfrac{\sigma_2^2}{n_2}}},$$

where the sample standard deviations s_1 and s_2 are substituted for the unknown population standard deviations σ_1 and σ_2.

$$z = \frac{(98.8 - 95.3) - 0}{\sqrt{\dfrac{8.9^2}{30} + \dfrac{7.4^2}{35}}} = \frac{3.5}{2.0506} = 1.71$$

Step 4: Rejection region.
Because the inequality $>$ in H_a "points" to the right, the rejection region is in the upper right-hand tail of the z-distribution. H_0 is rejected for

$$z > z_\alpha = z_{.05} = 1.645.$$

Step 5: Conclusion.
Since $z = 1.71$ is in the rejection region, H_0 is rejected, and the alternative hypothesis H_a is accepted. At the 5 percent significance level, the store manager can conclude that camera film sales at the front of the store are greater than those at the previous location.

In the preceding example, there was sufficient evidence at the 5 percent level to reject the null hypothesis. Consequently, the difference between the sample means is statistically significant. However, **statistical significance** should not be equated with **practical significance.** For the film example, the difference in mean sales for the samples was only 3.5 rolls per day, and from a financial viewpoint the relocation might not be practical. Permanent relocation may require costly renovations or perhaps use of the frontal space for a different item might be more financially rewarding for the store.

Example 11.3   Find the p-value for the hypothesis test in the previous example.

Solution

The computed value of the test statistic in Example 11.2 is $z = 1.71$. The p-value is the probability of obtaining a value of z that contradicts H_0 at least as much as 1.71. Such values of z are those for which $z \geq 1.71$. The p-value, then, is obtained from Table 3 of Appendix A and equals

$$p\text{-value} = P(z \geq 1.71) = 1 - 0.9564 = 0.0436.$$

The p-value is illustrated in Figure 11.2.
Note that since the p-value is smaller than $\alpha = 0.05$, the null hypothesis should be rejected. If H_0 were true, the probability would only be 0.0436 of obtaining a difference in the sample means as large (or larger) as that actually observed.

Figure 11.2
Determining the p-value for
Example 11.2.

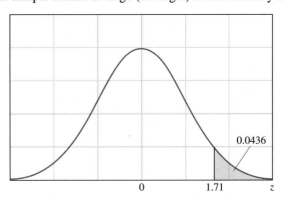

Currently, there is no MINITAB command using the z-distribution for performing a hypothesis test or constructing a confidence interval for $(\mu_1 - \mu_2)$. In the next section we will illustrate how the t-distribution can be used for these purposes.

Section 11.1 Exercises

Large-Sample Inferences for Two Means: Independent Samples

11.1   Independent random samples were selected from two populations, and the following sample results were obtained.

	Sample One	Sample Two
Sample size	$n_1 = 45$	$n_2 = 35$
Sample mean	$\bar{x}_1 = 257$	$\bar{x}_2 = 245$
Sample std. dev.	$s_1 = 26$	$s_2 = 19$

With $\alpha = 0.05$, test the following hypotheses.

$$H_0: (\mu_1 - \mu_2) = 0$$

$$H_a: (\mu_1 - \mu_2) > 0$$

11.2   Use the data in Exercise 11.1 to estimate $(\mu_1 - \mu_2)$ with a 95 percent confidence interval.

11.3   Independent random samples were selected from two populations, and the following sample results were obtained.

	Sample One	Sample Two
Sample size	$n_1 = 30$	$n_2 = 32$
Sample mean	$\bar{x}_1 = 60.8$	$\bar{x}_2 = 48.3$
Sample std. dev.	$s_1 = 3.6$	$s_2 = 3.1$

With $\alpha = 0.01$, test the following hypotheses.

$$H_0: (\mu_1 - \mu_2) = 10$$

$$H_a: (\mu_1 - \mu_2) \neq 10$$

11.4   Use the data in Exercise 11.3 to estimate $(\mu_1 - \mu_2)$ with a 99 percent confidence interval.

11.5   Independent random samples were selected from two populations, and the following sample results were obtained.

	Sample One	Sample Two
Sample size	$n_1 = 53$	$n_2 = 68$
Sample mean	$\bar{x}_1 = 949$	$\bar{x}_2 = 940$
Sample std. dev.	$s_1 = 23$	$s_2 = 17$

With $\alpha = 0.05$, test the following hypotheses.

$$H_0: (\mu_1 - \mu_2) = 5$$

$$H_a: (\mu_1 - \mu_2) > 5$$

11.6   Major physiological changes in body composition are a natural part of the aging process. A physiologist wants to compare the mean body percent of water for 30-year-old males (μ_1) with that of 60-year-old males (μ_2). Random samples from the two age groups produced the following results.

| | **Body % of Water** | | |
Age Group	n	Mean	Std. Dev.
30 year olds	39	59.8	1.3
60 year olds	32	55.2	0.9

Obtain a 95 percent confidence interval for $(\mu_1 - \mu_2)$ to estimate the difference in mean body percent of water for the 2 age groups.

11.7   To compare 2 dog training programs, an obedience school trained 43 dogs using Program A and 41 dogs using Program B. For Program A, the average number of training hours required was 24.8 with a standard deviation of 3.1 hours. For Program B, the mean was 22.9 hours with a standard deviation of 3.3 hours. Is the difference in the sample means statistically significant at the 0.05 level?

11.8   Use the data in Exercise 11.7 to estimate with 99 percent confidence the difference in the mean number of training hours required for Programs A and B.

11.9   A national computer retailer believes that average sales are greater for salespersons with a college degree. A random sample of 34 salespersons with a degree had an average weekly sales of $3,542 last year, while 37 salespersons without a degree averaged $3,301. The standard deviations were $468 for those with a degree and $642 for those without.
a. Is there sufficient evidence at the 0.05 level to support the retailer's belief?
b. Obtain the p-value for the test.

11.10 For Exercise 11.9, obtain a 90 percent confidence interval to estimate the difference in mean weekly sales for salespersons with and without a college degree.

11.11 Recently a Canadian scientist showed there is a danger that cardboard milk containers may contain the carcinogen dioxin, which is believed to enter paper products during the bleaching process. When this occurs, it is then possible for the dioxin to migrate into the milk (*Newsweek,* March 27, 1989). One solution being considered is to line the cartons with foil. Suppose a paper manufacturer measures the dioxin in a sample of 50 unlined cartons of milk and obtains a mean of 0.030 parts per trillion (ppt), with a standard deviation of 0.009 ppt. It also measures the dioxin in a sample of 50 lined cartons and obtains a mean of 0.006 ppt, with a standard deviation of 0.002. Would these results indicate a significant difference in the mean dioxin levels for the two types of containers? Test at the 0.01 level of significance.

11.12 An oral surgeon conducted an experiment to determine if background music tends to decrease a patient's anxiety during a tooth extraction. Over a 1-month period, 32 patients had a tooth removed while listening to music through headphones, and 36 patients had an extraction with no background music. After the tooth removal, each patient completed a questionnaire to measure their anxiety level during the procedure (higher scores indicate greater anxiety levels). For those with background music, the mean score was 4.2 with a standard deviation of 1.2, while the group with no music had a mean score of 5.9 with a standard deviation of 1.9. Is there sufficient evidence at the 0.05 level to conclude that music is effective in reducing anxiety during a tooth extraction?

11.13 With reference to Exercise 11.12, obtain a 95 percent confidence interval to estimate the difference in the mean anxiety level for the 2 methods.

11.14 An agronomist believes that a newly developed plant food will increase the mean yield of tomato plants by more than 5 pounds. Forty-eight plants were treated with the food and had a mean yield of 28.4 pounds and a standard deviation of 2.8 pounds. Forty-five identical plants were untreated and had a mean yield of 22.3 pounds and a standard deviation of 2.3 pounds.
   a. Do the data provide sufficient evidence to support the agronomist's theory? Test at the 0.01 significance level.
   b. Obtain the *p*-value for the test.

11.15 For Exercise 11.14, obtain a 90 percent confidence interval to estimate the mean increase in yield of tomato plants treated with the plant food.

11.2   Small-Sample Inferences for Two Means: Independent Samples and Equal Variances

In the preceding section, the two-sample *z*-test and its associated confidence interval were employed for inferences concerning the difference between two population means. These techniques should only be used when both $n_1 \geq 30$ and $n_2 \geq 30$. If one (or both) of the sample sizes is smaller than 30, then inferences are based on a *t*-statistic. Before discussing small-sample techniques, we should mention that generally one tries to avoid small-sample situations for several reasons. Small-sample analyses usually require more assumptions than their large-sample counterparts. For instance, in the small-sample case pertaining to $(\mu_1 - \mu_2)$, it is necessary to assume that the sampled populations have distributions that are normal (at least approximately). Also, as intuition dictates, inferences based on small samples are less precise (wider confidence interval) than when the sample sizes are large. Furthermore, as this section will show, the analysis may be more complex when small samples are used.

In the small-sample situation concerning $(\mu_1 - \mu_2)$, the *t*-statistic used for inferences depends on whether or not the sampled populations have the same variance. In this section we will consider the simpler situation in which the population variances are assumed to be equal. That is, suppose it is reasonable to make the assumption that

σ_1^2 and σ_2^2 have the same value, say σ^2. In nearly all practical applications, σ^2 is unknown and must be estimated by using s_1^2 and s_2^2, the variances of two independent random samples selected from the populations. The estimate used is called the **pooled sample variance** and is denoted by s_p^2. It is calculated as follows.

The Pooled Sample Variance:
$$s_p^2 = \frac{(n_1 - 1)s_1^2 + (n_2 - 1)s_2^2}{n_1 + n_2 - 2}$$

Note:
1. When two populations are assumed to have the same unknown variance, s_p^2 is used to estimate its value.
2. s_1^2 and s_2^2 are the variances of independent random samples of size n_1 and n_2.

Notice that the pooled sample variance combines (pools) the two sample variances to obtain an estimate of the common variance σ^2 shared by the sampled populations. More precisely, s_p^2 is just a weighted average of the two sample variances s_1^2 and s_2^2, where the weights are the degrees of freedom associated with each variance. In Exercise 11.29 the reader is asked to show that if the sample sizes are equal, then s_p^2 simplifies to the mean of s_1^2 and s_2^2, that is,

---

$$s_p^2 = \frac{(s_1^2 + s_2^2)}{2}$$

when $n_1 = n_2$.

Example 11.4   Assume that the variances of two populations are equal. To estimate this common value, independent random samples of size $n_1 = 6$ and $n_2 = 11$ were selected and yielded sample variances of $s_1^2 = 18.2$ and $s_2^2 = 20.6$. Estimate the common variance of the two populations.

Solution

The variance of each population is estimated by the pooled sample variance.

$$\begin{aligned}
s_p^2 &= \frac{(n_1 - 1)s_1^2 + (n_2 - 1)s_2^2}{n_1 + n_2 - 2} \\
&= \frac{(6 - 1)(18.2) + (11 - 1)(20.6)}{6 + 11 - 2} \\
&= \frac{297}{15} = 19.8
\end{aligned}$$

Notice that the value of s_p^2 is between $s_1^2 = 18.2$ and $s_2^2 = 20.6$. It is closer to the value of s_2^2 since n_2 is larger, and consequently, more weight is given to s_2^2 in computing s_p^2.

In Section 11.1 we pointed out that the sampling distribution of $(\bar{x}_1 - \bar{x}_2)$ has a standard deviation given by $\sigma_{(\bar{x}_1 - \bar{x}_2)} = \sqrt{\sigma_1^2/n_1 + \sigma_2^2/n_2}$. When the population variances are assumed to be equal, $\sigma_{(\bar{x}_1 - \bar{x}_2)}$ is estimated by replacing σ_1^2 and σ_2^2 with the pooled sample variance s_p^2. This gives

$$\sqrt{\frac{\sigma_1^2}{n_1} + \frac{\sigma_2^2}{n_2}} \approx \sqrt{\frac{s_p^2}{n_1} + \frac{s_p^2}{n_2}}$$

$$= \sqrt{s_p^2 \left(\frac{1}{n_1} + \frac{1}{n_2} \right)}.$$

By substituting this expression into the denominator of the large-sample z-statistic, the following small-sample t-statistic is obtained.

$$t = \frac{(\bar{x}_1 - \bar{x}_2) - (\mu_1 - \mu_2)}{\sqrt{s_p^2 \left(\frac{1}{n_1} + \frac{1}{n_2} \right)}}$$

The small-sample test of hypothesis and associated confidence interval are summarized below.

Pooled t-Test
Small-Sample Hypothesis Test for $(\mu_1 - \mu_2)$: Independent Samples and Equal Population Variances

For testing the null hypothesis H_0: $(\mu_1 - \mu_2) = d_0$ (d_0 is usually 0), the test statistic is

$$t = \frac{(\bar{x}_1 - \bar{x}_2) - d_0}{\sqrt{s_p^2 \left(\frac{1}{n_1} + \frac{1}{n_2} \right)}}. \tag{11.4}$$

The rejection region is given by the following.

For H_a: $(\mu_1 - \mu_2) < d_0$, H_0 is rejected when $t < -t_\alpha$.

For H_a: $(\mu_1 - \mu_2) > d_0$, H_0 is rejected when $t > t_\alpha$.

For H_a: $(\mu_1 - \mu_2) \neq d_0$, H_0 is rejected when $t < -t_{\alpha/2}$ or $t > t_{\alpha/2}$.

Assumptions:
1. Independent random samples.
2. Equal but unknown population variances.
3. Each population has a normal distribution.

Note:
1. The t-distribution has degrees of freedom $df = n_1 + n_2 - 2$.
2. $s_p^2 = \dfrac{(n_1 - 1)s_1^2 + (n_2 - 1)s_2^2}{n_1 + n_2 - 2}.$

> **Small-Sample $(1 - \alpha)$ Confidence Interval for $(\mu_1 - \mu_2)$: Independent Samples and Equal Population Variances**
>
> $$(\bar{x}_1 - \bar{x}_2) \pm t_{\alpha/2} \sqrt{s_p^2 \left(\frac{1}{n_1} + \frac{1}{n_2}\right)} \qquad (11.5)$$
>
> *Assumptions:*
> 1. Independent random samples.
> 2. Equal but unknown population variances.
> 3. Each population has a normal distribution.
>
> *Note:*
> 1. The t-distribution has degrees of freedom $df = n_1 + n_2 - 2$.
> 2. $s_p^2 = \dfrac{(n_1 - 1)s_1^2 + (n_2 - 1)s_2^2}{n_1 + n_2 - 2}$.

Example 11.5   In 1989, New York became the first state to limit the length of the work week to 80 hours for medical interns. Suppose the health commissioner of a neighboring state is investigating the number of hours worked by interns in the state's 2 largest cities. To determine if a difference exists in the mean work week for interns employed in these 2 cities, random samples of interns' work weeks are selected. For the first sample of 13 weeks, the mean and standard deviation are 93.1 and 5.7 hours. For the second sample of 16 weeks, the mean and standard deviation are 97.9 and 5.2 hours. At the 0.05 level, test if a difference exists in the mean length of the work week for interns in these cities. Assume that the distributions of length of work weeks are normal and have approximately the same variance.

Solution

We are given $n_1 = 13$, $\bar{x}_1 = 93.1$, $s_1 = 5.7$, $n_2 = 16$, $\bar{x}_2 = 97.9$, and $s_2 = 5.2$.

Step 1: Hypotheses.

$$H_0 : (\mu_1 - \mu_2) = 0$$
$$H_a : (\mu_1 - \mu_2) \neq 0$$

Step 2: Significance level.

$$\alpha = 0.05.$$

Step 3: Calculations.
Since the population variances are assumed equal, we first must estimate this common variance with the pooled sample variance s_p^2.

$$s_p^2 = \frac{(n_1 - 1)s_1^2 + (n_2 - 1)s_2^2}{n_1 + n_2 - 2} = \frac{12(5.7)^2 + 15(5.2)^2}{27} = 29.46$$

The value of the test statistic is given by

$$t = \frac{(\bar{x}_1 - \bar{x}_2) - d_0}{\sqrt{s_p^2 \left(\dfrac{1}{n_1} + \dfrac{1}{n_2}\right)}} = \frac{(93.1 - 97.9) - 0}{\sqrt{29.46 \left(\dfrac{1}{13} + \dfrac{1}{16}\right)}} = -2.37.$$

Step 4: Rejection region.
Since H_a involves $\neq$, the rejection region is two tailed with an area of $\alpha/2 = 0.05/2 = 0.025$ in each tail. The t-distribution has $df = n_1 + n_2 - 2 = 27$. From Table 4 in Appendix A, $t_{.025} = 2.052$, and H_0 is rejected for $t < -2.052$ or $t > 2.052$.

Step 5: Conclusion.
The null hypothesis is rejected since $t = -2.37$ is in the rejection region. Therefore, there is sufficient evidence at the 0.05 significance level to conclude that a difference exists in the mean length of the work week for interns in the 2 cities.

Example 11.6   For Example 11.5, use a 90 percent confidence interval to estimate the difference in the mean length of the work week for interns in the 2 cities.

Solution

From Example 11.5, we are given that $n_1 = 13$, $\bar{x}_1 = 93.1$, $s_1 = 5.7$, $n_2 = 16$, $\bar{x}_2 = 97.9$, and $s_2 = 5.2$. The pooled sample variance was found to be $s_p^2 = 29.46$. A 90 percent confidence interval for $(\mu_1 - \mu_2)$ is given by the following.

$$(\bar{x}_1 - \bar{x}_2) \pm t_{\alpha/2} \sqrt{s_p^2 \left(\frac{1}{n_1} + \frac{1}{n_2}\right)}$$

$$(93.1 - 97.9) \pm t_{.05} \sqrt{29.46 \left(\frac{1}{13} + \frac{1}{16}\right)}$$

The subscript of .05 for the t-value was obtained by computing

$$\frac{\alpha}{2} = \frac{1 - \text{confidence level}}{2} = \frac{1 - 0.90}{2} = .05$$

The value of $t_{.05}$ is 1.703, and this is found from Table 4 using $df = n_1 + n_2 - 2 = 27$. Substituting this into the above confidence interval formula, we obtain the interval -4.8 ± 3.45. Thus, with 90 percent confidence we estimate that $(\mu_1 - \mu_2)$ is between -8.25 and -1.35. That is, the mean number of hours for the first city is smaller than the mean for the second city by some amount between 1.35 and 8.25 hours.

Ⓜ Example 11.7   A coal-fired utility company has recently installed new pollution control equipment at one of its facilities. An environmental engineer wants to determine if there has been a change in air pollution levels at a nearby national park. Twelve randomly selected measurements of sulfate levels prior to installation and 10 readings after the equipment was installed appear below. All figures give the concentration of sulfates in micrograms per cubic meter.

With the Old Equipment	With the New Equipment
10.0	5.0
8.0	7.0
8.0	1.0
7.0	9.0
6.0	1.5
9.0	5.0
11.5	2.5
8.0	4.0
9.5	9.0
7.5	6.0
5.0	
10.0	

We will use MINITAB to test at the 1 percent significance level whether a change has occurred in the mean concentration levels of sulfates. We want to test

$$H_0 : (\mu_1 - \mu_2) = 0$$

$$H_a : (\mu_1 - \mu_2) \neq 0$$

We will place the sample of levels before installation in column C1 and those after installation in C2. The **SET** command rather than **READ** is used since n_1 and n_2 are not the same.

```
SET C1
10.0 8.0 8.0 7.0 6.0 9.0 11.5 8.0 9.5 7.5 5.0 10.0
END
SET C2
5.0 7.0 1.0 9.0 1.5 5.0 2.5 4.0 9.0 6.0
END
```

Next, type the following commands.

```
TWOSAMPLE T C1 C2;
ALTERNATIVE 0;
POOLED.
```

If you prefer, you can just type the following.

```
TWOS C1 C2;
ALTE 0;
POOL.
```

Concerning the commands, note the following.

1. The first line contains the main command **TWOSAMPLE T** and must be followed by the columns that contain the samples. It ends with a semicolon to indicate that a subcommand will appear on the next line.
2. The second line contains the subcommand **ALTERNATIVE** and is followed by the code number 0 for the relation $\neq$ that appears in the alternative hypothesis. Since another subcommand will appear on the next line, the second line must end with a semicolon.
3. The third line contains the subcommand **POOLED** and ends with a period, which informs MINITAB that it is the last command. **POOLED** indicates that we want the calculation of the t-statistic to be based on the pooled sample standard deviation s_p. This subcommand is used whenever one assumes that the two populations have the same standard deviation.

    The output produced by MINITAB appears in Exhibit 11.1. We observe that the value of the test statistic is $t = 3.29$. Since the p-value of 0.0037 is less than $\alpha = 0.01$,

---

Exhibit 11.1

```
MTB > TWOS C1 C2;
SUBC> ALTE 0;
SUBC> POOL.

TWOSAMPLE T FOR C1 VS C2
        N       MEAN      STDEV    SE MEAN
C1   12        8.29       1.83      0.53
C2   10        5.00       2.84      0.90

95 PCT CI FOR MU C1 - MU C2: (1.20, 5.38)

TTEST MU C1 = MU C2 (VS NE): T= 3.29   P=0.0037  DF=  20
POOLED STDEV =         2.34
```

---

we reject the null hypothesis and conclude that a change has occurred in the mean concentration levels of sulfates since the new equipment was installed.

Note from the MINITAB output above that a 95 percent confidence interval for $(\mu_1 - \mu_2)$ is given. We have

$$1.20 \le (\mu_1 - \mu_2) \le 5.38.$$

Ninety-five percent is the default value for the confidence level. A different value can be used by specifying it after the main command **TWOS.** For example, a 99 percent confidence interval can be obtained with the commands in Exhibit 11.2. Thus, a 99 percent confidence interval for the difference between the mean concentration levels of sulfates is

$$0.44 \le (\mu_1 - \mu_2) \le 6.14.$$

Exhibit 11.2	

```
MTB > TWOS 99 C1 C2;
SUBC> ALTE 0;
SUBC> POOL.

TWOSAMPLE T FOR C1 VS C2
         N       MEAN      STDEV    SE MEAN
C1   12         8.29       1.83      0.53
C2   10         5.00       2.84      0.90

99 PCT CI FOR MU C1 - MU C2: (0.44, 6.14)

TTEST MU C1 = MU C2 (VS NE): T= 3.29   P=0.0037   DF=   20
POOLED STDEV =          2.34
```

11.3   Small-Sample Inferences for Two Means: Independent Samples and Unequal Variances (Optional)

The small-sample hypothesis test and confidence interval in the previous section should only be used when one is reasonably certain that the variances of the sampled populations are equal (we will say more about this at the end of the section). When one is unsure about the validity of assuming that $\sigma_1^2 = \sigma_2^2$, then the following approximate test and confidence interval should be used. They are based on a test statistic t that uses the sample variances s_1^2 and s_2^2 to approximate the population variances σ_1^2 and σ_2^2.

Nonpooled t-Test

Small-Sample Hypothesis Test for $(\mu_1 - \mu_2)$: Independent Samples (no assumptions needed concerning σ_1^2 and σ_2^2)

For testing the null hypothesis H_0: $(\mu_1 - \mu_2) = d_0$ (d_0 is usually 0), the test statistic is

$$t = \frac{(\bar{x}_1 - \bar{x}_2) - d_0}{\sqrt{\dfrac{s_1^2}{n_1} + \dfrac{s_2^2}{n_2}}}. \tag{11.6}$$

The rejection region is given by the following.

For H_a: $(\mu_1 - \mu_2) < d_0$, H_0 is rejected when $t < -t_\alpha$.

For H_a: $(\mu_1 - \mu_2) > d_0$, H_0 is rejected when $t > t_\alpha$.

For H_a: $(\mu_1 - \mu_2) \neq d_0$, H_0 is rejected when $t < -t_{\alpha/2}$ or $t > t_{\alpha/2}$.

Assumptions:

1. Independent random samples.
2. Each population has a normal distribution.

Note: The degrees of freedom of the t-distribution are given by

$$df = \frac{\left(\dfrac{s_1^2}{n_1} + \dfrac{s_2^2}{n_2}\right)^2}{\dfrac{\left(\dfrac{s_1^2}{n_1}\right)^2}{n_1 - 1} + \dfrac{\left(\dfrac{s_2^2}{n_2}\right)^2}{n_2 - 1}},$$

where this value is *rounded down** to the nearest integer.

* The test is approximate and rounding down rather than rounding up is a more conservative approach in that the probability of a Type I error is reduced.

Small-Sample $(1 - \alpha)$ Confidence Interval for $(\mu_1 - \mu_2)$: Independent Samples (no assumptions needed concerning σ_1^2 and σ_2^2)

$$(\bar{x}_1 - \bar{x}_2) \pm t_{\alpha/2} \sqrt{\frac{s_1^2}{n_1} + \frac{s_2^2}{n_2}} \tag{11.7}$$

Assumptions:

1. Independent random samples.
2. Each population has a normal distribution.

Note: The degrees of freedom of the t-distribution are given by

$$df = \frac{\left(\dfrac{s_1^2}{n_1} + \dfrac{s_2^2}{n_2}\right)^2}{\dfrac{\left(\dfrac{s_1^2}{n_1}\right)^2}{n_1 - 1} + \dfrac{\left(\dfrac{s_2^2}{n_2}\right)^2}{n_2 - 1}},$$

where this value is *rounded down* to the nearest integer.

Example 11.8   To determine if a new additive improves the mileage performance of gasoline, seven test runs were conducted with the additive, and six runs were made without it. The test results appear below, with all figures in miles per gallon (mpg).

With Additive	Without Additive
32.6	31.3
30.1	29.7
29.8	29.1
34.6	30.3
33.5	30.9
29.6	29.9
33.8	

Is there sufficient evidence at the 0.05 level to conclude that the additive increases gasoline mileage?

Solution

To perform the test, we need each sample's mean and variance. These are easily computed to be

$$\bar{x}_1 = 32.0, \quad s_1^2 = 4.470, \quad \bar{x}_2 = 30.2, \quad s_2^2 = 0.652.$$

Since the sample variances differ so greatly, there is doubt about assuming that the population variances are equal. A procedure is presented in Section 11.7 for testing this assumption. Its application here would lead to the conclusion that $\sigma_1^2 \neq \sigma_2^2$. Consequently, we will use the nonpooled t-test that does not require equal population variances.

Step 1: Hypotheses.

$$H_0: (\mu_1 - \mu_2) = 0$$
$$H_a: (\mu_1 - \mu_2) > 0$$

Step 2: Significance Level.

$$\alpha = 0.05.$$

Step 3: Calculations.

$$t = \frac{(\bar{x}_1 - \bar{x}_2) - d_0}{\sqrt{\dfrac{s_1^2}{n_1} + \dfrac{s_2^2}{n_2}}} = \frac{(32.0 - 30.2) - 0}{\sqrt{\dfrac{4.47}{7} + \dfrac{0.652}{6}}} = 2.08.$$

Step 4: Rejection region.

The degrees of freedom are given by the following.

$$df = \frac{\left(\dfrac{s_1^2}{n_1} + \dfrac{s_2^2}{n_2}\right)^2}{\dfrac{\left(\dfrac{s_1^2}{n_1}\right)^2}{n_1 - 1} + \dfrac{\left(\dfrac{s_2^2}{n_2}\right)^2}{n_2 - 1}} = \frac{\left(\dfrac{4.47}{7} + \dfrac{0.652}{6}\right)^2}{\dfrac{\left(\dfrac{4.47}{7}\right)^2}{6} + \dfrac{\left(\dfrac{0.652}{6}\right)^2}{5}} = 7.94$$

Since the *df* value is always *rounded down,* we use $df = 7$. Because H_a involves the relation >, the rejection region is right tailed and H_0 is rejected for $t > t_{.05} = 1.895$.

Step 5: Conclusion.
The null hypothesis is rejected since $t = 2.08$ lies in the rejection region. Thus, with $\alpha = 0.05$, there is sufficient evidence to conclude that the additive tends to increase gasoline mileage.

Example 11.9   Use the data in Example 11.8 to obtain a 90 percent confidence interval for $(\mu_1 - \mu_2)$.

Solution

The confidence interval is given by Formula 11.7.

$$(\bar{x}_1 - \bar{x}_2) \pm t_{\alpha/2} \sqrt{\frac{s_1^2}{n_1} + \frac{s_2^2}{n_2}}$$

For 90 percent confidence, $t_{\alpha/2} = t_{.05} = 1.895$. This value is obtained from Table 4, using $df = 7$ from Example 11.8.

$$(32.0 - 30.2) \pm 1.895 \sqrt{\frac{4.47}{7} + \frac{0.652}{6}}$$

$$1.8 \pm 1.64$$

Thus, with 90 percent confidence we estimate the mileage increase from the additive as some value between 0.16 and 3.44 mpg.

Ⓜ**Example 11.10**   Use MINITAB to perform the hypothesis test in Example 11.8.

Solution

The **SET** command is used to store the sample measurements in columns C1 and C2.

```
SET C1
32.6 30.1 29.8 34.6 33.5 29.6 33.8
END
SET C2
31.3 29.7 29.1 30.3 30.9 29.9
END
```

Next, type the following commands.

```
TWOSAMPLE T C1 C2;
ALTERNATIVE 1.
```

In the above, the first line contains the main command **TWOSAMPLE T** and the columns in which the sample values are stored. It must terminate with a semicolon since a subcommand follows on the next line. On the second line appears the subcommand **ALTERNATIVE** and the code number 1 for the relation $>$ that appears in H_a (see Example 10.6 for a review of the codes). The line ends with a period since no other subcommands follow it.

Because only the first four letters of a command are necessary, you may prefer just to type **TWOS** and **ALTE** in place of **TWOSAMPLE T** and **ALTERNATIVE**.

The output produced by MINITAB from employing the preceding commands appears in Exhibit 11.3.

The value of the test statistic is $t = 2.08$. Since the p-value of 0.038 is less than the significance level $\alpha = 0.05$, the null hypothesis is rejected, and we conclude that mean mpg with the additive is greater than mean mpg without it.

From Exhibit 11.3 it is seen that in addition to performing a hypothesis test, the **TWOSAMPLE T** command also constructs a 95 percent confidence interval for $(\mu_1 - \mu_2)$. Ninety-five percent is the default value for the confidence level. A different level can be used by specifying it after **TWOSAMPLE T.** For instance, a 90 percent confidence interval could have been obtained in Exhibit 11.3 by typing the following.

```
TWOS 90 C1 C2;
ALTE 1.
```

Exhibit 11.3

```
MTB > SET C1
DATA> 32.6 30.1 29.8 34.6 33.5 29.6 33.8
DATA> END
MTB > SET C2
DATA> 31.3 29.7 29.1 30.3 30.9 29.9
DATA> END
MTB > TWOSAMPLE T C1 C2;
SUBC> ALTERNATIVE 1.

TWOSAMPLE T FOR C1 VS C2
        N       MEAN      STDEV    SE MEAN
C1      7       32.00      2.11      0.80
C2      6       30.200     0.807     0.33

95 PCT CI FOR MU C1 - MU C2: (-0.24, 3.84)

TTEST MU C1 = MU C2 (VS GT): T= 2.08   P=0.038   DF=   7
```

When Should the Pooled t-Test Be Used?

In this and the preceding section, two procedures have been presented for making small-sample inferences about $(\mu_1 - \mu_2)$. Both procedures assume independent random samples and normal populations. The pooled test that was presented first also requires that the populations have equal variances. When the variances are not equal, or when one is in doubt about their equality, the nonpooled test should be employed. In practice, some statisticians favor the use of the nonpooled procedure in all but a few instances. They point out that even when $\sigma_1^2 = \sigma_2^2$, the pooled test is only slightly more powerful (smaller Type II error probability), and when $\sigma_1^2 \neq \sigma_2^2$, use of the pooled test can greatly increase the chances of a false conclusion.

Some statisticians recommend that the experimenter first test the hypotheses $H_0: \sigma_1^2 = \sigma_2^2$ versus $H_a: \sigma_1^2 \neq \sigma_2^2$. If H_0 is rejected, then conclude that the population variances are not equal and proceed to use the nonpooled t-test for testing $(\mu_1 - \mu_2)$.

If the hypothesis $H_0: \sigma_1^2 = \sigma_2^2$ is not rejected, then they advise that the pooled procedure be applied for testing $(\mu_1 - \mu_2)$. This recommendation has opponents who point out that it is a misuse of the variance test, since it may result in accepting the null hypothesis that $\sigma_1^2 = \sigma_2^2$ without knowing the probability of a Type II error. (Recall that when we fail to reject a null hypothesis we should not conclude that it is true, but only that there is insufficient evidence to warrant its rejection.) In addition, the probability of committing a Type I error is altered by performing two tests.

In conclusion, there is no universal agreement on when to use the pooled versus the nonpooled procedure. The risk of misusing the pooled test is not as serious when the sample sizes n_1 and n_2 are equal, or at least nearly so. On the other hand, even when the population variances are equal, the pooled procedure is only slightly more powerful than the nonpooled method. For actual applications, our personal recommendation is given below. As far as the exercises are concerned, instructions will be given as to how to proceed.

RECOMMENDATION: For small-sample inferences concerning $(\mu_1 - \mu_2)$ that are based on independent random samples, the pooled procedure should be used only when there is sufficient experience with the phenomena being studied to indicate that the population variances are approximately equal. Use the nonpooled procedure if in doubt about the equality of σ_1^2 and σ_2^2.

Sections 11.2 and 11.3 Exercises

Small-Sample Inferences for Two Means: Independent Samples

In the following exercises, assume that independent random samples were selected from populations that have approximately normal distributions. Note that an asterisk (*) indicates that the exercise pertains to the optional Section 11.3.

11.16 Independent random samples produced the following results.

	Sample One	**Sample Two**
Sample size	$n_1 = 13$	$n_2 = 15$
Sample mean	$\bar{x}_1 = 65.7$	$\bar{x}_2 = 62.1$
Sample std. dev.	$s_1 = 2.4$	$s_2 = 2.8$

Assume that the populations have the same variance. Use the pooled t-test at the 0.01 level to test the following hypotheses.

$$H_0: (\mu_1 - \mu_2) = 0$$

$$H_a: (\mu_1 - \mu_2) \neq 0$$

11.17 Use the data in Exercise 11.16 to estimate $(\mu_1 - \mu_2)$ with a 95 percent confidence interval (assume that $\sigma_1 = \sigma_2$).

*11.18 The following results were obtained for two independent samples.

	Sample One	Sample Two
Sample size	$n_1 = 8$	$n_2 = 21$
Sample mean	$\bar{x}_1 = 134$	$\bar{x}_2 = 127$
Sample std. dev.	$s_1 = 9.3$	$s_2 = 21.8$

Use the nonpooled t-test with $\alpha = 0.05$ to test the following hypotheses.

$$H_0: (\mu_1 - \mu_2) = 11$$

$$H_a: (\mu_1 - \mu_2) < 11$$

*11.19 Use the data in Exercise 11.18 to estimate $(\mu_1 - \mu_2)$ with a 95 percent confidence interval (do not assume that $\sigma_1 = \sigma_2$).

11.20 As part of a study of engineering programs at two universities, a career counselor wanted to compare the mean number of years that it took engineering students to earn their doctoral degrees. Records of graduates from the programs were randomly selected and the following results obtained.

University A	University B
$n_1 = 13$	$n_2 = 11$
$\bar{x}_1 = 5.74$ yrs.	$\bar{x}_2 = 5.02$ yrs.
$s_1 = 0.87$ yr.	$s_2 = 0.75$ yr.

Assume that the population variances are equal and use the pooled t-test to determine if a difference exists in the mean number of years required to complete the doctorate for the two programs. Test at the 0.05 level.

11.21 For Exercise 11.20, construct a 90 percent confidence interval to estimate the difference in the mean number of years required to obtain an engineering doctorate from the two universities (assume that $\sigma_1 = \sigma_2$).

*11.22 Dangerous chemicals from industrial wastes linked to cancer and other diseases can enter the food chain through their presence in lake sediments. Some species of fish from the Great Lakes have been found to contain more than 100 compounds present in industrial wastes (*Newsweek*, March 27, 1989). The amount of contamination fluctuates considerably in different lakes and from year to year. Suppose the amounts of DDT were determined for samples of salmon from two lakes, and the following results were obtained.

First Lake	Second Lake
$n_1 = 10$	$n_2 = 15$
$\bar{x}_1 = 1.77$ ppm	$\bar{x}_2 = 0.99$ ppm
$s_1 = 0.68$ ppm	$s_2 = 0.27$ ppm

Use the nonpooled t-test to determine if the mean amounts of DDT in salmon from the two lakes differ at the 0.05 level.

*11.23 For Exercise 11.22, estimate with 95 percent confidence the difference in the mean amount of DDT in salmon from the two lakes (do not assume $\sigma_1 = \sigma_2$).

11.24 Pellet stoves use as fuel wood pellets made from lumber industry waste. A stove manufacturer believes that its pellet stoves are more efficient than those of a competing brand. An independent testing laboratory measured the efficiency ratings for 10 of the manufacturer's stoves and obtained a mean of 80.6 percent and a standard deviation of 5.7 percent. For 12 of the competitor's stoves, the mean efficiency was 76.1 percent with a standard deviation of 6.1 percent. Is there sufficient evidence at the 0.01 level to support the manufacturer's belief? Assume that the population variances are equal.

11.25 Determine the approximate p-value for the test in Exercise 11.24.

11.26 A nutritionist believes that the mean percentage of saturated fat in stick margarine exceeds that in liquid margarine by more than 9 percent. The percentages of saturated fat were determined for a sample of 7 brands of stick and 6 brands of liquid margarines.

Type of Margarine	Percentage of Saturated Fat						
Stick	25.8	26.9	26.2	25.3	26.7	26.1	26.9
Liquid	16.9	17.4	16.8	16.2	17.3	16.8	

Formulate a suitable set of hypotheses and test at the 0.05 level. Assume that the population variances are equal.

11.27 Determine the approximate p-value for the test in Exercise 11.26.

11.28 For Exercise 11.26, estimate with 90 percent confidence the difference in the mean percentages of saturated fat in stick and liquid margarines (assume $\sigma_1 = \sigma_2$).

11.29 The pooled sample variance s_p^2 is a weighted average of the sample variances s_1^2 and s_2^2, where the weights are $(n_1 - 1)$ and $(n_2 - 1)$, respectively. Show that when the sample sizes n_1 and n_2 are equal, s_p^2 is just $(s_1^2 + s_2^2)/2$.

*11.30 The test statistics for the pooled t-test and the nonpooled t-test are repeated below.

Pooled t-Test	Nonpooled t-Test
$t = \dfrac{(\bar{x}_1 - \bar{x}_2) - d_0}{\sqrt{s_p^2 \left(\dfrac{1}{n_1} + \dfrac{1}{n_2}\right)}}$	$t = \dfrac{(\bar{x}_1 - \bar{x}_2) - d_0}{\sqrt{\dfrac{s_1^2}{n_1} + \dfrac{s_2^2}{n_2}}}$

Show that these formulas are the same when the sample sizes n_1 and n_2 are equal.

MINITAB Assignments

Ⓜ 11.31 An economist wanted to compare the hourly labor rates of automobile mechanics in two states. Dealerships were randomly selected from each state, and the following hourly charges in dollars were obtained.

First State	Second State
40.00	35.00
38.00	37.00
38.00	31.00
37.00	39.00
36.00	31.50
39.00	35.00
41.50	32.50
38.00	34.00
39.50	39.00
37.50	36.00
35.00	
40.00	

Use MINITAB to test at the 0.05 significance level whether a difference exists in the mean hourly rates for these two states. Assume that the sampled population have the same standard deviation.

Ⓜ 11.32 For the data in Exercise 11.31, use MINITAB to construct a 99 percent confidence interval for the difference in the mean hourly rates of mechanics in the 2 states. Assume that the sampled populations have the same standard deviation.

*Ⓜ 11.33 To determine if a difference exists in the mean weight of apples from 2 orchards, 11 apples were randomly selected from Orchard One and 10 were selected at random from Orchard Two. Their weights in ounces are given below. Use MINITAB to determine if the difference in the sample means is statistically significant at the 0.05 level. Do not assume that the standard deviations of the sampled populations are equal.

Orchard	Weights in Ounces										
One	4.7	5.3	5.9	4.8	5.1	6.2	6.1	6.1	5.3	6.1	4.9
Two	6.3	5.7	5.8	4.9	6.9	6.8	7.2	6.9	6.8	7.3	

*Ⓜ 11.34 For the preceding exercise, estimate the difference in the mean weight of apples from the 2 orchards by using MINITAB to construct a 90 percent confidence interval. Do not assume that the standard deviations of the sampled populations are equal.

11.4 Inferences for Two Means: Paired Samples

In Sections 11.1 and 11.2, we considered two-sample z- and t-tests for comparing two population means. The assumption of **independent samples** is one of the requirements to use these methods. In such situations, the selection of one sample is neither affected by nor related to the selection of the other sample. There are, however, many occasions when the researcher prefers to design an experiment that is based on **dependent samples.** In this section we will consider mean comparisons that involve **paired samples** for which the sample measurements are grouped in pairs. A common use of paired samples is with **before-and-after** comparisons, for which some examples are given below.

First sample: Weights of 20 adults at the beginning of an exercise program.
Second sample: Their weights at the end of the program.

First sample: Reading speeds of 15 students at the start of a speed-reading course.
Second sample: Reading speeds of the students at the completion of the course.

First sample: Tensile strengths of 12 cotton fibers.
Second sample: Tensile strengths of the fibers after the application of a chemical treatment.

As the following examples illustrate, not all paired samples involve before-and-after comparisons.

First sample: Starting salaries of 25 male social workers.
Second sample: Starting salaries of 25 female social workers who were matched by gpa with the male workers.

First sample: IQs of 18 married women.
Second sample: IQs of the women's husbands.

First sample: Reaction times of 13 mice to a light stimulus.
Second sample: Reaction times of the 13 mice to a shock stimulus.

In the six examples of paired samples cited above, each value from sample one is related by some characteristic to a value from the second sample. This pairing of similar experimental units allows for the elimination of unwanted variation, thus resulting in dependent samples that can increase the inferential ability to distinguish between two population means. Pairing is a special case of an experimental design technique known as **blocking.** For this reason, each pair of values is sometimes referred to as a **block.**

The analysis of paired samples is easier than for independent samples. The difference between the measurements for each pair is calculated, and these differences are then analyzed using the one-sample z- or t-test, depending on whether the sample size is large or small. We will illustrate the methodology in the following example, and follow that with a summary of the general technique.

Example 11.11   Ten infants were involved in a study to compare the effectiveness of two medications for the treatment of diaper rash. For each baby, two areas of approximately the same size and rash severity were selected, and one area was treated with medication A and the other with B. The number of hours for the rash to disappear was recorded for each medication and each infant, and the values obtained are given below. At the 1% significance level, is there sufficient evidence to conclude that a difference exists in the mean time required for the elimination of the rash?

| | Medication | | Difference |
Infant	A	B	D = A − B
1	46	43	3
2	50	49	1
3	46	48	−2
4	51	47	4
5	43	40	3
6	45	40	5
7	47	47	0
8	48	44	4
9	46	41	5
10	48	45	3

Solution

The difference for each of the 10 pairs has been calculated and listed above. We will let μ_D denote the mean of the hypothetical population of all possible differences between healing times for medications A and B. μ_D equals $(\mu_1 - \mu_2)$, where μ_1 and μ_2 are the mean healing times for medications A and B, respectively. A difference will exist in the mean healing times if $\mu_D \neq 0$.

Step 1: Hypotheses.

$$H_0: \mu_D = 0$$

$$H_a: \mu_D \neq 0$$

Step 2: Significance level.

$$\alpha = 0.01.$$

Step 3: Calculations.

Since the number of pairs ($n = 10$) is small, the test is based on the t-statistic

$$t = \frac{\overline{D} - \mu_D}{\frac{s_D}{\sqrt{n}}}.$$

$\overline{D}$ and s_D denote the sample mean and the sample standard deviation of the n differences.

$$\overline{D} = \frac{\Sigma D}{n} = \frac{26}{10} = 2.6$$

$$s_D^2 = \frac{\Sigma D^2 - \frac{(\Sigma D)^2}{n}}{n - 1} = \frac{114 - \frac{(26)^2}{10}}{9} = 5.1556$$

$$s_D = \sqrt{5.1556} = 2.27$$

$$t = \frac{\overline{D} - \mu_D}{\frac{s_D}{\sqrt{n}}} = \frac{2.6 - 0}{\frac{2.27}{\sqrt{10}}} = 3.62$$

Step 4: Rejection region.
Since H_a involves $\neq$, the rejection region is two tailed and consists of $t < -t_{\alpha/2} = -t_{.005} = -3.250$ and $t > t_{\alpha/2} = 3.250$. The value of $t_{.005}$ is obtained from Table 4 in Appendix A using $df = n - 1 = 9$.

Step 5: Conclusion.
Because $t = 3.62$ is in the rejection region, H_0 is rejected. There is sufficient evidence at the 1 percent significance level to conclude that a difference exists in the mean healing times for medications A and B.

The hypothesis test and confidence interval for paired samples are summarized below.

Hypothesis Test for ($\mu_1 - \mu_2$): Paired Samples
For testing the null hypothesis $H_0: \mu_D = d_0$ ($\mu_D = \mu_1 - \mu_2$; d_0 is usually 0), the test statistic is

$$t = \frac{\overline{D} - d_0}{\frac{s_D}{\sqrt{n}}}. \tag{11.8}$$

The rejection region is given by the following.

For $H_a: \mu_D < d_0$, H_0 is rejected when $t < -t_\alpha$.

For $H_a: \mu_D > d_0$, H_0 is rejected when $t > t_\alpha$.

For $H_a: \mu_D \neq d_0$, H_0 is rejected when $t < -t_{\alpha/2}$ or $t > t_{\alpha/2}$.

Assumptions:
1. The population of differences is normally distributed.
2. The sample of differences represents a random sample from the population of differences.

Note:
1. In Formula 11.8, $\overline{D}$ and s_D are the mean and standard deviation of the n paired-sample differences.
2. The t-distribution has degrees of freedom $df = n - 1$.
3. When n is large the normality assumption above is not necessary, and the t-statistic is replaced by z.

(1 − α) Confidence Interval for ($\mu_1 - \mu_2$): Paired Samples

$$\overline{D} \pm t_{\alpha/2} \frac{s_D}{\sqrt{n}} \tag{11.9}$$

Assumptions:
1. The population of differences is normally distributed.
2. The sample of differences represents a random sample from the population of differences.

> *Note:*
> 1. In Formula 11.9, $\overline{D}$ and s_D are the mean and standard deviation of the n paired-sample differences.
> 2. The t-distribution has $df = n - 1$.
> 3. When n is large the normality assumption above is not necessary, and $t_{\alpha/2}$ is replaced by $z_{\alpha/2}$.

Example 11.12

Estimate with a 95 percent confidence interval the difference in the mean healing times of the diaper medications in Example 11.11.

Solution

From Example 11.11, $\overline{D} = 2.6$ and $s_D = 2.27$ for the sample of $n = 10$ differences. For $df = n - 1 = 9$ and confidence level 0.95, $t_{\alpha/2} = t_{.025} = 2.262$. A 95 percent confidence interval for $\mu_D = (\mu_1 - \mu_2)$ is as follows.

$$\overline{D} \pm t_{\alpha/2} \frac{s_D}{\sqrt{n}}$$

$$2.6 \pm (2.262) \frac{2.27}{\sqrt{10}}$$

$$2.6 \pm 1.62$$

Thus, with 95 percent confidence, we estimate that medication A requires an average of between 0.98 and 4.22 hours more than medication B for the elimination of a diaper rash.

Ⓜ Example 11.13

We will use the data in Example 11.11 to illustrate how MINITAB can be utilized to test hypotheses that are based on paired samples. The two samples are stored in columns C1 and C2, and then the **LET** command is used to calculate and place the differences in column C3.

```
READ C1 C2
46  43
50  49
46  48
51  47
43  40
45  40
47  47
48  44
46  41
48  45
END
LET C3 = C1 - C2
```

The **TTEST** command (see Example 10.8) is then used on the column of differences, C3. The MINITAB output is displayed in Exhibit 11.4. Note that the p-value in

Exhibit 11.4 is 0.0056, which is smaller than the specified significance level $\alpha = 0.01$. Therefore, the null hypothesis should be rejected.

Exhibit 11.4

```
MTB > LET C3 = C1 - C2
MTB > TTEST 0 C3;      # 0 is the value in the null hypothesis
SUBC> ALTERNATIVE 0.   # code is 0 for not equal in alt hyp

TEST OF MU = 0.000 VS MU N.E. 0.000

            N      MEAN     STDEV    SE MEAN        T    P VALUE
C3         10     2.600     2.271     0.718      3.62     0.0056
```

For paired samples, the **TINTERVAL** command discussed in Example 9.7 can be applied to the column of differences to obtain a confidence interval for $(\mu_1 - \mu_2)$.

Section 11.4 Exercises
Inferences for Two Means: Paired Samples

In each of the following exercises, assume that the population of differences has approximately a normal distribution and the sample of differences is random.

11.35 A paired samples experiment yielded the data below. Test the null hypothesis $H_0: \mu_D = 0$ against the alternative hypothesis $H_a: \mu_D < 0$, where $D = A - B$. Let $\alpha = 0.05$.

Pair	A	B
1	25	32
2	35	38
3	56	65
4	52	50
5	24	30

11.36 Use the data in Exercise 11.35 to obtain a 90 percent confidence interval for μ_D.

11.37 The data below are from a paired samples experiment. Test the null hypothesis $H_0: \mu_D = 7$ against the alternative hypothesis $H_a: \mu_D \neq 7$, where $D = A - B$. Let $\alpha = 1$ percent.

Pair	A	B
1	85	76
2	28	19
3	76	56
4	99	84
5	51	41
6	46	46

11.38 Use the data in Exercise 11.37 to estimate μ_D with a 95 percent confidence interval.

11.39 A manufacturer of platinum-tipped spark plugs believes that they last longer than conventional spark plugs. To show this, 1 platinum plug and 1 conventional plug were installed in each of six 2-cylinder engines. The effective life in hours was determined for each plug and appears below.

Engine	1	2	3	4	5	6
Platinum Plug	640	570	530	410	600	580
Conventional Plug	470	370	460	490	380	410

Is there sufficient evidence at the 0.05 level to support the manufacturer's contention that the platinum plugs last longer?

11.40 Use the data in Exercise 11.39 to estimate with 90 percent confidence the difference in mean length of life for the two types of plugs.

11.41 A shoe manufacturer has developed a new running shoe that purportedly enables one to run faster. Eight adults participated in an experiment in which each ran a mile with regular track shoes and then ran a mile the next day with the new shoes. Their running times in seconds appear below.

Runner	Track Shoe	New Shoe
1	321	318
2	307	299
3	397	401
4	269	260
5	285	285
6	364	363
7	295	289
8	302	296

Is there sufficient evidence at the 0.05 level to support the manufacturer's claim?

11.42 Find the approximate p-value for the test in Exercise 11.41.

11.43 Initial difficulties experienced by children in starting their formal education can often be attributed to a delay in the development of writing readiness skills. To enhance these skills, Carol E. Oliver devised an occupational therapy sensorimotor treatment program ("A Sensorimotor Program for Improving Writing Readiness Skills in Elementary-Age Children," *American Journal of Occupational Therapy,* February, 1990). As part of her study, 12 children with writing difficulties and normal intelligence were measured on a standardized test before and after completion of the program. The mean increase in test scores was 9.5 and the standard deviation of the differences was 4.61. Is there sufficient evidence at the 0.05 level to conclude that the program tends to increase test scores for students of this type?

11.44 As part of its admissions evaluation process, an actuarial science department administraters an examination to each applicant. To determine if there is a significant difference in the grading of the exams by 2 faculty members, 10 exams were graded by each professor, and the following grades were assigned.

Applicant	1	2	3	4	5	6	7	8	9	10
Professor A's Grade	75	87	89	63	93	54	83	71	88	71
Professor B's Grade	69	84	80	57	95	49	79	65	88	67

   a. Do the data provide sufficient evidence to indicate a difference in the mean grades assigned by the 2 professors? Test at the 0.01 significance level.
   b. Determine the approximate p-value for the test.

11.45 Using the data in Exercise 11.44, estimate with 95 percent confidence the difference in mean scores assigned by the 2 professors.

11.46 A weight reduction center advertises that participants in its program lose an average of at least 5 pounds during the first week of participation. Because of numerous complaints, the state's consumer protection agency doubts this claim. To test it, the records of 12 participants were randomly selected. Their initial weights and their weights after 1 week in the program appear below.

Member	Initially	One Week
1	197	195
2	153	151
3	174	170
4	125	123
5	149	144
6	152	149
7	135	131
8	143	147
9	139	138
10	198	192
11	215	211
12	153	152

   Is there sufficient evidence at the 0.01 level to reject the center's claim?

11.47 For Exercise 11.46, estimate with 95 percent confidence the mean weight loss after 1 week's participation in the program.

11.48 To test the claim that a certain herb lowers blood pressure, 36 people participated in an experiment. The systolic blood pressure of each was recorded initially and then 90 minutes after digesting the herb. The observed blood pressures are given in the table that follows.

Subject	Initial Recording	90 Minutes Later
1	117	114
2	125	129
3	108	106
4	145	143
5	139	140
6	156	151
7	143	144
8	127	124
9	143	141
10	132	129
11	103	104
12	124	125
13	142	140
14	187	182
15	105	103
16	117	114
17	121	119
18	142	139
19	165	160
20	134	131
21	111	116
22	125	121
23	118	118
24	138	139
25	132	130
26	142	135
27	128	134
28	120	117
29	154	149
30	141	140
31	119	115
32	121	117
33	153	149
34	138	139
35	167	171
36	182	175

Use a z-test and a 0.05 level of significance to determine if there is sufficient evidence to conclude that the herb is effective in lowering blood pressure.

MINITAB Assignments

Ⓜ 11.49 Use MINITAB to perform the hypothesis test in Exercise 11.48.

Ⓜ 11.50 Use MINITAB and the data in Exercise 11.48 to obtain a 95 percent confidence interval for estimating the mean change in blood pressure after digesting the herb.

Ⓜ 11.51 The following are the median purchase prices of existing 1-family houses in 32 metropolitan areas for the years 1980 and 1987. The figures are in thousands of dollars. Use MINITAB to test if a significant increase in metropolitan housing prices has occurred during this time period. Use a significance level of 1 percent.

Metropolitan Area	**1980**	**1987**
Atlanta, GA	81.8	131.1
Baltimore, MD	69.5	125.4
Boston-Lawrence-Lowell, MA-NH	70.5	186.2
Chicago-Gary, IL-IN	78.0	116.9
Cleveland-Akron-Lorain, OH	63.8	89.4
Columbus, OH	65.7	92.9
Dallas, TX	89.4	128.5
Denver-Boulder, CO	80.0	139.8
Detroit-Ann Arbor, MI	67.9	93.6
Greensboro-Winston Salem-High Point, NC	63.6	110.0
Honolulu, HI	119.3	177.6
Houston-Galveston, TX	85.7	106.8
Indianapolis, IN	60.3	94.7
Kansas City, MO-KS	64.2	105.8
Los Angeles-Long Beach-Anaheim, CA	110.7	167.7
Louisville, KY-IN	58.5	85.6
Miami-Fort Lauderdale, FL	75.3	103.9
Milwaukee-Racine, WI	84.9	95.7
Minneapolis-St. Paul, MN-WI	80.9	139.4
New York-Newark-Jersey City, NY-NJ-CT	93.8	181.8
Philadelphia-Wilmington-Trenton, PA-DE-NJ-MD	62.6	113.7
Phoenix, AZ	92.7	132.9
Pittsburgh, PA	64.6	82.8
Portland, OR-WA	75.5	108.5
Rochester, NY	57.5	79.6
St. Louis, MO-IL	56.5	92.1
Salt Lake City-Ogden, UT	73.2	121.0
San Diego, CA	105.7	156.7
San Francisco-Oakland-San Jose, CA	122.9	179.9
Seattle-Tacoma, WA	82.2	127.7
Tampa-St. Petersburg, FL	64.0	100.3
Washington, DC-MD-VA	100.1	159.5

Source: Federal Home Loan Bank Board, *Savings and Home Financing Source Book,* annual.

11.5   Large-Sample Inferences for Two Proportions

Many occasions arise for which a comparison of two proportions is needed.

- A sociologist might want to determine if the proportion of families below the poverty level differs for two geographical regions.

- A company's safety engineer might be interested in comparing the proportion of workers at two plants who have incurred a work-related accident.

- A political scientist may want to know if the proportion of voters who favor a gubernatorial candidate differs for men and women.

- A manufacturer might want to compare the defective rate for two machines used to produce its product.

Inferences concerning a single proportion were considered in Chapters 9 and 10. We have seen that to estimate a population proportion we have to consider the problem of estimating the parameter p, the probability of a success in a binomial experiment. The usual point estimate of p is the sample proportion, $\hat{p} = x/n$, where x is the number of successes in a sample of size n. For inferences concerning two population proportions, we need to assume that independent random samples were selected from the two populations, and we will use the notation that is illustrated in Figure 11.3.

Figure 11.3
Notation for population and sample proportions.

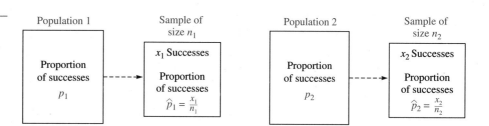

Comparisons of the population proportions p_1 and p_2 will be based on their difference $(p_1 - p_2)$. The usual **point estimate** of this is the difference between the corresponding sample proportions, $(\hat{p}_1 - \hat{p}_2)$, whose sampling distribution is summarized below.

The Sampling Distribution of $(\hat{p}_1 - \hat{p}_2)$ for Independent Samples:

1. The mean of $(\hat{p}_1 - \hat{p}_2)$ is $\mu_{(\hat{p}_1 - \hat{p}_2)} = (p_1 - p_2)$.
2. The standard deviation of $(\hat{p}_1 - \hat{p}_2)$ is

$$\sigma_{(\hat{p}_1 - \hat{p}_2)} = \sqrt{\frac{p_1 q_1}{n_1} + \frac{p_2 q_2}{n_2}}.$$

3. The shape of the sampling distribution is approximately normal for large samples (we will assume that each n_i satisfies $n_i \geq 30$, $n_i p_i \geq 5$, and $n_i q_i \geq 5$).

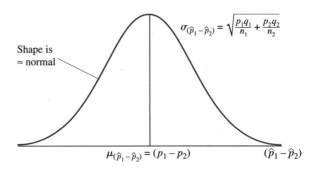

$$\sigma_{(\hat{p}_1 - \hat{p}_2)} = \sqrt{\frac{p_1 q_1}{n_1} + \frac{p_2 q_2}{n_2}}$$

Shape is ≈ normal

$$\mu_{(\hat{p}_1 - \hat{p}_2)} = (p_1 - p_2)$$ $(\hat{p}_1 - \hat{p}_2)$

From Formula 9.5, a large-sample confidence interval for a parameter θ is given by the following.

$$\hat{\theta} \pm z_{\alpha/2}\sigma_{\hat{\theta}}$$

$\longleftarrow$ $z_{\alpha/2}\sigma_{\hat{\theta}}$ $\longrightarrow$ $\longleftarrow$ $z_{\alpha/2}\sigma_{\hat{\theta}}$ $\longrightarrow$

$\hat{\theta} - z_{\alpha/2}\sigma_{\hat{\theta}}$ $\hat{\theta}$ $\hat{\theta} + z_{\alpha/2}\sigma_{\hat{\theta}}$

Lower Confidence Limit   Upper Confidence Limit

Using this, a large-sample confidence interval for $(p_1 - p_2)$ can be obtained by replacing $\hat{\theta}$ with $(\hat{p}_1 - \hat{p}_2)$ and $\sigma_{\hat{\theta}}$ with $\sigma_{(\hat{p}_1 - \hat{p}_2)}$. However, in $\sigma_{(\hat{p}_1 - \hat{p}_2)}$ we must use $\hat{p}_1\hat{q}_1$ and $\hat{p}_2\hat{q}_2$ in place of the unknown quantities $p_1 q_1$ and $p_2 q_2$. These substitutions result in the following approximate large-sample confidence interval for $(p_1 - p_2)$.

$$(\hat{p}_1 - \hat{p}_2) \pm z_{\alpha/2}\sqrt{\frac{\hat{p}_1\hat{q}_1}{n_1} + \frac{\hat{p}_2\hat{q}_2}{n_2}}$$

$\longleftarrow$ $z_{\alpha/2}\sqrt{\dfrac{\hat{p}_1\hat{q}_1}{n_1} + \dfrac{\hat{p}_2\hat{q}_2}{n_2}}$ $\longrightarrow$ $\longleftarrow$ $z_{\alpha/2}\sqrt{\dfrac{\hat{p}_1\hat{q}_1}{n_1} + \dfrac{\hat{p}_2\hat{q}_2}{n_2}}$ $\longrightarrow$

$(\hat{p}_1 - \hat{p}_2) - z_{\alpha/2}\sqrt{\dfrac{\hat{p}_1\hat{q}_1}{n_1} + \dfrac{\hat{p}_2\hat{q}_2}{n_2}}$ $\hat{p}_1 - \hat{p}_2$ $(\hat{p}_1 - \hat{p}_2) + z_{\alpha/2}\sqrt{\dfrac{\hat{p}_1\hat{q}_1}{n_1} + \dfrac{\hat{p}_2\hat{q}_2}{n_2}}$

Lower Confidence Limit   Upper Confidence Limit

Large-Sample $(1 - \alpha)$ Confidence Interval for $(p_1 - p_2)$:

$$(\hat{p}_1 - \hat{p}_2) \pm z_{\alpha/2}\sqrt{\frac{\hat{p}_1\hat{q}_1}{n_1} + \frac{\hat{p}_2\hat{q}_2}{n_2}} \qquad (11.10)$$

Assumptions:
1. Independent random samples.
2. Each sample size is large.

Example 11.14   The U.S. Department of Education estimates that nearly 17 percent of public school teachers moonlight during the school year (*Lebanon Daily News,* February 3, 1989). Suppose a school administrator wants to estimate the difference in the proportion of teachers who moonlight in 2 school districts. A random sample of 384 teachers from District I revealed 112 who moonlight, while a random sample of 432 teachers from District II found 91 who moonlight. Use a 95 percent confidence interval to estimate the true difference in the proportion of teachers in these 2 districts who moonlight during the school year.

Solution

For the 2 samples, we are given that $n_1 = 384$, $x_1 = 112$, $n_2 = 432$, and $x_2 = 91$. The sample proportions are $\hat{p}_1 = x_1/n_1 = 112/384 = 0.292$, and $\hat{p}_2 = x_2/n_2 = 91/432 = 0.211$. A 95 percent confidence interval for $(p_1 - p_2)$ is

$$(\hat{p}_1 - \hat{p}_2) \pm z_{\alpha/2} \sqrt{\frac{\hat{p}_1 \hat{q}_1}{n_1} + \frac{\hat{p}_2 \hat{q}_2}{n_2}}$$

$$(0.292 - 0.211) \pm z_{.025} \sqrt{\frac{(0.292)(0.708)}{384} + \frac{(0.211)(0.789)}{432}}$$

$$0.081 \pm 1.96(0.0304)$$

$$0.081 \pm 0.060$$

We are 95 percent confident that $(p_1 - p_2)$ is between 0.021 and 0.141. Thus, we estimate that the percentage of teachers who moonlight in District I exceeds the percentage in District II by some value between 2.1 percent and 14.1 percent.

To test a hypothesis concerning $(p_1 - p_2)$, the difference in sample proportions $(\hat{p}_1 - \hat{p}_2)$ is converted to standard units and used as the test statistic.

$$z = \frac{(\hat{p}_1 - \hat{p}_2) - \mu_{(\hat{p}_1 - \hat{p}_2)}}{\sigma_{(\hat{p}_1 - \hat{p}_2)}} = \frac{(\hat{p}_1 - \hat{p}_2) - (p_1 - p_2)}{\sqrt{\dfrac{p_1 q_1}{n_1} + \dfrac{p_2 q_2}{n_2}}} \qquad \textbf{(11.11)}$$

Usually the null hypothesis is $H_0: (p_1 - p_2) = 0$, which states that p_1 and p_2 have the same value, say p. In this case, the value of p is estimated by the **pooled sample proportion**

$$\hat{p} = \frac{x_1 + x_2}{n_1 + n_2}.$$

Notice that $\hat{p}$ combines (pools) the total number of successes in the two samples, $x_1 + x_2$, and divides this result by the total number of elements sampled, $n_1 + n_2$.

Substituting $\hat{p}$ into Formula 11.11, the following large-sample test statistic can be used for testing $H_0: (p_1 - p_2) = 0$.

$$z = \frac{(\hat{p}_1 - \hat{p}_2) - (p_1 - p_2)}{\sqrt{\dfrac{p_1 q_1}{n_1} + \dfrac{p_2 q_2}{n_2}}} = \frac{(\hat{p}_1 - \hat{p}_2) - 0}{\sqrt{\hat{p}\hat{q}\left(\dfrac{1}{n_1} + \dfrac{1}{n_2}\right)}}$$

The test is summarized below.

Large-Sample Hypothesis Test for $(p_1 - p_2)$:
For testing the null hypothesis $H_0\colon (p_1 - p_2) = 0$, the test statistic is

$$z = \frac{(\hat{p}_1 - \hat{p}_2)}{\sqrt{\hat{p}\hat{q}\left(\dfrac{1}{n_1} + \dfrac{1}{n_2}\right)}}. \qquad (11.12)$$

The rejection region is given by the following.

For $H_a\colon (p_1 - p_2) < 0$, H_0 is rejected when $z < -z_\alpha$.

For $H_a\colon (p_1 - p_2) > 0$, H_0 is rejected when $z > z_\alpha$.

For $H_a\colon (p_1 - p_2) \neq 0$, H_0 is rejected when $z < -z_{\alpha/2}$ or $z > z_{\alpha/2}$.

Assumptions:
1. Independent random samples.
2. Each sample size is large.

Note: In the test statistic,

$$\hat{p}_1 = \frac{x_1}{n_1}, \qquad \hat{p}_2 = \frac{x_2}{n_2}, \qquad \text{and} \quad \hat{p} = \frac{x_1 + x_2}{n_1 + n_2}$$

where x_1 and x_2 are the number of successes in the samples of size n_1 and n_2.

Example 11.15   With landfills quickly reaching their capacities, recycling household trash has assumed increased importance. A state legislator believes that a greater proportion of residents in the southern part of her state favor a mandatory recycling bill. To show this, 982 southern residents were randomly sampled, and 678 were found to be supportive of the proposal. A random sample of 952 residents in the northern region revealed 599 in favor of the bill. Formulate a suitable set of hypotheses and test at the 1 percent significance level.

Solution

Letting p_1 and p_2 denote the proportion of residents favoring the bill in the southern and northern parts, respectively, we are given that $n_1 = 982, x_1 = 678, n_2 = 952$, and $x_2 = 599$. The sample proportions are $\hat{p}_1 = 678/982 = 0.690$ and $\hat{p}_2 = x_2/n_2 = 599/952 = 0.629$.

Step 1: Hypotheses.
Since the legislator is attempting to show that p_1 exceeds p_2, the alternative hypothesis is $(p_1 - p_2) > 0$.

$$H_0\colon (p_1 - p_2) = 0$$

$$H_a\colon (p_1 - p_2) > 0$$

Step 2: Significance level.

$$\alpha = 0.01.$$

Step 3: Calculations.

$$z = \frac{(\hat{p}_1 - \hat{p}_2)}{\sqrt{\hat{p}\hat{q}\left(\dfrac{1}{n_1} + \dfrac{1}{n_2}\right)}},$$

where

$$\hat{p} = \frac{x_1 + x_2}{n_1 + n_2} = \frac{678 + 599}{982 + 952} = \frac{1{,}277}{1{,}934} = 0.660$$

$$z = \frac{(0.690 - 0.629)}{\sqrt{(0.66)(0.34)\left(\dfrac{1}{982} + \dfrac{1}{952}\right)}} = \frac{0.061}{0.0215} = 2.83.$$

Step 4: Rejection region.
The alternative hypothesis involves $>$, so the rejection region consists of values for which $z > z_{.01} = 2.326$.

Step 5: Conclusion.
The null hypothesis is rejected since $z = 2.83$ is in the rejection region. At the 1 percent significance level there is sufficient evidence to conclude that a larger proportion of residents in the southern part of the state favor the recycling proposal.

Example 11.16   Are VDTs a health risk? The beginning of this chapter referred to recent concerns about possible health risks from exposures to low level radiation sources such as video display terminals (VDTs). In March, 1991, the *New England Journal of Medicine* reported on a study conducted by the government's National Institute of Occupational Safety and Health. The project investigated 882 pregnancies among directory assistance telephone operators in several southeastern states. About one-half the operators used VDTs, and the other group used non-VDT equipment that emitted much lower levels of radiation. For the two groups, the investigators found no significant difference in the proportions of operators who had miscarriages during the first trimester of their pregnancies. For the operators who used VDTs the miscarriage rate was $\hat{p}_1 = 0.15$, compared to a rate of $\hat{p}_2 = 0.16$ for the non-VDT group.

Section 11.5 Exercises
Large-Sample Inferences for Two Proportions

11.52 Independent random samples of size $n_1 = 400$ and $n_2 = 300$ produced $x_1 = 320$ and $x_2 = 210$ successes. Test the following hypotheses at the 1 percent significance level.

$$H_0: p_1 = p_2 \text{ against } H_a: p_1 > p_2$$

11.53 Use the data in Exercise 11.52 to construct a 90 percent confidence interval for $(p_1 - p_2)$.

11.54 Independent random samples were selected from two populations, and the following sample results were obtained.

	Sample One	**Sample Two**
Sample size	$n_1 = 1{,}200$	$n_2 = 1{,}000$
Sample proportion	$\hat{p}_1 = 0.64$	$\hat{p}_2 = 0.56$

With $\alpha = 0.01$, test the following hypotheses.
$$H_0: p_1 = p_2$$
$$H_a: p_1 \neq p_2$$

11.55 Obtain a 99 percent confidence interval for the difference in the population proportions in Exercise 11.54.

11.56 Before pharmaceutical drugs are approved for general use, they must undergo extensive evaluations in controlled clinical trials. During the testing of a drug now used for the treatment of hypertension, a group of 644 patients received the drug, and 28 reported headaches as a side effect. A control group of 207 patients received a placebo, and 4 reported headaches as a side effect. For the 2 groups in the study, is the difference in the proportions of patients who reported headaches as a side effect statistically significant at the 0.05 level?

11.57 Obtain the *p*-value for the hypothesis test in Exercise 11.56.

11.58 In a recent survey, about two-thirds of the men and half the women classified themselves as aggressive drivers (The *Wall Street Journal,* February 20, 1990). Assume that 1,200 men and 1,200 women were interviewed (the actual numbers were not given), and that the samples can be regarded as representative of all men and women drivers. Estimate with 95 percent confidence the difference in the proportions of men and women who think they drive aggressively.

11.59 Using the data in Exercise 11.58, is there sufficient evidence at the 0.05 level to conclude that a larger proportion of men than women consider themselves aggressive drivers?

11.60 A recent study indicated that a drug designed to rid animals of worms greatly improves the chances of survival for advanced Stage C colon cancer patients (The *Wall Street Journal,* February 8, 1990). The drug (levamisole) was combined with a standard cancer medicine (fluorouracil) to treat 304 patients, and 103 suffered relapses. A control group of 315 Stage C colon cancer patients had 155 relapses.
a. Test at the 0.05 level if a significant difference exists in the recurrence rates for the treatment and control groups.
b. Determine the *p*-value for the test.

11.61 Estimate with 95 percent confidence the difference in the recurrence rates referred to in Exercise 11.60.

11.62 A manufacturer of lawn mowers has 2 assembly plants. A random sample of 2,000 mowers produced at the first plant revealed that 58 were returned to the dealer for service within 30 days of purchase. For a random sample of 1,800 mowers from the second plant, 45 were returned within this time period. Is there sufficient evidence to conclude that the return rate is greater for the first plant? Test at the 0.01 level of significance.

11.63 Determine the *p*-value for the test in Exercise 11.62.

11.64 After a frost, the owner of 2 orange groves sampled 100 trees from each grove to estimate the proportion of trees in each grove that had sustained damage. The sample from the first grove contained 38 damaged trees, while the second grove had 22 damaged trees.

    a. Estimate with 90 percent confidence the proportion of trees in the first grove that sustained damage.

    b. Obtain a 90 percent confidence interval to estimate the proportion of damaged trees in the second grove.

    c. Construct a 90 percent confidence interval to estimate the difference in the proportions of damaged trees in the 2 groves.

11.65 A public and a private university are located in the same city. For the private university, 1,046 alumni were surveyed, and 653 said that they had attended at least 1 class reunion. For the public university, 791 of 1,327 sampled alumni claimed they had attended a class reunion. Test at the 0.05 level whether the difference in the sample proportions is statistically significant.

11.66 Obtain the *p*-value for the test in Exercise 11.65.

11.6 *F* **Probability Distributions**

In the next section a procedure is given for testing the null hypothesis that two populations have the same variance. The test is based on an *F*-statistic whose distribution is introduced in this section.

In some respects, **F-distributions** are similar to chi-square distributions. Their graphs are nonsymmetrical, skewed to the right, and the distributions are positive only for positive values of the random variable. There are an infinite number of *F*-distributions. While a chi-square distribution is determined by only one parameter, two are needed to specify a particular *F*-distribution. For reasons that will become apparent in the following section, these parameters are referred to as **numerator degrees of freedom** and **denominator degrees of freedom.** We will denote them by *ndf* and *ddf*, respectively. Each of the degrees of freedom can be any positive integer, and each possible pair of parameters determines a different distribution. Figure 11.4 shows the graph of an *F*-distribution with numerator degrees of freedom *ndf* = 5 and denominator degrees of freedom *ddf* = 9.

Figure 11.4
F-distribution; *ndf* = 5 and *ddf* = 9.

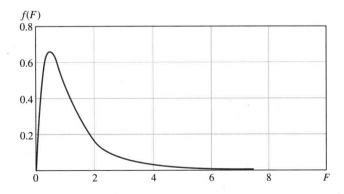

Since a pair of parameters is required to specify a particular F-distribution, tables of areas are more complex than those for previously considered distributions. Table 6 of Appendix A consists of 5 parts: a, b, c, d, and e. Part a pertains to a right-tail area of 0.10, while Parts b, c, d, and e apply to right-tail areas of 0.05, 0.025, 0.01, and 0.005, respectively. Thus, Parts a, b, c, d, and e provide values of F_a for right-tail areas of $a = 0.10, 0.05, 0.025, 0.01,$ and 0.005.

A portion of the table from Part b is given in Table 11.1. In using the table, one selects the column that is headed by the numerator degrees of freedom (ndf) and

Table 11.1

A Portion of Part B of Table 6 from Appendix A

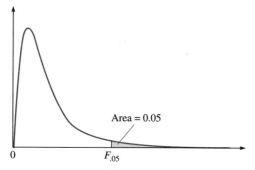

Each table entry is the F-value with an area of 0.05 to the right, and whose degrees of freedom are given by the column and row numbers.

Area = 0.05

$F_{.05}$

Denominator df (ddf)	Numerator df (ndf)								
	1	**2**	**3**	**4**	**5**	**6**	**7**	**8**	**9**
⋮	⋮	⋮	⋮	⋮	⋮	⋮	⋮	⋮	⋮
5	6.61	5.79	5.41	5.19	5.05	4.95	4.88	4.82	4.77
6	5.99	5.14	4.76	4.53	4.39	4.28	4.21	4.15	4.10
7	5.59	4.74	4.35	4.12	3.97	3.87	3.79	3.73	3.68
8	5.32	4.46	4.07	3.84	3.69	3.58	3.50	3.44	3.39
9	5.12	4.26	3.86	3.63	3.48	3.37	3.29	3.23	3.18
10	4.96	4.10	3.71	3.48	3.33	3.22	3.14	3.07	3.02
11	4.84	3.98	3.59	3.36	3.20	3.09	3.01	2.95	2.90
12	4.75	3.89	3.49	3.26	3.11	3.00	2.91	2.85	2.80
13	4.67	3.81	3.41	3.18	3.03	2.92	2.83	2.77	2.71
14	4.60	3.74	3.34	3.11	2.96	2.85	2.76	2.70	2.65
15	4.54	3.68	3.29	3.06	2.90	2.79	2.71	2.64	2.59
⋮	⋮	⋮	⋮	⋮	⋮	⋮	⋮	⋮	⋮

chooses the row that corresponds to the denominator degrees of freedom (*ddf*). The intersection of the chosen row and column gives the *F*-value that has an area to the right of it equal to 0.05.

Example 11.17   Determine the 95th percentile for the *F*-distribution having *ndf* = 7 and *ddf* = 15.

Solution

P_{95} is the *F*-value that has a cumulative probability to its left equal to 0.95. Consequently, the area to the right of it is $1 - 0.95 = 0.05$, and thus Part b of Table 6 applies. Using the column headed by numerator *df* = 7 and the row corresponding to denominator *df* = 15, we obtain $F_{.05} = 2.71$. Thus,

$$P_{95} = F_{.05} = 2.71.$$

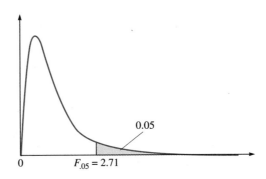

0.05

0          $F_{.05} = 2.71$

ⓂExample 11.18   For the *F*-distribution with *ndf* = 12 and *ddf* = 48, use MINITAB to determine:

1. $P(F > 1.23)$,
2. the *F*-value that is exceeded by 3 percent of the distribution.

Solution

1. Using the **CDF** command, we first obtain $P(F \le 1.23)$, the cumulative probability for 1.23. Type the following.

```
CDF 1.23;
F 12 48.
```

The second line contains the subcommand that specifies the *F*-distribution and the values of *ndf* and *ddf* in that order. From Exhibit 11.5, $P(F \le 1.23) = 0.7088$.

Exhibit 11.5

```
MTB > CDF 1.23;
SUBC> F 12 48.
    1.2300     0.7088
MTB > INVCDF 0.97;
SUBC> F 12 48.
    0.9700     2.1580
```

From this, it follows that

$$P(F > 1.23) = 1 - P(F \leq 1.23) = 0.2912.$$

2. The F-value that is exceeded by 3 percent of the distribution is the value that has a right-tail area of 0.03, that is, $F_{.03}$. This F-value has a cumulative probability to the left of it equal to $1 - 0.03 = 0.97$. The **INVCDF** command with the **F** subcommand can be used to obtain the F-value with this cumulative area of 0.97.

```
INVCDF 0.97;
F 12 48.
```

From Exhibit 11.5, the desired value is 2.158.

Ⓜ**Example 11.19**   Use MINITAB to sketch the graph of the F-distribution with numerator degrees of freedom equal to 20 and denominator degrees of freedom equal to 6.

Solution

First we instruct MINITAB to generate a set of (x, y) points that lie on the curve. We will arbitrarily use x-values from 0 to 3 with a spacing of 0.1. If the sketch is incomplete, then additional points can be added.

```
SET C1
0:3/0.1
END
```

The **PDF** command with the **F** subcommand below will produce the corresponding y-values and store them in column C2.

```
PDF C1 C2;
F 20 6.
```

The x-values are stored in C1 and the corresponding y-values are located in C2. To graph these, we use the **PLOT** command.

```
PLOT C2 C1
```

The resulting output appears in Exhibit 11.6.

Exhibit 11.6

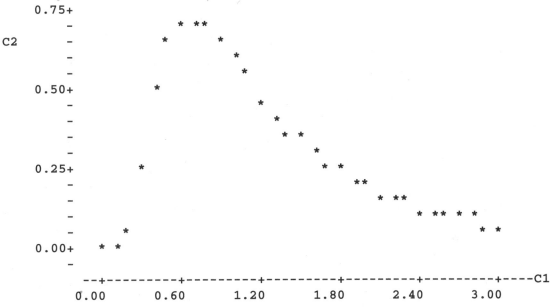

Section 11.6 Exercises

F Probability Distributions

For each of the Exercises 11.67–11.71, find the *F*-value with a right-tail area equal to the specified subscript.

11.67 $F_{.01}$ with $ndf = 9$, $ddf = 15$

11.68 $F_{.10}$ with $ndf = 20$, $ddf = 4$

11.69 $F_{.05}$ with $ndf = 4$, $ddf = 21$

11.70 $F_{.005}$ with $ndf = 30$, $ddf = 16$

11.71 $F_{.025}$ with $ndf = 10$, $ddf = 30$

Determine the following percentiles.

11.72 P_{90} with $ndf = 15$, $ddf = 12$

11.73 P_{95} with $ndf = 10$, $ddf = 21$

11.74 P_{99} with $ndf = 5$, $ddf = 40$

11.75 P_{99} with $ndf = 8$, $ddf = 7$

11.76 P_{95} with $ndf = 7$, $ddf = 9$

Use Table 6 in Appendix A to determine the following probabilities.

11.77 $P(F > 2.11)$ with $ndf = 6$, $ddf = 19$

11.78 $P(F > 2.67)$ with $ndf = 10$, $ddf = 23$

11.79 $P(F < 3.59)$ with $ndf = 2$, $ddf = 17$

11.80 $P(F < 7.19)$ with $ndf = 6$, $ddf = 7$

11.81 $P(3.49 < F < 5.85)$ with $ndf = 2$, $ddf = 20$

11.82 $P(2.35 < F < 4.94)$ with $ndf = 9$, $ddf = 10$

For each of the Exercises 11.83–11.86, find the value of c.

11.83 $P(F > c) = 0.01$ with $ndf = 5$, $ddf = 29$

11.84 $P(F > c) = 0.05$ with $ndf = 9$, $ddf = 10$

11.85 $P(F < c) = 0.975$ with $ndf = 20$, $ddf = 8$

11.86 $P(F < c) = 0.90$ with $ndf = 15$, $ddf = 4$

MINITAB Assignments

Ⓜ 11.87 For the F-distribution with $ndf = 11$ and $ddf = 39$, use MINITAB to find
   a. $P(F < 1.12)$,
   b. $F_{.07}$.

Ⓜ 11.88 For the F-distribution with $ndf = 18$ and $ddf = 33$, use MINITAB to find $F_{.10}$, $F_{.05}$, and $F_{.01}$.

Ⓜ 11.89 Use MINITAB to sketch the graph of the F-distribution with $ndf = 9$ and $ddf = 20$.

11.7 Inferences for Two Variances

In the last two chapters we saw that there are many practical situations where one is primarily interested in a population's variance. This is also true when dealing with two populations. An engineer may want to compare the precision of two manufacturing processes by performing a test of hypothesis to determine if they differ in their variability. A teacher might want to check whether the scores on a new standardized test are more variable than scores on a previously used test.

An additional need for comparing the variances of two populations occurred in Section 11.2. There we saw that before making small-sample inferences about two population means, one needs to decide whether or not the population variances differ.

NOTE: For inferences concerning σ_1^2 and σ_2^2, it will be convenient always to label (or relabel) as **population one** that which has the **larger sample variance.** Thus, s_1^2 will be used to denote the **larger sample variance.**

A test of the null hypothesis that the variances of two populations are equal is based on a statistic involving the ratio of the sample variances. For testing H_0: $\sigma_1^2 = \sigma_2^2$, the test statistic is the quantity s_1^2/s_2^2. When H_0 is true and the sampled populations are normally distributed, the sampling distribution of s_1^2/s_2^2 can be shown to have an F-distribution with numerator degrees of freedom $ndf = n_1 - 1$ and denominator degrees of freedom $ddf = n_2 - 1$. A value of $F = s_1^2/s_2^2$ that is close to 1 is consistent with the null hypothesis that the population variances are equal. A value of F that differs significantly from 1 would provide evidence against H_0. For example, suppose samples of size $n_1 = 8$ and $n_2 = 10$ yield the sample variances $s_1^2 = 20$ and $s_2^2 = 5$. With $\alpha = 0.05$, do these results provide sufficient evidence to conclude that the population variances differ? To answer this, we perform the following hypothesis test.

Step 1: Hypotheses.

$$H_0: \sigma_1^2 = \sigma_2^2$$

$$H_a: \sigma_1^2 \neq \sigma_2^2$$

Step 2: Significance level.

$$\alpha = 0.05.$$

Step 3: Calculations.

$$F = \frac{s_1^2}{s_2^2} = \frac{20}{5} = 4.00.$$

Note that the F-value of 4.00 indicates that the first sample variance is 4 times the size of the second sample variance. Although this ratio appears considerably different from 1, it might not be statistically significant in light of the very small sample sizes.

Step 4: Rejection region.
The rejection region is two tailed since the alternative hypothesis involves the $\neq$ relation. Each tail has an area of $\alpha/2 = 0.025$. The upper tail of the rejection region is determined from Part C of Table 6, where the numerator degrees of freedom $ndf = n_1 - 1 = 7$ and the denominator degrees of freedom $ddf = n_2 - 1 = 9$. The right-tail rejection region consists of $F > F_{.025} = 4.20$ and is shown in Figure 11.5.

Figure 11.5
Rejection region of H_0: $\sigma_1^2 = \sigma_2^2$.

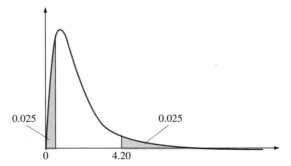

Notice in Figure 11.5 that although the rejection region is two tailed, we have only indicated the portion in the upper tail. Since we will always label s_1^2 as the larger sample variance, it is not necessary to know the lower tail of the rejection region. With the larger value of s_1^2 in the numerator of F, the upper tail is the only part of the rejection region that F could lie in. (This is convenient since it means that F-tables with lower tail areas are not needed.)

Step 5: Conclusion.
Since the value of the test statistic is $F = 4.00$, and this does not lie in the rejection region, we do not reject the null hypothesis. Thus, although the sample variance s_1^2 is 4 times the size of s_2^2, there isn't quite enough evidence at the 0.05 level to conclude that the population variances differ. In light of the small sample sizes, the difference in the sample variances can reasonably be attributed to chance.

The hypothesis test for two population variances is summarized below and is applicable for both large and small samples, as long as the specified assumptions of the test are satisfied.

Large- or Small-Sample Hypothesis Test for Two Population Variances:
For testing the null hypothesis $H_0: \sigma_1^2 = \sigma_2^2$, the test statistic is

$$F = \frac{s_1^2}{s_2^2},\qquad\qquad (11.13)$$

where s_1^2 is the larger sample variance. The rejection region is given by the following.

For $H_a: \sigma_1^2 > \sigma_2^2$, H_0 is rejected when $F > F_\alpha$.

For $H_a: \sigma_1^2 \neq \sigma_2^2$, H_0 is rejected when $F > F_{\alpha/2}$ (rejection region is two tailed with $\alpha/2$ in each tail, but it is not necessary to find the lower tail).

Note:
1. For $H_a: \sigma_1^2 < \sigma_2^2$, H_0 is never rejected since $F = s_1^2/s_2^2$ exceeds 1 and thus could not fall in a left-tail rejection region.
2. The rejection region is determined from Table 6 of Appendix A with numerator degrees of freedom given by $ndf = n_1 - 1$ and with denominator degrees of freedom given by $ddf = n_2 - 1$.

Assumptions:
1. Independent random samples of size n_1 and n_2.
2. Each population has a normal distribution.
3. The populations have been labeled (or relabeled) so that population one is that with the larger sample variance.

Example 11.20   A pharmaceutical company manufactures a soaking solution for contact lenses. The product contains a wetting agent to reduce friction, and the company wants to determine if the variability of its amount is different from that of a competitor's brand. The amount of wetting agent was determined for a sample of 25 bottles of its product, and the mean was 28.7 ml and the standard deviation was 1.3 ml. For a sample of 21 similar bottles of the competitor's brand, the mean was 27.2 ml and the standard deviation equaled 1.9 ml. With $\alpha = 0.10$, is there sufficient evidence to conclude that a difference exists in variability of the quantity of wetting agent for the two brands?

Solution

The population of the competitor's bottles will be labeled as population one, since its sample produced the larger sample variance. Thus, $n_1 = 21$, $\bar{x}_1 = 27.2$, $s_1 = 1.9$, $n_2 = 25$, $\bar{x}_2 = 28.7$, and $s_2 = 1.3$.

Step 1: Hypotheses.

$$H_0: \sigma_1^2 = \sigma_2^2$$

$$H_a: \sigma_1^2 \neq \sigma_2^2$$

Step 2: Significance level.

$$\alpha = 0.10.$$

Step 3: Calculations.

$$F = \frac{s_1^2}{s_2^2} = \frac{1.9^2}{1.3^2} = \frac{3.61}{1.69} = 2.14.$$

Note that the sample standard deviations must be squared, and the values of the sample means are not used in the calculation of the test statistic F.

Step 4: Rejection region.
Since H_a involves the $\neq$ relation, the rejection region is two tailed. The area of each tail is $\alpha/2 = 0.10/2 = 0.05$. Only the upper tail is needed, and it is obtained from Part B of Table 6, using $ndf = n_1 - 1 = 20$ and $ddf = n_2 - 1 = 24$. H_0 is rejected for $F > F_{.05} = 2.03$. The rejection region is shown in Figure 11.6.

Figure 11.6
Rejection region for Example 11.20.

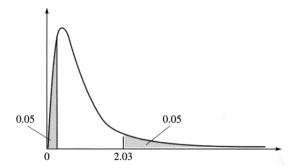

0.05   0.05

0   2.03

Step 5: Conclusion.
The value of the test statistic, $F = 2.14$, falls in the rejection region. Thus, H_0 is rejected, and we conclude that there is a difference in variability of the quantity of wetting agent for the two brands.

Example 11.21   A pizza shop is considering the purchase of a new oven that the manufacturer claims will maintain the temperature setting better than the shop's present oven. To check this, each oven is set at 425°F, and a series of temperature measurements is taken for each oven. Based on the sample results given below, is there sufficient evidence at the 0.05 level to conclude that the temperature of the new oven varies less than that of the pizza shop's oven?

	Present Oven	New Oven
Sample size	11	12
Sample mean	429.9	428.1
Sample std. dev.	3.7	2.1

(© *Richard Hutchings/Photo Researchers, Inc.*)

Solution

Since the pizza shop's oven has the larger sample standard deviation, the population of temperature readings for this oven will be labeled as population one. Therefore, $n_1 = 11$, $\bar{x}_1 = 429.9$, $s_1 = 3.7$, $n_2 = 12$, $\bar{x}_2 = 428.1$, and $s_2 = 2.1$.

Step 1: Hypotheses.

The temperature of the new oven will vary less than that of the present oven if the variance σ_1^2 of the present oven is more than the variance σ_2^2 of the new oven. Therefore, the alternative hypothesis is $\sigma_1^2 > \sigma_2^2$.

$$H_0: \sigma_1^2 = \sigma_2^2$$

$$H_a: \sigma_1^2 > \sigma_2^2$$

Step 2: Significance level.

$$\alpha = 0.05.$$

Step 3: Calculations.

$$F = \frac{s_1^2}{s_2^2} = \frac{3.7^2}{2.1^2} = \frac{13.69}{4.41} = 3.10.$$

Step 4: Rejection region.

Since the relation in H_a is $>$, the rejection region is only an upper tail and has an area of $\alpha = 0.05$. Using Part B of Table 6 with $ndf = n_1 - 1 = 10$ and $ddf = n_2 - 1 = 11$, we reject H_0 if $F > F_{.05} = 2.85$. The rejection region is shown in Figure 11.7.

Figure 11.7
Rejection region for Example 11.21.

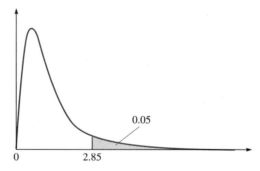

Step 5: The null hypothesis is rejected since $F = 3.10$ lies in the rejection region. Thus, for a temperature setting of 425°F, the temperature of the new oven appears to vary less than the pizza shop's present oven.

Ⓜ **Example 11.22**   Consider the hypothesis test in Example 11.21.

1. Use Table 6 to determine the approximate *p*-value.
2. Use MINITAB to obtain the exact *p*-value.

Solution

1. The observed value of the test statistic is $F = 3.10$. The *p*-value is the probability of obtaining an *F*-value at least this large. That is, the *p*-value equals $P(F \geq 3.10)$.

To determine this, Table 6 is used with $ndf = n_1 - 1 = 10$ and $ddf = n_2 - 1 = 11$. Since the table only has tail areas equal to $0.10, 0.05, 0.025, 0.01$, and 0.005, the value of $P(F \geq 3.10)$ can only be approximated. $F = 3.10$ is between $F_{.05} = 2.85$ and $F_{.025} = 3.53$ (see Figure 11.8). Therefore,

$$0.025 < p\text{-value} < 0.05.$$

Figure 11.8
Determining the p-value for
Example 11.21.

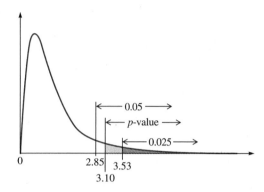

2. First, the command **CDF** (cumulative distribution function) with the subcommand **F** is used to find the cumulative probability to the left of 3.10. Type the following.

```
CDF 3.10;
F 10 11.      #   10 is ndf and 11 is ddf
```

MINITAB will produce the output in Exhibit 11.7. From Exhibit 11.7, the cumulative probability to the left of 3.10 is 0.9616. Thus, the p-value is given by the following.

$$
\begin{aligned}
p\text{-value} &= P(F > 3.10) \\
&= 1 - P(F \leq 3.10) \\
&= 1 - 0.9616 = 0.0384
\end{aligned}
$$

Exhibit 11.7

```
MTB > CDF 3.10;
SUBC> F 10 11.
    3.1000    0.9616
```

When a statistical procedure is insensitive to moderate deviations from an assumption upon which it is based, the procedure is said to be **robust** with respect to that assumption. For example, the small-sample t-test concerning a population mean that was discussed in the previous chapter is quite robust concerning the assumption that the sampled population is normally distributed. The F-test of this section, however, is not robust with regard to the assumption that the populations have, at least approximately, normal distributions. When the normality assumption is not reasonable, then the probability of a Type I error may actually be considerably larger than the value specified by the researcher.

Section 11.7 Exercises

Inferences for Two Variances

In the following exercises, assume that independent random samples were selected from populations that have approximately normal distributions.

11.90   Independent random samples were selected from two populations, and the following sample results were obtained.

	Sample One	**Sample Two**
Sample size	$n_1 = 9$	$n_2 = 22$
Sample std. dev.	$s_1 = 29$	$s_2 = 17$

With $\alpha = 0.10$, test the following hypotheses.

$$H_0: \sigma_1^2 = \sigma_2^2$$
$$H_a: \sigma_1^2 > \sigma_2^2$$

11.91   Independent random samples were selected from two populations, and the following sample results were obtained.

	Sample One	**Sample Two**
Sample size	$n_1 = 13$	$n_2 = 15$
Sample std. dev.	$s_1 = 2.89$	$s_2 = 1.38$

With $\alpha = 0.05$, test the following hypotheses.

$$H_0: \sigma_1^2 = \sigma_2^2$$
$$H_a: \sigma_1^2 > \sigma_2^2$$

11.92   Determine the approximate p-value for the hypothesis test in Exercise 11.90.

11.93   Determine the approximate p-value for the hypothesis test in Exercise 11.91.

11.94   Independent random samples were selected from two populations, and the following sample results were obtained.

	Sample One	**Sample Two**
Sample size	$n_1 = 25$	$n_2 = 21$
Sample std. dev.	$s_1 = 183$	$s_2 = 105$

With $\alpha = 0.01$, test the following hypotheses.

$$H_0: \sigma_1^2 = \sigma_2^2$$
$$H_a: \sigma_1^2 \neq \sigma_2^2$$

11.95   Independent random samples were selected from two populations, and the sample results in the following table were obtained.

	Sample One	Sample Two
Sample size	$n_1 = 7$	$n_2 = 9$
Sample std. dev.	$s_1 = 23.7$	$s_2 = 10.8$

With $\alpha = 0.05$, test the following hypotheses.

$$H_0: \sigma_1^2 = \sigma_2^2$$

$$H_a: \sigma_1^2 \neq \sigma_2^2$$

11.96   Water hardness is mainly caused by limestone in the earth that dissolves and accumulates in water as it seeps through the ground. A certain city draws its water supply from 2 reservoirs, A and B. A sample of 10 analyses of hardness was conducted on water from reservoir A, and the mean and standard deviation were 9.7 ppm and 1.2 ppm, respectively. For a sample of 9 analyses from B, the mean and standard deviation were 11.2 ppm and 1.9 ppm, respectively. Is there sufficient evidence at the 0.10 significance level to indicate a difference in the variance of water hardness for the 2 reservoirs?

11.97   A company uses 2 machines to fill bags of dog food. For a sample of 14 bags filled by the newer machine, the mean was 81.7 ounces with a standard deviation of 1.3 ounces. Sixteen bags filled by the older machine had a mean of 81.0 ounces and a standard deviation of 2.1 ounces. Is there sufficient evidence at the 5 percent level to conclude that the variability in the amounts dispensed by the older machine is greater?

11.98   Determine the approximate p-value for the test in Exercise 11.97.

11.99   A national real estate agency randomly sampled 15 of its Pennsylvania agents and 16 of its agents from New York. Last year, the Pennsylvania agents had a mean income of $38,653 and a standard deviation of $2,786, while the mean and standard deviation for the New York agents were $43,659 and $3,153, respectively. Do the data provide sufficient evidence to indicate a difference in the variability of incomes for agents in these 2 states? Test at the 10 percent level.

11.100   A state university has satellite campuses in the northern and southern parts of the state. A random sample of 31 student applicants for the northern branch had a standard deviation of 93 on a national aptitude test, while 25 randomly selected applicants for the southern branch had a standard deviation of 59. Do the data provide sufficient evidence to indicate that test scores of applicants for the northern branch have a larger standard deviation than that of applicants for the southern branch? Test at the 5 percent level.

MINITAB Assignments

Ⓜ 11.101   For the hypothesis test in Exercise 11.99,
a. use Table 6 to obtain the approximate p-value,
b. use the **CDF** command and the subcommand **F** in MINITAB to obtain the p-value.

Ⓜ 11.102 For the hypothesis test in Exercise 11.100,
   a. obtain the approximate *p*-value by using Table 6,
   b. obtain the *p*-value by using the **CDF** command and the **F** subcommand in MINITAB.

Ⓜ 11.103 Solid fats (fats that are solid at room temperature) are more likely to raise blood cholesterol levels than liquid fats. Suppose a nutritionist analyzed the percentage of saturated fat for a sample of 7 brands of stick margarine (solid fat) and for a sample of 6 brands of liquid margarine, and obtained the following results:

Type of Margarine	Percentage of Saturated Fat						
Stick	25.8	26.9	26.2	25.3	26.7	26.1	26.9
Liquid	16.9	17.4	16.8	16.2	17.3	16.8	

   a. Use MINITAB to calculate each sample variance.
   b. Using the results from Part a, test at the 0.05 level whether the difference in the sample variances is statistically significant.
   c. Use MINITAB to obtain the *p*-value for the test in Part b.

 Looking Back

Chapter 11 is concerned with inferences about two populations. The methodology of confidence intervals and hypothesis testing introduced in Chapters 9 and 10 is extended to two population parameters. Specifically, this chapter considers the comparison of two populations with regard to their means, μ_1 and μ_2; variances, σ_1^2 and σ_2^2; and proportions, p_1 and p_2.

The following tests of hypotheses and confidence intervals are discussed.
1. Large-sample ($n_1 \geq 30$ and $n_2 \geq 30$) inferences for $(\mu_1 - \mu_2)$: independent samples.
   a. Hypothesis test of $H_0: (\mu_1 - \mu_2) = d_0$.

> The test statistic is
> $$z = \frac{(\bar{x}_1 - \bar{x}_2) - d_0}{\sqrt{\dfrac{\sigma_1^2}{n_1} + \dfrac{\sigma_2^2}{n_2}}}.$$

   b. Confidence interval for $(\mu_1 - \mu_2)$.

> $$(\bar{x}_1 - \bar{x}_2) \pm z_{\alpha/2} \sqrt{\frac{\sigma_1^2}{n_1} + \frac{\sigma_2^2}{n_2}}$$

Note: Nearly always the population variances σ_1^2 and σ_2^2 are unknown and must be approximated by s_1^2 and s_2^2.

2. Small-sample ($n_1 < 30$ or $n_2 < 30$) inferences for ($\mu_1 - \mu_2$): independent samples and equal population variances (normal populations).

a. Hypothesis test of H_0: ($\mu_1 - \mu_2$) = d_0.

> The test statistic is
> $$t = \frac{(\bar{x}_1 - \bar{x}_2) - d_0}{\sqrt{s_p^2 \left(\dfrac{1}{n_1} + \dfrac{1}{n_2}\right)}}$$

b. Confidence interval for ($\mu_1 - \mu_2$).

> $$(\bar{x}_1 - \bar{x}_2) \pm t_{\alpha/2} \sqrt{s_p^2 \left(\frac{1}{n_1} + \frac{1}{n_2}\right)}$$

Note: 1. The t-distribution has $df = n_1 + n_2 - 2$.

2. $s_p^2 = \dfrac{(n_1 - 1)s_1^2 + (n_2 - 1)s_2^2}{n_1 + n_2 - 2}$.

3. Small-sample ($n_1 < 30$ or $n_2 < 30$) inferences for ($\mu_1 - \mu_2$): independent samples (no assumptions needed for σ_1^2 and σ_2^2; normal populations).

a. Hypothesis test of H_0: ($\mu_1 - \mu_2$) = d_0.

> The test statistic is
> $$t = \frac{(\bar{x}_1 - \bar{x}_2) - d_0}{\sqrt{\dfrac{s_1^2}{n_1} + \dfrac{s_2^2}{n_2}}}.$$

b. Confidence interval for ($\mu_1 - \mu_2$).

> $$(\bar{x}_1 - \bar{x}_2) \pm t_{\alpha/2} \sqrt{\frac{s_1^2}{n_1} + \frac{s_2^2}{n_2}}$$

Note: The degrees of freedom of the t-distribution are given by

$$df = \frac{\left(\dfrac{s_1^2}{n_1} + \dfrac{s_2^2}{n_2}\right)^2}{\dfrac{\left(\dfrac{s_1^2}{n_1}\right)^2}{n_1 - 1} + \dfrac{\left(\dfrac{s_2^2}{n_2}\right)^2}{n_2 - 1}},$$

where this value is *rounded down* to the nearest integer.

4. Inferences for $(\mu_1 - \mu_2)$: paired samples (normal differences).
   a. Hypothesis test of H_0: $\mu_D = d_0$, where μ_D is $(\mu_1 - \mu_2)$ and d_0 is usually 0.

The test statistic is

$$t = \frac{\overline{D} - d_0}{\dfrac{s_D}{\sqrt{n}}}$$
---

   b. Confidence interval for $\mu_D = (\mu_1 - \mu_2)$

$$\overline{D} \pm t_{\alpha/2} \frac{s_D}{\sqrt{n}}$$

   Note: 1. $\overline{D}$ and s_D are the mean and standard deviation of the n paired-sample differences.
   2. The t-distribution has $df = n - 1$.

5. Large-sample inferences for two population proportions (independent samples).
   a. Hypothesis test of H_0: $(p_1 - p_2) = 0$.

The test statistic is

$$z = \frac{(\hat{p}_1 - \hat{p}_2)}{\sqrt{\hat{p}\hat{q}\left(\dfrac{1}{n_1} + \dfrac{1}{n_2}\right)}},$$
---

   b. Confidence interval for $(p_1 - p_2)$.

$$(\hat{p}_1 - \hat{p}_2) \pm z_{\alpha/2} \sqrt{\frac{\hat{p}_1\hat{q}_1}{n_1} + \frac{\hat{p}_2\hat{q}_2}{n_2}}$$

   Note: $\hat{p}_1 = \dfrac{x_1}{n_1}$, $\hat{p}_2 = \dfrac{x_2}{n_2}$, and $\hat{p} = \dfrac{x_1 + x_2}{n_1 + n_2}$, where x_1 and x_2 are the number of successes in the samples of size n_1 and n_2.

6. Inferences for two population variances (independent samples, normal populations).
   Hypothesis test of H_0: $\sigma_1^2 = \sigma_2^2$.

The test statistic is

$$F = \frac{s_1^2}{s_2^2}$$

where s_1^2 is the larger sample variance.

   Note: The F-distribution has $ndf = n_1 - 1$, $ddf = n_2 - 1$.

Key Words

In reviewing this chapter, you should be able to define, explain, and illustrate each of the following.

difference between two population means ($\mu_1 - \mu_2$) *(page 428)*

independent samples *(page 428)*

pooled sample variance *(page 437)*

pooled *t*-test *(page 438)*

nonpooled *t*-test *(page 444)*

dependent samples *(page 452)*

paired samples *(page 452)*

difference between two population proportions ($p_1 - p_2$) *(page 461)*

pooled sample proportion *(page 463)*

F probability distribution *(page 467)*

numerator degrees of freedom *(page 467)*

denominator degrees of freedom *(page 467)*

robustness *(page 477)*

Ⓜ *MINITAB Commands*

SET _ *(page 441)*

END *(page 441)*

TWOSAMPLE T _ _ _; *(page 446)*
ALTERNATIVE _.

TWOSAMPLE T _ _ _; *(page 442)*
POOLED;
ALTERNATIVE _.

READ _ _ *(page 455)*

LET _ *(page 455)*

TTEST _ _; *(page 455)*
ALTERNATIVE _.

CDF _; *(page 469)*
F _ _.

INVCDF _; *(page 470)*
F _ _.

PDF _ _; *(page 470)*
F _ _.

PLOT _ _ *(page 470)*

Review Exercises

11.104 Independent random samples were selected from two populations, and the following sample results were obtained.

	Sample One	Sample Two
Sample size	$n_1 = 1,500$	$n_2 = 1,800$
Number of successes	$x_1 = 630$	$x_2 = 612$

With $\alpha = 0.01$, test the following hypotheses.

$$H_0: p_1 = p_2$$

$$H_a: p_1 \neq p_2$$

11.105 Claiming that health care costs more for smokers and that they file a larger percentage of claims than nonsmokers, some companies are now charging smokers more for their health insurance (The *Wall Street Journal,* March 6,

1990). Suppose a large national company wanted to investigate its employees in this regard. It randomly sampled 600 smokers and 1,200 nonsmokers and found that 125 smokers and 192 nonsmokers filed medical claims last year.
   a. Would this provide sufficient evidence at the 0.05 level to conclude that smokers file a larger percentage of claims?
   b. Find the p-value for the test in Part a.

11.106 Use the data in Exercise 11.105 to estimate with 90 percent confidence the difference in the proportions of claims filed by smokers and nonsmokers.

11.107 The following data are from a paired samples experiment. Test the null hypothesis $H_0: \mu_D = 0$ against the alternative hypothesis $H_a: \mu_D < 0$, where $D = \text{I} - \text{II}$. Use $\alpha = 0.05$.

Pair	I	II
1	195	199
2	198	205
3	196	196
4	189	197
5	191	195

11.108 For the data in Exercise 11.107, estimate μ_D with a 95 percent confidence interval.

11.109 Independent random samples were selected from two normal populations, and the following sample results were obtained.

	Sample One	**Sample Two**
Sample size	$n_1 = 10$	$n_2 = 12$
Sample mean	$\bar{x}_1 = 348$	$\bar{x}_2 = 335$
Sample std. dev.	$s_1 = 19.6$	$s_2 = 16.8$

Test to determine if sufficient evidence exists at the $\alpha = 0.10$ level to conclude that the population variances differ.

11.110 Refer to Exercise 11.109 and assume that the population variances are equal. Is there sufficient evidence to conclude that the population means differ? Test at the 5 percent level.

11.111 Use the data in Exercise 11.109 to estimate $(\mu_1 - \mu_2)$ with a 90 percent confidence interval (assume that $\sigma_1 = \sigma_2$).

11.112 Some automobile dealers are able to sell a new car at a higher price than that advertised by adding extra charges that are often called "documentary fees." In 1990, Illinois became the first state to limit the amount of these extra charges. Suppose a state's attorney general wants to determine if a difference exists between the mean documentary fees charged by dealers in her state and in a neighboring state. A random sample of 16 dealers in her state charged an average fee of $173.83 with a standard deviation of $55.32, while a random sample of 14 dealers from the neighboring state charged a mean fee of $142.53 with a standard deviation of $31.05.

a. Test at the 10 percent level whether the variances of the sampled populations differ.

b. Is there sufficient evidence at the 5 percent level to conclude that a difference exists in the mean documentary fee charged by dealers in the two states?

11.113 Determine the approximate p-value for the hypothesis test concerning the population means in Part b of Exercise 11.112.

11.114 Marine farms must closely monitor temperatures of fish ponds since a significant temperature change could result in the death of the fish and a large financial loss. To monitor the temperature of a particular pond, 35 readings are taken at the eastern end, and the mean and standard deviation are 72.6 and 0.8 degrees. For 38 readings at the western end, the mean and standard deviation are 71.9 and 0.4 degrees. Estimate with 99 percent confidence the true difference in the mean temperatures at the two sites.

11.115 Independent random samples of $n_1 = 400$ and $n_2 = 500$ observations resulted in $x_1 = 280$ and $x_2 = 275$ successes. Estimate $(p_1 - p_2)$, the difference between the population proportions, with a 99 percent confidence interval.

11.116 A psychologist conducted a color experiment in which 34 four-year-old children were given a red peg board and 35 children of the same age received a black board. The psychologist believed that children would play with the red board for a longer time. Those who received the red board played with it an average of 828 seconds with a standard deviation of 96 seconds, while the children with the black board played an average of 781 seconds with a standard deviation of 85 seconds. Is there sufficient evidence at the 0.05 level to support the psychologist's belief?

11.117 Determine the p-value for the test in Exercise 11.116.

11.118 Independent random samples were selected from two normal populations, and the following sample results were obtained.

	Sample One	**Sample Two**
Sample size	$n_1 = 9$	$n_2 = 15$
Sample mean	$\bar{x}_1 = 86.9$	$\bar{x}_2 = 50.2$
Sample std. dev.	$s_1 = 12.6$	$s_2 = 3.7$

Test to determine if sufficient evidence exists at the $\alpha = 0.10$ level to conclude that the population variances are not equal.

11.119 Refer to Exercise 11.118 and assume that the population variances differ. Test the following hypotheses at the 0.05 level.

$$H_0: \mu_1 - \mu_2 = 20$$

$$H_a: \mu_1 - \mu_2 > 20$$

11.120 Use the data in Exercise 11.118 to estimate $(\mu_1 - \mu_2)$ with a 99 percent confidence interval (assume that $\sigma_1 \neq \sigma_2$).

11.121 A manufacturer of air pollution controls believes that a chemical treatment of its air filters will increase the number of airborne particles collected by the filters. To test this theory, a treated and an untreated filter are installed at 16 sites. The following data are the number of airborne particles collected over a one-hour period.

Site	Treated	Untreated
1	58	50
2	27	19
3	38	29
4	45	34
5	27	22
6	16	13
7	87	70
8	97	82
9	25	29
10	36	30
11	45	29
12	39	39
13	56	43
14	43	29
15	47	49
16	54	51

Is there sufficient evidence to conclude that the chemical treatment is effective? Test at the 0.01 significance level.

MINITAB Assignments

Ⓜ 11.122 Use MINITAB to test the hypotheses in Exercise 11.121.

Ⓜ 11.123 For the data in Exercise 11.121, use MINITAB to construct a 99 percent confidence interval for the difference in the mean number of particles removed by the two types of filters.

Ⓜ 11.124 To compare the amount of fat in two brands of canned gravy, cans of each brand were randomly selected and analyzed. The grams of fat for each sample are given below.

Brand 1	Brand 2
37.3	31.6
35.7	33.9
39.2	32.8
38.2	31.9
32.9	32.1
36.2	32.2
38.4	33.3
39.2	30.9
41.9	32.7
30.6	33.0
	32.5
	32.4

    a. Use MINITAB to calculate each sample variance.

    b. Using the results from Part a, test at the 0.05 level whether the difference in the samples variances is statistically significant.

    c. Use MINITAB to obtain the p-value for the test in Part b.

Ⓜ 11.125 Use MINITAB to perform a nonpooled t-test on the data in Exercise 11.124 in order to determine if the two brands differ in the mean fat content. Test at the 0.01 level.

Ⓜ 11.126 With reference to the data in Exercise 11.124, use MINITAB to construct a 90 percent confidence interval for the difference in the mean amount of fat for the two brands. (Do not assume that the population variances are equal.)

Ⓜ 11.127 Use MINITAB to sketch the graph of the F-distribution with $ndf = 13$ and $ddf = 17$.

Ⓜ 11.128 For the F-distribution with $ndf = 14$ and $ddf = 34$, use MINITAB to find the following.

    a. $P(F < 1.23)$ b. The 83rd percentile

(© *Grant Heilman/Grant Heilman Photography, Inc.*)

12 Chi-Square Tests for Analyzing Count Data

◀ *Consumer Groups' ''Beef'' with Beef*
In 1990, Congress mandated nutritional information for nearly all types of food except meats. The following year, a coalition of consumer and health groups lobbied hard to extend the requirement to meats. The consumer group, Public Voice for Food and Health Policy, checked nearly 1,600 meat products in 1991 for nutritional labels (The Wall Street Journal, March 5, 1991). They found that the presence of nutritional labeling was greatly dependent on the type of meat product examined. For instance, only 20 of 49 hot dog brands contained nutritional labels, while 61 of 65 dried meats provided nutritional information. For canned and microwave products, 50 of 75 multiserve items had nutritional labels, but only 2 of 42 single-serve products provided such.

This chapter is concerned with the analysis of data that consist of counts such as the values stated above. After learning how to analyze counts for different experimental settings, we will revisit this nutritional labeling study.

Looking Ahead

A **count variable** indicates the number of times that some event of interest has occurred. Since it represents a number of occurrences, its possible values are zero and the positive integers. This chapter is concerned with the **analysis of count data** that have been generated under different types of experimental settings. First, we will consider counts produced from **multinomial experiments.** A special case of a multinomial experiment was discussed in Chapter 6. There we considered binomial experiments, and we saw that they pertain to situations that involve the repetition of n independent trials, where each trial can result in one of two possibilities—success or failure. In a multinomial experiment, the number of possible outcomes for each trial is not limited to two.

In Section 12.2 we will see that a multinomial experiment can be viewed as the classification of count data with respect to **one characteristic of interest.**

Then, in Section 12.3, counts of data will be classified along **two dimensions of interest.** The results will be displayed in a **contingency table** in which the rows represent one method of classification and the columns denote the other.

The last section will focus on the analysis of counts that have been obtained by taking independent random samples from binomial distributions.

All analyses in this chapter will utilize the same test statistic. It is called the **chi-square statistic** because its sampling distribution has been shown for large samples to be approximately chi-square. The statistic is applied to situations that consist of several observations of some phenomenon of interest, where the outcome of each observation is classified into one of several categories, called **cells.** Interest is usually focused on the number of observations that fall within each cell. Often an experimenter assumes a theory concerning the cell counts, and then proceeds to check whether the **observed cell counts** actually obtained from a sample agree with the set of **expected cell counts** that are dictated by that theory. The chi-square test statistic is used to measure the discrepancies between the observed and the expected cell counts.

12.1   Count Data and the Chi-Square Statistic

Discrete random variables were the primary consideration of Chapters 5 and 6. Although there are exceptions, a discrete random variable is usually a **count variable** that indicates the number of times something is observed. Numerous examples of practical situations can be cited that involve **count data,** the main concern of this chapter.

- Governmental agencies at the local, state, and national levels maintain counts of the number of births, the number of deaths, the number of newly diagnosed cases of AIDS, the number of registered voters, and so on.

- Manufacturers keep counts on the number of employee absences, the number of work-related accidents, the number of times that a production line must be shut down, and the number of items produced.

- Educational institutions are interested in counts such as the number of applicants for next year's freshmen class, the number of full-time students and faculty members, the number of class days canceled each year due to inclement weather, and the number of alumni who contribute to their annual fund.

Notice that count data consist of the number of occurrences of something and, thus, count data can only have the nonnegative whole number values 0, 1, 2, 3, ...

In many research studies a phenomenon of interest is observed, and then its outcome is classified into one of several categories, called **cells.** For instance:

- A poultry farmer might classify eggs according to the categories small, medium, large, extra large, or jumbo. Each classified egg can be thought of as falling into exactly one of the five cells that represent the different sizes.

- A highway safety engineer might classify each city intersection according to its volume of traffic. In this case, the cells could be low, normal, high, or extremely high volume.

(© *David R. Frazier/Photo Researchers, Inc.*)

- An ice cream shop might be interested in the number of scoops per cone that its customers purchase. If the shop offers a choice of one, two, or three scoops, then each purchased cone would be classified into one of three possible cells.

In problems, such as the above, in which a phenomenon is observed and then classified into a particular cell, interest is usually focused on the number of items falling within each cell. Often one assumes a theory concerning the cell counts, and then proceeds to check whether the observed counts obtained from a sample agree with the set of expected (theoretical) counts that are dictated by the theory. One of the simplest illustrations of this technique involves a procedure for testing the fairness of a coin. Each toss of a coin can be thought of as falling into one of two cells—heads or tails. One might formulate the null hypothesis that the coin is fair, and then toss it a large number of times, say 100. One would then compare how closely the observed cell counts (the number of heads and the number of tails actually tossed) agree with the expected cell counts under the null hypothesis. For 100 tosses of a fair coin, each expected cell count is 50. If the observed numbers of heads and tails differ significantly from 50, then the null hypothesis is rejected, and it is concluded that the coin is biased.

   At the beginning of this century, the eminent statistician, Karl Pearson, published a statistical procedure for testing if a set of observed cell counts is consistent with a set of theoretical counts. The testing procedure is known as the **chi-square goodness of fit test,** because for large samples, the sampling distribution of the test statistic is approximately chi-square. At the time of its introduction in 1900 to the statistical literature, it is doubtful that Pearson could have foreseen the prominent role that the test would later play in the statistical decision-making process. Today chi-square goodness of fit tests are frequently employed to analyze count data for a multitude of applications in a wide spectrum of disciplines.

   Pearson's chi-square test statistic measures the discrepancies between the observed counts for a set of cells and the corresponding expected counts under the assumption that the null hypothesis is true. Suppose that n observations have been made, and the outcome of each observation is classified as falling into exactly one of k cells. The number of outcomes that have been placed in each of the cells will be denoted by $o_1, o_2, o_3, \ldots, o_k$, and these are called the **observed counts.** The expected number of outcomes for each cell, if the null hypothesis were true, will be denoted by $e_1, e_2, e_3, \ldots, e_k$, and are called the **expected counts.** As part of the analysis, it is customary to summarize the counts in the form of a table such as Table 12.1 below. There, the expected count for each cell appears in parentheses. For the

Early in his career, Karl Pearson (1857–1936) lectured and wrote about geometry. His initial fame came from work in statistical graphics. Pearson's development of the chi-square statistic was partially inspired by investigations of the fairness of Monte Carlo roulette. *(Photo courtesy FPG)*

Table 12.1

Observed Counts and Expected Counts for *k* Cells

Cell 1	Cell 2	Cell 3	Cell 4		Cell *k*
$o_1\ (e_1)$	$o_2\ (e_2)$	$o_3\ (e_3)$	$o_4\ (e_4)$	. . .	$o_k\ (e_k)$

Table 12.2

A Coin Is Tossed 100 Times to Test H_0: The Coin is Fair

Heads	Tails
62 (50)	38 (50)

Observed counts are $o_1 = 62$, $o_2 = 38$
Expected counts are $e_1 = 50$, $e_2 = 50$

---

coin example, if 100 tosses resulted in 62 heads and 38 tails, then these results could be summarized as in Table 12.2.

While the observed counts must be whole numbers, the expected counts often will not be, since they are long-run average values. The sum of the observed counts, as well as the sum of the expected counts, must equal n, the total number of observations ($\Sigma o = \Sigma e = n$).

If the null hypothesis is true, then an observed cell count generally should be reasonably close to the corresponding expected count, since the latter was determined under the assumption of the null hypothesis. On the other hand, large differences between the observed and the expected counts are inconsistent with the null hypothesis and provide evidence against it. The **chi-square test statistic** measures these disparities by calculating a weighted value of the squared differeces between o and e for each cell. More specifically, the test statistic is as follows.

$$\chi^2 = \Sigma \frac{(o - e)^2}{e}$$

$$= \frac{(o_1 - e_1)^2}{e_1} + \frac{(o_2 - e_2)^2}{e_2} + \frac{(o_3 - e_3)^2}{e_3} + \cdots + \frac{(o_k - e_k)^2}{e_k}$$

Each squared difference between o and e is weighted by $1/e$, the reciprocal of the cell's expected count. For the coin illustration, the test statistic equals the following:

$$\chi^2 = \frac{(62 - 50)^2}{50} + \frac{(38 - 50)^2}{50}$$

$$= \frac{144}{50} + \frac{144}{50} = 5.76$$

Notice that if there were perfect agreement between the observed and the expected counts for each cell, then χ^2 would equal zero. χ^2 becomes larger as the discrepancies between the observed and the expected counts increase. Consequently, the null hypothesis will be rejected only if χ^2 is sufficiently large, and thus, the rejection region will always be a right-hand upper tail. To determine the rejection region, Pearson showed that for large values of n, χ^2 has approximately a chi-square distribution with degrees of freedom $df = k - 1$, where k is the number of cells. Generally, n is sufficiently large if all the expected counts are 5 or more (each $e_i \geq 5$).

The following sections will illustrate some of the various types of problems that can be analyzed with the chi-square goodness of fit statistic.

Hypothesis Test for a Multinomial Experiment

Binomial experiments were discussed in Section 6.1. These pertain to situations that involve the repetition of n independent trials, where each trial can result in one of two possibilities. A commonly cited example of a binomial experiment is the tossing of a coin n times; a trial is a single toss, and each trial can result in either a head or a tail. There are many situations that involve the repetition of n independent trials, but where each trial can result in one of several possibilities. An example of such would be the tossing of a die $n = 15$ times, where each toss will produce either one, two, three, four, five, or six spots. In the context of the previous section, the outcome of each trial (toss) is classified as falling into one of k cells, where the cells are the different possibilities on each trial. This type of problem can be thought of as a generalization of a binomial experiment, and it is referred to as a **multinomial experiment.**

A **multinomial experiment** is an experiment that satisfies the following conditions:

1. The experiment consists of the repetition of n identical trials.
2. Each trial will result in exactly one of k possible outcomes (cells).
3. The probabilities of the k possible outcomes (cells) are denoted by p_1, p_2, p_3, . . . , p_k, and these values remain unchanged from trial to trial. Also, $\Sigma p = 1$.
4. The trials are independent, that is, what happens on a trial is not affected by what occurs on any other trial.
5. The random variables of interest are o_1, o_2, o_3, . . . , o_k, where o_i is the observed number of times that the ith outcome occurs, and $\Sigma o = n$, the number of trials.

For the tossing of a die 15 times, suppose we are only interested in the number of times that either 3 or 5 spots are tossed. This problem can be thought of as a multinomial experiment, since the following 5 conditions are satisfied.

1. The experiment consists of the repetition of $n = 15$ identical trials, where 1 trial is a single toss.
2. Each trial will result in exactly 1 of the following $k = 3$ outcomes.

    Outcome (cell) 1: 3 spots show.

    Outcome (cell) 2: 5 spots show.

    Outcome (cell) 3: another number (1, 2, 4, 6) of spots show.

3. For each toss, $p_1 = 1/6$, $p_2 = 1/6$, and $p_3 = 4/6$, where p_1, p_2, and p_3 are the probabilities of tossing 3 spots, 5 spots, and any other number of spots, respectively. Note that $p_1 + p_2 + p_3 = 1$.
4. The trials are independent. What happens on a toss is not affected by what occurs on any other toss.
5. o_1 is the observed number of times that 3 spots are tossed, o_2 is the number of

occurrences of 5 spots, and o_3 is the number of times that either 1, 2, 4, or 6 spots show. $o_1 + o_2 + o_3$ must equal 15 (value of n).

Often one is interested in testing a set of hypotheses concerning the probabilities p_1, p_2, p_3, . . . , p_k in a multinomial experiment. This section illustrates how the chi-square statistic can be used for this purpose.

Example 12.1   Offering several versions of the same car model is a common practice among automobile manufacturers. For example, a popular sports car can be purchased in a standard, deluxe, or luxurious version. Suppose the manufacturer claims that 38 percent of sales are for the standard line, while 26 percent and 36 percent are for the deluxe and luxurious versions, respectively. To check on the veracity of this claim, an independent testing organization surveyed 580 new owners of this model. They found that 239, 143, and 198 had purchased the standard, deluxe, and luxurious versions, respectively. Is there sufficient evidence at the 0.05 significance level to conclude that the manufacturer's claim is false?

Solution

This problem can be thought of as a multinomial experiment with $n = 580$ trials, where a single trial is a purchase of this model and each trial has $k = 3$ possible outcomes.

> Outcome 1: the model is the standard version.
>
> Outcome 2: the model is the deluxe version.
>
> Outcome 3: the model is the luxurious version.

Step 1: Hypotheses.

$$H_0: p_1 = 0.38, \ p_2 = 0.26, \ p_3 = 0.36$$

$$H_a: \text{Not all of the above are true.}$$

where p_1, p_2, and p_3 are the proportions of sales for the standard, deluxe, and luxurious versions, respectively.

Step 2: Significance level.

$$\alpha = 0.05.$$

Step 3: Calculations.
To calculate the value of the chi-square test statistic, the theoretical expected count e_i first needs to be determined for each of the 3 lines. If the null hypothesis were true, then over the long run, we would expect about 38 percent of the 580 cars to be the standard version, 26 percent of 580 to be the deluxe, and 36 percent of 580 to be the luxurious line. Thus, we have the following.

$$e_1 = np_1 = 580(0.38) = 220.4$$

$$e_2 = np_2 = 580(0.26) = 150.8$$

$$e_3 = np_3 = 580(0.36) = 208.8$$

Table 12.3

Observed and Expected Counts for Example 12.1

Standard	Deluxe	Luxurious
239 (220.4)	143 (150.8)	198 (208.8)

Note that like the observed counts o_i, the expected counts e_i sum to n. However, since they are long-run average values, they do not have to be whole numbers. Also note that the expected counts are calculated under the assumption that the null hypothesis is true. In Table 12.3, the observed count and the expected count are given for each of the 3 versions. The expected counts appear in parentheses. To measure the discrepancies between the observed and the expected counts, the value of the chi-square statistic is calculated.

$$\chi^2 = \Sigma \frac{(o - e)^2}{e}$$

$$= \frac{(239 - 220.4)^2}{220.4} + \frac{(143 - 150.8)^2}{150.8} + \frac{(198 - 208.8)^2}{208.8}$$

$$= 1.57 + 0.40 + 0.56 = 2.53$$

Step 4: Rejection region.
The greater the differences between the observed and the expected counts, the larger will be the value of chi-square. Consequently, the null hypothesis is rejected only if χ^2 is sufficiently large. Thus, the rejection region is the right tail of a chi-square distribution. The applicable chi-square curve is determined by its degrees of freedom, $df = k - 1 = 2$. For $\alpha = 0.05$, Table 5 of Appendix A reveals that the null hypothesis should be rejected when $\chi^2 > \chi^2_{.05} = 5.99$. The rejection region is shown in Figure 12.1.

Figure 12.1
Rejection region for Example 12.1.

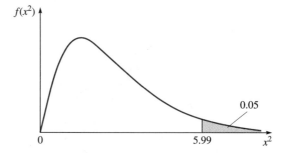

Step 5: Conclusion.
Since the calculated value of χ^2 is 2.53, and this is not in the rejection region, we do not reject the null hypothesis. Thus, there is insufficient evidence to refute the manufacturer's sales percentages for the 3 versions of the model.

Lincoln cents were first pro-
duced in 1909 at the U.S.
mint in Philadelphia and San
Francisco. The coin replaced
the "Indian head" penny.
(Photos courtesy of Coins
Magazine, 700 E. State St.
Iola, WI 54990.)

Example 12.2   A coin dealer has a large bin that contains tens of thousands of old Lincoln pennies.
The numismatist advertises that the coins can be purchased by the pound in unsorted
bags. She claims that the ratios of coins with dates in the 1950s, 1940s, 1930s, 1920s,
and 1910s are $5:5:2:2:1$. A young collector purchased a bag containing 1,065 coins,
and he found the dates distributed as follows.

Date	1950s	1940s	1930s	1920s	1910s
Number	398	372	129	110	56

Is there sufficient evidence at the 0.05 level of significance to reject the dealer's claim
concerning the ratios of the dates?

Solution

Step 1: Hypotheses.
If the dates are in the ratios $5:5:2:2:1$, the corresponding proportions are $p_1 = 5/15$,
$p_2 = 5/15$, $p_3 = 2/15$, $p_4 = 2/15$, and $p_5 = 1/15$. Thus, we'll test the following.

$$H_0: p_1 = \frac{5}{15}, p_2 = \frac{5}{15}, p_3 = \frac{2}{15}, p_4 = \frac{2}{15}, p_5 = \frac{1}{15}$$

H_a: Not all of the above are true.

Step 2: Significance level.

$$\alpha = 0.05.$$

Step 3: Calculations.
Under the assumption that the null hypothesis is true, the expected counts for the
categories of dates are as follows.

$$e_1 = np_1 = 1{,}065 \left(\frac{5}{15}\right) = 355$$

$$e_2 = np_2 = 1{,}065 \left(\frac{5}{15}\right) = 355$$

$$e_3 = np_3 = 1{,}065 \left(\frac{2}{15}\right) = 142$$

Table 12.4

Observed and Expected Counts for the Date Categories

1950s	1940s	1930s	1920s	1910s
398 (355)	372 (355)	129 (142)	110 (142)	56 (71)

$$e_4 = np_4 = 1{,}065 \left(\frac{2}{15}\right) = 142$$

$$e_5 = np_5 = 1{,}065 \left(\frac{1}{15}\right) = 71$$

The observed and the expected counts appear in Table 12.4. The discrepancies between the observed and the expected counts in Table 12.4 are measured by calculating the value of the chi-square statistic.

$$\chi^2 = \Sigma \frac{(o - e)^2}{e}$$

$$= \frac{(398 - 355)^2}{355} + \frac{(372 - 355)^2}{355} + \frac{(129 - 142)^2}{142} + \frac{(110 - 142)^2}{142} + \frac{(56 - 71)^2}{71}$$

$$= 5.21 + 0.81 + 1.19 + 7.21 + 3.17$$

$$= 17.59$$

Step 4: Rejection region.
The rejection region of the null hypothesis is based on the chi-square distribution having $df = k - 1 = 4$. From Table 5, for $\alpha = 0.05$ we should reject H_0 if $\chi^2 > \chi^2_{.05} = 9.49$. The rejection appears in Figure 12.2.

Figure 12.2
Rejection region for Example 12.2.

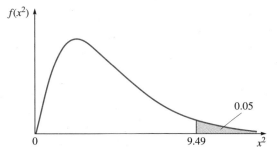

Step 5: Conclusion.
Since the value of the test statistic is $\chi^2 = 17.59$, and this is in the rejection region, the null hypothesis is rejected. Thus, there is sufficient evidence at the 0.05 level to reject the dealer's claim that the coin dates are in the ratios $5:5:2:2:1$. A comparison of each observed count with its expected count in Table 12.4 indicates that there is an excess of dates in the 1940s and 1950s, and a shortage of coins with older dates.

When using the chi-square test statistic, the reader should remember that it is based on the assumption that the sample size n is large.

A conservative rule of thumb in deciding if n is sufficiently large is to check that each expected count e_i is 5 or larger.

Section 12.2 Exercises

Hypothesis Test for a Multinomial Experiment

12.1   A multinomial experiment consisting of $n = 200$ trials with $k = 4$ possible outcomes produced the following observed counts.

Outcomes			
1	2	3	4
54	29	83	34

Is there sufficient evidence at the 0.10 level to reject the following null hypothesis?

$$H_0: p_1 = 0.30, \ p_2 = 0.10, \ p_3 = 0.40, \ p_4 = 0.20$$

12.2   A gambler believes that a die is not balanced. To show this, he tossed it 600 times and obtained the following counts.

Number of Spots					
1	2	3	4	5	6
108	111	78	105	97	101

With $\alpha = 0.05$, do the results provide sufficient evidence to support the gambler's belief?

12.3   Determine the approximate p-value for the test in Exercise 12.2.

12.4   A coin was tossed 1,000 times, and 570 heads and 430 tails were obtained. Use the chi-square statistic to determine if this constitutes sufficient evidence at the 0.05 level to conclude that the coin is not fair.

12.5   Find the approximate p-value for the test in Exercise 12.4.

12.6   A shop has found that 57 percent of its ice cream cone sales are for 1 scoop, 32 percent are for 2 scoops, and 11 percent are for 3 scoops. Recently, the shop added frozen yogurt to its line, and a random sample of 250 cone sales of this product revealed the following distribution.

Number of scoops	1	2	3
Number of sales	169	70	11

Is there sufficient evidence that the pattern of yogurt sales is not the same as that for ice cream sales? Test at the 1 percent significance level.

12.7 In a large city, an insurance company has claims offices in Center City and on the East and West sides. A random sample of 800 claims submitted in this city revealed that 247, 269, and 284 were processed at these 3 locations, respectively. Test the null hypothesis that the proportions of claims handled at the 3 locations are the same. Use $\alpha = 0.10$.

12.8 Most computer software companies provide a telephone number that customers can use to obtain technical assistance. A particular manufacturer recently changed from a toll number to a free 800 number. The distribution of the lengths of calls (in seconds) when the number was a toll call is given below.

Length of call	0–119	120–239	240–359	360–479	≥480
% of calls	13%	21%	40%	19%	7%

After installing the 800 number, a random sample of 300 calls had the following durations.

Length of call	0–119	120–239	240–359	360–479	≥480
Number of calls	27	33	93	88	59

Is there sufficient evidence to conclude that a change has occurred in the distribution of the lengths of calls now that the number is toll free? Test at the 0.05 level.

12.9 Determine the approximate p-value for the test in Exercise 12.8.

12.10 A state's department of motor vehicles is open Monday through Friday for driver examinations. Examination days were checked for a random sample of 2,500 drivers, and the distribution obtained was as follows:

Exam Day	Frequency
Monday	536
Tuesday	509
Wednesday	472
Thursday	460
Friday	523

Do the data provide sufficient evidence that the days are not chosen with equal frequency? Test at the 0.05 level.

12.11 Obtain the approximate p-value for the test in Exercise 12.10.

12.12 Based on a recent survey of several thousand government workers, the *News Digest*, a federal employees' newsletter, gave the following distribution of the highest levels of education completed by the respondents.

Highest Education Level	Proportion of Workers
High school graduate	0.146
Some college	0.292
College graduate	0.243
Some postgraduate	0.138
Postgraduate degree	0.181

Suppose a similar survey of 500 state employees was conducted, and the numbers of workers classified in the above categories (in the order given) were 95, 196, 128, 43, and 38. Would the data provide sufficient evidence of a difference in the proportions for state workers compared to federal employees? Test at the 0.05 significance level.

12.13 Twenty years ago, an introductory statistics course was added to a college's curriculum. Since then, the proportions of A, B, C, D, and F grades given in the course are 0.15, 0.24, 0.32, 0.20, and 0.09, respectively. At a neighboring institution, a random sample of 600 grades in a similar course during the same time period produced the following distribution.

Course grade	A	B	C	D	F
Number	103	153	189	111	44

Is there sufficient evidence to conclude that the grade distributions are not the same for the two institutions? Test at the 5 percent level.

12.14 *In Health* (January/February, 1990) reported that a study of the residents in Maryland's Montgomery County found that 5 percent are afflicted with full-fledged SAD (seasonal affective disorder, but usually referred to as the winter blues). It was found that an additional 13 percent are afflicted with a milder form of SAD. Suppose a psychologist interviewed 375 residents of a second county and obtained the following results:

Form of SAD	Number
Strong	21
Mild	60
None	294

Would this provide sufficient evidence to indicate that the proportions of people with these conditions of the disorder differ from the proportions for Montgomery County? Use $\alpha = 0.01$.

MINITAB Assignments

Ⓜ 12.15 Use MINITAB to obtain the p-value for the test in Exercise 12.10.

Ⓜ 12.16 Use MINITAB to obtain the p-value for the test in Exercise 12.12.

Ⓜ 12.17 Use MINITAB to obtain the p-value for the test in Exercise 12.13.

Ⓜ 12.18 Use MINITAB to obtain the p-value for the test in Exercise 12.14.

12.3 Hypothesis Test for a Contingency Table

The previous section was concerned with analyzing data counts that have been classified with respect to 1 characteristic. In Example 12.2, for instance, the purchaser of a bag of 1,065 pennies classified each coin according to its date. Of course, the classification of data does not have to be restricted to only 1 dimension of interest. Suppose the purchaser decides also to grade each coin according to its physical condition, and each coin will be classified as falling into 1 of the following 3 grade categories.

A: very fine or better

B: very good–fine

C: good or below

The results of classifying each coin by its grade and also by its date are exhibited in Table 12.5.

Table 12.5

3 × 5 Contingency Table

Grade\Date	1950s	1940s	1930s	1920s	1910s	Totals
A	207	91	24	16	6	344
B	159	131	44	38	13	385
C	32	150	61	56	37	336
Totals	398	372	129	110	56	1,065

Displays, such as Table 12.5, in which the rows and the columns represent 2 different methods of classifying count data, are called **contingency tables.** In particular, Table 12.5 is a 3 × 5 (3 by 5) contingency table, since there are 3 classifications for the row dimension and 5 for the column dimension. The same chi-square statistic employed in the previous section for testing 1-dimensional data can also be used for the 2-dimensional case. It allows us to test the following hypotheses.

H_0: The row classifications are independent of the column classifications.

H_a: The row classifications are dependent on the column classifications.

Using the chi-square statistic to test these hypotheses is called a **chi-square test of independence,** since one is testing the null hypothesis that a contingency table's row and column classifications are independent.

To compute the value of the test statistic, it is necessary first to determine the expected count for each cell of the contingency table. To illustrate how these are obtained, consider e_{23}, the expected count for the cell in row 2 and column 3. This equals np_{23}, where p_{23} is the probability of an item falling in the cell in row 2 and column 3.

$$e_{23} = np_{23}$$

But if the null hypothesis is true (row and column classifications are independent), then $p_{23} = (p_2 \cdot)(p \cdot _3)$, where $p_2 \cdot$ is the probability of an item falling in the 2nd row, and $p \cdot _3$ is the probability of an item falling in the 3rd column. This follows from the fact that for independent events A and B, $P(AB) = P(A)P(B)$. Therefore, assuming that the null hypothesis is true, we have

$$e_{23} = np_{23} = n(p_2 \cdot)(p \cdot _3).$$

The probability of an item falling in row 2, $p_2 \cdot$, is approximated by computing the proportion of the sample that lies in row 2, namely, R_2/n, where R_2 is the total of row 2. Similarly, $p \cdot _3$, the probability of an item falling in the 3rd column, is estimated by C_3/n, where C_3 is the total of column 3. Thus, we have

$$e_{23} = np_{23} = n(p_2 \cdot)(p \cdot _3) \approx n\left(\frac{R_2}{n}\right)\left(\frac{C_3}{n}\right) = \frac{R_2 C_3}{n}$$

$$= \frac{(385)(129)}{1,065} = 46.63.$$

Estimating the Expected Cell Counts:
The expected count e_{ij} for the contingency table cell in row i and column j is estimated by

$$e_{ij} \approx \frac{R_i C_j}{n}, \tag{12.1}$$

where

$$R_i \text{ is the total count for row } i,$$

$$C_j \text{ is the total count for column } j,$$

$$n \text{ is the total count for the table.}$$

Using Equation 12.1, the expected count for the cell in row 3 and column 5 is

$$e_{35} \approx \frac{R_3 C_5}{n} = \frac{(336)(56)}{1,065} = 17.67.$$

Estimates of the remaining expected counts appear in parentheses in Table 12.6.

Example 12.3   Based on the data in Table 12.6, is there sufficient evidence at the 1 percent level to indicate a dependency between the condition of a coin and its date?

Solution
Step 1: Hypotheses.

H_0: A coin's condition is independent of its date.

H_a: There is a dependency between condition and date.

Table 12.6

Contingency Table for Coins Classified by Grade and Date

Grade\Date	1950s	1940s	1930s	1920s	1910s	Totals
A	207(128.56)	91(120.16)	24(41.67)	16(35.53)	6(18.09)	344
B	159(143.88)	131(134.48)	44(46.63)	38(39.77)	13(20.24)	385
C	32(125.57)	150(117.36)	61(40.70)	56(34.70)	37(17.67)	336
Totals	398	372	129	110	56	1,065

Step 2: Significance level.

$$\alpha = 0.01.$$

Step 3: Calculations.
For the 15 cells in Table 12.6, the discrepancies between the observed and the expected counts are measured by calculating the value of the chi-square statistic.

$$\chi^2 = \Sigma \frac{(o - e)^2}{e}$$

$$= \frac{(207 - 128.56)^2}{128.56} + \frac{(91 - 120.16)^2}{120.16} + \frac{(24 - 41.67)^2}{41.67} + \frac{(16 - 35.53)^2}{35.53}$$

$$+ \frac{(6 - 18.09)^2}{18.09} + \frac{(159 - 143.88)^2}{143.88} + \frac{(131 - 134.48)^2}{134.48} + \frac{(44 - 46.63)^2}{46.63}$$

$$+ \frac{(38 - 39.77)^2}{39.77} + \frac{(13 - 20.24)^2}{20.24} + \frac{(32 - 125.57)^2}{125.57} + \frac{(150 - 117.36)^2}{117.36}$$

$$+ \frac{(61 - 40.70)^2}{40.70} + \frac{(56 - 34.70)^2}{34.70} + \frac{(37 - 17.67)^2}{17.67}$$

$$= 47.86 + 7.08 + 7.49 + 10.74 + 8.08$$

$$+ 1.59 + 0.09 + 0.15 + 0.08 + 2.59$$

$$+ 69.72 + 9.08 + 10.13 + 13.07 + 21.15$$

$$= 208.90$$

Notice that the contribution to the size of χ^2 is small for a cell in which there is relatively little difference between the observed and the expected counts. For instance, the cell in row 2 and column 4 has an observed count and expected count of 38 and 39.77, respectively, and its portion of the χ^2 value is only 0.08.

Figure 12.3
Rejection region for Example
12.3.

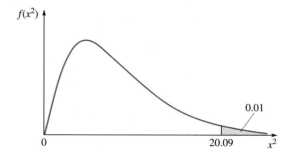

Step 4: Rejection region.
The degrees of freedom associated with the rejection region for a contingency table
consisting of r rows and c columns are given by the formula

$$df = (r - 1)(c - 1)$$
$$= (3 - 1)(5 - 1)$$
$$= 8.$$

From Table 5, with $\alpha = 0.01$, the null hypothesis is rejected if the value of
$\chi^2 > \chi^2_{.01} = 20.09$. The rejection region is given in Figure 12.3.

Step 5: Conclusion.
Since $\chi^2 = 208.90$ exceeds the table value of 20.09, we reject H_0 and conclude that the
condition of the coins is related to their dates.

Ⓜ**Example 12.4**   Use MINITAB to calculate the value of χ^2, the test statistic, for Example 12.3.

Solution

The **CHISQUARE** command can be used to obtain a contingency table containing
the observed and the expected cell counts. The value of χ^2 and its degrees of freedom
are also displayed. First, the table's columns of observed counts must be entered into
MINITAB.

```
READ C1-C5
207   91   24   16    6
159  131   44   38   13
 32  150   61   56   37
END
```

Next, type the following command.

```
CHISQUARE C1-C5
```

MINITAB will produce the output given in Exhibit 12.1.

Exhibit 12.1

```
MTB > CHISQUARE C1-C5

Expected counts are printed below observed counts

            C1        C2        C3        C4        C5      Total
    1      207        91        24        16         6        344
          128.56    120.16    41.67     35.53     18.09

    2      159       131        44        38        13        385
          143.88    134.48    46.63     39.77     20.24

    3       32       150        61        56        37        336
          125.57    117.36    40.70     34.70     17.67

Total      398       372       129       110        56       1065

ChiSq = 47.866 +   7.075 +   7.491 + 10.736 +   8.079 +
         1.589 +   0.090 +   0.149 +   0.078 +   2.592 +
        69.721 +   9.076 +  10.127 + 13.068 +  21.154 = 208.891

df = 8
```

Example 12.5   In the February 1987 *Notices* of the American Mathematical Society, a survey of new doctorates granted by mathematical sciences departments revealed the following distribution of fields of specialization (E. A. Connors, "Employment of New Mathematical Science Doctorates, Faculty Mobility, Employment Trends, Enrollments and Departmental Size, Fall 1986.")

	Year Surveyed			
Specialty	**1982–83**	**1983–84**	**1984–85**	**1985–86**
Applied math	103	110	115	149
Statistics	188	173	189	171
Operations research	63	66	41	62
Computer science	18	20	15	16
Others	420	420	409	403

Is there sufficient evidence at the 0.05 level to indicate that the distributions of specialties differ for the 4 time periods?

Solution
Step 1: Hypotheses.

H_0: The distribution of specialties is independent of the time period.

H_a: The distribution of specialties is dependent on the time period.

Step 2: Significance level.

$$\alpha = 0.05.$$

Step 3: Calculations.
Before calculating χ^2, it is necessary to obtain the estimated expected cell counts. They are calculated on the next page and appear in parentheses in Table 12.7.

Table 12.7

Specialty Fields of New Doctorates

| | Year Surveyed | | | | |
Specialty	1982–83	1983–84	1984–85	1985–86	Totals
Applied math	103(119.89)	110(119.44)	115(116.41)	149(121.26)	477
Statistics	188(181.22)	173(180.54)	189(175.96)	171(183.28)	721
Operations research	63(58.31)	66(58.09)	41(56.62)	62(58.98)	232
Computer science	18(17.34)	20(17.28)	15(16.84)	16(17.54)	69
Others	420(415.23)	420(413.66)	409(403.17)	403(419.95)	1,652
Totals	792	789	769	801	3,151

$$e_{11} \approx \frac{R_1C_1}{n} = \frac{(477)(792)}{3,151} = 119.89$$

$$e_{12} \approx \frac{R_1C_2}{n} = \frac{(477)(789)}{3,151} = 119.44$$

$$e_{13} \approx \frac{R_1C_3}{n} = \frac{(477)(769)}{3,151} = 116.41$$

$$\vdots \quad \vdots \qquad \vdots \qquad \vdots$$

$$e_{53} \approx \frac{R_5C_3}{n} = \frac{(1,652)(769)}{3,151} = 403.17$$

$$e_{54} \approx \frac{R_5C_4}{n} = \frac{(1,652)(801)}{3,151} = 419.95$$

The value of the test statistic is given by the following.

$$\chi^2 = \Sigma \frac{(o - e)^2}{e}$$

$$= \frac{(103 - 119.89)^2}{119.89} + \frac{(110 - 119.44)^2}{119.44} + \frac{(115 - 116.41)^2}{116.41} + \frac{(149 - 121.26)^2}{121.26}$$

$$+ \frac{(188 - 181.22)^2}{181.22} + \frac{(173 - 180.54)^2}{180.54} + \frac{(189 - 175.96)^2}{175.96} + \frac{(171 - 183.28)^2}{183.28}$$

$$+ \frac{(63 - 58.31)^2}{58.31} + \frac{(66 - 58.09)^2}{58.09} + \frac{(41 - 56.62)^2}{56.62} + \frac{(62 - 58.98)^2}{58.98}$$

$$+ \frac{(18 - 17.34)^2}{17.34} + \frac{(20 - 17.28)^2}{17.28} + \frac{(15 - 16.84)^2}{16.84} + \frac{(16 - 17.54)^2}{17.54}$$

$$+ \frac{(420 - 415.23)^2}{415.23} + \frac{(420 - 413.66)^2}{413.66} + \frac{(409 - 403.17)^2}{403.17} + \frac{(403 - 419.95)^2}{419.95}$$

$$= 2.38 + 0.75 + 0.02 + 6.35$$
$$+ 0.25 + 0.31 + 0.97 + 0.82$$
$$+ 0.38 + 1.08 + 4.31 + 0.15$$
$$+ 0.03 + 0.43 + 0.20 + 0.14$$
$$+ 0.05 + 0.10 + 0.08 + 0.68$$

$$\chi^2 = 19.48$$

With few exceptions, each cell's contribution to χ^2 is very small, indicating that the observed and expected counts are in close agreement. The notable exceptions are the cell in row 1, column 4 (applied math in 1985–86), and the cell in row 3, column 3 (operations research in 1984–85). Their contributions are 6.35 and 4.31, respectively, and these constitute more than half the chi-square value of 19.48.

Step 4: Rejection region.

$$df = (r - 1)(c - 1) = (4)(3) = 12$$

From Table 5, using $\alpha = 0.05$, the null hypothesis is rejected if the calculated value of χ^2 exceeds the table value of $\chi^2_{.05}$, namely, 21.03 (see Figure 12.4).

Figure 12.4
Rejection region for Example 12.5.

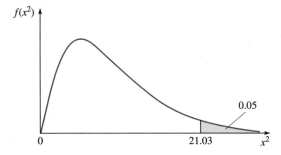

Step 5: Conclusion.
Since the computed value of chi-square is 19.48, which is less than 21.03, we do not reject the null hypothesis. Thus, there is not sufficient evidence to indicate that the distributions of doctoral specialties are dependent on the given time periods.

Section 12.3 Exercises
Hypothesis Test for a Contingency Table

12.19 For the following table of counts, test the null hypothesis that the row and column classifications are independent. Use a significance level of 5 percent.

	Column		
Row	**1**	**2**	**3**
1	20	75	25
2	25	60	45

12.20 Two hundred university students were randomly selected and asked if they favor a revision of the university's student government constitution. The following results were obtained:

	Class Standing			
Opinion	Freshman	Sophomore	Junior	Senior
Favor	21	24	25	19
Against	41	35	21	14

Do the data suggest that opinion on this issue is dependent on class standing? Test at the 0.05 level.

12.21 Determine the approximate *p*-value for the test in Exercise 12.20.

12.22 Many personal computers have built-in hard disk drives to store large quantities of information and programs. The speed of a hard disk drive is usually measured by its access time. A computer magazine conducted a performance evaluation of hard disk drives used in a newly designed laptop computer. It measured access times for a sample of 100 machines. Since the computer maker had utilized 2 suppliers for its disk drives, these were also recorded. The results of the evaluation appear below.

	Access Time in Milliseconds		
Supplier	< 18	18–20	21–23
A	21	33	7
B	8	21	10

Is there sufficient evidence to indicate that access time is dependent on the particular supplier? Test at the 1 percent level.

12.23 The following table pertains to 840 claims processed by an insurance company last year for fire damage that originated in the kitchen.

Kitchen Fire Extinguisher?	Dollar Amount of Damage			
	< 250	250–499	500–999	≥ 1,000
Yes	124	110	40	16
No	170	175	114	91

Is there a dependency between the amount of fire damage sustained and whether or not the kitchen had a fire extinguisher? Test with $\alpha = 0.01$.

12.24 Durl O'Neil and colleagues recently investigated the type and prevalence of dental injuries seen in a hospital emergency room (''Oral Trauma in Children: A Hospital Survey,'' *Oral Surgery, Oral Medicine, and Oral Pathology,* December, 1989). The study involved 765 cases of injuries to structures of the oral cavity. One aspect of the study investigated the type of dental injury sustained and patient gender. Do the following results suggest that there is a relationship between the type of dental injury and gender? Test at the 5 percent significance level.

Type of Dental Injury	Females	Males
Lip laceration	189	291
Internal structures	35	62
Extrusion/intrusion	29	30
Tooth fracture	17	32
Maxillary mandibular fracture	13	33
Burn	5	13
Avulsion	4	12

12.25 Another purpose of the study referred to in Exercise 12.24 was to investigate seasonal variation of dental injuries and gender of the patient. Results pertaining to these appear below.

Month of Injury Occurrence

Gender	Jan.	Feb.	Mar.	Apr.	May	June	Jul.	Aug.	Sep.	Oct.	Nov.	Dec.	Totals
Female	22	23	46	29	34	49	23	21	10	17	11	7	292
Male	39	26	53	59	59	74	35	28	31	26	25	18	473
Totals	61	49	99	88	93	123	58	49	41	43	36	25	765

Do the data provide sufficient evidence to conclude that the month in which a dental injury occurs is dependent on the injured's gender? Test at the 0.05 level.

12.26 Determine the approximate p-value in Exercise 12.25.

12.27 To satisfy graduation requirements, a college requires that students complete a research project sometime during their last 3 semesters. A professor believes that the semester in which a student satisfies the requirement is dependent on his/her grade point average. To test this theory, records of 300 graduates were selected and produced the following results. Is there sufficient evidence at the 0.05 level to support the professor's belief?

Cumulative GPA	Semester Project Completed		
	6th	7th	8th
2.5 and below	10	22	36
2.6–3.0	26	47	54
3.1 and above	22	52	31

12.28 Determine the approximate p-value in Exercise 12.27.

12.29 Patricia Ostrow and colleagues conducted a study of 193 acute care patients in 6 Midwestern hospitals (''Functional Outcomes and Rehabilitation: An Acute Care Field Study,'' *Journal of Rehabilitation Research and Development*, Summer, 1989). One hundred and thirty-two patients received occupational therapy services because of requests from the referring physicians, while the other 61 did not receive such services. All patients were classified according to the severity of disability, and 1 aspect of the study investigated the relationship between this and whether or not the patient had received occupational therapy. Do the following data suggest a dependency between these 2 methods of classification? Test at the 5 percent level.

Received Occupational Therapy?	Severity of Disability		
	Mild	**Moderate**	**Severe**
Yes	26	66	40
No	30	24	7

12.30 Obtain the approximate p-value for the test in Exercise 12.29.

MINITAB Assignments

Ⓜ 12.31 Use MINITAB to perform the chi-square test in Exercise 12.29.

Ⓜ 12.32 Use MINITAB to obtain the exact p-value for the test in Exercise 12.29.

Ⓜ 12.33 Use MINITAB to perform the chi-square test in Exercise 12.24.

Ⓜ 12.34 The personnel director of a large corporation wants to investigate the relationship between the level of physical fitness and the annual salaries of its professional employees. A random sample of 905 employees resulted in the following counts.

Fitness	1992 Salary (in thousands)			
	20–29	**30–39**	**40–49**	**≥ 50**
Fair	22	25	76	103
Good	89	93	78	82
Excellent	98	103	75	61

Use MINITAB to determine if there is a relationship between the physical fitness and the salary of professional employees at this company. Test at the 0.01 significance level.

Ⓜ 12.35 Use MINITAB to obtain the p-value for the chi-square test in Exercise 12.34.

12.4 Hypothesis Test for Two or More Proportions

The previous section was concerned with the analysis of counts for which n items were randomly selected and classified with respect to 2 dimensions of interest. In these situations, n, the total number sampled, was predetermined, but the row and column totals vary according to the particular items sampled. In many situations, it is desirable to fix in advance the totals of either the rows or columns. For example, the column classifications might represent 4 different brands of televisions, and the experimenter may want to sample 100 sets of each make. In this case, each column would have a predetermined total of 100. Fortunately, the analysis of an $r \times c$ table with fixed row or column totals is exactly the same as that for the $r \times c$ contingency table in the previous section.

In this section we will focus on a particular situation involving fixed totals that arises frequently in applications. Specifically, we will use a chi-square test of independence to test the null hypothesis that k populations have the same proportion of successes.

$$H_0: p_1 = p_2 = \cdots = p_k,$$

where p_j $(j = 1, 2, \ldots, k)$ is the proportion of successes in the jth population. Since the null hypothesis states that the populations are homogeneous with regard to their proportions of successes, the test procedure is called a **test of homogeneity.** In the following work, the k populations will be represented by the k column classifications of a $2 \times k$ table, and the column totals will be $n_1, n_2, \ldots, n_k$, the sample sizes for independent random samples selected from the k populations. The 2 row classifications will be success and failure. These ideas are illustrated in Example 12.6.

Example 12.6   In 1989, the U.S. Department of Health and Human Services issued the Surgeon General's report titled *Reducing the Health Consequences of Smoking: 25 Years of Progress*. It discussed developments in smoking prevalence and mortality since the Surgeon General's 1964 landmark report. One of its 1990 national health objectives was to have employer-supported smoking cessation programs available to at least 35 percent of all workers.

Suppose a researcher is investigating labor practices in 3 states. He wants to determine if a difference exists in the proportions of employees in the 3 states for whom such a program is available. Independent random samples of 250, 225, and 275 workers are selected from the 3 states, and 70, 76, and 71, respectively, are found to have access to such a program. Are the differences in the sample proportions statistically significant at the 0.05 level?

Solution

Let p_1, p_2, and p_3 denote the actual proportions of workers in the 3 states with availability to an employer-supported smoking cessation program.

Step 1: Hypotheses.

$$H_0: p_1 = p_2 = p_3$$

$$H_a: \text{Not all proportions are equal.}$$

Step 2: Significance level.

$$\alpha = 0.05.$$

Step 3: Calculations.
The sample results are arranged in Table 12.8, along with the expected cell counts in parentheses.

$$e_{11} \approx \frac{R_1 C_1}{n} = \frac{(217)(250)}{750} = 72.33$$

Table 12.8

Sample Results and Expected Counts for the Three States

Program	1st State	2nd State	3rd State	Totals
Yes	70(72.33)	76(65.10)	71(79.57)	217
No	180(177.67)	149(159.90)	204(195.43)	533
Totals	250	225	275	750

$$e_{12} \approx \frac{R_1 C_2}{n} = \frac{(217)(225)}{750} = 65.10$$

$$e_{13} \approx \frac{R_1 C_3}{n} = \frac{(217)(275)}{750} = 79.57$$

$$e_{21} \approx \frac{R_2 C_1}{n} = \frac{(533)(250)}{750} = 177.67$$

$$e_{22} \approx \frac{R_2 C_2}{n} = \frac{(533)(225)}{750} = 159.90$$

$$e_{23} \approx \frac{R_2 C_3}{n} = \frac{(533)(275)}{750} = 195.43$$

The value of the chi-square test statistic is as follows.

$$\chi^2 = \Sigma \frac{(o - e)^2}{e}$$

$$= \frac{(70 - 72.33)^2}{72.33} + \frac{(76 - 65.10)^2}{65.10} + \frac{(71 - 79.57)^2}{79.57}$$

$$+ \frac{(180 - 177.67)^2}{177.67} + \frac{(149 - 159.90)^2}{159.90} + \frac{(204 - 195.43)^2}{195.43}$$

$$= 0.08 + 1.83 + 0.92 + 0.03 + 0.74 + 0.38$$

$$= 3.98$$

Step 4: Rejection region.

$$df = (r - 1)(c - 1) = (1)(2) = 2$$

From Table 5, with $\alpha = 0.05$, the null hypothesis is rejected for values of $\chi^2 > \chi^2_{.05} = 5.99$. The rejection region appears in Figure 12.5.

Step 5: Conclusion.
Since $\chi^2 = 3.98$ is not in the rejection region, we do not reject the null hypothesis. Thus, there is not enough evidence to conclude that a difference exists in the proportions of workers in the three states with access to an employer-sponsored smoking cessation program.

Figure 12.5
Rejection region for Example
12.6.

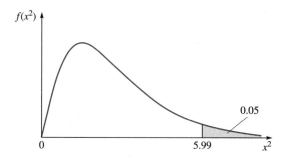

Ⓜ**Example 12.7**

For the previous example, use MINITAB to

1. calculate the value of the chi-square statistic,
2. determine the *p*-value for the test.

Solution

1. First, the columns of observed counts that appear in Table 12.8 are placed in columns C1, C2, and C3, and then the **CHISQUARE** command is applied to the columns.

```
READ C1-C3
 70  76  71
180 149 204
END
CHISQUARE C1-C3
```

The MINITAB output appears in Exhibit 12.2. From this, $\chi^2 = 3.972$.

Exhibit 12.2

```
MTB > READ C1-C3
DATA>  70  76  71
DATA> 180 149 204
DATA> END
         2 ROWS READ
MTB > CHISQUARE C1-C3
```

Expected counts are printed below observed counts

	C1	C2	C3	Total
1	70	76	71	217
	72.33	65.10	79.57	
2	180	149	204	533
	177.67	159.90	195.43	
Total	250	225	275	750

```
ChiSq =  0.075 +  1.825 +  0.922 +
         0.031 +  0.743 +  0.376 = 3.972
df = 2
```

2. To obtain the p-value, the **CDF** command and the **CHISQUARE** subcommand are used to find the cumulative probability up to (to the left of) 3.972.

```
CDF 3.972;
CHISQUARE 2.
```

The p-value is the probability to the right of 3.972. Using the results from Exhibit 12.3, this is given by the following.

$$
\begin{aligned}
p\text{-value} &= P(\chi^2 > 3.972) \\
&= 1 - P(\chi^2 \leq 3.972) \\
&= 1 - 0.8628 \\
&= 0.1372
\end{aligned}
$$

Exhibit 12.3

```
MTB > CDF 3.972;
SUBC> CHIS 2.
      3.9720      0.8628
MTB >
```

Example 12.8

We mentioned at the beginning of the chapter that in 1990, Congress mandated nutritional information for nearly all types of food except meats. A consumer group, Public Voice for Food and Health Policy, checked several hundred meat products in 1991 and found that the presence of a nutrition label varied greatly according to the type of product. For instance, the following results were obtained for samples of 341 frozen single-serve, 224 frozen multiserve, 42 unfrozen single-serve, and 75 unfrozen multiserve meat products.

	Type of Meat Product			
	Frozen Single Serve	**Frozen Multiserve**	**Unfrozen Single Serve**	**Unfrozen Multiserve**
Nutritional label	134	194	2	50
Sample size	341	224	42	75

At the 5 percent level, do the data provide sufficient evidence to indicate a difference in the proportions of product types that have a nutritional label?

Solution

Let p_1, p_2, p_3, and p_4 denote the actual proportions of the four product types that contain nutritional labels.

Step 1: Hypotheses.

$$H_0: p_1 = p_2 = p_3 = p_4$$

$$H_a: \text{Not all proportions are equal.}$$

Step 2: Significance level.

$$\alpha = 0.05.$$

Table 12.9

Sample Results and Expected Counts for the Four Meat Products

Nutritional Label	Type of Meat Product				
	Frozen Single Serve	Frozen Multiserve	Unfrozen Single Serve	Unfrozen Multiserve	Totals
Yes	134(190.00)	194(124.81)	2(23.40)	50(41.79)	380
No	207(151.00)	30(99.19)	40(18.60)	25(33.21)	302
Totals	341	224	42	75	682

Step 3: Calculations.

The sample results and the expected cell counts are given in Table 12.9.

$$e_{11} \approx \frac{R_1 C_1}{n} = \frac{(380)(341)}{682} = 190.00$$

$$e_{12} \approx \frac{R_1 C_2}{n} = \frac{(380)(224)}{682} = 124.81$$

$$e_{13} \approx \frac{R_1 C_3}{n} = \frac{(380)(42)}{682} = 23.40$$

$$e_{14} \approx \frac{R_1 C_4}{n} = \frac{(380)(75)}{682} = 41.79$$

$$e_{21} \approx \frac{R_2 C_1}{n} = \frac{(302)(341)}{682} = 151.00$$

$$e_{22} \approx \frac{R_2 C_2}{n} = \frac{(302)(224)}{682} = 99.19$$

$$e_{23} \approx \frac{R_2 C_3}{n} = \frac{(302)(42)}{682} = 18.60$$

$$e_{24} \approx \frac{R_2 C_4}{n} = \frac{(302)(75)}{682} = 33.21$$

The value of the chi-square test statistic is as follows.

$$\chi^2 = \Sigma \frac{(o - e)^2}{e}$$

$$= \frac{(134 - 190.00)^2}{190.00} + \frac{(194 - 124.81)^2}{124.81} + \frac{(2 - 23.40)^2}{23.40} + \frac{(50 - 41.79)^2}{41.79}$$

$$+ \frac{(207 - 151.00)^2}{151.00} + \frac{(30 - 99.19)^2}{99.19} + \frac{(40 - 18.60)^2}{18.60} + \frac{(25 - 33.21)^2}{33.21}$$

$$= 16.51 + 38.36 + 19.57 + 1.61 + 20.77 + 48.26 + 24.62 + 2.03$$

$$= 171.73$$

Step 4: Rejection region.

$$df = (r - 1)(c - 1) = (1)(3) = 3$$

From Table 5, with $\alpha = 0.05$, the null hypothesis is rejected for values of $\chi^2 > \chi^2_{.05} = 7.81$. The rejection region is shown in Figure 12.6.

Figure 12.6
Rejection region for Example 12.8.

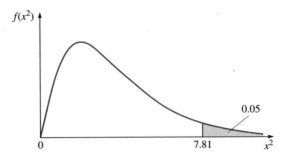

Step 5: Conclusion.

$\chi^2 = 171.73$ does lie in the rejection region, and H_0 is rejected. There is overwhelming evidence to conclude that a difference exists in the proportions of product types that provide a nutritional label.

Section 12.4 Exercises

Hypothesis Test for Two or More Proportions

12.36 Independent random samples were selected from three populations, and the following sample results were obtained.

Sample One	Sample Two	Sample Three
$n_1 = 1,500$	$n_2 = 1,800$	$n_3 = 1,200$
$x_1 = 630$	$x_2 = 612$	$x_3 = 600$

With $\alpha = 0.01$, test the null hypotheses $H_0: p_1 = p_2 = p_3$.

12.37 Independent random samples of size $n_1 = 500$, $n_2 = 400$, and $n_3 = 600$ produced $x_1 = 192$, $x_2 = 168$, and $x_3 = 270$ successes. Is there sufficient evidence at the 0.05 level to conclude that the population proportions are not all equal?

12.38 Find the approximate p-value for the test in Exercise 12.37.

12.39 Four coins were each tossed 500 times. Based on the following results, is there sufficient evidence at the 10 percent level to conclude that the probability of a head is not the same for the 4 coins?

	Coin			
Toss Result	**1st**	**2nd**	**3rd**	**4th**
Heads	236	259	239	261
Tails	264	241	261	239

12.40 Nicholas Fiebach and colleagues conducted a study of 332 women and 790 men after myocardial infarction to examine the influence of gender on survival (*Journal of the American Medical Association,* February 23, 1990). During hospitalization, 47 women and 70 men died. Is the difference in the sample mortality rates for men and women during hospitalization statistically significant? Test at the 5 percent level.

12.41 Find the approximate p-value for the test in Exercise 12.40.

12.42 For the myocardial infarction study described in Exercise 12.40, 285 women and 720 men were hospital survivors, and a follow up of each was conducted 3 years after hospital release. During the 3-year period, there were 48 female and 112 male deaths. Are the gender 3-year mortality rates significantly different at the 5 percent level?

12.43 Use the z-test in Section 11.5 to perform the hypothesis test in Exercise 12.40.

12.44 Perform the test of hypothesis in Exercise 12.42 by using the z-test in Section 11.5.

12.45 A company has manufacturing plants in New York, Michigan, Georgia, Colorado, and California. One hundred employees from each state were randomly selected and asked if they favor the suspension of manufacturing operations for a 2-week period each August. The survey results appear below.

	Plant Location				
Survey Results	**NY**	**MI**	**GA**	**CO**	**CA**
Favor	39	47	32	33	49
Against	61	53	68	67	51

At the 0.05 level, do the data provide sufficient evidence to indicate a difference in the proportions of workers in the 5 states who favor the proposal?

12.46 Obtain the approximate p-value for the test in Exercise 12.45.

12.47 Five hundred adults participated in a comparison of the effectiveness of 3 arthritic pain relievers. Each participant used 1 of the 3 medications for 1 month and then was asked if the product was effective. The results were as follows.

	Pain Reliever		
Effective	**A**	**B**	**C**
Yes	115	78	140
No	60	72	35
Totals	175	150	175

Do the sample proportions differ significantly at the 1 percent level?

12.48 In Exercise 12.47, suppose one is only interested in a comparison of pain relievers A and B. Is there sufficient evidence to conclude that the proportion of people for whom A is effective differs from the proportion for whom B is effective? Test at the 0.05 level.

MINITAB Assignments

Ⓜ 12.49 Five hundred adults were interviewed from each of the 6 New England states to determine the support for a legislative proposal under consideration by the U.S. Senate. The numbers favoring the proposal are given below.

State	Number
Maine	375
New Hampshire	321
Massachusetts	315
Rhode Island	362
Connecticut	358
Vermont	347

Use MINITAB to test at the 5 percent level the null hypothesis that there is no difference in the proportions of adults in each of the 6 New England states who favor the legislation.

Ⓜ 12.50 Use MINITAB to find the p-value for the test in Exercise 12.49.

Ⓜ 12.51 For the data in Exercise 12.49, use MINITAB to determine if a difference exists among the proportions of adults in Maine, New Hampshire, and Vermont who favor the legislation. Test at the 5 percent significance level.

Ⓜ 12.52 Use MINITAB to find the p-value for the test in Exercise 12.51.

Looking Back

Chapter 12 is concerned with the analysis of **count data,** the number of times that some event occurs. An experiment may consist of several observations of some phenomenon of interest, and the outcome of each observation is classified into one of several categories, called **cells.** Interest is usually centered on the number of observations that fall within each cell. Often a researcher assumes a theory concerning the cell counts, and then proceeds to check whether the **observed cell counts** from a sample agree with the set of **expected (theoretical) counts** that are dictated by that theory. Karl Pearson developed the **chi-square test statistic** to measure the discrepancies between the observed counts for a set of cells and the corresponding expected counts if the null hypothesis were true.

$$\chi^2 = \Sigma \frac{(o - e)^2}{e},$$

where o and e are a cell's observed and expected counts, respectively.

Although a small value of χ^2 is consistent with the null hypothesis, a large value is indicative of major discrepancies between the observed and the expected counts for some of the cells. Thus, a large χ^2 value provides evidence against the null hypothesis and, consequently, the null hypothesis is rejected if the value of χ^2 is sufficiently large.

Section 12.2 is concerned with analyzing data counts that have been classified with respect to one characteristic. Specifically, the chi-square statistic is used to test a null hypothesis concerning the cell probabilities in a **multinomial experiment.** A multinomial experiment, a generalization of a binomial experiment, is one in which each trial can admit of k possible outcomes.

In Section 12.3, the chi-square statistic is applied to the analysis of counts of data that have been classified with respect to two characteristics of interest. The count data are displayed in a **contingency table** in which the rows represent one method of classification and the columns denote the other. χ^2 is used to test the null hypothesis that the row classifications are independent of the column classifications.

A third application of the chi-square statistic is illustrated in Section 12.4. It is used to test the null hypothesis that k populations have the same proportion of successes (**test of homogeneity** for proportions). The statistical analysis is identical to that for a contingency table. K column classifications are used to represent the different populations, and two row classifications are used to denote success and failure.

Key Words

In reviewing this chapter, you should be able to define, explain, and illustrate each of the following.

count data *(page 490)*

cell *(page 490)*

chi-square goodness of fit test *(page 491)*

observed cell count *(page 491)*

expected (theoretical) cell count *(page 491)*

chi-square test statistic *(page 492)*

multinomial experiment *(page 493)*

contingency table *(page 501)*

chi-square test of independence *(page 501)*

test of homogeneity *(page 511)*

Ⓜ ## MINITAB Commands

READ _ *(page 504)*

END *(page 504)*

CHISQUARE _ _ *(page 504)*

CDF _; *(page 514)*

CHISQUARE _.

Review Exercises

12.53 H. Stattin and D. Magnusson investigated the relationship between aggressiveness of Swedish children at early school age and their later criminal activities ("The Role of Early Aggressive Behavior in the Frequency, Seriousness, and Types of Later Crime," *Journal of Consulting and Clinical Psychology,* December, 1989). A portion of the study involved 514 boys who were assigned an aggressiveness rating at age 10 by their teachers. For each boy, the number of incidences of registered law breaking was tracked through age 26.

Number of Crimes thru Age 26	Aggressiveness Score at Age 10 (Boys)		
	1–2	3–5	6–7
0	100	197	29
1	15	52	13
≥ 2	7	71	30

Do the data suggest that there is a relationship between the aggressiveness ratings at age 10 and adult delinquency for boys? Test at the 0.05 significance level.

12.54 Determine the approximate *p*-value for the test in Exercise 12.53.

12.55 In the study by Stattin and Magnusson, 507 girls were also assigned an aggressiveness rating at age 10, and the number of incidences of registered law breaking was tracked through age 26. Their results appear below.

Number of Crimes thru Age 26	Aggressiveness Score at Age 10 (Girls)		
	1–2	3–5	6–7
0	152	291	21
1	6	17	3
≥ 2	3	11	3

Explain why a chi-square test of independence should not be applied to this 3 × 3 contingency table.

12.56 For each of the 7 aggressiveness categories in the study referred to in Exercise 12.53, the following table gives the number of boys classified at age 10 who had no criminal record through age 26.

	Aggressiveness Score at Age 10 (Boys)						
	1	2	3	4	5	6	7
No record	41	59	66	95	36	22	7
Sample size	48	74	94	153	73	51	21

For the 7 aggressiveness categories, are the sample proportions significantly different at the 0.05 level?

12.57 For the study referred to in Exercise 12.55, the following table gives the numbers of girls classified at age 10 who had no criminal record through age 26.

	Aggressiveness Score at Age 10 (Girls)		
	1–3	4	5–7
No record	241	158	65
Sample size	256	178	73

Do the sample proportions for the 3 aggressiveness categories differ significantly at the 5 percent level?

12.58 Determine the approximate p-value for the test in Exercise 12.56.

12.59 Determine the approximate p-value for the test in Exercise 12.57.

12.60 A supermarket sells brown and white eggs in sizes small, medium, large, and extra large. The table below shows the numbers of cartons sold for the various sizes and colors during a one-week period.

Type of Egg	**Egg Size**			
	Small	**Medium**	**Large**	**X-large**
Brown	123	217	286	114
White	208	359	406	197

Is egg color preference dependent on the size purchased? Test at the 0.05 significance level.

12.61 Obtain the approximate p-value for the test in Exercise 12.60.

12.62 A spinning wheel is divided into 5 colored sectors of the same size. The wheel is spun 1,000 times, and the number of occurrences for each of the 5 colors appears below.

	Color of Sector				
	Red	**White**	**Blue**	**Green**	**Black**
Number of occurrences	228	207	164	188	213

Do the data indicate that the wheel is out of balance? Test at the 0.01 significance level.

12.63 Obtain the approximate p-value for the test in Exercise 12.62.

12.64 A home mortgage company classifies each applicant into 1 of 3 income levels: low, average, and high. The company offers 4 types of interest rate plans:

A: fixed rate for the life of the mortgage,

B: annually adjustable rate,

C: 3-year adjustable rate,

D: fixed rate for 5 years and then adjustable yearly.

To see if there is a relationship between income level and type of plan chosen, 843 loans were randomly selected and classified as follows. Formulate a suitable set of hypotheses and test at the 5 percent level.

	Type of Mortgage Selected			
Income Level	**A**	**B**	**C**	**D**
Low	21	34	39	49
Average	131	127	128	119
High	79	31	37	48

12.65 Obtain the approximate p-value for the test in Exercise 12.64.

12.66 Independent random samples were selected from 4 populations, and the following sample results were obtained.

Sample One	**Sample Two**	**Sample Three**	**Sample Four**
$n_1 = 500$	$n_2 = 400$	$n_3 = 800$	$n_4 = 300$
$x_1 = 220$	$x_2 = 180$	$x_3 = 260$	$x_4 = 90$

Is there sufficient evidence to conclude that the population proportions are not all the same? Use $\alpha = 0.01$.

12.67 A television manufacturer has assembly plants in 3 countries. A random sample of 1,000 sets was selected from each plant, and each television was tested to determine if it complied with FCC radiation regulations. The numbers of sets from the 3 countries that failed to satisfy the regulations were 231, 329, and 259. Is there sufficient evidence at the 0.05 level to indicate a difference in the noncompliance rates?

12.68 Obtain the approximate p-value for the test in Exercise 12.67.

12.69 A city has 4 exit ramps off an interstate highway. Historically, the percentage of exiting cars that use exits 1 through 4 are 24 percent, 38 percent, 21 percent, and 17 percent, respectively. After completion of a major urban renewal program, 10,000 cars were observed exiting the highway. The numbers using exits 1 through 4 were 2,250, 3,980, 2,040, and 1,730, respectively. Do these results indicate that a change has occurred in the pattern of exiting? Test at the 5 percent level.

12.70 Determine the approximate p-value for the test in Exercise 12.69.

MINITAB Assignments

Ⓜ 12.71 A prominent newspaper surveyed voters in 3 geographical regions of a state in order to assess voter support for the state's governor. A poll of 1,200 voters from each region revealed that 697, 603, and 571 expressed satisfaction with the governor's administration. Formulate a suitable set of hypotheses, and use MINITAB to test at the 0.01 level of significance.

Ⓜ 12.72 Use MINITAB to determine the p-value for the test in Exercise 12.71.

Ⓜ 12.73 In the manufacture of a certain product, a worker must perform a difficult operation. To investigate if the skill with which this work can be done is dependent on one's gender, the performances of 485 workers were judged. The following table summarizes the results.

	Level of Performance			
Gender	**Inadequate**	**Adequate**	**Very Good**	**Excellent**
Males	27	108	79	35
Females	18	99	71	48

Is there sufficient evidence to say that the skill with which this operation can be performed is dependent on the gender of the worker? Use MINITAB to test at the 0.05 level.

Ⓜ 12.74 Use MINITAB to determine the p-value for the test in Exercise 12.73.

Ⓜ 12.75 The produce buyer for a chain of supermarkets is considering 3 suppliers for peaches. He samples 400 peaches from each supplier and finds that 60, 47, and 32 peaches from the 3 suppliers have major blemishes. Is there sufficient evidence that a difference exists in the proportions with serious blemishes for the 3 suppliers? Use MINITAB to test with $\alpha = 0.05$.

Ⓜ 12.76 Use MINITAB to determine the p-value for the test in Exercise 12.75.

(© Barry L. Runk/Grant Heilman Photography, Inc.)

13 Analysis of Variance

◀ *Comparing Two or More Population Means*

Earlier we learned how to conduct mean comparisons between two groups, and we saw several illustrations of their practical application. In this chapter we will learn how the methodology for two comparisons can be extended to any number. For instance, we will compare three different experimental processes for the treatment of water hardness caused by the presence of calcium and magnesium salts in the soil. Next, a comparison of four brands of spaghetti sauce will be made to determine if they differ in their amounts of beef. In the exercises, you will be asked to perform a variety of mean comparisons such as the

- *weight gains of puppies fed three brands of dog food,*
- *quality ratings of bond paper made with three sources of pulp,*
- *performance ratings for four computer word processing programs,*
- *amounts of DDT in lake trout from four lakes,*
- *beginning salaries of education majors in three states,*
- *drying times of four brands of furniture stain,*
- *radon levels in four school districts,*
- *prices of a camcorder at three types of retail outlets,*
- *costs of milk in three geographical regions of a state,*
- *life lengths of three brands of flashlight batteries,*
- *hourly rates of automobile mechanics in three states,*
- *ascorbic acid (vitamin C) levels in Cortland apples grown in three geographical regions.*

Looking Ahead

The title of this chapter, Analysis of Variance, might suggest that our primary intent will be analyzing variability in data. While we are concerned with this aspect, it is only a means to an end. The focus of the chapter is the **testing of hypotheses concerning several population means.** It will be seen that analysis of variance is simply a technique that allows one to generalize the pooled two-sample *t*-test for two means that was considered in Chapter 11. With this generalization, we will be able to test the null hypothesis that *k* population means are equal, where *k* may be two or more. The required assumptions for this procedure are the same as those for the pooled *t*-test, namely, that **independent random samples** have been selected from **normal populations** that have a **common variance σ^2.**

Analysis of variance is actually a general term that applies to several different types of statistical analyses. This chapter is concerned with that form called a **one-way analysis of variance.** To understand the meaning of ''one-way'' recall from our study of regression that the **response variable** (dependent variable) is the variable of primary interest in the experiment. For instance, it might be the length of life of a dishwasher. We are usually concerned with how the response variable is affected by one or more independent variables. In analysis of variance, an independent variable is called a **factor,** and the different values of the factor are called its **levels.** For the dishwasher example, our primary concern might be to determine if there is a difference in the mean life of three dishwasher brands, such as A, B, and C. In this experiment, we would be interested in whether the means (μ_1, μ_2, μ_3) of the response variable (length of life) differ for the three levels (A, B, C) of

The field of analysis of variance was developed by R. A. Fisher (1890–1962). Fisher brought the methodology to India, where it was used to analyze experimental designs applied to agricultural studies. *(Photo courtesy UPI/Bettmann)*

the factor (dishwasher brand). The term **"one-way"** in a one-way analysis of variance indicates that the experiment is concerned with the effects of only **one factor** on the response variable.

In a one-way analysis of variance, the different levels (values) of the single factor are called the **treatments.** This term has been adopted because analysis of variance had its roots in agricultural experiments for which the factor levels were often different combinations of soil treatments.

Terminology used in a one-way analysis of variance is summarized below.

The **response variable** is the dependent variable and is the variable of primary interest in the experiment.

A **factor** is an independent variable.

The **levels** of a factor are its different values, and the levels are called the **treatments.**

The term **one-way** denotes that the experiment is concerned with the effects of only one factor on the response variable.

13.1 The Analysis of Variance Technique

In Section 11.2, the pooled two-sample t-statistic was used to test the null hypothesis that two population means are equal. By using a technique referred to as **analysis of variance,** this type of problem can be generalized to testing hypotheses concerning the equality of two or more population means. The procedure is based on the same assumptions as were required for the pooled two-sample t-test. Throughout this chapter we will assume the following (Figure 13.1).

Assumptions for the One-Way Analysis of Variance:
1. The samples are random and independent.
2. Each population has a normal distribution.
3. The populations have the same variance σ^2.

Figure 13.1
Assumptions for testing
$H_0: \mu_1 = \mu_2 = \mu_3 = \ldots = \mu_k$

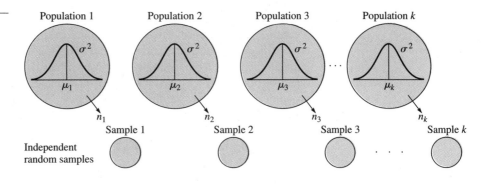

Population 1   Population 2   Population 3   Population k

σ^2 σ^2 σ^2 σ^2

μ_1 μ_2 μ_3 μ_k

n_1 n_2 n_3 n_k

Sample 1   Sample 2   Sample 3   Sample k

Independent
random samples

Each day, the United States uses about 40 billion gallons of water just for normal household purposes (*U.S. Geological Survey*).

(© G. V. Faint/The Image Bank)

Table 13.1

Amounts (ppm) of Salts Removed by the Three Treatments

Process 1	Process 2	Process 3
14	16	18
12	14	16
13	15	16
15		17
		19

$n_1 = 4 \qquad n_2 = 3 \qquad n_3 = 5 \qquad n = 12$

$\bar{x}_1 = 13.5 \qquad \bar{x}_2 = 15 \qquad \bar{x}_3 = 17.2 \qquad \bar{x} = 15.417$

$s_1^2 = 1.667 \qquad s_2^2 = 1 \qquad s_3^2 = 1.7$

Because the experiment is designed on the premise that independent random samples are used, the experiment is known as a **completely randomized design.**

We will use the following example to introduce the basic concept that underlies the analysis of variance methodology. Water hardness is principally caused by the presence of calcium and magnesium salts. The research and development division of a water conditioning company wants to evaluate the effectiveness of three different experimental processes for the treatment of water hardness. For each treatment process, a random sample of treated water was selected and analyzed for calcium and magnesium salts. Table 13.1 gives the amounts (in parts per million, ppm) of these salts that were removed by the three treatment methods. The company wants to test the null hypothesis that the mean amount of salts removed is the same for each process.

$$H_0: \mu_1 = \mu_2 = \mu_3$$

H_a: Not all the means are equal.

In the above we have denoted the total number of observations $(n_1 + n_2 + n_3)$ by n, and the mean of the entire group has been denoted by $\bar{x}$ and is called the **grand mean.** To illustrate graphically the three samples, we had MINITAB construct the dotplots that appear in Exhibit 13.1.

We have seen that for two populations, the equality of their means is tested by measuring the difference between their sample means, $(\bar{x}_1 - \bar{x}_2)$. This procedure must be modified when dealing with more than two populations. To determine if the three population means in our example differ, we will analyze the variability in the samples (analysis of variance). First, think of the three samples as being combined into a single group of $n = 12$ values as illustrated in Figure 13.2, where each value is plotted by using its sample number.

A frequently used measure of the variability in a set of values is $SS(x)$, the sum of the squared deviations of the values from their mean. This quantity is called the **total**

Exhibit 13.1   Dotplots for the Three Samples of Measurements

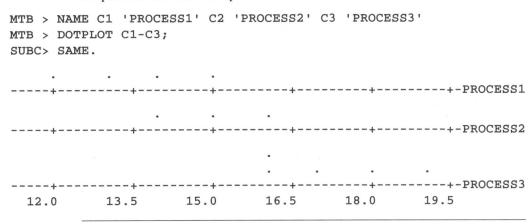

```
MTB > NAME C1 'PROCESS1' C2 'PROCESS2' C3 'PROCESS3'
MTB > DOTPLOT C1-C3;
SUBC> SAME.
```

sum of squares and is denoted by **SST.** For the three samples, the total sum of squares is

$$SST = SS(x) = \Sigma x^2 - \frac{(\Sigma x)^2}{n}$$

$$= (14^2 + 12^2 + \cdots + 17^2 + 19^2) - \frac{(14 + 12 + \cdots + 17 + 19)^2}{12}$$

$$= 2{,}897 - \frac{(185)^2}{12} = 44.92.$$

In the analysis of variance procedure, the total sum of squares is partitioned into two parts that are called the **treatment sum of squares (SSTr)** and the **error sum of squares (SSE).** The treatment sum of squares is a measure of the variability in the $n = 12$ measurements that can be attributed to differences among the three sample means. We will see that SSTr *measures the variability between the sample means* and equals 31.12.

The error sum of squares (also called the **residual sum of squares**) is the difference between the total sum of squares (SST) and the treatment sum of squares (SSTr). It can be thought of as that portion of SST that is left over (residual) after accounting for the differences between the treatment means.

The partitioning of the total sum of squares into the treatment and the error sum of squares is illustrated in Figure 13.3.

To determine if the three population means differ, a test statistic can be derived by obtaining two independent estimates of σ^2, the assumed common variance of the

Figure 13.2
Plot of three samples combined (the sample number of
each value is shown).

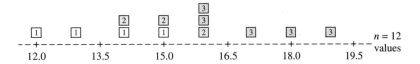

Figure 13.3

Partitioning of the total sum of squares.

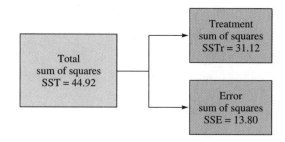

three populations. One such estimate is based on SSE, and is just the extension to three samples of the pooled sample variance, s_p^2, that was used with the pooled two-sample t-test. Recall that s_p^2 is a weighted average of the sample variances, where the weights are the degrees of freedom associated with each variance. For three samples, the pooled sample variance is

$$s_p^2 = \frac{(n_1 - 1)s_1^2 + (n_2 - 1)s_2^2 + (n_3 - 1)s_3^2}{(n_1 - 1) + (n_2 - 1) + (n_3 - 1)}.$$

For the given samples, s_p^2 is obtained by multiplying the sample variances ($s_1^2 = 1.667$, $s_2^2 = 1$, and $s_3^2 = 1.7$) by their weights [$(4 - 1)$, $(3 - 1)$, and $(5 - 1)$], and dividing the result by the sum of the weights.

$$s_p^2 = \frac{(4 - 1)(1.667) + (3 - 1)(1) + (5 - 1)(1.7)}{(4 - 1) + (3 - 1) + (5 - 1)} = \frac{13.80}{9} = 1.533$$

Notice that the numerator of s_p^2 is 13.80, the value of the error sum of squares shown in Figure 13.3.

Since the pooled sample variance is a weighting of the three sample variances, s_p^2 (and its numerator SSE) can be thought of as a *measure of the variability within the samples.* An examination of the three dotplots in Exhibit 13.1 reveals that the observations within each sample are quite close to each sample mean, and this is why s_p^2 is small.

A second estimate of the three populations' common variance σ^2 can be obtained by dividing SSTr, the treatment sum of squares, by $k - 1$, where k is the number of sampled populations. This ratio is called the **mean square for treatments,** or more simply, the **treatment mean square.** It will be denoted by **MSTr** and can be calculated as follows:

$$\text{MSTr} = \frac{\text{SSTr}}{k - 1} = \frac{n_1(\bar{x}_1 - \bar{x})^2 + n_2(\bar{x}_2 - \bar{x})^2 + n_3(\bar{x}_3 - \bar{x})^2}{k - 1},$$

where $\bar{x}_1$, $\bar{x}_2$, and $\bar{x}_3$ denote the three sample means, and $\bar{x}$ is the grand mean of the entire group of $n = n_1 + n_2 + n_3$ sample values. For the three samples, $\bar{x}_1 = 13.5$, $\bar{x}_2 = 15$, $\bar{x}_3 = 17.2$, and $\bar{x} = 15.417$ (see Table 13.1). Thus, the treatment mean square is

$$\text{MSTr} = \frac{4(13.5 - 15.417)^2 + 3(15 - 15.417)^2 + 5(17.2 - 15.417)^2}{3 - 1}$$

$$= \frac{31.12}{2} = 15.56.$$

The treatment mean square *measures the variability between the samples* by calculating the square of the deviation of each sample mean from the grand mean $\bar{x}$. If the three population means are equal, then we would expect that the three sample means will not differ much in value and will be close to the grand mean. Thus, a small value of MSTr is consistent with the null hypothesis. On the other hand, a large value of MSTr indicates considerable variation in the sample means and, thus, evidence against the null hypothesis of equal population means. For our example, the MSTr value of 15.56 is large compared to the pooled sample variance $s_p^2 = 1.533$, and the large value of MSTr reflects the presence of substantial variability between the three sample means. This is also evident by examining Figure 13.2. Although the *variation within* each sample is small, there is considerable *variability between* the three samples. The analysis of variance technique is based on a comparison of these two sources of variation. In particular, the test statistic is the following ratio, which can be shown to have an *F*-distribution.

$$F = \frac{\text{MSTr}}{s_p^2} = \frac{\text{variation between the samples}}{\text{variation within the samples}}$$

If the null hypothesis is true (and the stated assumptions hold), then the numerator and the denominator of the *F*-ratio are each estimates of σ^2 and, consequently, the *F*-ratio is expected to be near 1. When the null hypothesis is false, s_p^2 still estimates σ^2, but the numerator has an expected value larger than σ^2 and, thus, the *F*-ratio is expected to be greater than 1. The weight of evidence against H_0 is directly related to the size of *F* and, therefore, the rejection region is always upper tailed. It consists of values for which $F > F_\alpha$, where α is the specified significance level. For our example, the *F*-ratio is

$$F = \frac{\text{MSTr}}{s_p^2} = \frac{15.56}{1.533} = 10.15.$$

In the next section we will see that this *F*-value is sufficiently large to warrant the rejection of the null hypothesis that the three population means are equal. We will thus conclude that the mean amount of salts removed is not the same for all three water treatment processes.

13.2 The Analysis of Variance Table and Computing Formulas

In performing an analysis of variance, various components involved in the calculation of the *F*-statistic are usually displayed in tabular form, referred to as the **ANOVA (ANalysis Of VAriance) table.** For the water treatment example in the previous section, the ANOVA table appears in Table 13.2.

The first column of the table lists the sources of variation in the data. The *SS* column contains the sum of squares for the three sources, and the second column gives their associated degrees of freedom. The *MS* column contains the **mean square for treatments (MSTr)** and the **mean square for error (MSE).** A mean square is obtained by dividing a sum of squares by its degrees of freedom. The last column gives the *F*-ratio that is obtained by dividing the mean square for treatments by the mean

Table 13.2

ANOVA Table for Testing $H_0 : \mu_1 = \mu_2 = \mu_3$

Source of Variation	df	SS	MS	F
Treatments	2	31.12	15.56	10.15
Error	9	13.80	1.533	
Total	11	44.92		

square for error. The general form of the ANOVA table based on k samples appears in Table 13.3.

To begin the construction of the ANOVA table, the sums of squares for the different sources of variation must be obtained. SSTr and SSE are seldom found by using the formulas of the previous section. Preferably, the sums of squares are obtained as part of the computer output from a statistical software package such as MINITAB. When this is not feasible, they can be obtained by employing shortcut formulas that we will now illustrate. We will construct the ANOVA table for the water treatment data considered earlier and reproduced below.

Amounts (ppm) of Salts Removed by the Three Treatments

Process 1	Process 2	Process 3
14	16	18
12	14	16
13	15	16
15		17
		19

$$T_1 = 54 \quad T_2 = 45 \quad T_3 = 86 \quad T = 185$$

$$n_1 = 4 \quad n_2 = 3 \quad n_3 = 5 \quad n = 12$$

In the above, the totals for the samples have been calculated and are denoted by T_1, T_2, and T_3. The combined total of the three samples is also needed and is denoted by T,

Table 13.3

ANOVA Table for Comparing k Population Means

Source of Variation	df	SS	MS	F
Treatments	$k - 1$	SSTr	$\text{MSTr} = \dfrac{\text{SSTr}}{k - 1}$	$\dfrac{\text{MSTr}}{\text{MSE}}$
Error	$n - k$	SSE	$\text{MSE} = \dfrac{\text{SSE}}{n - k}$	
Total	$n - 1$	SST		

where

$$T = \Sigma x = T_1 + T_2 + T_3 = 54 + 45 + 86 = 185.$$

To construct the ANOVA table, follow these steps.

1. **Calculate the total sum of squares.**

$$SST = \Sigma x^2 - \frac{T^2}{n}$$

$$= (14^2 + 12^2 + \cdots + 17^2 + 19^2) - \frac{185^2}{12}$$

$$= 2,897 - 2,852.08 = 44.92$$

2. **Calculate the treatment sum of squares.**

$$SSTr = \left(\frac{T_1^2}{n_1} + \frac{T_2^2}{n_2} + \frac{T_3^2}{n_3}\right) - \frac{T^2}{n}$$

$$= \left(\frac{54^2}{4} + \frac{45^2}{3} + \frac{86^2}{5}\right) - \frac{185^2}{12}$$

$$= 2,883.2 - 2,852.08$$

$$= 31.12$$

3. **Calculate the error sum of squares.**

$$SSE = SST - SSTr$$

$$= 44.92 - 31.12$$

$$= 13.80$$

The computational formulas are given below for obtaining the sums of squares for the general situation involving k samples.

Computing Formulas for the Sums of Squares in the ANOVA Table:

1. Calculate the total sum of squares.

$$SST = \Sigma x^2 - \frac{T^2}{n} \qquad (13.1)$$

2. Calculate the treatment sum of squares.

$$SSTr = \left(\frac{T_1^2}{n_1} + \frac{T_2^2}{n_2} + \cdots + \frac{T_k^2}{n_k}\right) - \frac{T^2}{n} \qquad (13.2)$$

3. Calculate the error sum of squares.

$$SSE = SST - SSTr \qquad (13.3)$$

In the above formulas,

k is the number of samples;

$n_1, n_2, \ldots, n_k$ are the sample sizes, and $n = \Sigma n_i$;

$T_1, T_2, \ldots, T_k$ are the sample totals, and $T = \Sigma T_i$.

After obtaining the treatment sum of squares SSTr and the error sum of squares SSE, the ANOVA table is completed by computing the associated mean squares and their F-ratio.

$$MSTr = \frac{SSTr}{df} = \frac{31.12}{2} = 15.56$$

$$MSE = \frac{SSE}{df} = \frac{13.80}{9} = 1.533$$

$$F = \frac{MSTr}{MSE} = \frac{15.56}{1.533} = 10.15$$

As discussed in the previous section, the F-ratio is used as the test statistic for deciding whether or not the null hypothesis H_0 is rejected. For instance, if we are testing at the 0.05 significance level, then H_0 would be rejected for values of $F > F_{.05} = 4.26$. This value is obtained from Part b of Table 6 of Appendix A using $ndf = 2$ (df for SSTr) and $ddf = 9$ (df for SSE). In this instance H_0 would be rejected, since the F-value in the ANOVA table is $F = 10.15$, and this exceeds the table value 4.26. Consequently, we would conclude that there is a difference in the mean amounts of salts removed by the 3 water treatment processes.

Sections 13.1 and 13.2 Exercises
The Analysis of Variance Technique;
The Analysis of Variance Table and Computing Formulas

In Exercises 13.1 through 13.4, independent random samples were selected from populations having a common variance σ^2. Use the given sample variances to calculate the pooled variance s_p^2.

13.1

Sample 1	Sample 2
$n_1 = 21$	$n_2 = 16$
$s_1^2 = 46$	$s_2^2 = 38$

13.2

Sample 1	Sample 2	Sample 3
$n_1 = 16$	$n_2 = 6$	$n_3 = 11$
$s_1^2 = 80$	$s_2^2 = 72$	$s_3^2 = 75$

13.3

Sample 1	Sample 2	Sample 3	Sample 4
$n_1 = 8$	$n_2 = 5$	$n_3 = 10$	$n_4 = 6$
$s_1^2 = 36$	$s_2^2 = 40$	$s_3^2 = 35$	$s_4^2 = 42$

13.4

Sample 1	Sample 2	Sample 3	Sample 4	Sample 5
$n_1 = 13$	$n_2 = 11$	$n_3 = 7$	$n_4 = 16$	$n_5 = 11$
$s_1^2 = 8$	$s_2^2 = 10$	$s_3^2 = 9$	$s_4^2 = 12$	$s_5^2 = 8$

13.5   Independent random samples were selected from three populations, each with variance σ^2. The results were as follows.

Sample 1	Sample 2	Sample 3
$\bar{x}_1 = 10$	$\bar{x}_2 = 22$	$\bar{x}_3 = 5$
$s_1^2 = 20$	$s_2^2 = 16$	$s_3^2 = 15$
$n_1 = 6$	$n_2 = 6$	$n_3 = 8$

   a. Obtain the grand mean $\bar{x}$ by computing a weighted average of the three sample means.
   b. Use the grand mean and the three sample means to calculate SSTr, the treatment sum of squares.
   c. Calculate the pooled sample variance s_p^2.
   d. Using the results from Parts b and c, construct the ANOVA table.

13.6   Independent random samples were selected from four populations, each with variance σ^2. The results were as follows.

Sample 1	Sample 2	Sample 3	Sample 4
$n_1 = 10$	$n_2 = 6$	$n_3 = 5$	$n_4 = 9$
$\bar{x}_1 = 50$	$\bar{x}_2 = 20$	$\bar{x}_3 = 80$	$\bar{x}_4 = 10$
$s_1^2 = 28$	$s_2^2 = 42$	$s_3^2 = 25$	$s_4^2 = 30$

   a. Obtain the grand mean $\bar{x}$ by computing a weighted average of the three sample means.
   b. Use the grand mean and the three sample means to calculate SSTr, the treatment sum of squares.
   c. Calculate the pooled sample variance s_p^2.
   d. Use the results from Parts b and c to construct the ANOVA table.

13.7   A completely randomized design produced the following sample results.

Sample 1	Sample 2	Sample 3
19	10	11
23	11	6
20	15	9
18		6
20		

$n_1 = 5$	$n_2 = 3$	$n_3 = 4$	$n = 12$
$\bar{x}_1 = 20$	$\bar{x}_2 = 12$	$\bar{x}_3 = 8$	$\bar{x} = 14$
$s_1^2 = 3.5$	$s_2^2 = 7$	$s_3^2 = 6$	

   a. Use the grand mean $\bar{x}$ and the three sample means to calculate SSTr, the treatment sum of squares.
   b. Use the shortcut Formula 13.2 to calculate SSTr.

   c. Calculate the pooled sample variance s_p^2, and use it to obtain SSE, the error sum of squares.

   d. Use the shortcut Formula 13.3 to calculate SSE.

   e. Construct the analysis of variance table.

13.8 A completely randomized design produced the following sample results.

Sample 1	Sample 2	Sample 3	Sample 4	
50	34	70	18	
49	38	67	15	
54	32	68	21	
	31	67	16	
	35		15	
$n_1 = 3$	$n_2 = 5$	$n_3 = 4$	$n_4 = 5$	$n = 17$
$\bar{x}_1 = 51$	$\bar{x}_2 = 34$	$\bar{x}_3 = 68$	$\bar{x}_4 = 17$	$\bar{x} = 40$
$s_1^2 = 7$	$s_2^2 = 7.5$	$s_3^2 = 2$	$s_4^2 = 6.5$	

   a. Use the grand mean $\bar{x}$ and the four sample means to calculate SSTr, the treatment sum of squares.

   b. Use the shortcut Formula 13.2 to calculate SSTr.

   c. Calculate the pooled sample variance s_p^2, and use it to obtain SSE, the error sum of squares.

   d. Use the shortcut Formula 13.3 to calculate SSE.

   e. Construct the analysis of variance table.

13.9 Use the computational Formulas 13.1 through 13.3 to construct the ANOVA table for the following samples.

Sample 1	Sample 2	Sample 3
8	13	11
9	16	12
6	16	19
11	19	18
8		

13.10 Use the computational Formulas 13.1 through 13.3 to construct the ANOVA table for the following samples.

Sample 1	Sample 2	Sample 3	Sample 4
4.7	6.3	2.1	5.8
3.2	7.8	1.1	5.9
6.9	3.9	5.2	2.5
1.7	6.7	1.3	8.3
	3.2		5.3

13.11 Use the computational Formulas 13.1 through 13.3 to construct the ANOVA table for the following samples.

Sample 1	Sample 2	Sample 3	Sample 4	Sample 5
33	28	39	21	19
30	21	35	20	18
25	22	38	24	33
28	20	39	22	36
30		35	21	27

13.12 Consider the following partially completed ANOVA table for a completely randomized design.

Source of Variation	df	SS	MS	F
Treatments	5	—	200.2	—
Error	—	—	—	
Total	24	1,555		

a. How many treatments are involved in the experiment?
b. What is the total number of sample observations?
c. Complete the ANOVA table by filling in the missing blanks.

13.13 Consider the following partially completed ANOVA table for a completely randomized design.

Source of Variation	df	SS	MS	F
Treatments	—	21.7	3.1	—
Error	28	—	1.7	
Total	—	—		

a. Complete the ANOVA table by filling in the missing blanks.
b. How many treatments are involved in the experiment?
c. What is the total number of sample observations?

13.14 In Exercise 11.26, a pooled two-sample t-test was performed on the following samples.

Sample One	25.8	26.9	26.2	25.3	26.7	26.1	26.9
Sample Two	16.9	17.4	16.8	16.2	17.3	16.8	

Construct the ANOVA table.

13.3   Applications of the One-Way Analysis of Variance

We will now consider some applications of the one-way analysis of variance technique for testing the null hypothesis that k population means are equal. The procedure discussed earlier is summarized on the following page.

The One-Way Analysis of Variance for a Completely Randomized Design:
To test the null hypothesis

$$H_0: \mu_1 = \mu_2 = \cdots = \mu_k,$$

obtain the ANOVA table from computer output or by using the following formulas to calculate the sum of squares.

$$\text{SST} = \Sigma x^2 - \frac{T^2}{n}$$

$$\text{SSTr} = \left(\frac{T_1^2}{n_1} + \frac{T_2^2}{n_2} + \cdots + \frac{T_k^2}{n_k} \right) - \frac{T^2}{n}$$

$$\text{SSE} = \text{SST} - \text{SSTr}$$

The test statistic is

$$F = \frac{\text{MSTr}}{\text{MSE}} = \frac{\dfrac{\text{SSTr}}{k - 1}}{\dfrac{\text{SSE}}{n - k}}$$

The null hypothesis is rejected when $F > F_\alpha$, where α is the significance level of the test.

Assumptions:
1. The samples are random and independent.
2. Each population has a normal distribution.
3. The population variances are equal.

Note:
1. The F-distribution has $ndf = k - 1$, $ddf = n - k$.
2. In the above formulas,
   k is the number of sampled populations, n_i and T_i are the ith sample size and total, $n = \Sigma n_i$, and $T = \Sigma T_i$.

Example 13.1 To determine if there is a difference in the average amount of beef contained in four brands of spaghetti sauce, jars of each brand were randomly selected, and the amount of beef contained in each jar was determined. The results appear below, where the units are grams per jar.

Brand 1	Brand 2	Brand 3	Brand 4
27	29	33	23
24	27	29	24
27	31	34	23
25	32	32	25
27	30	31	24
	31		

Is there sufficient evidence at the 0.05 significance level to conclude that there is a difference in the mean beef content for the four brands? Assume that the sampled populations are normally distributed with a common variance σ^2.

Solution

We will let μ_i denote the true mean amount of beef for the population of all jars of brand i, where $i = 1, 2, 3, 4$.

Step 1: Hypotheses.

$$H_0: \mu_1 = \mu_2 = \mu_3 = \mu_4.$$

H_a: Not all the means are equal.

Step 2: Significance level.

$$\alpha = 0.05.$$

Step 3: Calculations.

We need the total for each of the four samples, as well as the grand total of all the measurements.

$$T_1 = 130 \qquad T_2 = 180 \qquad T_3 = 159 \qquad T_4 = 119 \qquad T = 588$$

$$n_1 = 5 \qquad n_2 = 6 \qquad n_3 = 5 \qquad n_4 = 5 \qquad n = 21$$

Next, the sum of squares SST, SSTr, and SSE are obtained.

$$\text{SST} = \Sigma x^2 - \frac{T^2}{n}$$

$$= (27^2 + 24^2 + \cdots + 25^2 + 24^2) - \frac{588^2}{21}$$

$$= 16{,}710 - 16{,}464 = 246$$

$$\text{SSTr} = \left(\frac{T_1^2}{n_1} + \frac{T_2^2}{n_2} + \frac{T_3^2}{n_3} + \frac{T_4^2}{n_4} \right) - \frac{T^2}{n}$$

$$= \left(\frac{130^2}{5} + \frac{180^2}{6} + \frac{159^2}{5} + \frac{119^2}{5} \right) - \frac{588^2}{21}$$

$$= 16{,}668.4 - 16{,}464 = 204.4$$

$$SSE = SST - SSTr$$
$$= 246 - 204.4 = 41.6$$

With the sums of squares, the ANOVA table can now be constructed.

Source	df	SS	MS	F
Treatments	3	204.4	68.13	27.84
Error	17	41.6	2.447	
Total	20	246.0		

From the ANOVA table, the value of the test statistic is $F = 27.84$.

Step 4: Rejection region.
The rejection region is right tailed and consists of values for which $F > F_\alpha = F_{.05} = 3.20$. This value is obtained from Part b of Table 6 of Appendix A using $ndf = k - 1 = 3$ and $ddf = n - k = 17$. Note that these are the degrees of freedom for the numerator (MSTr) and for the denominator (MSE) of the test statistic F. The rejection region is shown in Figure 13.4.

Figure 13.4
Rejection region for Example 13.1.

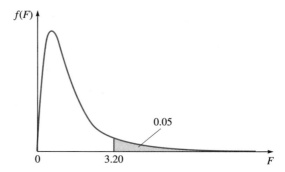

Step 5: Conclusion.
Since the value of the test statistic in the ANOVA table is $F = 27.84$, and this exceeds the table value 3.20, H_0 is rejected. Thus, at the 5 percent significance level we conclude that at least 2 of the brands differ in the mean amount of beef content.

Ⓜ **Example 13.2**   Use MINITAB to obtain the analysis of variance table for the four samples in Example 13.1.

Solution

The measurements for the four samples are placed in columns C1 through C4.

```
SET C1
27 24 27 25 27
END
SET C2
29 27 31 32 30 31
END
```

```
SET C3
33 29 34 32 31
END
SET C4
23 24 23 25 24
END
```

To have MINITAB produce the analysis of variance table, use the command **AOVONEWAY**.

```
AOVONEWAY C1-C4
```

MINITAB will respond by producing the output in Exhibit 13.2. In the ANOVA table, MINITAB gives the p-value for the value of the test statistic F. Since the p-value is smaller than the specified α value of 0.05, the null hypothesis would be rejected.

In addition to the ANOVA table, MINITAB gives the sample means and standard deviations for the four treatments (the levels of the factor—brand of sauce). It also visually displays 95 percent confidence intervals for the treatment means. Consider, for example, the confidence interval for the mean amount of beef in brand 3. Each tick mark on the horizontal axis represents 0.3 (10 ticks cover a spread of 3 units). The interval for μ_3 appears to extend from about 30.3 to 33.3. In the next section we will show how this interval is constructed and that the actual interval is from 30.32 to 33.28 grams.

Exhibit 13.2

```
MTB > AOVONEWAY C1-C4

ANALYSIS OF VARIANCE
SOURCE      DF        SS        MS        F         p
FACTOR       3     204.40     68.13     27.84     0.000
ERROR       17      41.60      2.45
TOTAL       20     246.00
                                    INDIVIDUAL 95 PCT CI'S FOR MEAN
                                    BASED ON POOLED STDEV
LEVEL        N      MEAN      STDEV   ------+---------+---------+---------+-
C1           5     26.000     1.414            (----*----)
C2           6     30.000     1.789                       (---*---)
C3           5     31.800     1.924                              (----*----)
C4           5     23.800     0.837   (----*----)
                                    ------+---------+---------+---------+-
POOLED STDEV =     1.564              24.0      27.0      30.0      33.0
```

The one-way analysis of variance can be used instead of the pooled two-sample t-test for determining if a difference exists between two population means. This is illustrated in the following example.

Example 13.3   In Example 13.1, suppose one were only interested in comparing the amount of beef for brands 1 and 2. The samples for these 2 brands are copied below. At the 0.05

significance level, test to determine if a difference exists in the mean beef content for these 2 brands by using

I. a pooled two-sample t-test,
II. a one-way analysis of variance.

Brand 1	Brand 2
27	29
24	27
27	31
25	32
27	30
	31

Solution

For each test we need to assume independent random samples from normal populations having a common variance σ^2.

I. Pooled two-sample t-test.

Step 1: Hypotheses.

$$H_0: \mu_1 = \mu_2$$

$$H_a: \mu_1 \neq \mu_2$$

Step 2: Significance level.

$$\alpha = 0.05.$$

Step 3: Calculations.

$$\bar{x}_1 = \frac{T_1}{n_1} = \frac{130}{5} = 26, \bar{x}_2 = \frac{T_2}{n_2} = \frac{180}{6} = 30$$

$$s_1^2 = \frac{(27^2 + 24^2 + 27^2 + 25^2 + 27^2) - \frac{130^2}{5}}{5 - 1} = 2$$

$$s_2^2 = \frac{(29^2 + 27^2 + 31^2 + 32^2 + 30^2 + 31^2) - \frac{180^2}{6}}{6 - 1} = 3.2$$

The assumed common variance of the 2 populations is estimated by the pooled sample variance s_p^2.

$$s_p^2 = \frac{(n_1 - 1)s_1^2 + (n_2 - 1)s_2^2}{n_1 + n_2 - 2} = \frac{4(2) + 5(3.2)}{9} = 2.667$$

The value of the test statistic is

$$t = \frac{(\bar{x}_1 - \bar{x}_2) - (\mu_1 - \mu_2)}{\sqrt{s_p^2 \left(\frac{1}{n_1} + \frac{1}{n_2}\right)}} = \frac{(26 - 30) - 0}{\sqrt{2.667 \left(\frac{1}{5} + \frac{1}{6}\right)}} = -4.04.$$

Figure 13.5
Rejection region for Example
13.3 (I).

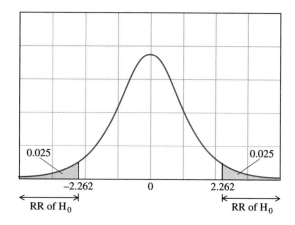

Step 4: Rejection region.

Since the alternative hypothesis is two sided, the rejection region is two tailed. The degrees of freedom are $df = n_1 + n_2 - 2 = 9$, and H_0 is rejected for $t < -t_{\alpha/2} = -t_{.025} = -2.262$ and $t > t_{.025} = 2.262$. The rejection region appears in Figure 13.5.

Step 5: Conclusion.

Since $t = -4.04$ is in the rejection region, H_0 is rejected, and we conclude that the mean beef content for brands 1 and 2 do differ.

II.　One-way analysis of variance.

Step 1: Hypotheses.

$$H_0: \mu_1 = \mu_2$$
$$H_a: \mu_1 \neq \mu_2$$

Step 2: Significance level.

$$\alpha = 0.05.$$

Step 3: Calculations.

$$T = T_1 + T_2 = 130 + 180 = 310, \, n = n_1 + n_2 = 11$$

$$\mathrm{SST} = \Sigma x^2 - \frac{T^2}{n}$$

$$= (27^2 + 24^2 + \ldots + 30^2 + 31^2) - \frac{310^2}{11}$$

$$= 8{,}804 - 8{,}736.36 = 67.64$$

$$\mathrm{SSTr} = \left(\frac{T_1^2}{n_1} + \frac{T_2^2}{n_2}\right) - \frac{T^2}{n}$$

$$= \left(\frac{130^2}{5} + \frac{180^2}{6}\right) - \frac{310^2}{11}$$

$$= 8{,}780 - 8{,}736.36 = 43.64$$

$$SSE = SST - SSTr$$
$$= 67.64 - 43.64 = 24$$

The sums of squares are now used to form the ANOVA table.

Source	df	SS	MS	F
Treatments	1	43.64	43.64	16.36
Error	9	24	2.667	
Total	10	67.64		

Step 4: Rejection region.
The rejection region is based on the F-distribution with $ndf = 1$ and $ddf = 9$. From Part b of Table 6, H_0 is rejected for $F > F_\alpha = F_{.05} = 5.12$ (see Figure 13.6).

Figure 13.6
Rejection region for Example 13.3 (II).

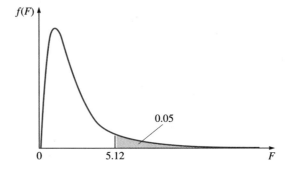

Step 5: Conclusion.
From the ANOVA table, the value of the test statistic is $F = 16.36$. Since this exceeds the table value of 5.12, H_0 is rejected and we conclude that the mean beef contents for the 2 brands are different.

The pooled two-sample t-test and the one-way analysis of variance used in Example 13.3 are equivalent for determining if two population means differ. Each is based on the same assumptions of independent random samples from normal populations with the same variance. Also, except for round off error, the square of the value of the t-statistic equals the value of the F-statistic ($4.04^2 \approx 16.36$). The same relation between t and F applies to the table values used to determine the rejection regions ($2.262^2 = 5.12$). Furthermore, the MSE value in the ANOVA table equals the pooled sample variance used in the t-statistic. Of course, the analysis of variance is more versatile in that it can be used to compare more than two population means, while the pooled t-test is restricted to two means.

13.4 Estimation of Means

In addition to conducting an analysis of variance to determine if the population means differ, one may want to estimate either a single mean or the difference between two means. The form of each confidence interval is very similar to those used in earlier chapters. Because the mean square error (MSE) in the ANOVA table is a pooled sample variance (s_p^2), it is used to estimate the assumed common variance of the populations.

$(1 - \alpha)$ Confidence Interval for One Population Mean μ_i:

$$\bar{x}_i \pm t_{\alpha/2} \frac{s_p}{\sqrt{n_i}} \tag{13.4}$$

Assumptions:
1. The samples are random and independent.
2. Each population has a normal distribution.
3. The population variances are equal.

Note:
1. $\bar{x}_i = T_i/n_i$ and $s_p = \sqrt{\text{MSE}}$.
2. The t-distribution is based on $df = n - k$, where k is the number of samples and $n = n_1 + n_2 + \cdots + n_k$.

$(1 - \alpha)$ Confidence Interval for $(\mu_i - \mu_j)$:

$$(\bar{x}_i - \bar{x}_j) \pm t_{\alpha/2} \sqrt{s_p^2 \left(\frac{1}{n_i} + \frac{1}{n_j} \right)} \tag{13.5}$$

Assumptions:
1. The samples are random and independent.
2. Each population has a normal distribution.
3. The population variances are equal.

Note:
1. $\bar{x}_i = T_i/n_i$, $\bar{x}_j = T_j/n_j$, and $s_p^2 = \text{MSE}$.
2. The t-distribution is based on $df = n - k$, where k is the number of samples and $n = n_1 + n_2 + \cdots + n_k$.

Example 13.4 Use the results of Example 13.1 to obtain a 95 percent confidence interval for

1. μ_3, the mean beef content of brand 3,
2. $(\mu_3 - \mu_2)$, the difference in the mean beef content for brands 3 and 2.

Solution

From Example 13.1, we have

$$T_2 = 180, \ T_3 = 159, \ n_2 = 6, \ n_3 = 5, \text{ and MSE} = 2.447,$$

$$\bar{x}_2 = \frac{T_2}{n_2} = \frac{180}{6} = 30,$$

$$\bar{x}_3 = \frac{T_3}{n_3} = \frac{159}{5} = 31.80,$$

$$s_p = \sqrt{\text{MSE}} = \sqrt{2.447} = 1.564,$$

and

$$t_{\alpha/2} = t_{.025} = 2.110$$

from Table 4 using $df = n - k = 21 - 4 = 17$.

1. A 95 percent confidence interval for μ_3 is

$$\bar{x}_3 \pm t_{\alpha/2} \frac{s_p}{\sqrt{n_3}}$$

$$31.80 \pm 2.11 \frac{1.564}{\sqrt{5}}$$

$$31.80 \pm 1.48.$$

Thus, with 95 percent confidence, we estimate that the mean amount of beef for brand 3 is between 30.32 and 33.28 grams.

2. A 95 percent confidence interval for $(\mu_3 - \mu_2)$ is

$$(\bar{x}_3 - \bar{x}_2) \pm t_{\alpha/2} \sqrt{s_p^2 \left(\frac{1}{n_3} + \frac{1}{n_2} \right)}$$

$$(31.80 - 30) \pm 2.11 \sqrt{(2.447) \left(\frac{1}{5} + \frac{1}{6} \right)}$$

$$1.80 \pm 2.00.$$

Therefore, we are 95 percent confident that $(\mu_3 - \mu_2)$ lies in the interval from -0.20 to 3.80. Since this interval contains both negative and positive values, we cannot conclude with 95 percent confidence that μ_3 is larger than μ_2.

In conducting an analysis of variance, we are testing the null hypothesis that k population means are equal. When the test results indicate that the null hypothesis should be rejected, then we conclude that at least two of the population means are not the same. At this point we may want to pursue the problem further and determine which means are different. For instance, in Example 13.1 we performed an analysis of variance to determine if there is a difference in the mean amount of beef contained in four brands of spaghetti sauce. At the completion of the test we concluded that at least two brands differ in their average beef content. We could begin our exploration of which means differ by visually inspecting the confidence intervals for the four population means that were produced by MINITAB in Exhibit 13.2. The visual display of the intervals from Exhibit 13.2 is reproduced in Figure 13.7.

Figure 13.7
The 95 percent confidence intervals for each mean.

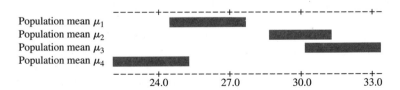

An examination of Figure 13.7 indicates that the confidence intervals for μ_1 and μ_4 overlap, and there is also an overlap between the intervals for μ_2 and μ_3. This separation of the confidence intervals into the two groups

$$\mu_1, \mu_4 \qquad \text{and} \qquad \mu_2, \mu_3$$

suggests that means one and four are similar, as are means two and three. That there is no interval overlapping between the two groups further suggests that μ_1 and μ_4 may be different from μ_2 and μ_3.

We emphasize that the above analysis is only exploratory and does not conclusively indicate which population means are similar and which are different. To support our initial and tentative assessments, a formal evaluation needs to be conducted such as the performance of multiple comparisons between all possible pairs of population means. Several multiple comparison procedures are available and can readily be found in books on analysis of variance and experimental design.

In concluding our discussion of the one-way analysis of variance procedure, it should be emphasized that its validity is dependent on its underlying assumptions. The assumption concerning independent random samples is critical and must not be violated. As with the pooled two-sample t-test, it is also important that the variances of the populations are approximately equal, particularly when the sample sizes are different. The analysis of variance procedure, however, is quite robust with regard to the normality assumption.

When one has doubts concerning the reasonableness of the normality or variance assumptions, an alternative method known as the Kruskal-Wallis H-test can be used. This is a nonparametric test discussed in Chapter 15.

Sections 13.3 and 13.4 Exercises

Applications of the One-Way Analysis of Variance;
Estimation of Means

In the following exercises, assume that independent random samples were selected from populations that have approximately normal distributions with the same variance.

13.15 Consider the following three samples.

Sample 1	Sample 2	Sample 3
17	16	12
21	12	7
18	11	10
16		11
18		

a. Construct a dotplot of each sample and form an opinion as to whether or not the sample means differ significantly.

b. Construct the analysis of variance table and test the null hypothesis H_0: $\mu_1 = \mu_2 = \mu_3$. Use a significance level of 5 percent.

13.16 Consider the following four samples.

Sample 1	Sample 2	Sample 3	Sample 4
9.9	4.3	3.1	5.3
6.2	5.8	2.1	5.9
7.7	1.9	6.2	2.8
4.7	4.7	2.3	8.5
	1.2		5.3

    a. Construct a dotplot of each sample and form an opinion as to whether or not the sample means differ significantly.

    b. Construct the analysis of variance table and test the null hypothesis H_0: $\mu_1 = \mu_2 = \mu_3 = \mu_4$. Use $\alpha = 0.05$.

13.17 Refer to Exercise 13.15 and obtain a 95 percent confidence interval for the following.

    a. μ_1 b. $(\mu_1 - \mu_2)$

13.18 Refer to Exercise 13.16 and obtain a 95 percent confidence interval for the following.

    a. μ_3 b. $(\mu_4 - \mu_3)$

13.19 Determine the approximate p-value for the analysis of variance test in Exercise 13.15.

13.20 Determine the approximate p-value for the analysis of variance test in Exercise 13.16.

13.21 Exercise 11.31 considered an economist's investigation of the hourly labor rates of automobile mechanics in two states. A pooled t-test was used on the following samples to determine if a difference exists in the mean hourly rates for the two states. Use the analysis of variance procedure to perform this test at the 0.05 significance level.

First State	Second State
$40.00	$35.00
$38.00	$37.00
$38.00	$31.00
$37.00	$39.00
$36.00	$31.50
$39.00	$35.00
$41.50	$32.50
$38.00	$34.00
$39.50	$39.00
$37.50	$36.00
$35.00	
$40.00	

13.22 Use the F-table to obtain the approximate p-value for the analysis of variance test in Exercise 13.21.

13.23 Use the results of Exercise 13.21 to construct a 99 percent confidence interval for the difference in the mean hourly rates of automobile mechanics in the two states.

13.24 A kennel owner studied the effects of 3 different brands of dog food on the weight gain of puppies. From a litter of 12 puppies, 4 were randomly selected and fed brand A, 4 were randomly selected from the remaining 8 and fed brand B, and the remaining 4 were fed brand C. Their weight gains after 6 weeks are given below.

Brand of Dog Food		
A	**B**	**C**
1.3	3.1	3.9
1.9	2.8	3.4
1.8	3.0	4.3
1.4	2.9	4.7

Test at the 0.01 significance level if there is a difference in the mean weight gains for the 3 brands of dog food.

13.25 Determine the approximate p-value for the analysis of variance test in Exercise 13.24.

13.26 For Exercise 13.24, estimate with 99 percent confidence the mean weight gain for brand C.

13.27 In Exercise 13.24, construct a 99 percent confidence interval to estimate the difference in the mean weight gains for brands C and A.

13.28 Radon, the second largest cause of lung cancer after smoking, is a radioactive gas produced by the natural decay of radium in the ground. Radon can seep into a building through openings in the foundation, and the EPA recommends that corrective measures be taken if levels reach 4 or more picocuries per liter (pc/l). To investigate radon levels in 4 public schools, an official took a sample of 6 readings at each school. The results, in pc/l, appear below.

School 1	School 2	School 3	School 4
1.7	5.3	5.1	1.2
1.3	4.2	4.1	2.9
0.8	3.9	5.2	1.5
1.5	5.7	3.3	2.3
0.9	3.8	2.6	2.1
1.1	5.2	4.9	1.8

Do the data suggest that there is a difference in the mean radon levels at the 4 schools? Test with $\alpha = 5$ percent.

13.29 Estimate with 95 percent confidence the mean radon level for school 3 in Exercise 13.28.

13.30 In Exercise 13.28, construct a 95 percent confidence interval to estimate the difference in the mean radon levels for schools 2 and 4.

13.31 An agronomist wanted to determine if the amount of ascorbic acid (vitamin C) in Cortland apples depends on the soil. The amounts of ascorbic acid were measured for samples of apples from 3 geographical regions, and the results are given below. The figures are in milligrams of ascorbic acid per 100 grams.

Region 1	Region 2	Region 3
9	13	10
8	10	12
11	9	13
10	10	10
11	11	11
9	8	11
10	14	13
10	13	

Is there sufficient evidence to indicate a difference among mean ascorbic acid levels for Cortland apples grown in the 3 regions? Test at the 0.05 significance level.

13.32 Determine the approximate p-value for the analysis of variance test in Exercise 13.31.

13.33 For Exercise 13.31, estimate with 90 percent confidence the difference in mean ascorbic acid levels for Cortland apples grown in regions 1 and 3.

13.34 Dangerous chemicals from industrial wastes linked to cancer and other diseases can enter the food chain through their presence in lake sediments. The amounts of DDT were determined for samples of trout from 4 lakes and are given below.

Lake	Levels of DDT in Parts per Million					
One	1.7	1.4	1.9	1.1	2.1	1.8
Two	0.3	0.7	0.5	0.1	1.1	0.9
Three	2.7	1.9	2.0	1.5	2.6	
Four	1.2	3.1	1.9	3.7	2.8	3.5

Do the data provide sufficient evidence at the 0.05 level to conclude that there is a difference in mean DDT levels for the 4 lakes?

13.35 Determine the approximate p-value for the analysis of variance test in Exercise 13.34.

13.36 For Exercise 13.34, compare the mean DDT levels in lakes 2 and 4 by constructing a 90 percent confidence interval.

13.37 Suppose in Exercise 13.34 the experimenter hypothesized before collecting the data that the mean DDT levels were not all the same for lakes 1, 3, and 4. Is there sufficient evidence with $\alpha = 0.05$ to support the researcher's belief?

13.38 A university's career development director conducted a survey of starting salaries offered to education majors in three states. Ten offers were recorded for each state, and the analysis of variance output produced by MINITAB appears below.

```
ANALYSIS OF VARIANCE
SOURCE     DF         SS         MS        F        P
FACTOR      2   17470016    8735008     9.99    0.001
ERROR      27   23611000     874482
TOTAL      29   41081016
                                INDIVIDUAL 95 PCT CI'S FOR MEAN
                                BASED ON POOLED STDEV
LEVEL       N    MEAN STDEV   ----------+---------+---------+------
STATE 1    10   23958   968   (------*-------)
STATE 2    10   25512   682                         (-------*------)
STATE 3    10   23836  1105   (-------*-------)
                             ----------+---------+---------+------
POOLED STDEV = 935                  24000     24800     25600
```

a. Do the data provide sufficient evidence to indicate that the mean starting salary is different for at least 2 states? Test at the 0.01 significance level.

b. Based on the visual display of the confidence intervals in the above output, form a tentative opinion as to which population means may be equal and which may differ.

13.39 Use the MINITAB output in Exercise 13.38 to estimate with 95 percent confidence the difference in mean starting salaries in states 1 and 2.

MINITAB Assignments

Ⓜ 13.40 A furniture manufacturer wanted to compare the mean drying times for 4 brands of stains. Each stain was applied to 10 chairs, and the drying times in minutes appear below.

Brand 1	Brand 2	Brand 3	Brand 4
80.6	91.6	90.5	86.7
81.3	83.5	98.5	75.4
82.8	83.4	97.5	79.7
81.5	88.6	99.9	76.5
80.4	96.7	96.9	75.7
79.7	84.8	90.5	84.7
82.3	88.4	96.7	74.5
81.7	89.5	93.8	83.3
80.6	84.4	97.8	84.2
81.5	85.1	96.8	75.3

a. Use MINITAB to test at the 0.05 level if there is sufficient evidence to conclude that a difference exists in the mean drying times for the 4 brands.

b. Based on the visual display of the confidence intervals in the MINITAB output, form a tentative opinion as to which population means may be equal and which may differ.

Ⓜ 13.41 Use the MINITAB output produced in Exercise 13.40 to estimate with 95 percent confidence the difference in mean drying times for brands 2 and 4.

Ⓜ 13.42 Use MINITAB to perform an analysis of variance to determine if there is a difference in the mean drying times for brands 3 and 4 in Exercise 13.40. Use $\alpha = 0.05$.

Ⓜ 13.43 Use MINITAB to obtain the ANOVA table for the data in Exercise 13.28.

Ⓜ 13.44 Use MINITAB to obtain the analysis of variance table for the data in Exercise 13.31.

Ⓜ 13.45 Use MINITAB to obtain the ANOVA table for the samples in Exercise 13.34.

Looking Back

Analysis of variance is a technique for testing the null hypothesis that the means of two or more populations are equal. Chapter 13 is concerned with the **one-way analysis of variance** in which the experiment investigates the effects of only one factor on the response variable. A one-way analysis of variance is a generalization of the two-sample t-test that was discussed earlier in Chapter 11. Both methods are based on the **assumptions** that **independent random samples** have been selected from **normal populations** that have the **same variance.**

In performing an analysis of variance, the various components involved in the calculation of the test statistic F are displayed in tabular form, referred to as the **ANOVA table.**

ANOVA Table for Comparing k Population Means

Source of Variation	df	SS	MS	F
Treatments	$k - 1$	SSTr	MSTr = SSTr/$(k - 1)$	MSTr/MSE
Error	$n - k$	SSE	MSE = SSE/$(n - k)$	
Total	$n - 1$	SST		

The first step in constructing the table is to calculate the three **sums of squares.** These are obtained by using the following shortcut formulas.

Total SS:

$$SST = \Sigma x^2 - \frac{T^2}{n}$$

Treatment SS:

$$SSTr = \left(\frac{T_1^2}{n_1} + \frac{T_2^2}{n_2} + \cdots + \frac{T_k^2}{n_k} \right) - \frac{T^2}{n}$$

Error SS:

$$SSE = SST - SSTr$$

In the above formulas, n_i and T_i denote the sample sizes and totals, respectively, $n = \Sigma n_i$, and $T = \Sigma T_i$.

Confidence intervals for one mean or the difference between two population means can be constructed by using the following formulas.

$$\bar{x}_i \pm t_{\alpha/2} \frac{s_p}{\sqrt{n_i}}$$

$$(\bar{x}_i - \bar{x}_j) \pm t_{\alpha/2} \sqrt{s_p^2 \left(\frac{1}{n_i} + \frac{1}{n_j} \right)}$$

The degrees of freedom for determining $t_{\alpha/2}$ are $df = n - k$, and $s_p^2 = \text{MSE}$ from the ANOVA table.

Key Words

In reviewing this chapter, you should be able to define, explain, and illustrate each of the following.

analysis of variance *(page 527)*

one-way analysis of variance *(page 527)*

response variable *(page 527)*

factor *(page 527)*

levels *(page 527)*

treatments *(page 527)*

completely randomized design *(page 528)*

total sum of squares (SST) *(page 529)*

treatment sum of squares (SSTr) *(page 529)*

error sum of squares (SSE) *(page 529)*

mean square for treatments (MSTr) *(page 530)*

F-ratio *(page 531)*

mean square for error (MSE) *(page 531)*

ANOVA table *(page 531)*

Ⓜ *MINITAB Commands*

NAME _ _ *(page 529)*

DOTPLOT _ _; *(page 529)*

SAME.

AOVONEWAY _ _ *(page 541)*

SET _ *(page 540)*

END *(page 540)*

Review Exercises

In the following exercises, assume that independent random samples were selected from populations that have approximately normal distributions with the same variance.

13.46 A particular camcorder is sold by mail order companies, camera shops, and general merchandise outlets. The manufacturer is interested in determining if the mean selling price differs for these three sources. Seven stores of each type were randomly selected, and the following prices were obtained:

Type of Store	Selling Price in Dollars						
Mail order	899	929	900	979	925	950	959
Camera shop	995	935	950	979	979	995	929
General merch.	979	999	950	995	925	975	989

Formulate a suitable set of hypotheses and test at the 0.01 significance level.

13.47 Determine the approximate p-value for the test in Exercise 13.46.

13.48 With reference to Exercise 13.46, construct a 90 percent confidence interval to estimate the mean price charged by mail order companies.

13.49 With reference to Exercise 13.46, construct a 90 percent confidence interval to estimate the difference in mean prices charged by mail order and general merchandise companies.

13.50 Calculate the pooled sample variance for the three samples whose variances are given below.

Sample 1	Sample 2	Sample 3
$n_1 = 7$	$n_2 = 10$	$n_3 = 9$
$s_1^2 = 20$	$s_2^2 = 16$	$s_3^2 = 22$

13.51 MSE, the mean square error in a one-way analysis of variance table, is a weighted average of the sample variances $s_1^2, s_2^2, s_3^2, \ldots, s_k^2$, where the weights are $(n_1 - 1), (n_2 - 1), (n_3 - 1), \ldots, (n_k - 1)$, respectively. Show that when all the sample sizes are the same, MSE is just

$$\frac{(s_1^2 + s_2^2 + s_3^2 + \cdots + s_k^2)}{k}.$$

13.52 Independent random samples from four populations produced the following results:

Sample 1	Sample 2	Sample 3	Sample 4
$n_1 = 9$	$n_2 = 10$	$n_3 = 5$	$n_4 = 6$
$\bar{x}_1 = 42$	$\bar{x}_2 = 39$	$\bar{x}_3 = 54$	$\bar{x}_4 = 50$
$s_1^2 = 18$	$s_2^2 = 25$	$s_3^2 = 22$	$s_4^2 = 20$

a. Calculate the treatment sum of squares, SSTr.
b. Calculate the pooled sample variance, s_p^2.
c. Construct the analysis of variance table.
d. With $\alpha = 0.05$, test the null hypothesis $H_0: \mu_1 = \mu_2 = \mu_3 = \mu_4$.

13.53 In Exercise 13.52, estimate μ_2 by obtaining a 90 percent confidence interval.

13.54 In Exercise 13.52, construct a 90 percent confidence interval for $(\mu_3 - \mu_1)$.

13.55 To determine if the means of four populations differ, the following independent random samples were selected:

Sample 1	Sample 2	Sample 3	Sample 4
6	9	12	13
7	10	11	11
9	9	9	10
8	13	10	13
5			15

a. Construct a dotplot of each sample and form an opinion as to whether the sample means differ significantly.

b. Construct the analysis of variance table and test the null hypothesis H_0: $\mu_1 = \mu_2 = \mu_3 = \mu_4$. Use $\alpha = 0.05$.

13.56 A partially completed ANOVA table is given below for a completely randomized design.

Source of Variation	df	SS	MS	F
Treatments	—	493	70.43	—
Error	22	—	—	
Total	—	721		

Fill in the missing values in the analysis of variance table, and test at the 0.01 level the null hypothesis that the sampled populations have the same mean.

13.57 In a state for which the price of milk is not regulated, 8 stores were randomly selected in the western, central, and eastern parts of the state. The prices charged by the 24 stores for a quart of milk appear below. Do the data provide sufficient evidence to conclude that there is a difference in the mean prices for a quart of milk in the 3 geographical regions? Test at the 0.05 significance level.

Region	Cost in Cents for a Quart of Milk							
Western	63	64	63	60	55	62	60	62
Central	67	62	69	68	65	65	65	66
Eastern	68	68	73	64	69	72	69	68

13.58 Determine the approximate p-value for the test in Exercise 13.57.

13.59 For Exercise 13.57, estimate with 95 percent confidence the mean price of a quart of milk in the central part of the state.

13.60 For Exercise 13.57, construct a 95 percent confidence interval to estimate the difference in the mean price of a quart of milk in the east and in the west.

13.61 A study was conducted to compare the four leading computer word processing programs. Each of four groups of students received training in one program for a four-week period. At the conclusion of the instruction, all participants were rated on their ability to perform a given task. Their ratings are given on the following page.

Program 1	Program 2	Program 3	Program 4
24	29	24	25
22	27	38	23
21	21	25	24
20	25	24	25
21	22	29	23
	28		

Is there sufficient evidence at the 0.05 level to conclude that the mean performance ratings for the four word processing programs are not all the same?

13.62 Obtain the approximate p-value for the analysis of variance test in Exercise 13.61.

13.63 For Exercise 13.61, estimate with 90 percent confidence the difference in mean performance ratings for programs 3 and 1.

13.64 A consumer products evaluation magazine tested three brands of flashlight batteries. Each brand was used in five flashlights, and the lights were left on until the batteries failed. The life lengths in hours are given below.

Brand	Length of Life in Hours
A	7.3 6.9 5.8 7.9 8.2
B	6.7 7.1 6.0 6.5 5.9
C	7.5 8.3 7.9 8.4 8.3

Is there a difference in the mean length of life among the three brands of batteries? Test at the 0.05 level.

MINITAB Assignments

Ⓜ 13.65 In Exercise 13.21, a test of hypothesis was conducted to determine if there is a difference in the mean hourly labor rates of automobile mechanics in two states. Suppose it was decided to include a third state in the comparison. For each of the three states, dealerships were randomly selected, and the following hourly charges in dollars were obtained.

First State	Second State	Third State
40.00	35.00	40.00
38.00	37.00	42.00
38.00	31.00	40.00
37.00	39.00	39.50
36.00	31.50	37.50
39.00	35.00	45.00
41.50	32.50	43.00
38.00	34.00	42.50
39.50	39.00	39.50
37.50	36.00	41.00
35.00		44.00
40.00		

a. Use MINITAB to determine if a difference exists in the mean hourly rates for these three states. Test at the 0.01 significance level.
b. Based on the visual display of the confidence intervals in the MINITAB output, form a tentative opinion as to which population means may be equal and which may differ.

Ⓜ 13.66 Use the output produced by MINITAB in Exercise 13.65 to estimate with 95 percent confidence the difference in the mean hourly rates for mechanics in the first and third states.

Ⓜ 13.67 A manufacturer of bond writing paper conducted a study to compare the quality of paper for 3 different sources of pulp used in the manufacturing process. For each type of pulp, 16 sheets of paper were randomly selected and assigned a quality rating. The results were as follows:

Pulp A		Pulp B		Pulp C	
78	74	73	62	79	91
69	89	65	80	87	90
75	71	78	76	84	87
89	80	76	72	98	92
76	88	73	56	87	85
78	84	65	69	92	78
65	82	69	61	91	86
78	91	74	71	87	92

Is there sufficient evidence to conclude that a difference exists in the mean quality ratings for the 3 sources of pulp? Use MINITAB to test at the 0.05 significance level.

Ⓜ 13.68 With reference to Exercise 13.67, use MINITAB to perform an analysis of variance to determine if there is a difference in the mean quality ratings for sources A and C. Test at the 5 percent level.

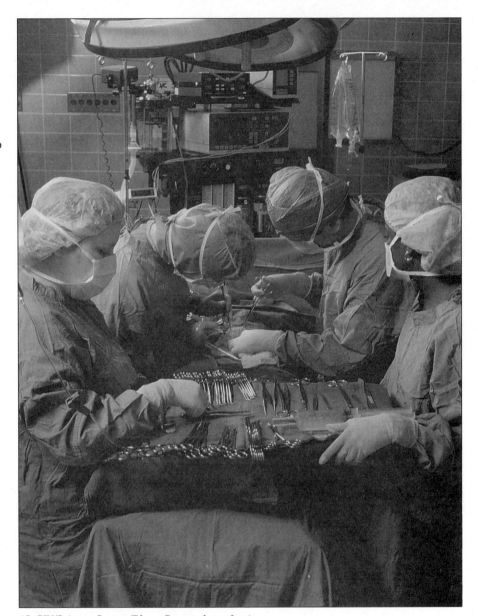

(© SIV/Science Source/Photo Researchers, Inc.)

14 Linear Regression Analysis

◀ *An Objective Approach for Doctors' Fees?*

Harvard University researchers published in 1988 a 1,000-page study referred to as the RBRVS (Resource-Based Relative Value Scale) report. Commissioned by Congress, the report presented radical changes in the way that doctors are reimbursed for their services by Medicare. Instead of payments being based on historical charges by physicians, the study suggested that a relative value scale be used. This would take into account several factors such as the time, skill, and effort expended by the physician during a service.

*In the formulation of their proposal, the Harvard researchers developed a **mathematical model** to measure the relative value associated with a procedure. The model consisted of an equation involving many variables such as time, technical skill, mental effort, stress, training, and overhead expenses. Knowledge of these variables would be used to determine an appropriate reimbursement for a service.*

This chapter is concerned with investigating the relation between two variables and using a mathematical model to describe the hypothesized relationship.

Looking Ahead

Chapter 3 provided an introduction to the topic of modeling bivariate data. Many problems that are subjected to statistical analyses are concerned with investigating the relationship between two variables x and y.

- A social scientist might be concerned with how a city's crime rate (y) is related to its unemployment rate (x).

- A nutritionist might try to relate the quantity (y) of breakfast cereal consumed to the cereal's percentage of sugar (x).

- An economist might be interested in studying the relationship between consumer spending (y) and the prime interest rate (x).

- An investment analyst might investigate how the price (y) of a stock is related to its price earnings ratio (x).

Seldom will an exact mathematical relationship exist between two variables x and y and, thus, the researcher must be content with obtaining a model (equation) that will describe it approximately. In such situations it is important to have some measure of how well the data support the hypothesized model.

In fitting a bivariate model to a set of data, one can proceed systematically by first making a scatter diagram to discern what type of relationship might exist between x and y. This chapter will focus on describing relationships that can be modeled by a straight line. These linear models are called **simple** because they involve only one predictor variable x. The process of fitting a simple linear model to a set of data is called a **simple linear regression analysis.**

If the scatter diagram indicates that the data points tend to lie near a straight line, then the **least squares line** will be obtained and considered as a possible model. Before adopting it, however, it is necessary to evaluate the model's potential usefulness. In Chapter 3, the correlation coefficient r was considered as a means of evaluating the utility of a model. We saw that this is a measure of the strength of the linear relationship between x and y. A value of r that is "near" -1 or $+1$ indicates the existence of a good fit between the data points and the least squares line. In Chapter 3, we had no way of determining if r was sufficiently "near" -1 or $+1$. Now that we have a background in hypothesis testing, this issue can be addressed. In fact, we will see that additional criteria can be used to assess the utility of a model.

Finally, once we have obtained supporting evidence of the usefulness of a model, we will then show how it can be utilized for estimation and prediction.

14.1   The Simple Linear Model and Related Assumptions

In Chapter 3, we considered an example in which the mathematics department at a liberal arts college used a placement test to assist in assigning appropriate math courses to incoming freshmen. A 25-point test is administered during freshman orientation to measure quantitative and analytical skills. The department believes that the test is a good predictor of a student's final numerical grade in its introductory statistics course. Table 14.1 contains the placement test scores and final course grades for a sample of 15 students who took the course. In Chapter 3, these 15 data points were used to build a model that uses a student's placement test score (x) to predict his or her final statistics grade (y). We will review and, in the process, elucidate the steps involved.

To determine what type of relationship might exist between x and y, we begin by constructing a **scatter diagram,** which is a graph of the data points. This appears in Figure 14.1. From the scatter diagram it appears that there might be approximately a straight line relationship between placement test score x and course grade y. Based on the appearance of the scatter diagram, we will tentatively assume as our model that the mean course grade μ_y is related to test score x by a straight line. Mathematically, we write this as

$$\mu_y = \beta_0 + \beta_1 x, \qquad \textbf{(14.1)}$$

Table 14.1

Placement Scores and Statistics Grades for Fifteen Freshmen

Student	Placement Test Score x	Numerical Grade y
1	21	69
2	17	72
3	21	94
4	11	61
5	15	62
6	19	80
7	15	65
8	23	88
9	13	54
10	19	75
11	16	80
12	25	93
13	8	55
14	14	60
15	17	64

where β_0 and β_1 are unknown constants (parameters) that denote the y-intercept and the slope, respectively, of the line. This line of means is called the **regression line** and is illustrated in Figure 14.2.

It is important to note that the model given in Equation 14.1 describes the relationship between a placement score x and the **average** course grade of all students with this score. To illustrate this, assume temporarily that the unknown y-intercept and

Figure 14.1
Scatter diagram for the 15 points.

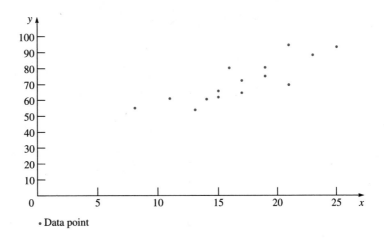

• Data point

Figure 14.2
Assumed model relating
mean grade to test score x.

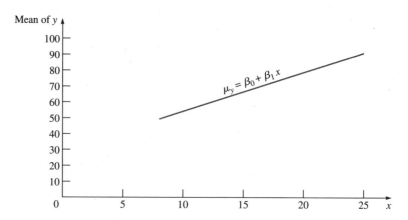

slope are actually $\beta_0 = 30$ and $\beta_1 = 2.5$ (in reality, these are unknown and must be estimated). Then for an arbitrarily chosen value of x, say 20, the model states that the mean course grade μ_y of all those who scored 20 on the placement test is given by

$$\mu_y = 30 + 2.5(20) = 80.$$

To write the model so that it describes an individual's course grade y, we must add a **random error component** ϵ to the straight line given in Equation 14.1. The random error component is a random variable that represents the deviation of each person's grade from the mean grade of all who have a placement score of x. Expressing the assumed model in terms of an individual's grade, we have

$$y = \beta_0 + \beta_1 x + \epsilon. \qquad \textbf{(14.2A)}$$

To illustrate let's temporarily assume that the unknown y-intercept is $\beta_0 = 30$ and the unknown slope is $\beta_1 = 2.5$. Then for a randomly selected individual with a placement score of 20, for example, his or her course grade is given by the following.

$y = 30 + 2.5(20) + \epsilon$

$\quad = 80 + \epsilon$

$\quad = $ (the mean grade of all who score 20) + (a random error component)

Without the random error component in the equation, the model would state that all who score $x = 20$ on the placement test will also obtain the same course grade of 80. The inclusion of ϵ in the model allows for individual grades to vary from the mean grade of 80. For the general model in Equation 14.2A the random error component (ϵ) allows for variation in individual grades (y) from the mean grade ($\beta_0 + \beta_1 x$) of all who score x on the placement test.

$$\overbrace{y = \beta_0 + \beta_1 x}^{\mu_y} + \epsilon \qquad \textbf{(14.2B)}$$

In Equation 14.2B, $\mu_y = \beta_0 + \beta_1 x$ is called the **deterministic component** of the model, since it always gives the same value for a given score of x (it gave a value of 80 with $x = 20$ for the above illustration). The model in Equation 14.2B is obtained by

Figure 14.3
Probabilistic linear model
$y = \beta_0 + \beta_1 x + \epsilon$.

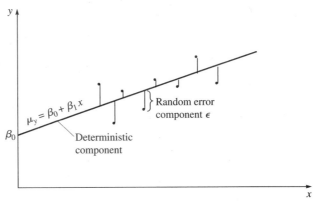

adding a random error component (ϵ) to the deterministic component ($\beta_0 + \beta_1 x$), and the resulting model is called a **probabilistic model.**

$$
y = \boxed{\;\beta_0 + \beta_1 x\;} + \boxed{\;\epsilon\;}
$$

Deterministic component   Random error component

The essential distinguishing feature between a deterministic and a probabilistic model is that a deterministic model assumes that the phenomenon of interest can be predicted precisely, while a probabilistic model allows for variation in the predictions through the inclusion of a random error component. (See Figure 14.3)

In the sections that follow, inferences will be made in the form of confidence intervals and tests of hypotheses. These inferences require certain assumptions about the behavior of the random error component ϵ in the model. The assumed model and its related assumptions are described in the following box and Figure 14.4.

Simple Linear Probabilistic Model and Assumptions:
$$y = \beta_0 + \beta_1 x + \epsilon, \qquad (14.3)$$
y is the **dependent (response) variable** being modeled,
x is the **independent (predictor) variable,**
β_0 is the **y-intercept** of the line,
β_1 is the **slope** of the line,
ϵ is a random variable called the **random error component.**

Assumptions:
1. For each possible setting of x, the random error component ϵ has
   a. a normal probability distribution,
   b. a mean of 0,
   c. a constant variance that is denoted by σ^2.
2. For every possible pair of observations y_i and y_j, the associated random errors ϵ_i and ϵ_j are independent. That is, the error associated with one y-value does not affect the error associated with another y-value.

Note: The mean of y is given by $\mu_y = \beta_0 + \beta_1 x$, and individual values of y deviate about this straight line with a variance of σ^2.

Figure 14.4
Probability distribution of ϵ for different settings of x.

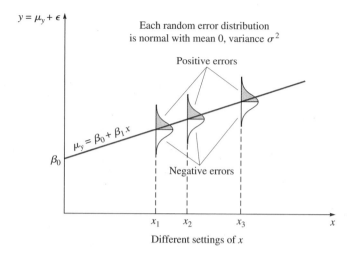

14.2 Fitting the Model by the Method of Least Squares

In the previous section, we began our investigation of the relationship between placement score x and course grade y by constructing a scatter diagram. Since the pattern of the data points suggested a straight line, we tentatively assumed that the

mean course grade μ_y is related to test score x by a straight line. This is the deterministic component of the model, written in Equation 14.1 as

$$\mu_y = \beta_0 + \beta_1 x.$$

The coefficients β_0 and β_1 denote the y-intercept and the slope of the regression line, and they are unknown parameters that must be estimated from the sample data. In Chapter 3 we discussed the method of least squares for estimating the slope and y-intercept when fitting a straight line to a set of data points. As discussed there, the **least squares line** is generally considered to be the line of best fit. The formulas for the least squares estimates of β_0 and β_1 are repeated below.

The Least Squares Estimates of β_0 and β_1:
The slope is estimated by

$$\hat{\beta}_1 = \frac{SS(xy)}{SS(x)} \qquad (14.4)$$

and the y-intercept by

$$\hat{\beta}_0 = \bar{y} - \hat{\beta}_1 \bar{x}, \qquad (14.5)$$

where

$$SS(xy) = \Sigma xy - \frac{(\Sigma x)(\Sigma y)}{n},$$

$$SS(x) = \Sigma x^2 - \frac{(\Sigma x)^2}{n},$$

$$\bar{x} = \frac{\Sigma x}{n}, \qquad \text{and}$$

$$\bar{y} = \frac{\Sigma y}{n}.$$

Example 14.1   Obtain the estimated model for describing the relationship between test score x and course grade y.

Solution

The assumed model is $\mu_y = \beta_0 + \beta_1 x$, which will be estimated by the least squares line. This estimated model will be denoted by $\hat{y} = \hat{\beta}_0 + \hat{\beta}_1 x$. To find $\hat{\beta}_0$ and $\hat{\beta}_1$, we must first determine $SS(xy)$ and $SS(x)$.

Test Score x	Stat Grade y	xy	x^2	y^2
21	69	1,449	441	4,761
17	72	1,224	289	5,184
21	94	1,974	441	8,836
11	61	671	121	3,721
15	62	930	225	3,844
19	80	1,520	361	6,400
15	65	975	225	4,225
23	88	2,024	529	7,744
13	54	702	169	2,916
19	75	1,425	361	5,625
16	80	1,280	256	6,400
25	93	2,325	625	8,649
8	55	440	64	3,025
14	60	840	196	3,600
17	64	1,088	289	4,096
254	1,072	18,867	4,592	79,026

$$SS(x) = \Sigma x^2 - \frac{(\Sigma x)^2}{n} = 4,592 - \frac{(254)^2}{15} = 290.9333$$

$$SS(xy) = \Sigma xy - \frac{(\Sigma x)(\Sigma y)}{n} = 18,867 - \frac{(254)(1,072)}{15} = 714.4667$$

The slope of the least squares line is

$$\hat{\beta}_1 = \frac{SS(xy)}{SS(x)} = \frac{714.4667}{290.9333} = 2.4558.$$

The y-intercept equals

$$\hat{\beta}_0 = \bar{y} - \hat{\beta}_1 \bar{x} = \frac{1,072}{15} - (2.4558)\frac{254}{15} = 29.88.$$

Thus, the estimated model is the least squares line

$$\hat{y} = 29.88 + 2.46x.$$

The least squares line and the 15 data points are shown in Figure 14.5. For each data point, the difference between the observed value y and the predicted value $\hat{y}$ is called the **error** (or **residual**). The size of an error equals the vertical distance from the point to the line. In Chapter 3, the **sum of the squares of the errors, $\Sigma(y - \hat{y})^2$,** was denoted by SSE, and we will continue with the use of this notation.

In Figure 14.5, the least squares line crosses the y-axis at the y-intercept of 29.88. The slope of 2.46 indicates that each 1-point increase in placement score x will result in a predicted increase of 2.46 in the final course grade.

Figure 14.5
Least squares line for the data.

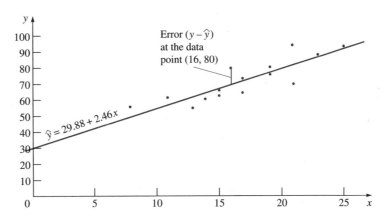

Using SSE to Estimate the Variance σ^2 of the Random Error Components

The least squares line is usually considered the "best" choice as an estimated model for describing a linear relationship between two variables x and y. Recall from Chapter 3 that the least squares line is unique in that it is *the line for which the sum of the squares of the errors, SSE, is minimized.* No other line would give a smaller value of SSE for the given data points. The importance of SSE in regression analysis extends beyond characterizing the least squares line. We will see that SSE is an essential part of all hypotheses tests and confidence intervals in this chapter.

In our investigation of the relationship between placement test scores x and final statistics grades y, we first constructed a scatter diagram to discern what type of model might be appropriate. A plot of the 15 data points suggested that a straight line relationship should be considered, so we assumed the linear model

$$y = \beta_0 + \beta_1 x + \epsilon,$$

where $(\beta_0 + \beta_1 x)$ is the deterministic part and equals μ_y, the mean course grade for a score of x. The probabilistic part of the model is ϵ, which is a random error component assumed to have a mean of 0 and a variance of σ^2. We proceeded to obtain as our estimated model the least squares line

$$\hat{y} = 29.88 + 2.46x.$$

Before this can be used to predict course grades, we need to test its utility, that is, how well it describes the relationship between x and y. Testing the usefulness of the model requires that we first obtain an estimate of σ^2, the variance of the random error components. The best estimate of σ^2 is based on SSE, the sum of the squared deviations of the y-values from the least squares line. This estimate is obtained by dividing SSE by its degrees of freedom df, where $df = (n - 2)$. In Chapter 3, SSE was calculated by using its definition. However, a much easier shortcut formula is given on the following page.

> **Estimate of the Variance σ^2 of the Random Error Components:**
>
> $$\hat{\sigma}^2 = s^2 = \frac{SSE}{df} = \frac{SSE}{n-2} \qquad (14.6)$$
>
> SSE is most easily calculated by the formula
>
> $$SSE = SS(y) - \hat{\beta}_1 SS(xy).$$
>
> *Note:*
> σ is called the **standard deviation of the model,** and s is the **estimated standard deviation of the model.**

In calculating SSE, try to retain several significant places in intermediate calculations, since excessive round off in the values of $SS(y)$, $\hat{\beta}_1$, or $SS(xy)$ can seriously affect the accuracy of SSE.

Example 14.2   Estimate σ, the standard deviation of the model, for the data pertaining to placement test scores (x) and final statistics grades (y).

Solution

In Example 14.1 we obtained

$$SS(xy) = 714.4667 \text{ and } \hat{\beta}_1 = 2.4558.$$

We also found that $\Sigma y^2 = 79{,}026$ and $\Sigma y = 1{,}072$. These two quantities are needed to find $SS(y)$.

$$SS(y) = \Sigma y^2 - \frac{(\Sigma y)^2}{n} = 79{,}026 - \frac{(1{,}072)^2}{15} = 2{,}413.7333$$

The error sum of squares can now be computed.

$$\begin{aligned} SSE &= SS(y) - \hat{\beta}_1 SS(xy) \\ &= 2{,}413.7333 - (2.4558)(714.4667) \\ &= 659.146 \end{aligned}$$

The estimated variance of the random error components is

$$\hat{\sigma}^2 = s^2 = \frac{SSE}{n-2} = \frac{659.146}{13} = 50.704.$$

Thus, the estimated standard deviation of the model is

$$s = \sqrt{50.704} = 7.12.$$

In the following sections we will see that s, the estimated standard deviation of the model, is involved in all inferential processes. At this stage, we can give a simple illustration of its significance. The unknown standard deviation of the random error components, σ, measures the variability of y with respect to the regression line

$\mu_y = \beta_0 + \beta_1 x$. Similarly, s (which estimates σ) measures the variability of the sample data points with respect to the least squares line (which estimates the regression line). By applying the Empirical rule, we would expect that roughly 95 percent of the sample y-values will lie within $2s$ units of the least squares line, that is, few data points would be expected to have an error $(y - \hat{y})$ that is larger than $2s$. For the placement scores example, $2s = 2(7.12) = 14.24$, and a check of Figure 14.5 reveals that each of the 15 points lies within this range of 2 standard deviations from the least squares line.

Sections 14.1 and 14.2 Exercises

The Simple Linear Model and Related Assumptions;
Fitting the Model by the Method of Least Squares

14.1 Explain the difference between a deterministic model and a probabilistic model.

14.2 Give an example for which a deterministic model would be more appropriate than a probabilistic model.

14.3 Explain why a probabilistic model will produce different values of y for the same value of x. Is this also true for a deterministic model? Explain.

14.4 For one day's use of a compact car, a rental company charges a fixed fee of $35.00 plus 30 cents per mile. Let x denote the number of miles traveled on a given day, and let y be the rental cost. Give the model that relates y to x, and state whether it is deterministic or probabilistic.

14.5 The service department of an appliance store charges $50 for a service call plus an hourly rate of $45.00 to repair a dishwasher in the home. Let x denote the number of hours required for such a repair and let y denote its cost. Give the model that relates y to x, and state whether it is deterministic or probabilistic.

14.6 Suppose one assumes a simple linear model to describe the relationship between x and y, and the following sample of four data points is observed.

x	2	3	4	5
y	4	6	12	10

   a. Construct a scatter diagram.
   b. Explain why the model must be probabilistic.
   c. If the model were deterministic, what would have to be the value of y when x is 4?

14.7 Consider the following data points:

x	1	3	4	6	9
y	8	9	5	1	0

   a. Make a scatter diagram.
   b. Find the least squares line for the data, and draw its graph on the scatter diagram. Does the line appear to provide a good fit to the data points?

14.8 For the data in Exercise 14.7, determine the following:
 a. SSE, the sum of the squares of the errors;
 b. s^2, the estimated variance of the random error components;
 c. s, the estimated standard deviation of the model.

14.9 Consider the following five data points:

x	0	2	5	8	10
y	1	3	4	6	10

 a. Construct a scatter diagram.
 b. Find the least squares line for the data, and draw its graph on the scatter diagram. Does the line appear to provide a good fit to the data points?
 c. Calculate the sum of the squares of the errors.
 d. Determine the estimated variance of the random error components.
 e. Obtain the estimated standard deviation of the model.

14.10 Five data points are such that $SS(y) = 218$ and $SS(xy) = 68$. The least squares line is fit to these points and its slope is 1.7. Determine SSE and s^2.

14.11 The least squares line is fit to 100 data points for which $\Sigma y^2 = 1{,}750$, $\Sigma y = 280$, $SS(x) = 310$, and $SS(xy) = -155$. Determine SSE and s^2.

14.12 The Jebsen-Taylor Hand Function Test is used to measure the recovery of coordination after traumatic injury. The following are the times after injury (in weeks) and the scores on one subtest for eight patients with similar medial nerve injuries.

Time after injury	x	3	2	5	6	2	4	10	5
Subtest score	y	6	8	5	3	7	6	3	4

 a. Construct a scatter diagram of the data.
 b. Obtain the least squares line to use as an estimated model of the relationship between x and y.
 c. Determine the estimated variance of the random error components.
 d. Determine the estimated standard deviation of the model and interpret its value.

14.13 The average prices (in dollars) of one ounce of gold and one ounce of silver for the years 1980 through 1987 are given below (U.S. Bureau of Mines, *Minerals Yearbook*).

Year	Gold (x)	Silver (y)
80	613	20.63
81	460	10.52
82	376	7.95
83	424	11.44
84	361	8.14
85	318	6.14
86	368	5.47
87	448	7.01

For the above data, $\Sigma x = 3{,}368$, $\Sigma y = 77.3$, $\Sigma x^2 = 1{,}476{,}094$, $\Sigma y^2 = 913.3636$, and $\Sigma xy = 35{,}369.65$.

a. Make a scatter diagram of the data points.

b. Obtain the least squares line to model the relationship between the price of gold and the price of silver.

c. Graph the least squares line on your scatter diagram.

14.14 For the data given in Exercise 14.13, calculate the estimated standard deviation of the model. What is the largest deviation from the least squares line that you would expect in the average price of silver for a given year?

14.15 Can physical exercise extend a person's life? This popular belief was supported in a recent study by Paffenbarger, Hyde, Wing, and Hsieh. They examined the physical activity and other life-style characteristics of 16,936 Harvard alumni for relationships to lengths of life ("Physical Activity, All-Cause Mortality, and Longevity of College Alumni," *New England Journal of Medicine*, 1986, 314). The table below gives estimates of years of added life gained by men expending 2,000 or more kcal per week on exercise, as compared with those expending less than 500 kcal.

Age at the Start of Followup (x)	Estimated Years of Added Life (y)
37	2.51
42	2.34
47	2.10
52	2.11
57	2.02
62	1.75
67	1.35
72	0.72
77	0.42

For the above data, $\Sigma x = 513$, $\Sigma y = 15.32$, $\Sigma x^2 = 30{,}741$, $\Sigma y^2 = 30.298$, and $\Sigma xy = 797.84$.

a. Construct a scatter diagram of the data points, and observe that there appears to be approximately a linear relationship between x and y.

b. Model the relationship between y and x by obtaining the least squares line, and graph this on the scatter diagram.

c. Determine SSE and s^2.

14.16 Consider the data in Exercise 14.15.

a. Calculate the estimated standard deviation of the model.

b. For each of the given x-values, determine the interval $\hat{y} \pm 2s$ and check if the observed value of y lies within this interval.

14.3 Inferences for the Slope to Assess the Usefulness of the Model

Before using an estimated model, it is necessary to check how well it describes the relationship between x and y. There are several ways that this can be tested. The procedure to be described is based on the slope β_1 of the regression line. The assumed model is

$$y = \beta_0 + \beta_1 x + \epsilon.$$

The model implies that knowledge of x contributes information for predicting y if the coefficient of x, β_1, is not 0. If β_1 is 0, then this coefficient of x indicates that knowledge of x is irrelevant for predicting y (in which case we would just use $\bar{y}$, the sample mean of the y-values). Consequently, we can test the usefulness of the model by testing the null hypothesis

$$H_0: \beta_1 = 0.$$

The test statistic is based on the least squares estimate $\hat{\beta}_1$ of the slope β_1. Under the assumptions that have been made concerning the random error components, the sampling distribution of $\hat{\beta}_1$ is normal, with a mean of β_1 and a standard deviation of $\dfrac{\sigma}{\sqrt{SS(x)}}$ (see Figure 14.6). By transforming $\hat{\beta}_1$ to standard units, we obtain the test statistic

$$\frac{\hat{\beta}_1 - \beta_1}{\dfrac{\sigma}{\sqrt{SS(x)}}}.$$

However, σ is unknown and must be replaced by its estimate s, thus resulting in the test summarized below.

Testing the Slope for Model Usefulness:
For testing the null hypothesis

$$H_0: \beta_1 = 0,$$

the test statistic is

$$t = \frac{\hat{\beta}_1 - \beta_1}{\dfrac{s}{\sqrt{SS(x)}}} = \frac{\hat{\beta}_1 \sqrt{SS(x)}}{s} \qquad (14.7)$$

The rejection region is given by the following.

For $H_a: \beta_1 < 0$, H_0 is rejected for $t < -t_\alpha$;

For $H_a: \beta_1 > 0$, H_0 is rejected for $t > t_\alpha$;

For $H_a: \beta_1 \neq 0$, H_0 is rejected for $t < -t_{\alpha/2}$ or $t > t_{\alpha/2}$.

Assumptions:
1. The model is $y = \beta_0 + \beta_1 x + \epsilon$.
2. For each possible setting of x, the random errors ϵ have a normal distribution with mean 0 and variance σ^2. Also, the errors are independent.

Note:
1. The t-distribution is based on $df = n - 2$.
2. $\hat{\beta}_1 = \dfrac{SS(xy)}{SS(x)}$.
3. $s = \sqrt{\dfrac{SSE}{n-2}}$ and $SSE = SS(y) - \hat{\beta}_1 SS(xy)$.

Figure 14.6
Sampling distribution of the
estimated slope $\hat{\beta}_1$.

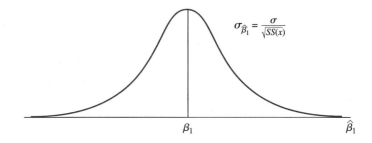

Example 14.3   For the placement scores and course grades example, is there sufficient evidence at the 5 percent level to conclude that the assumed linear model is useful and that knowledge of the placement test score x contributes information for predicting the final course grade y?

Solution

The assumed model is $y = \beta_0 + \beta_1 x + \epsilon$, and the estimated model was found to be $\hat{y} = 29.88 + 2.46x$. We also determined that $SS(x) = 290.93$ and $s = 7.12$.

Step 1: Hypotheses.

$$H_0: \beta_1 = 0$$
$$H_a: \beta_1 \neq 0$$

Step 2: Significance level.

$$\alpha = 0.05.$$

Step 3: Calculations.

$$t = \frac{\hat{\beta}_1 \sqrt{SS(x)}}{s} = \frac{2.46\sqrt{290.93}}{7.12} = 5.89$$

Step 4: Rejection region.
Since H_a is two sided, the rejection region is two tailed. Using Table 4 of Appendix A with $df = n - 2 = 13$, H_0 is rejected for $t < -t_{.025} = -2.16$ and $t > t_{.025} = 2.16$.

Step 5: Conclusion.
Since $t = 5.89$ is in the rejection region, H_0 is rejected and H_a is accepted. Thus, the data indicate that knowledge of the placement test score is useful for predicting the final course grade.

If we conclude, as in the previous example, that the slope of the linear model is not zero, then we may want to estimate it with a confidence interval. The confidence interval formula for β_1 is summarized on the following page.

A $(1 - \alpha)$ Confidence Interval for the Slope β_1:

$$\hat{\beta}_1 \pm t_{\alpha/2} \frac{s}{\sqrt{SS(x)}} \tag{14.8}$$

Assumptions:
1. The model is $y = \beta_0 + \beta_1 x + \epsilon$.
2. For each possible setting of x, the random errors ϵ have a normal distribution with mean 0 and variance σ^2. Also, all pairs of errors are independent.

Note:
1. The t-distribution is based on $df = n - 2$.
2. $\hat{\beta}_1 = \dfrac{SS(xy)}{SS(x)}$.
3. $s = \sqrt{\dfrac{SSE}{n - 2}}$ and $SSE = SS(y) - \hat{\beta}_1 SS(xy)$.

Example 14.4 Obtain a 99 percent confidence interval for the true slope of the regression line for the placement test and course grade example.

Solution

For 99 percent confidence, $t_{\alpha/2} = t_{.005}$. From Table 4 using $df = n - 2 = 13$, this value is found to be 3.012. The 99 percent confidence interval is given by the following.

$$\hat{\beta}_1 \pm t_{\alpha/2} \frac{s}{\sqrt{SS(x)}}$$

$$2.46 \pm (3.012) \frac{7.12}{\sqrt{290.93}}$$

$$2.46 \pm 1.26$$

Thus, with 99 percent confidence we estimate that the true slope of the regression line is between 1.20 and 3.72.

14.4 The Coefficients of Correlation and Determination to Measure the Usefulness of the Model

In Chapter 3, the **correlation coefficient** r was considered as a means of evaluating the usefulness of a linear model. It was defined as

$$r = \frac{SS(xy)}{\sqrt{SS(x)SS(y)}}. \tag{14.9}$$

We saw that r is a measure of the strength of the linear relationship between x and y, and r will have a value near its extremes of -1 or $+1$ when the data points lie near the least squares line (Figure 14.7 A, B). A value of r near 0 indicates the lack of a linear relationship between x and y (see Fig. 14.7C).

Figure 14.7
Correlation coefficients for
different scatter diagrams.
A, Large positive correlation;
B, Large negative correlation;
C, Lack of correlation.

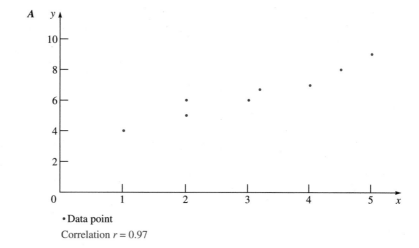

• Data point
Correlation $r = 0.97$

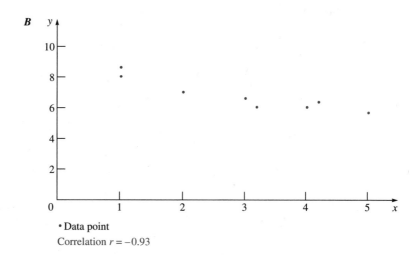

• Data point
Correlation $r = -0.93$

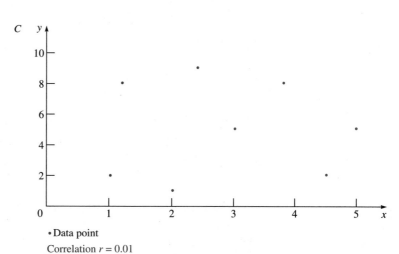

• Data point
Correlation $r = 0.01$

The correlation coefficient r between x and y is for the sample of points. The corresponding correlation coefficient for the population from which the points were selected is denoted by the Greek letter ρ (rho). In Chapter 3, we had to leave unanswered the question of how one determines if the sample correlation coefficient r is sufficiently close to -1 or $+1$ to conclude that the linear model is useful. Now that we have a background in hypothesis testing, we could address this issue by testing the following null hypothesis concerning the population's correlation coefficient ρ.

$$H_0: \rho = 0$$

A test statistic can be derived that is based on the sample correlation coefficient r. It is given by

$$t = \frac{r\sqrt{n-2}}{\sqrt{1-r^2}}. \tag{14.10}$$

This test is redundant, however, because it is exactly equivalent to the t-test of the preceding section concerning the slope β_1 of the regression line. There we tested

$$H_0: \beta_1 = 0$$

for the placement scores and test grades example. To illustrate the equivalency of the two tests, recall that in Example 14.3 the following value of the test statistic t was obtained.

$$t = \frac{\hat{\beta}_1 \sqrt{SS(x)}}{s} = \frac{2.46\sqrt{290.93}}{7.12} = 5.89$$

We could have found t by performing an equivalent calculation based on the correlation coefficient r. To do this, r must first be determined.

$$r = \frac{SS(xy)}{\sqrt{SS(x)SS(y)}} = \frac{714.47}{\sqrt{(290.93)(2{,}413.73)}} = 0.853$$

Now using Equation 14.10 to calculate t, we have

$$t = \frac{r\sqrt{n-2}}{\sqrt{1-r^2}} = \frac{0.853\sqrt{15-2}}{\sqrt{1-0.853^2}} = 5.89.$$

Because of the redundancy of this correlation test, we will continue to use the slope test for testing the existence of a linear relationship between x and y.

The Coefficient of Determination

Although the correlation coefficient r is a well-known measure of the strength of the linear relation between x and y, the significance of its value is difficult to interpret. For instance, what does a correlation of $r = 0.8$ signify, and how does it compare to a value of $r = 0.4$? A more meaningful measure of the linear relationship is given by the **coefficient of determination.** This concept can be developed in a manner similar to the partitioning of the total sum of squares that was used in Chapter 13 to introduce analysis of variance.

$SS(y)$ measures the variability in the y-values with respect to their mean $\bar{y}$, since $SS(y)$ equals the sum of the squared deviations of the y-values from $\bar{y}$. In analysis of variance terminology, $SS(y)$ is called the **total sum of squares,** and for our example we found this to be $SS(y) = 2{,}413.73$.

We have already seen that the **error sum of squares** SSE measures the variability in the y-values with respect to the least squares line, since SSE equals the sum of the squared deviations of the points from the line. For the 15 data points, we obtained a value of SSE $= 659.15$. If knowledge of x could predict y perfectly, then all the data points would lie on the least squares line and SSE would be 0. At the other extreme, if knowledge of x were completely worthless for predicting y, then SSE would equal the total sum of squares $SS(y)$. The difference between these two quantities,

$$SS(y) - \text{SSE}$$

is the reduction in the total sum of squares that can be attributed to the use of x in the model. This reduction is called the **explained variation in y,** that is, the variation in y that is being explained by the model.* For our example,

$$SS(y) - \text{SSE} = 2{,}413.73 - 659.15 = 1{,}754.58.$$

This partitioning of the total variation in y is illustrated in Figure 14.8.

Figure 14.8
Partitioning of the total variation in the y values.

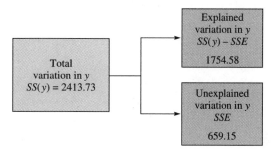

If the reduction in the total sum of squares, $SS(y) - \text{SSE}$, is divided by the total sum of squares, $SS(y)$, we obtain a ratio that gives the **proportion of the total variation in y that is explained by the model.** This ratio is called the **coefficient of determination.** It is denoted by r^2 because it can be shown to equal the square of the correlation coefficient.

The Coefficient of Determination r^2:

$$r^2 = \frac{SS(y) - \text{SSE}}{SS(y)} = 1 - \frac{\text{SSE}}{SS(y)} \qquad \textbf{(14.11)}$$

Note:
1. r^2 is the proportion of $SS(y)$, the total variation in y, that is explained by the least squares line.
2. r^2 equals the square of the correlation coefficient and, therefore, $0 \le r^2 \le 1$.

---

* Many statistical computer programs call this **regression sum of squares.**

Example 14.5

For the placement test and course grade example, find the coefficient of determination and interpret its value.

Solution

$$r^2 = 1 - \frac{SSE}{SS(y)}$$

$$= 1 - \frac{659.15}{2,413.73}$$

$$= 0.73$$

Thus, the least squares line accounts for 73 percent of the total variation in the y-values.

Note that since the correlation coefficient r had been determined earlier, we could have squared it to obtain the coefficient of determination ($r^2 = 0.853^2 = 0.73$).

At the beginning of the section we posed the question of how a correlation coefficient of $r = 0.8$ compares to a value of $r = 0.4$. We can now address this question by obtaining the corresponding coefficients of determination, which are $0.8^2 = 0.64$ and $0.4^2 = 0.16$. Thus, a least squares line with $r = 0.8$ accounts for 64 percent of the total variation in the y-values, while only 16 percent of the variation in the y-values is explained by a least squares line with $r = 0.4$.

Sections 14.3 and 14.4 Exercises

Inferences for the Slope to Assess the Usefulness of the Model; The Coefficients of Correlation and Determination to Measure the Usefulness of the Model

14.17 In Exercise 14.7, a scatter diagram was constructed for the following data. The graph suggested a linear relation between x and y.

x	1	3	4	6	9
y	8	9	5	1	0

a. Measure the strength of the linear relation by calculating the correlation coefficient r.
b. Measure the strength of the linear relation by calculating the coefficient of determination r^2.
c. Interpret the value of r^2 obtained in Part b.

14.18 Consider the following five data points:

x	7	2	6	1	5
y	8	2	5	0	4

a. Construct a scatter diagram.
b. Obtain the least squares line, and draw its graph on the scatter diagram.
c. Find the value of r^2.
d. Is there evidence of a linear relationship between x and y? Test at the 0.05 significance level.

14.19 Construct a 90 percent confidence interval to estimate the slope of the assumed model in Exercise 14.18.

14.20 The tread depth (in hundredths of an inch) and the number of miles of usage (in thousands) are given below for a sample of 10 tires of the same brand. Determine the coefficients of correlation and determination.

Miles (x)	39	37	35	15	25	38	18	60	36	40
Tread (y)	10	15	16	30	21	14	34	3	10	14

14.21 With reference to Exercise 14.20, is there sufficient evidence at the 1 percent level to conclude that x and y are negatively correlated? (Formulate the hypotheses in terms of the slope β_1.)

14.22 Find the approximate p-value for the test in Exercise 14.21.

14.23 For 100 data points, $\Sigma x = 387$, $\Sigma y = 588$, $\Sigma x^2 = 1,981$, $\Sigma y^2 = 5,023$, and $\Sigma xy = 2,530$. Find the coefficient of determination and interpret its value.

14.24 The least squares line was fit to a sample of data points for which $SS(y) = 952$ and $SSE = 207$. Determine r^2 and interpret its value.

14.25 The Jebsen-Taylor Hand Function Test is used to measure the recovery of coordination after traumatic injury. The following are the times after injury (in weeks) and the scores on one subtest for eight patients with similar medial nerve injuries.

Time after injury	x	3	2	5	6	2	4	10	5
Subtest score	y	6	8	5	3	7	6	3	4

For the above data, $\Sigma x = 37$, $\Sigma y = 42$, $\Sigma x^2 = 219$, $\Sigma y^2 = 244$, and $\Sigma xy = 165$. A scatter diagram and the least squares line were obtained in Exercise 14.12, and the scatter diagram suggested a linear relationship between x and y.
a. Measure the strength of the linear relationship by calculating the correlation coefficient r.
b. What is signified by the fact that r is negative?
c. Measure the strength of the linear relation between x and y by calculating the coefficient of determination.
d. Interpret the value of r^2 in Part c.

14.26 With reference to Exercise 14.25, is there sufficient evidence to conclude that knowledge of x contributes information for predicting y? Test at the 5 percent level.

14.27 Obtain the approximate p-value for Exercise 14.26.

14.28 Construct a 99 percent confidence interval to estimate the slope of the assumed model in Exercise 14.25.

14.29 The average prices (in dollars) of one ounce of gold and one ounce of silver for the years 1980 through 1987 are given below (U.S. Bureau of Mines, *Minerals Yearbook*).

Year	Gold (x)	Silver (y)
80	613	20.63
81	460	10.52
82	376	7.95
83	424	11.44
84	361	8.14
85	318	6.14
86	368	5.47
87	448	7.01

For the above data, $\Sigma x = 3,368$, $\Sigma y = 77.3$, $\Sigma x^2 = 1,476,094$, $\Sigma y^2 = 913.3636$, and $\Sigma xy = 35,369.65$. A scatter diagram and the least squares line were obtained in Exercise 14.13, and the scatter diagram suggested a linear relationship between x and y.

a. Calculate r^2 and interpret its value.

b. Is there sufficient evidence to conclude that knowledge of the gold price contributes information for predicting the price of silver? Test with $\alpha = 0.01$.

14.30 Find the approximate p-value for the test in Exercise 14.29.

14.31 Construct a 95 percent confidence interval to estimate the slope of the assumed model in Exercise 14.29.

14.32 Many baseball experts consider pitching as the key to a successful season. A frequently used measure of a pitching staff's effectiveness is its earned run average (ERA), which a pitcher strives to keep as low as possible. The following table gives the numbers of wins and the earned run averages of the American League teams for the 1990 season.

Team	ERA (x)	Number of Wins (y)
Oakland	3.18	103
Chicago	3.61	94
Seattle	3.69	77
Boston	3.72	88
California	3.79	80
Texas	3.83	83
Toronto	3.84	86
Kansas City	3.93	75
Baltimore	4.04	76
Milwaukee	4.08	74
Minnesota	4.12	74

(Table continues on next page)

Team	ERA (x)	Number of Wins (y)
New York	4.21	67
Cleveland	4.26	77
Detroit	4.39	79

For the above data, $\Sigma x = 54.69$, $\Sigma y = 1{,}133$, $\Sigma x^2 = 214.9087$, $\Sigma y^2 = 92{,}815$, and $\Sigma xy = 4{,}395.19$.

a. Construct a scatter diagram of the 14 data points.

b. Model the relationship between the number of wins and earned run average by obtaining the least squares line.

c. Calculate the coefficient of determination and interpret its value.

d. Is there sufficient evidence at the 1 percent level to conclude that knowledge of a team's ERA contributes information for predicting total wins?

14.33 Obtain the approximate p-value for Part d of Exercise 14.32.

14.34 Construct a 95 percent confidence interval to estimate the slope of the assumed model in Exercise 14.32.

14.5   Using the Model to Estimate and Predict

We began our study of placement test scores x and final statistics grades y by constructing a scatter diagram to determine what type of mathematical relationship might exist between x and y. The graph of the 15 data points suggested that a straight line model might be appropriate. We then assumed the linear model

$$y = \beta_0 + \beta_1 x + \epsilon$$

and we proceeded to obtain an estimated model of this by finding the least squares line

$$\hat{y} = 29.88 + 2.46x.$$

Before this estimated model could be used, we needed to check its usefulness by determining how well it describes the relation between x and y for the sample data. A test of the model's utility was conducted in Section 14.3 by performing a hypothesis test concerning the slope β_1. Further support of the model's usefulness was obtained in the previous section when the coefficient of determination was found to be $r^2 = 0.73$. We saw that this value indicates that 73 percent of the total variation in the y-values is being accounted for by the least squares line. We are now at the stage where we can use the estimated model to make inferences about final course grades. There are two types of inferences for which we want to use the least squares line:

1. **Estimating the mean value of y, μ_y, for a given value of x.**

   For example, we might want to estimate the mean course grade for all students who achieve a score of 20 on the math placement test.

2. **Predicting a single value of y for a given value of x.**

   As an illustration of this situation, suppose we have a student who scored 20 on the placement test, and we want to predict his or her final statistics grade.

In each of the two examples above, the least squares line is utilized by substituting $x = 20$ into it and evaluating $\hat{y}$. This gives the value

$$\hat{y} = 29.88 + 2.46x$$
$$= 29.88 + 2.46(20)$$
$$= 79.08$$

The same value of 79.08 is used in both cases to estimate the mean grade of all students who have a test score of 20, as well as to predict what a single student's course grade will be. The two types of problems will differ in regard to their maximum error. In other words, if a 95 percent confidence interval is used to estimate the mean grade, and a 95 percent prediction interval is used to predict the single grade, the two intervals will have different widths. Which interval do you think will be wider? Before answering, keep in mind that the size of the maximum error increases as the width of the interval increases. In Chapter 8 we saw that sample means fluctuate less than individual values and, consequently, the error associated with a mean score will be smaller than the associated error for an individual score. We can estimate more precisely the mean grade of all students with a score of 20 and predict less precisely the grade of a single student who achieves this score. A general confidence interval and a prediction interval for these two situations are given below.

A $(1 - \alpha)$ Confidence Interval for Estimating the Mean of y at $x = x_0$:

$$\hat{y} \pm t_{\alpha/2} s \sqrt{\frac{1}{n} + \frac{(x_0 - \bar{x})^2}{SS(x)}} \qquad (14.12)$$

Note:
1. The t-distribution is based on $df = n - 2$.
2. x_0 is the given value of x, and $\hat{y}$ is calculated by substituting x_0 into the least squares equation.
3. n is the number of data points, and $\bar{x}$ is the mean of their x values.
4. $s = \sqrt{\dfrac{SSE}{n - 2}}$ where $SSE = SS(y) - \hat{\beta}_1 SS(xy)$.

Assumptions:
These are the same as stated in the box for "Simple Linear Probabilistic Model and Assumptions," page 563.

A $(1 - \alpha)$ Prediction Interval for Predicting One y-Value at $x = x_0$:

$$\hat{y} \pm t_{\alpha/2} s \sqrt{1 + \frac{1}{n} + \frac{(x_0 - \bar{x})^2}{SS(x)}} \qquad (14.13)$$

Note:
1. The t-distribution is based on $df = n - 2$.
2. x_0 is the given value of x, and $\hat{y}$ is calculated by substituting x_0 into the least squares equation.

3. n is the number of data points, and $\bar{x}$ is the mean of their x-values.

4. $s = \sqrt{\dfrac{SSE}{n-2}}$, where $SSE = SS(y) - \hat{\beta}_1 SS(xy)$.

Assumptions:
These are the same as stated in the box for "Simple Linear Probabilistic Model and Assumptions," page 563.

Before illustrating the preceding formulas, we wish to point out that the term **estimate** is used in reference to a **parameter,** while the term **predict** is used when relating to a **random variable.** The value of a parameter is estimated, while the value of a random variable is predicted. Formula 14.12 above is used to estimate a mean (parameter) and is called a confidence interval. Formula 14.13 is used to predict a single value of y (random variable) and is referred to as a prediction interval. Notice that the only difference in the formulas is the inclusion of a 1 under the radical in the error bound for the prediction interval.

Example 14.6 Obtain a 95 percent confidence interval to estimate the mean statistics grade for all students who achieve a score of 20 on the placement exam.

Solution

The confidence interval for estimating the mean of y at a given value x_0 is

$$\hat{y} \pm t_{\alpha/2} s \sqrt{\frac{1}{n} + \frac{(x_0 - \bar{x})^2}{SS(x)}}.$$

The least squares line was found to be $\hat{y} = 29.88 + 2.46x$. For $x = x_0 = 20$, $\hat{y} = 29.88 + 2.46(20) = 79.08$. For 95 percent confidence, $t_{\alpha/2} = t_{.025} = 2.16$. This is obtained from Table 4 using $df = n - 2 = 13$. Earlier, we obtained $s = 7.12$, $SS(x) = 290.93$, and $\bar{x} = 254/15 = 16.93$.

$$79.08 \pm (2.16)(7.12) \sqrt{\frac{1}{15} + \frac{(20 - 16.93)^2}{290.93}}$$

$$79.08 \pm 4.84$$

Thus, with 95 percent confidence we estimate that the mean course grade for all students who score 20 on the placement test is between 74.24 and 83.92.

Example 14.7 Determine a 95 percent prediction interval for the purpose of predicting the final course grade of a student who obtains a placement test score of 20.

Solution

For predicting the course grade of a single student, the prediction interval in Formula 14.13 is used.

$$\hat{y} \pm t_{\alpha/2} s \sqrt{1 + \frac{1}{n} + \frac{(x_0 - \bar{x})^2}{SS(x)}}$$

$$79.08 \pm (2.16)(7.12) \sqrt{1 + \frac{1}{15} + \frac{(20 - 16.93)^2}{290.93}}$$

$$79.08 \pm 16.12$$

Therefore, we predict that the student's course grade will be between 62.96 and 95.20. The large width of this prediction interval renders it of little value from a practical point of view.

In the last example, note that the error bound for the prediction interval is 16.12, while the error bound for the confidence interval obtained in Example 14.6 is only 4.84. For the same amount of confidence and the same x_0, the error bound of the confidence interval for the mean of y will always be smaller than that of the prediction interval for a single y.

By examining Formulas 14.12 and 14.13 used to obtain confidence and prediction intervals, we see that in each case the error bound is smallest when the given value of x_0 is $\bar{x}$. At $x_0 = \bar{x}$ the term $\dfrac{(x_0 - \bar{x})^2}{SS(x)}$ in the error bound becomes 0. However, this term increases, and thus does the error of estimation and prediction, as x_0 is chosen farther away from $\bar{x}$. For instance, if we had chosen x_0 to be 24 instead of 20, then the error bound would increase from 4.84 to 7.51 for the mean grade and from 16.12 to 17.12 for an individual's grade. For this reason, one should be cautious in using the least squares line to estimate and predict at values of x_0 that are distant from the mean of the sample x-values. Another reason for caution is that while a straight line might be a good model near $\bar{x}$, it could be a very poor choice for values of x outside the extremes of the x-values for which the model was developed. Outside the extremes of x, curvature might be present in the actual relationship between x and y. For instance, suppose the true (but unknown) model relating y to x was the parabolic curve in Figure 14.9, but only data points with x-values between a and b were used to obtain an

Figure 14.9
Excessive curvature outside the range of the data points.

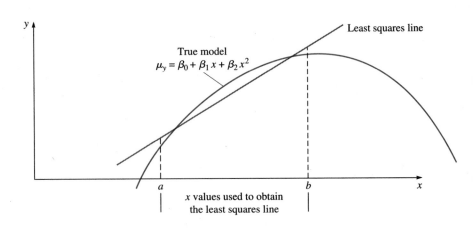

y

True model
$\mu_y = \beta_0 + \beta_1 x + \beta_2 x^2$

Least squares line

a b x

x values used to obtain
the least squares line

estimated model. Although a least squares line might suffice for values of x within this interval, the straight line would be a very poor choice for values of x that are distant from the interval.

Section 14.5 Exercises
Using the Model to Estimate and Predict

14.35 Consider the following data points.

x	-2	0	3	4	5
y	0	2	3	3	4

a. Construct a scatter diagram.
b. Fit the least squares line to the data.
c. Calculate the coefficient of determination and interpret its value.
d. Test at the 0.05 level if knowledge of x contributes information for predicting y.
e. Obtain a 95 percent confidence interval to estimate the mean of y when x is 3.
f. Obtain a 95 percent prediction interval for a single value of y when x is 3.
g. Compare the widths of the intervals obtained in Parts e and f.

14.36 A sample of 25 data points resulted in the following values: $\Sigma x = 100$, $\Sigma y = 375$, $\Sigma x^2 = 500$, $\Sigma y^2 = 6{,}125$, and $\Sigma xy = 1{,}700$.

a. Determine the least squares line.
b. Calculate the coefficient of determination and interpret its value.
c. Does knowledge of x contribute information for predicting y? Test at the 0.05 level.
d. Predict with 90 percent probability the value of y when x is 6.
e. Estimate with 90 percent confidence the mean value of y when x is 6.

14.37 A least squares line was fit to 20 data points, and the following results were obtained.

$$\text{Least squares line } \hat{y} = 21.7 + 10.8x$$

$$SS(x) = 22.2, \qquad \bar{x} = 20, \qquad SSE = 44.8$$

a. Obtain a 90 percent confidence interval for the average value of y when $x = 25$.
b. If x had been chosen as 21 in Part a, would the confidence interval be wider or narrower? Why?
c. What choice of x in Part a would result in a confidence interval of smallest width?

14.38 Refer to Exercise 14.37 and obtain a 90 percent prediction interval for y when $x = 25$.

14.39 Exercise 14.32 gave the numbers of wins and the earned run averages of the American League teams for the 1990 season. These figures are repeated on the following page.

Team	ERA (x)	Number of Wins (y)
Oakland	3.18	103
Chicago	3.61	94
Seattle	3.69	77
Boston	3.72	88
California	3.79	80
Texas	3.83	83
Toronto	3.84	86
Kansas City	3.93	75
Baltimore	4.04	76
Milwaukee	4.08	74
Minnesota	4.12	74
New York	4.21	67
Cleveland	4.26	77
Detroit	4.39	79

For the above data, $\Sigma x = 54.69$, $\Sigma y = 1{,}133$, $\Sigma x^2 = 214.9087$, $\Sigma y^2 = 92{,}815$, and $\Sigma xy = 4{,}395.19$. From Exercise 14.32, the estimated standard deviation of the model is $s = 5.583$, and the least squares line is

$$\hat{y} = 175.94 - 24.32x.$$

Estimate with 95 percent confidence the mean number of wins for teams with an earned run average of 4.08.

14.40 Refer to Exercise 14.39 and obtain a 95 percent prediction interval for the number of wins by a team whose earned run average is 4.08.

14.41 In Exercises 14.13 and 14.29, we gave the following average prices (in dollars) of one ounce of gold and one ounce of silver for the years 1980 through 1987.

Year	Gold (x)	Silver (y)
80	613	20.63
81	460	10.52
82	376	7.95
83	424	11.44
84	361	8.14
85	318	6.14
86	368	5.47
87	448	7.01

For the above data, $\Sigma x = 3{,}368$, $\Sigma y = 77.3$, $\Sigma x^2 = 1{,}476{,}094$, $\Sigma y^2 = 913.3636$, and $\Sigma xy = 35{,}369.65$. The estimated standard deviation of the model is $s = 2.203$ and the least squares line is

$$\hat{y} = -10.79 + 0.0486x.$$

For a year in which the average cost of gold is \$400 per ounce, predict with probability 0.95 the cost of an ounce of silver.

14.42 With reference to Exercise 14.41, estimate with 95 percent confidence the mean cost of an ounce of silver for years in which the average cost of gold is $400 per ounce.

14.43 The time required for a factory worker to install a certain component in a video camcorder appears to be related to the number of days of experience with this procedure. Installment times (in seconds) and the numbers of days of experience appear below for a sample of 10 workers.

Experience	x	6	8	10	8	1	3	2	5	5	7
Time required	y	32	30	25	28	39	35	40	30	33	38

a. Construct a scatter diagram of the data points.
b. Obtain the least squares line to approximate the relationship between y and x.
c. Calculate the coefficient of determination and interpret its value.
d. Test at the 5 percent level the utility of the model.
e. For a randomly selected worker with 5 days experience, predict with probability 0.95 the time required to install the component.

14.44 In Exercise 14.43, estimate with 95 percent confidence the average installation time for all workers with 5 days experience.

14.6 Conducting a Regression Analysis with MINITAB

In this section we will show how MINITAB can relieve the burden of performing many tedious and laborious calculations associated with the regression analysis of the preceding sections. In addition, the illustrations will serve to review and tie together the various steps involved in the model building process, from the initial formulation of the model through its ultimate use.

In our investigation of the relationship between placement test scores x and final statistics grades y, the first step was to discern what type of mathematical relationship might exist between the two variables. This was done by making a scatter diagram of the 15 data points. To have MINITAB accomplish this, the x- and y-values of the points are first stored in columns C1 and C2. The **PLOT** command is then applied to the two columns. In the **PLOT** command, the column of y-values is specified first, followed by the column that contains the x-values.

```
READ C1 C2
21 69
17 72
21 94
11 61
15 62
19 80
15 65
23 88
13 54          (Continued on following page)
```

```
(Continued)   19  75
              16  80
              25  93
               8  55
              14  60
              17  64
              END
              NAME C1 'X' C2 'Y'
              PLOT C2 C1
```

In response to the above commands, MINITAB will produce the output that appears in Exhibit 14.1.

Exhibit 14.1

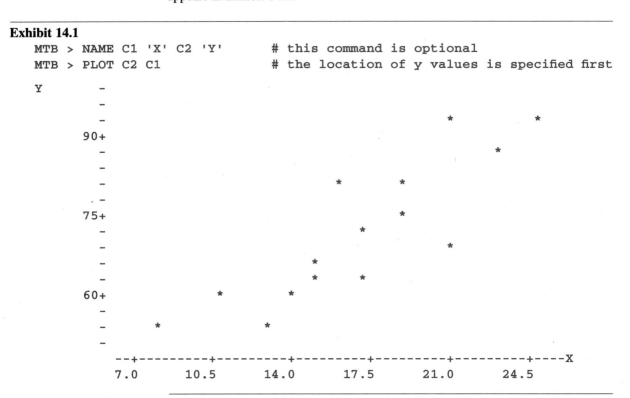

```
MTB > NAME C1 'X' C2 'Y'        # this command is optional
MTB > PLOT C2 C1                # the location of y values is specified first
```

The scatter diagram in Exhibit 14.1 suggests that test score x and course grade y might be approximately linearly related. Therefore, we tentatively assume the simple linear model

$$y = \beta_0 + \beta_1 x + \epsilon,$$

where ϵ is a random error component.

The second step is to estimate the assumed model by obtaining the least squares line. To have MINITAB produce the estimated model, type the command **REGRESS** as follows.

```
REGRESS C2 1 C1
```

In the above, the **REGRESS** command is followed by the column of y-values; then the number 1 (the number of predictor variables); and lastly, by the column of x-values. The output produced by the **REGRESS** command appears in Exhibit 14.2. In Exhibit 14.2, the least squares line is labeled as the regression equation and is

$$y = 29.9 + 2.46x.$$

Exhibit 14.2

```
MTB > REGRESS C2 1 C1

The regression equation is
Y = 29.9 + 2.46 X

Predictor        Coef        Stdev      t-ratio          p
Constant       29.882        7.304         4.09      0.001
X              2.4558       0.4175         5.88      0.000

s = 7.121        R-sq = 72.7%       R-sq(adj) = 70.6%

Analysis of Variance

SOURCE          DF          SS            MS          F          p
Regression       1       1754.6        1754.6      34.60      0.000
Error           13        659.2          50.7
Total           14       2413.7
```

If desired, additional decimal digits in the coefficients can be obtained from the "Coef" column just below the regression equation. If these values are used, then the estimated model is

$$y = 29.882 + 2.4558x.$$

The estimate of σ, the standard deviation of the random error components (standard deviation of the model), is given as $s = 7.121$. The value of s^2 also appears in the analysis of variance table. Recall that $s^2 = \text{SSE}/(n - 2)$. From the "SS" (sum of squares) column for the "Error" row, we have SSE $= 659.2$. Next to this in the "MS" (mean square) column appears the mean square for error, which is $s^2 = 50.7$ (note that this equals 7.121^2).

After constructing the scatter diagram (to assist in formulating an assumed model) and obtaining an estimated model (the least squares line), the next step in the model-building process is to assess the usefulness of the model. The coefficient of determination appears in Exhibit 14.2 as "R-sq," and its value is $r^2 = 72.7$ percent. This can also be determined from the analysis of variance table by calculating

$$r^2 = \frac{\text{Regression } SS}{\text{Total } SS} = \frac{1,754.6}{2,413.7} = 0.727.$$

Since nearly 73 percent of the variation in the y-values is being explained by the model, r^2 suggests that the fit of the least squares line to the data points is quite good. However, we can formally check this by testing the slope β_1 of the assumed model.

For testing

$$H_0: \beta_1 = 0$$

$$H_a: \beta_1 \neq 0,$$

the value of the test statistic is given in the "t-ratio" column for the row labeled "X". This is $t = 5.88$. The p-value appears to the right and equals 0.000. Since this is less than 0.05, we would reject H_0 at the 5 percent level and conclude that a straight line model is useful for describing the relationship between placement test score x and final course grade y.*

Concluding that the assumed linear model is useful, we can now utilize the estimated model (the least squares line) to estimate the mean of y or to predict an individual y for a given value of x. By using the subcommand **PREDICT** with the **REGRESS** command, MINITAB will construct a 95 percent confidence interval for μ_y and a 95 percent prediction interval for y. For example, suppose as in Section 14.5 we want to obtain a 95 percent confidence interval for the mean statistics grade of all students who achieve a score of 20 on the placement exam. In addition, we want a 95 percent prediction interval to predict the final course grade of a student who obtains a placement test score of 20. These intervals are obtained by the following commands.

```
REGRESS C2 1 C1;
PREDICT 20.
```

The main command is the same as used earlier, except that it ends with a semicolon since a subcommand will appear on the next line. The subcommand **PREDICT** is followed by the value of x. As many as 10 **PREDICT** subcommands can be used with the same **REGRESS** command. In addition to the output given earlier in Exhibit 14.2, MINITAB will also produce 95 percent confidence and prediction intervals for each of the specified x-values. At present, MINITAB has no provision for changing the confidence level from 95 percent. The output for the above commands appears in Exhibit 14.3.

The value of 79.00 in the "Fit" column is obtained by substituting $x = 20$ into the estimated model.

$$y = 29.882 + 2.4558x$$

In Section 14.5, we obtained a value of 79.08 since we used 29.88 and 2.46 for the coefficients. Because of this rounding off, MINITAB's confidence interval (74.16 to 83.84) and prediction interval (62.87 to 95.13) differ slightly from those obtained earlier.

MINITAB has also produced some output in Exhibit 14.3 that pertains to topics reserved for a more advanced course.

There are three levels of output that can be obtained with the **REGRESS** command. The maximum amount will be given by typing the following command prior to using the **REGRESS** command.

```
BRIEF 3
```

---

* The F-ratio in the analysis of variance table offers an equivalent alternative for testing the usefulness of the model. In Chapter 13, it was pointed out that the square of the value of a t-statistic equals the value of an F-statistic ($5.88^2 = 34.6$). Since $F = 34.6$ has a p-value of 0.000, we would reject H_0 and conclude that the assumed model is useful.

Exhibit 14.3

```
MTB > REGRESS C2 1 C1;
SUBC> PREDICT 20.

The regression equation is
Y = 29.9 + 2.46 X

Predictor        Coef       Stdev      t-ratio          p
Constant       29.882       7.304         4.09      0.001
X              2.4558      0.4175         5.88      0.000

s = 7.121       R-sq = 72.7%      R-sq(adj) = 70.6%

Analysis of Variance

SOURCE           DF          SS          MS          F         p
Regression        1      1754.6      1754.6      34.60     0.000
Error            13       659.2        50.7
Total            14      2413.7

      Fit   Stdev.Fit        95% C.I.            95% P.I.
    79.00        2.24   ( 74.16,  83.84)   ( 62.87,  95.13)
```

The minimum amount is achieved by using 1 in place of 3 in the command above. The output that appears in our exhibits corresponds to the default level of 2.

Section 14.6 Exercises
Conducting a Regression Analysis with MINITAB

Ⓜ 14.45 Exercise 14.15 referred to a study of the effects of physical exercise on the length of a person's life. The table below gives estimates of years of added life gained by men expending 2,000 or more kcal per week on exercise, as compared with those expending less than 500 kcal.

Age at the Start of Followup (x)	Estimated Years of Added Life (y)
37	2.51
42	2.34
47	2.10
52	2.11
57	2.02
62	1.75
67	1.35
72	0.72
77	0.42

MINITAB was used to fit a least squares line to the data, and the computer output appears on the following page.

```
MTB > NAME C1 'X' C2 'Y'
MTB > REGRESS C2 1 C1;
SUBC> PREDICT 50.

The regression equation is
Y = 4.57 - 0.0503 X

Predictor        Coef      Stdev      t-ratio         p
Constant       4.5674     0.3739        12.21     0.000
X            -0.050267   0.006398       -7.86     0.000

s = 0.2478      R-sq = 89.8%     R-sq(adj) = 88.4%

Analysis of Variance

SOURCE        DF         SS         MS         F        p
Regression     1      3.7901     3.7901     61.72    0.000
Error          7      0.4298     0.0614
Total          8      4.2200

     Fit   Stdev.Fit        95% C.I.           95% P.I.
  2.0541      0.0940   ( 1.8318, 2.2763)  ( 1.4272, 2.6809)
```

a. Give the estimated model, that is, the least squares line.
b. Give the coefficient of determination and interpret its value.
c. For testing the utility of the model, give the value of the test statistic and its p-value. Is there sufficient evidence at the 0.05 level to conclude that the model is useful?
d. Give and interpret the 95 percent confidence interval for the mean number of years of added life for men who are 50 at the start of followup.

Ⓜ 14.46 Refer to the MINITAB output in Exercise 14.45.
a. Use the analysis of variance table to determine the coefficient of determination, and compare the result with the value from Part b of Exercise 14.45.
b. Find the estimated standard deviation of the model and give an interpretation of its value.
c. Find the value of SSE. What can be said about this value when compared to the values of SSE for other straight lines that could be fit to the data?

Ⓜ 14.47 In Exercise 14.41 we gave the following average prices (in dollars) of one ounce of gold and one ounce of silver for the years 1980 through 1987.

Year	Gold (x)	Silver (y)
80	613	20.63
81	460	10.52
82	376	7.95
83	424	11.44
84	361	8.14
85	318	6.14
86	368	5.47
87	448	7.01

MINITAB was used to fit a least squares line to the data, and the computer output appears below. Use this to solve the following.

a. What is the least squares line?

b. What is the coefficient of determination? Do you think this value indicates that the model fits the data well?

c. For testing the utility of the model, give the value of the test statistic and its p-value. Is there sufficient evidence at the 0.05 level to conclude that the model is useful?

d. Find and interpret the 95 percent prediction interval for the price of an ounce of silver for a year in which the average cost of gold is $400.

e. Why would the interval obtained in Part d probably be of little practical value?

```
MTB > REGRESS C2 1 C1;
SUBC> PREDICT 400.

The regression equation is
Y = - 10.8 + 0.0486 X
```

Predictor	Coef	Stdev	t-ratio	p
Constant	-10.794	3.924	-2.75	0.033
X	0.048591	0.009134	5.32	0.002

```
s = 2.203        R-sq = 82.5%      R-sq(adj) = 79.6%
```

Analysis of Variance

SOURCE	DF	SS	MS	F	p
Regression	1	137.34	137.34	28.30	0.002
Error	6	29.12	4.85		
Total	7	166.45			

Unusual Observations

Obs.	X	Y	Fit	Stdev.Fit	Residual	St. Resid
1	613	20.630	18.992	1.919	1.638	1.51 X

X denotes an obs. whose X value gives it large influence.

Fit	Stdev.Fit	95% C.I.	95% P.I.
8.642	0.802	(6.679, 10.605)	(2.904, 14.380)

Ⓜ **14.48** Refer to the MINITAB output in Exercise 14.47.

a. What is the estimated standard deviation of the model?

b. Use the analysis of variance table to determine s, and check your result with that obtained in Part a.

Ⓜ **14.49** Refer to the MINITAB output in Exercise 14.47.

a. Obtain the coefficient of determination and interpret its value.

b. Use the analysis of variance table to determine r^2, and compare your result with that obtained in Part a.

Ⓜ **14.50** Exercise 14.32 gave the numbers of wins and the earned run averages of the American League teams for the 1990 season. The data are repeated on the following page.

Team	ERA (x)	Number of Wins (y)
Oakland	3.18	103
Chicago	3.61	94
Seattle	3.69	77
Boston	3.72	88
California	3.79	80
Texas	3.83	83
Toronto	3.84	86
Kansas City	3.93	75
Baltimore	4.04	76
Milwaukee	4.08	74
Minnesota	4.12	74
New York	4.21	67
Cleveland	4.26	77
Detroit	4.39	79

Use MINITAB to solve the following:
a. Construct a scatter diagram of the 14 data points.
b. Model the relationship between the number of wins and earned run average by obtaining the least squares line.
c. Find the coefficient of determination and interpret its value.
d. Test at the 5 percent level if there is sufficient evidence to conclude that knowledge of a team's ERA contributes information for predicting total wins.
e. Obtain a 95 percent prediction interval for the number of wins by a team whose earned run average is 4.08.

Ⓜ 14.51 *Money* magazine (March, 1990) had 50 tax professionals complete a 1040 Federal income tax return for a hypothetical family. The results were very surprising and also somewhat discouraging. Only 2 of the 50 pros came up with the correct amount of tax due. The tax assessments of the 50 preparers ranged from a low of $9,806 to a high of $21,216 (the correct tax was $12,038). In addition to considerable differences in the amounts of tax due, there was also a great deal of variability in the fees charged by the preparers. The fee for each preparer and the amount by which their result differed from the correct tax are given below. All figures are in dollars. (Adapted from Topolnicki, DM. ''The Pros Flub Our Third Annual Tax-Return Test.'' *Money,* Vol 19, No. 3, March 1990, p. 90.)

Fee	Error	Fee	Error	Fee	Error	Fee	Error
990	2,232	300	53	1,015	1,484	750	4,228
795	2,019	450	189	600	1,659	2,500	4,307
750	1,665	950	256	1,100	1,831	276	5,064
1,950	1,400	800	457	422	1,852	900	5,188
400	1,345	1,450	494	550	2,051	280	5,454
960	1,111	1,425	609	1,200	2,074	4,000	5,473

Fee	Error	Fee	Error	Fee	Error	Fee	Error
1,300	875	650	614	2,000	2,098	750	5,518
850	651	770	737	1,500	2,198	970	5,973
1,450	618	750	808	1,100	2,788	1,150	8,103
1,360	215	975	912	1,100	2,867	400	8,288
1,285	0	720	1,124	1,750	3,079	520	9,178
750	0	960	1,300	1,350	3,240		
271	54	450	1,407	640	3,338		

One might expect that the performance of a tax preparer as measured by the accuracy of the return is related to the fee charged. To check on this, use MINITAB to

a. Construct a scatter diagram of the preparations fees (x) and the errors (y).
b. Does the appearance of the scatter diagram suggest that fee and error size are correlated? Support your opinion by using MINITAB to obtain the coefficients of correlation and determination.
c. Formally check your opinion in Part (b) by testing at the 5 percent level if a linear relationship exists between the fee charged and the size of the error.

Looking Back

This chapter is concerned with investigating the relationship between two variables x and y. The ultimate goal is to obtain a mathematical model (equation) that will use knowledge of x to make inferences and predictions concerning y. To achieve this, a sample of n pairs of x and y values is obtained, and then we proceed systematically as follows to fit a model to the data.

First, a **scatter diagram** is constructed to discern what type of relationship might exist between x and y. We have only considered relationships that can be modeled by a straight line. If the scatter diagram indicates that a more complicated model is required, then one could consider a multiple regression analysis, which is an extension of the methodology of this chapter.

If the scatter diagram suggests that the data points tend to lie near a straight line, then we assume that x and y are related by the **simple linear model**

$$y = \beta_0 + \beta_1 x + \epsilon.$$

We then proceed to obtain an estimated model of this by finding the **least squares line.**

$$\hat{y} = \hat{\beta}_0 + \hat{\beta}_1 x$$

where

$$\hat{\beta}_1 = \frac{SS(xy)}{SS(x)}, \quad \text{and } \hat{\beta}_0 = \bar{y} - \hat{\beta}_1 \bar{x}.$$

The next step is to check how well the estimated model describes the relationship between x and y. The **coefficient of determination,** r^2, gives an informal indication of this, since it is the proportion of the total variation in y that is explained by the least squares line. This is given by the following.

$$r^2 = \frac{SS(y) - SSE}{SS(y)} = 1 - \frac{SSE}{SS(y)}$$

A formal test of the usefulness of the model can be accomplished by performing a hypothesis test concerning the slope of the line. To test $H_0: \beta_1 = 0$, the test statistic is

$$t = \frac{\hat{\beta}_1 \sqrt{SS(x)}}{s},$$

where $s = \sqrt{\dfrac{SSE}{n-2}}$, $SSE = SS(y) - \hat{\beta}_1 SS(xy)$, and the t-distribution has $df = n - 2$.

If H_0 is rejected, then we conclude that the linear model is useful for describing the relation between x and y.

After concluding that the model is useful, it can then be utilized to do the following.

1. Estimate the mean of y, μ_y, for a given value x_0 of x.

$$\hat{y} \pm t_{\alpha/2} s \sqrt{\frac{1}{n} + \frac{(x_0 - \bar{x})^2}{SS(x)}}$$

2. Predict a single value of y for a given value x_0 of x.

$$\hat{y} \pm t_{\alpha/2} s \sqrt{1 + \frac{1}{n} + \frac{(x_0 - \bar{x})^2}{SS(x)}}$$

Key Words

In reviewing this chapter, you should be able to define, explain, and illustrate each of the following.

simple linear model *(page 560)*

regression line *(page 561)*

scatter diagram *(page 560)*

estimated standard deviation of the model (s) *(page 568)*

correlation coefficient (r) *(page 574)*

random error component (ϵ) *(page 562)*

deterministic component *(page 562)*

probabilistic model *(page 563)*

least squares line *(page 565)*

error (residual) *(page 566)*

error sum of squares (SSE) *(page 566)*

standard deviation of the model
(σ) *(page 568)*

coefficient of determination (r^2)
(page 576)

explained variation in y *(page 577)*

confidence interval for the mean of
y *(page 582)*

prediction interval for an individual
y *(page 582)*

Ⓜ️ *MINITAB Commands*

`READ _ _` *(page 587)*

`END` *(page 588)*

`PLOT _ _` *(page 588)*

`NAME _ _` *(page 588)*

`BRIEF _` *(page 590)*

`REGRESS _ _ _` *(page 588)*

`REGRESS _ _ _;` *(page 590)*

`PREDICT _.`

Review Exercises

14.52 The following data are given.

x	10	13	8	15	5
y	14	10	15	4	20

a. Construct a scatter diagram.
b. Obtain the least squares line.
c. Calculate the coefficients of correlation and determination.
d. Does knowledge of x contribute information for the prediction of y? Test at the 5 percent level.
e. Determine the estimated standard deviation of the model.
f. Estimate the slope of the model with a 90 percent confidence interval.
g. Construct a 95 percent confidence interval to estimate the mean of y when x is 11.
h. Predict the value of y when x is 11 by constructing a 95 percent prediction interval.

14.53 A local automobile dealership pays each salesperson a weekly salary of $135 plus $75 for each car that he/she sells that week. Let x denote the number of cars sold in a week, and let y denote the salesperson's salary for that week.
a. What is the model that gives the relation between y and x?
b. Is the model deterministic or probabilistic? Explain.

14.54 The least squares line was fit to 40 points, and the following results were obtained: $\Sigma y = 680$, $\Sigma y^2 = 12,560$, and $r^2 = 0.76$. Determine:
a. the error sum of squares,
b. the estimated variance of the random error components,
c. the estimated standard deviation of the model.

14.55 The following figures give the daily attendance and the number of hot dog sales for a sample of 10 games of a minor league baseball team.

Attendance (x)	Hot Dog Sales (y)
8,747	6,845
5,857	4,168
8,360	5,348
6,945	5,687
8,688	6,007
4,534	3,216
7,450	5,018
5,874	4,652
9,821	7,002
5,873	3,897

For the data, $\Sigma x = 72,149$, $\Sigma y = 51,840$, $SS(x) = 25,378,309$, $SS(y) = 13,893,508$, and $SS(xy) = 17,604,101$.

a. Construct a scatter diagram.

b. Obtain the least squares line to model the relationship between hot dog sales and attendance.

c. Measure the strength of the linear relationship by finding the coefficient of determination, and interpret its value.

14.56 For Exercise 14.55, test the utility of the model at the 5 percent significance level, and determine the p-value for the test.

14.57 Refer to Exercise 14.55 and obtain a 95 percent confidence interval to estimate mean hot dog sales for days when the attendance is 6,000.

14.58 Refer to Exercise 14.55 and predict with probability 0.95 hot dog sales for a day when the attendance is 6,000.

14.59 The following data are the heights (inches) and weights (pounds) of a sample of 12 members of a high school football team.

Player	Height (x)	Weight (y)
1	62	135
2	68	182
3	69	168
4	73	198
5	70	174
6	68	159
7	75	221
8	72	197
9	71	182
10	70	170
11	66	154
12	77	234

For the above data, $\Sigma x = 841$, $\Sigma y = 2,174$, $\Sigma x^2 = 59,117$, $\Sigma y^2 = 402,480$, and $\Sigma xy = 153,547$.

a. Construct a scatter diagram of the data points.

b. Model the relationship between weight and height by obtaining the least squares line.

c. Calculate the coefficient of determination and interpret its value.

d. Test at the 5 percent level the utility of the model.

e. For a randomly selected player with a height of 70 inches, predict his weight with a 95 percent prediction interval.

14.60 In Wall Street terminology, the "January effect" refers to a historic pattern in which low-priced stocks such as those traded on the NASDAQ exchange often outperform the Dow Jones Industrial Average during January (The *Wall Street Journal,* December 4, 1989). The table below gives the percentage change during January for the DJIA and the NASDAQ composite during the years 1980 to 1989.

Year	DJIA % Change (x)	NASDAQ % Change (y)
80	4.4	7.0
81	− 1.7	− 2.2
82	− 0.4	− 3.8
83	2.8	6.9
84	− 3.0	− 3.6
85	6.2	12.7
86	1.6	3.3
87	13.8	12.4
88	1.0	4.3
89	7.2	4.7

a. Construct a scatter diagram of the data points.

b. Obtain the least squares line to model the relationship between the January percentage changes in the DJIA and the NASDAQ composite.

c. Calculate the correlation coefficient for x and y.

d. Test at the 0.05 level the usefulness of the model.

e. What proportion of the total variation in the y-values is being explained by the model?

14.61 Refer to Exercise 14.60 and estimate with 95 percent confidence the average percentage change in the NASDAQ composite for Januarys in which the DJIA increases by 2 percent.

14.62 Refer to Exercise 14.60 and predict with 95 percent probability the percentage change in the NASDAQ composite for a January in which the DJIA increases by 2 percent.

MINITAB Assignments

Ⓜ 14.63 Exercise 14.32 gave the number of wins and the earned run average for each of the 14 American League teams during the 1990 season. The data suggest that the number of games won by a major league team is related to the team's earned run average. Is there a relationship between the number of wins and a team's batting average? These figures are given below for the American League teams during the 1990 season.

Team	Batting Average (x)	No. of Wins (y)
Oakland	.254	103
Chicago	.258	94
Seattle	.259	77
Boston	.272	88
California	.260	80
Texas	.259	83
Toronto	.265	86
Kansas City	.267	75
Baltimore	.245	76
Milwaukee	.256	74
Minnesota	.265	74
New York	.241	67
Cleveland	.267	77
Detroit	.259	79

Use MINITAB to solve the following.
a. Construct a scatter diagram.
b. Fit a least squares line to the 14 points.
c. Find the coefficient of determination. Do you think its value indicates that the model is useful?
d. Test the utility of the model at the 5 percent significance level.

Ⓜ 14.64 The owner of an expensive gift shop believes that her weekly sales are related to the performance of the stock market for that week. To explore this possibility, she determines the amounts of sales and the means of the Dow Jones Industrial Average for 16 randomly selected weeks. These figures are given below, with the sales figures in units of $1,000. Refer to the MINITAB output to solve the following.

DJIA (x)	Sales (y)	DJIA (x)	Sales (y)
2215	58.3	1879	52.7
2518	62.9	1713	39.3
1781	46.3	2122	58.7
1823	48.2	2346	60.9
2117	58.2	1629	40.5
2703	65.8	2609	70.3
1423	36.7	1515	39.1
1532	32.3	1687	45.9

```
MTB > NAME C1 'X' C2 'Y'
MTB > REGRESS C2 1 C1;
SUBC> PREDICT 2300.

The regression equation is
Y = - 2.39 + 0.0270 X

Predictor        Coef        Stdev      t-ratio         p
Constant       -2.393        4.270        -0.56     0.584
X             0.027027     0.002119       12.75     0.000

s = 3.362       R-sq = 92.1%      R-sq(adj) = 91.5%

Analysis of Variance

SOURCE          DF          SS          MS          F         p
Regression       1        1839.1      1839.1     162.69     0.000
Error           14         158.3        11.3
Total           15        1997.3

Unusual Observations
Obs.     X        Y      Fit Stdev.Fit   Residual    St.Resid
  8    1532    32.300   39.013    1.261     -6.713      -2.15R
R denotes an obs. with a large st. resid.

    Fit    Stdev.Fit        95% C.I.            95% P.I.
  59.770     1.086    ( 57.441, 62.099)  ( 52.190, 67.350)
```

a. What is the estimated model for relating sales to the DJIA?
b. Give the coefficient of determination and interpret its value in the context of this problem.

Ⓜ 14.65 Refer to Exercise 14.64 and test at the 1 percent level if the model is useful. Give the *p*-value for the test.

Ⓜ 14.66 With reference to Exercise 14.64, give and interpret the 95 percent confidence interval for the mean amount of sales for weeks when the average DJIA is 2300.

(© Michael Salas/The Image Bank)

15 Nonparametric Tests

Looking Ahead

Most of the statistical tests in the previous chapters are based on the assumption that the sampled populations have normal distributions. Some have additional requirements pertaining to the variances of the populations. For instance, the pooled two-sample t-test for comparing two population means is quite restrictive, since it assumes that each population has a normal distribution and that the variances of the two populations are equal. In situations where little is known about the nature of a population, a researcher may be reluctant to use a test that is based on such assumptions. As an alternative, one may prefer to employ a procedure from an area of methodology that is referred to as **nonparametric statistics.** These techniques usually require few assumptions about the sampled populations. Moreover, compared to their parametric counterparts from the previous chapters, they involve simpler and fewer calculations. In fact, in many nonparametric tests the sample values are replaced by their relative ranks, which are generally whole numbers, and may thus be easier to work with than the original measurements. Some non-parametric tests, such as the sign test discussed in Sections 15.1 and 15.2, are even simpler and require little in the way of calculations.

Recall that the **power of a test** pertains to its ability to correctly reject a false null hypothesis (correctly accept a true alternative hypothesis). The major draw-back to the use of nonparametric tests is that they are less powerful than their parametric counterparts in situations for which the assumptions of the latter are true. Loosely speaking, a nonparametric test requires more evidence to "prove" the alternative hypothesis. However, the difference in power is often small. Fur-thermore, when the assumptions of a parametric test are not satisfied, a nonpara-metric test may actually be more powerful.

In this chapter we will discuss nonparametric tests that can be used in place of the one-sample test for a population mean, the independent samples test for two means, the paired-sample test for two means, and the one-way analysis of variance for several population means. In addition, we will consider a nonparametric corre-lation coefficient that offers an alternative to Pearson's product-moment correla-tion coefficient. We will also discuss a test for checking whether a sequence of observations is random.

15.1   The One-Sample Sign Test

The sign test for a single sample is a nonparametric alternative to the one-sample z- and t-tests that were discussed in Sections 10.2 and 10.3. The sign test does not require that the sampled population have a normal distribution. Its derivation only assumes that the sample is random and that the population has a continuous distribution (in practice, it is often applied to discrete distributions as well). While the one-sample z- and t-tests are used to test a hypothesis concerning a population's mean, the one-sample sign test applies to its median.* The basis for the test is the binomial distribu-tion, and it is one of the easiest nonparametric tests to apply. The one-sample sign test is illustrated in the following example.

Example 15.1   For the purpose of conducting future experiments, a psychologist is designing a maze for laboratory mice. The contention is made that the maze can be traversed in an average time of less than 8 seconds. To test this belief, 11 test runs are conducted, and the following times in seconds are recorded.

| 7.8 | 7.7 | 7.8 | 8.1 | 7.6 | 7.8 | 7.9 | 8.2 | 7.6 | 8.0 | 7.7 |

Is there sufficient evidence at the 0.05 level to support the hypothesis that the median time required to traverse the maze is less than 8 seconds?

Solution

The median of a population will be denoted by $\tilde{\mu}$ (read "mu tilde"). For this problem, $\tilde{\mu}$ represents the true median time required to complete the maze.

---
* If the population has a symmetric distribution, then the mean and the median are equal. In this case, the hypotheses in the sign test can be stated in terms of the population mean.

(© Eric Kamp/Phototake, NYC)

Step 1: Hypotheses.

$$H_0: \tilde{\mu} = 8$$

$$H_a: \tilde{\mu} < 8$$

Step 2: Significance level.

$$\alpha = 0.05.$$

Step 3: Calculations.

The test statistic is based on x, the number of values in the sample that fall above the assumed median value of 8 in the null hypothesis. To determine x, each sample value is replaced by a plus sign $(+)$ if it exceeds 8 and by a minus sign $(-)$ if it is below 8. If a value equals 8, it is discarded and the sample size n is reduced by 1. The sample of 11 measurements is thus replaced by the following signs (we have used * to indicate that the measurement has been deleted).

sample value:	7.8	7.7	7.8	8.1	7.6	7.8	7.9	8.2	7.6	8.0	7.7
sign replacement:	−	−	−	+	−	−	−	+	−	*	−

The number of plus signs is $x = 2$, and the sample size is reduced to $n = 10$ since there is 1 value that equals 8.

Step 4: Rejection region.

Under the assumption that H_0 is true, the random variable x has a binomial distribution with $p = 0.5$ and $n = 10$. Since H_a involves the $<$ relation, an unusually small value of x will provide evidence against H_0 and in support of H_a. Therefore, the rejection region is left tailed, and it is determined from the table of binomial probabilities (Table 1, Appendix A). The rejection region is shown in Figure 15.1, and it consists of the values of x in the left tail for which the cumulative probability is as close as possible to $\alpha = 0.05$. From Table 1, using $n = 10$ and $p = 0.5$, the values $x \leq 2$ have a total probability of 0.055.

Figure 15.1
Rejection region for Example 15.1.

Probability 0.055

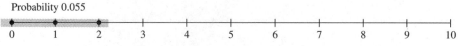

From Figure 15.1, we should reject H_0 for $x \leq 2$ since this will result in a value of $\alpha = 0.055$. This value is very close to the desired value of 0.05, and it is preferred to a rejection region of $x \leq 1$ for which α would be only 0.011. Thus, we are actually conducting the test at the $\alpha = 0.055$ significance level.

Step 5: Conclusion.
Since the value of the test statistic in Step 3 is $x = 2$, and this is in the rejection region, H_0 is rejected. Therefore, there is sufficient evidence at the $\alpha = 0.055$ significance level to conclude that the median time required to run the maze is less than 8 seconds.

For sample sizes larger than 20, Table 1 cannot be used to determine the rejection region. However, for n this large, the normal approximation to the binomial distribution can be used to produce the following test statistic.

$$z = \frac{x - \mu}{\sigma} = \frac{x - np}{\sqrt{npq}} = \frac{x - n(0.5)}{\sqrt{n(0.5)(0.5)}} = \frac{x - 0.5n}{0.5\sqrt{n}}$$

The large-sample test is summarized below.

Large-Sample Sign Test for a Population Median $\tilde{\mu}$:
For testing the null hypothesis $H_0 \colon \tilde{\mu} = \tilde{\mu}_0$ ($\tilde{\mu}_0$ is a constant), the test statistic is

$$z = \frac{x - 0.5n}{0.5\sqrt{n}}.$$ **(15.1)**

The rejection region is given by the following.

For $H_a \colon \tilde{\mu} < \tilde{\mu}_0$, H_0 is rejected when $z < -z_\alpha$;
For $H_a \colon \tilde{\mu} > \tilde{\mu}_0$, H_0 is rejected when $z > z_\alpha$;
For $H_a \colon \tilde{\mu} \neq \tilde{\mu}_0$, H_0 is rejected when $z < -z_{\alpha/2}$ or $z > z_{\alpha/2}$.

Assumption: The sample is random from a continuous population.

Note:
1. x is the number of sample observations that exceed $\tilde{\mu}_0$ (x is the number of $+$ signs).
2. Sample values equal to $\tilde{\mu}_0$ are discarded, and n is the number of values that are retained.

Example 15.2   In Example 10.6 we were given a sample of 40 dining charges per person at an expensive restaurant. These amounts (in dollars) are repeated below.

38.00	36.25	37.75	38.62	38.50	40.50	41.25	43.75	44.50	42.50
43.00	42.25	42.75	42.88	44.25	45.75	45.50	35.75	35.12	39.95
35.88	35.50	36.75	35.50	36.25	36.62	37.25	37.75	37.62	36.80
38.50	38.25	37.88	38.88	38.25	46.00	46.50	45.25	34.60	39.50

Is there sufficient evidence at the 1 percent level to conclude that the median cost per person exceeds $38.00?

Solution

Step 1: Hypotheses.

$$H_0: \tilde{\mu} = 38.00$$

$$H_a: \tilde{\mu} > 38.00$$

Step 2: Significance level.

$$\alpha = 0.01.$$

Step 3: Calculations.

Each sample observation is replaced by $+$ if it is greater than 38 and by $-$ if it falls below 38.

*	−	−	+	+	+	+	+	+	+
+	+	+	+	+	+	+	−	−	+
−	−	−	−	−	−	−	−	−	−
+	+	−	+	+	+	+	+	−	+

The number of plus signs is $x = 23$, and $n = 39$ since the first sample value equals 38 and is discarded.

$$z = \frac{x - 0.5n}{0.5\sqrt{n}} = \frac{23 - 0.5(39)}{0.5\sqrt{39}} = 1.12$$

Step 4: Rejection region.

The rejection region is right tailed since H_a involves the $>$ relation. From the z-table (Table 3, Appendix A), H_0 is rejected for $z > z_{.01} = 2.326$.

Step 5: Conclusion.

$z = 1.12$ is not in the rejection region, and H_0 is not rejected. Therefore, there is insufficient evidence at the 0.01 level to conclude that the median cost exceeds $38.00 per person.

It is interesting to note that the use of the sign test in the previous example did not result in the rejection of H_0. However, when the large-sample z-test was applied in Example 10.6 to the data, there was sufficient evidence to reject the null hypothesis. This reinforces the statement made earlier that nonparametric tests usually require more evidence than their parametric counterparts to reject H_0. In this example, the sign test only considers whether a measurement falls above or below $38.00 and it ignores the magnitude of the difference. Thus, the sign test is not fully using all the relevant information contained in the sample. In a sense, this is the "price" that we are paying for the simplicity and fewer assumptions associated with this nonparametric test.

15.2   The Two-Sample Sign Test: Paired Samples

The sign test can also be applied as a nonparametric alternative to the paired t- and z-tests that were considered in Section 11.4. Although the tested hypotheses do not involve means, the sign test can be used to determine if one population tends to yield larger or smaller values than the other population. When the sign test is applied to data consisting of paired samples, the difference is calculated for each pair, and this is replaced by either a $+$ or $-$ sign, depending on whether the difference is positive or negative. If the difference equals zero, then that pair is discarded and n, the number of pairs, is reduced by one. The test procedure is then the same as in the previous section, except that it now tests the null hypothesis that the median of the population of differences is zero.

Example 15.3   In Example 11.11, 10 infants were involved in a study to compare the effectiveness of 2 medications for the treatment of diaper rash. For each baby, 2 areas of approximately the same size and rash severity were selected, and 1 area was treated with medication A and the other with B. The number of hours until the rash disappeared was recorded for each medication and each infant. The results are repeated below.

Infant	Medication A	Medication B	Difference $D = A - B$
1	46	43	3
2	50	49	1
3	46	48	-2
4	51	47	4
5	43	40	3
6	45	40	5
7	47	47	0
8	48	44	4
9	46	41	5
10	48	45	3

Is there sufficient evidence at the 5 percent level to conclude that a difference exists in the effectiveness of the 2 medications, that is to say, that one of the medications tends to heal more quickly than the other?

Solution

The difference for each of the 10 pairs has been calculated and listed above. We will let $\tilde{\mu}_D$ denote the median of the hypothetical population of all possible differences. A value of $\tilde{\mu}_D$ that is different from 0 is supportive of the statement that 1 of the medications tends to heal more quickly than the other.

Step 1: Hypotheses.

$$H_0: \tilde{\mu}_D = 0$$

$$H_a: \tilde{\mu}_D \neq 0$$

Step 2: Significance level. $\alpha = 0.05$.

Step 3: Calculations.

The differences for the 10 pairs are replaced by the following signs:

The value of the test statistic is $x = 8$, since there are 8 plus signs. The number of pairs is reduced to $n = 9$, because the difference for pair 7 is 0, and thus is discarded.

Step 4: Rejection region.

Since the alternative hypothesis is 2 sided, the rejection region is 2 tailed. H_0 is rejected for values of x that are unusually small or large. The rejection region is shown in Figure 15.2, and is obtained from Table 1 using $n = 9$ and $p = 0.5$. It consists of those values of x in the 2 tails for which the cumulative probability is as close as possible to $\alpha = 0.05$. From Figure 15.2, by choosing 0, 1, 8, and 9 for the rejection region, α will equal 0.04, and the test will be conducted at this significance level.

Figure 15.2
Rejection region for Example 15.3.

Step 5: Conclusion.

In Step 3, the value of the test statistic is $x = 8$, and this lies in the rejection region. H_0 is thus rejected, and we conclude with $\alpha = 0.04$ that there is a difference in the healing effectiveness of the 2 medications. From an inspection of the data, it appears that medication B tends to heal more quickly.

Ⓜ**Example 15.4**   Use MINITAB to perform the sign test for the data in Example 15.3.

Solution

The two samples are stored in columns C1 and C2, and the **LET** command is used to determine the column of differences.

```
READ C1 C2
46 43
50 49
46 48
51 47
43 40
45 40
47 47
48 44
46 41
48 45
END
LET C3 = C1 - C2
```

We now apply the **STEST** command to the differences that are stored in column C3.

```
STEST 0 C3;        # 0 is the value specified in Ho
ALTERNATIVE 0.     # 0 is the code for not equal in Ha
```

The output generated from the above commands is shown in Exhibit 15.1.

The *p*-value in Exhibit 15.1 is 0.0391. Since this is less than the specified significance value of 0.05, the null hypothesis would be rejected and, thus, the alternative hypothesis accepted.

Exhibit 15.1

```
MTB > LET C3 = C1 - C2
MTB > STEST 0 C3;        # 0 is the value specified in Ho
SUBC> ALTERNATIVE 0.     # 0 is the code for not equal in Ha

SIGN TEST OF MEDIAN = 0.00000 VERSUS  N.E.   0.00000

                 N   BELOW   EQUAL   ABOVE   P-VALUE     MEDIAN
C3              10     1       1       8     0.0391      3.000
```

Sections 15.1 and 15.2 Exercises

The One-Sample Sign Test;
The Two-Sample Sign Test: Paired Samples

15.1   The following data are from a paired samples experiment. Use the sign test to determine if there is sufficient evidence at the 0.05 level to conclude that population I tends to yield smaller values than population II.

Pair	I	II
1	390	395
2	256	258
3	190	199
4	189	189
5	395	391
6	121	126
7	254	253
8	189	197
9	243	254

15.2   A random sample of size $n = 15$ produced the following values.

352 354 365 358 338 358 364 341
369 353 363 360 342 358 376

Use the sign test with $\alpha = 0.05$ to determine if sufficient evidence exists to say that the median of the sampled population differs from 350.

15.3   An administrator of a large community hospital claims that the average age of blood donors during the last 12 months has fallen below the previous value of 40 years. To check on this belief, 20 records were randomly selected from those of all donors during the last 12 months. The ages of the 20 donors are given below.

32 43 51 43 21 42 23 29 35 37
39 28 27 64 27 31 36 28 31 22

Use the sign test to determine if there is sufficient evidence at the 5 percent level to support the administrator's claim.

15.4   Determine the p-value for the test in Exercise 15.3.

15.5   A consumer prefers a particular brand of potato chips that is packaged with a label weight of 15 ounces. However, he believes that the bags tend to be underweight. In an attempt to prove this belief, 15 bags were purchased over a period of several weeks. The resulting weights are given below.

14.1 14.0 14.3 15.1 14.2 14.7 15.3 14.9
15.1 14.2 13.7 14.1 14.2 14.3 14.2

Use the sign test to determine if the data present sufficient evidence at the 0.05 level to support the consumer's belief.

15.6   The balances of a random sample of 10 savings accounts at a credit union appear below.

$1,500.89 $995.98 $1,258.87 $598.32 $1,282.58
$1,929.64 $303.76 $1,372.56 $429.76 $1,075.12

At the 10 percent level, use the sign test to determine if sufficient evidence exists to conclude that the average balance of all savings accounts at this institution exceeds $1,000.00.

15.7   Determine the p-value for the test in Exercise 15.6.

15.8   A manufacturer of air pollution controls believes that a chemical treatment of its air filters will increase the number of airborne particles collected by the filters. To test this theory, a treated and an untreated filter are installed at 16 sites. The following data are the number of airborne particles collected over a one-hour period.

Site	Treated	Untreated
1	58	50
2	27	19
3	38	29
4	45	34
5	27	22

(continued on following page)

(continued)

Site	Treated	Untreated
6	16	13
7	87	70
8	97	82
9	25	29
10	36	30
11	45	29
12	39	39
13	56	43
14	43	29
15	47	49
16	54	51

Is there sufficient evidence that the chemical treatment is effective? Use the sign test at the 0.01 significance level.

15.9 A manufacturer of platinum-tipped spark plugs believes that they last longer than conventional spark plugs. To show this, one platinum plug and one conventional plug were installed in each of six 2-cylinder engines. The effective life in hours was determined for each plug and appears below.

Engine	1	2	3	4	5	6
Platinum plug	640	570	530	410	600	580
Conventional plug	470	370	460	490	380	410

Is there sufficient evidence to support the manufacturer's contention that the platinum plugs do tend to last longer? Use the sign test and a significance level of 10 percent.

15.10 To test the claim that a certain herb lowers blood pressure, 36 people participated in an experiment. The systolic blood pressure of each was recorded initially and 90 minutes after digesting the herb. The observed blood pressures are given below.

Subject	Initial Blood Pressure	90 Minutes Later
1	117	114
2	125	129
3	108	106
4	145	143
5	139	140
6	156	151
7	143	144
8	127	124
9	143	141
10	132	129
11	103	104

Subject	Initial Blood Pressure	90 Minutes Later
12	124	125
13	142	140
14	187	182
15	105	103
16	117	114
17	121	119
18	142	139
19	165	160
20	134	131
21	111	116
22	125	121
23	118	118
24	138	139
25	132	130
26	142	135
27	128	134
28	120	117
29	154	149
30	141	140
31	119	115
32	121	117
33	153	149
34	138	139
35	167	171
36	182	175

Use the sign test and a 0.01 level of significance to determine if there is sufficient evidence to conclude that the herb is effective in lowering blood pressure.

15.11 Determine the p-value for the test in Exercise 15.10.

15.12 A certain candy bar is supposed to weigh 75 grams. Fifty bars were randomly selected from a production run, and the following weights in grams were recorded.

75.4	68.8	71.3	74.4	77.0	73.0	70.8	75.2	70.4	73.6
73.4	75.2	73.7	77.2	74.9	73.1	68.8	73.5	77.9	76.2
75.2	74.1	73.9	75.8	74.1	74.3	70.5	74.0	73.5	71.6
75.6	76.1	71.9	71.1	75.4	74.0	77.4	76.4	78.8	74.2
73.9	74.5	72.5	71.6	73.0	75.2	71.0	75.9	73.5	74.1

Use the sign test and $\alpha = 5$ percent to determine if there is sufficient evidence to conclude that the median weight of all bars in this production run differs from 75 grams.

15.13 Determine the p-value for the test in Exercise 15.12.

15.14 A shoe manufacturer has developed a new running shoe that purportedly enables an athlete to run faster. Eight adults participated in an experiment in which each ran a mile with regular track shoes and then ran a mile the next day with the new shoes. Their running times in seconds appear below.

Runner	Track Shoe	New Shoe
1	321	318
2	307	299
3	397	401
4	269	260
5	285	285
6	364	363
7	295	289
8	302	296

Is there sufficient evidence to support the manufacturer's claim? Use the sign test with $\alpha = 0.05$.

15.15 Find the approximate p-value for the test in Exercise 15.14.

15.16 As part of its admissions evaluation process, an actuarial science department administers an examination to each applicant. To determine if there is a significant difference in the grading of the exams by 2 faculty members, 10 exams were graded by both professors and the following scores were assigned:

Applicant	1	2	3	4	5	6	7	8	9	10
Professor A's score	75	87	89	63	93	54	83	71	88	71
Professor B's score	69	84	80	57	95	49	79	65	88	67

Do the data provide sufficient evidence to indicate that one professor tends to assign higher grades? Use the sign test at the 0.05 significance level.

MINITAB Assignments

Ⓜ 15.17 The following are the heights in inches of 36 blue spruce trees randomly selected from a 10-acre tree farm.

84	79	67	86	75	89	76	91	83
74	87	78	86	90	84	79	73	88
87	79	73	76	87	86	80	82	74
85	87	69	84	83	74	81	85	79

Use MINITAB to perform the sign test to determine if sufficient evidence exists at the 0.05 level to conclude that the median height of all trees on the farm exceeds 80 inches.

Ⓜ 15.18 A fish wholesaler has a catch of several thousand lobsters. A prospective buyer selected 50 at random and obtained the following weights in ounces.

21.3	21.1	21.4	18.9	20.2	19.3	19.1	18.3	19.9	22.0
20.6	20.7	21.9	20.1	17.1	19.3	21.2	18.4	21.0	21.6
16.5	18.9	17.4	20.8	18.5	18.1	21.1	19.3	21.5	20.1
21.8	20.2	19.7	18.9	19.5	20.0	18.7	21.6	20.9	21.5
17.5	16.1	20.1	21.8	19.4	21.6	23.1	20.5	22.0	20.6

The prospective buyer will purchase the entire catch if it can be shown that the average weight exceeds 19.9 ounces. Formulate a suitable set of hypotheses, and use MINITAB to perform the sign test at the 1 percent significance level.

Ⓜ 15.19   Use MINITAB to perform the sign test for the data in Exercise 15.10.

15.3   The Mann-Whitney U-Test: Independent Samples

The pooled two-sample t-test for comparing two population means was discussed in Section 11.2. The test is quite restrictive because it assumes that each population has a normal distribution and the variances of the populations are equal. The Mann-Whitney U-test offers a nonparametric alternative that is less restrictive in that the populations do not have to be normal. The test does assume, however, that the populations have continuous distributions that are identical in shape. (Like the sign test, the populations are assumed to be continuous, but in practice the Mann-Whitney U-test is frequently applied to discrete distributions as well.) The Mann-Whitney U-procedure is used to test the null hypothesis that the populations also have identical locations. The alternative hypothesis is that one of the distributions is shifted to the left or right of the other (Figure 15.3).

Figure 15.3
Identical distributions with shifted locations.

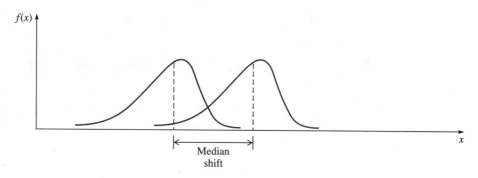

Like the pooled t-test, the Mann-Whitney U-test also requires independent random samples. The test, and all the remaining procedures in this chapter, are based on replacing the sample measurements with their relative ranks. To illustrate this concept, consider the following samples of the actual weights of "eight-ounce" bags of two brands of potato chips (the units are ounces).

Brand One: $n_1 = 10$					**Brand Two: $n_2 = 11$**				
8.47	8.32	8.20	8.41	8.38	8.13	8.28	8.31	8.24	8.17
8.42	8.29	8.45	8.45	8.29	8.10	8.27	8.26	8.04	8.21
					8.29				

* The two samples are combined into a single group of $n_1 + n_2 = 21$ values, and these are arranged in order of size from smallest to largest (to distinguish between the samples, we have underlined the values from sample one). The 21 measurements are then ranked from 1 (smallest value) to 21 (largest value). The assigned ranks appear in parentheses below each measurement.

8.04	8.10	8.13	8.17	8.20	8.21	8.24
(1)	(2)	(3)	(4)	(5)	(6)	(7)

8.26	8.27	8.28	8.29	8.29	8.29	8.31
(8)	(9)	(10)	(12)	(12)	(12)	(14)

8.32	8.38	8.41	8.42	8.45	8.45	8.47
(15)	(16)	(17)	(18)	(19.5)	(19.5)	(21)

When there are ties, the measurements are assigned the mean of the ranks that they jointly share. For instance, there are 3 values of 8.29, and these values occupy positions 11, 12, and 13 in the array. Consequently, each value of 8.29 is assigned a mean rank of $(11 + 12 + 13)/3 = 12$. There are also 2 values of 8.45, which occupy positions 19 and 20. Each value is thus assigned a mean rank of $(19 + 20)/2 = 19.5$.

For each sample, the sum of the ranks (**rank sum**) is determined. For sample one (its measurements are underlined) the rank sum is as follows.

$$R_1 = 5 + 12 + 12 + 15 + 16 + 17 + 18 + 19.5 + 19.5 + 21$$
$$= 155$$

The rank sum of sample two is as follows.

$$R_2 = 1 + 2 + 3 + 4 + 6 + 7 + 8 + 9 + 10 + 12 + 14$$
$$= 76$$

If the two samples actually come from populations with the same location, then we would expect that the ranks as determined above should be randomly dispersed among the two samples. However, if one of the rank sums is unusually small compared to the other, then this is construed as evidence that one population is shifted relative to the other and that the medians of the populations are not the same. A test statistic can be developed that is based on R_1 and R_2. This is accomplished by utilizing the following quantities.

$$U_1 = n_1 n_2 + \frac{n_2(n_2 + 1)}{2} - R_2$$

$$U_2 = n_1 n_2 + \frac{n_1(n_1 + 1)}{2} - R_1$$

Either U_1 or U_2 can be used as a test statistic. Often the test statistic is chosen to be the smaller of these two, and it is denoted by U. Special tables have been constructed that give the values of U for which the null hypothesis should be rejected. However, by standardizing U to its z-value, one can use a simpler approximate test based on the z-distribution. It works well even for values of n_1 and n_2 as small as 10. Consequently, we shall use this procedure, which is summarized below. Also, when using a z-statistic there is no advantage in calculating both U_1 and U_2 and then selecting the smaller. For the sake of simplicity, we will only calculate U_1 and use its z-value as the test statistic.

Mann-Whitney U-Test for $(\tilde{\mu}_1 - \tilde{\mu}_2)$ $(n_1 \geq 10$ and $n_2 \geq 10)$:
For testing the null hypothesis $H_0: (\tilde{\mu}_1 - \tilde{\mu}_2) = 0$, the test statistic is

$$z = \frac{U_1 - \dfrac{n_1 n_2}{2}}{\sqrt{\dfrac{n_1 n_2 (n_1 + n_2 + 1)}{12}}} \tag{15.2}$$

The rejection region is given by the following.

For $H_a: (\tilde{\mu}_1 - \tilde{\mu}_2) < 0$, H_0 is rejected when $z < -z_\alpha$;

For $H_a: (\tilde{\mu}_1 - \tilde{\mu}_2) > 0$, H_0 is rejected when $z > z_\alpha$;

For $H_a: (\tilde{\mu}_1 - \tilde{\mu}_2) \neq 0$, H_0 is rejected when $z < -z_{\alpha/2}$ or $z > z_{\alpha/2}$.

Assumptions:
1. Independent random samples.
2. The sampled populations have continuous probability distributions with the same shape.

Note:

$U_1 = n_1 n_2 + \dfrac{n_2(n_2 + 1)}{2} - R_2$, where R_2 is the rank sum for sample 2.

Example 15.5   Use the two samples of weights considered at the beginning of this section to determine if a difference exists in the median weights of the two brands of potato chips. Test at the 0.05 significance level.

Solution

Let $\tilde{\mu}_1$ and $\tilde{\mu}_2$ denote the population medians for brands 1 and 2, respectively.

Step 1: Hypotheses.

$$H_0: (\tilde{\mu}_1 - \tilde{\mu}_2) = 0$$
$$H_a: (\tilde{\mu}_1 - \tilde{\mu}_2) \neq 0$$

Step 2: Significance Level.

$$\alpha = 0.05.$$

Step 3: Calculations.
The two samples, after they have been combined and ranked, are repeated below. The underlined values are from sample 1, $n_1 = 10$, and $n_2 = 11$.

8.04	8.10	8.13	8.17	<u>8.20</u>	8.21	8.24
(1)	(2)	(3)	(4)	(5)	(6)	(7)

8.26	8.27	8.28	8.29	8.29	8.29	8.31
(8)	(9)	(10)	(12)	(12)	(12)	(14)

8.32	8.38	8.41	8.42	8.45	8.45	8.47
(15)	(16)	(17)	(18)	(19.5)	(19.5)	(21)

As determined earlier, the rank sum for the second sample is $R_2 = 76$. Next, we need the value of U_1.

$$U_1 = n_1 n_2 + \frac{n_2(n_2 + 1)}{2} - R_2$$

$$= 10(11) + \frac{11(11 + 1)}{2} - 76 = 176 - 76 = 100$$

The z-value for U_1 is used as the test statistic.

$$z = \frac{U_1 - \frac{n_1 n_2}{2}}{\sqrt{\frac{n_1 n_2 (n_1 + n_2 + 1)}{12}}}$$

$$= \frac{100 - \frac{10(11)}{2}}{\sqrt{\frac{10(11)(10 + 11 + 1)}{12}}} = \frac{45}{14.201} = 3.17$$

Step 4: Rejection region.
Since H_a contains the $\neq$ relation, the rejection region is two tailed and consists of those values for which $z < -z_{\alpha/2} = -z_{.025} = -1.96$, and those values for which $z > z_{\alpha/2} = z_{.025} = 1.96$.

Step 5: Conclusion.
$Z = 3.17$ is in the rejection region. Therefore, H_0 is rejected, and we conclude that the median weights of the 2 brands of potato chips do differ.

Ⓜ**Example 15.6**   For the two samples in Example 15.5, have MINITAB perform the Mann-Whitney U-test.

Solution

We begin by storing samples one and two in columns C1 and C2, respectively.

```
SET C1
8.47 8.32 8.20 8.41 8.38 8.42 8.29 8.45 8.45 8.29
END
SET C2
8.13 8.28 8.31 8.24 8.17 8.29 8.10 8.27 8.26 8.04 8.21
END
```

Now type the following commands.

```
MANN-WHITNEY C1 C2;
ALTERNATIVE 0.              # 0 is the code for ≠ in alt hyp
```

The resulting output appears in Exhibit 15.2.

Exhibit 15.2

```
MTB > MANN C1 C2;
SUBC> ALTE 0.

Mann-Whitney Confidence Interval and Test

C1         N =  10     Median =      8.3950
C2         N =  11     Median =      8.2400
Point estimate for ETA1-ETA2 is      0.1600
95.5 pct c.i. for ETA1-ETA2 is (0.0700,0.2400)
W = 155.0
Test of ETA1 = ETA2 vs. ETA1 n.e. ETA2 is significant at 0.0017
The test is significant at 0.0017 (adjusted for ties)
```

MINITAB uses as the test statistic the rank sum R_1, and this is labeled W in Exhibit 15.2 (W = 155). In addition to calculating the *p*-value for the test statistic, MINITAB also provides a 95 percent (approximately) confidence interval for the difference between the population medians (MINITAB denotes the population medians by ETA1 and ETA2). Since the *p*-value of 0.0017 is less than $\alpha = 0.05$, the null hypothesis would be rejected. We conclude that there is a difference in the median weights of the two brands.

For the **MANN-WHITNEY** command used in the above example, 95 percent is the default value that MINITAB uses for the confidence interval. A different value can be used by specifying it after **MANN-WHITNEY**. For instance, a 90 percent confidence interval could have been obtained in Exhibit 15.2 by typing the following.

```
MANN 90 C1 C2;
ALTE 0.
```

As we mentioned at the beginning of this section, the Mann-Whitney *U*-test assumes that the populations have an identical shape (this implies that the standard deviations must be the same). If this common shape is that of a normal distribution, then the pooled *t*-test could be used instead, and it would be slightly more powerful than the Mann-Whitney *U*-test. However, in situations for which the populations have distributions with the same nonnormal shape, the Mann-Whitney test may be considerably more powerful.

Section 15.3 Exercises

The Mann-Whitney U *Test: Independent Samples*

15.20 Independent random samples from 2 populations produced the following observations. Use the Mann-Whitney U-test to determine if the median of population 1 exceeds that of population 2. Use $\alpha = 0.10$.

Sample 1	Sample 2
158	150
127	170
138	129
145	134
127	122
116	113
187	119
197	182
150	167
127	155
149	137

15.21 Apply the Mann-Whitney U-test to the following independent random samples to determine if the medians of the sampled populations differ. Test at the 0.05 significance level.

Sample 1	56	47	52	67	58	59	51	62	63	64	70	57
Sample 2	45	63	51	48	49	51	69	57	46	71	65	

15.22 To determine if a difference exists in the average weight of apples from 2 orchards, 11 apples were randomly selected from Orchard One and 10 were selected at random from Orchard Two. Their weights in ounces are given below. Use the Mann-Whitney U-test to determine if the medians of the sampled populations differ. Test with $\alpha = 0.05$.

Orchard	Weights in Ounces										
One	4.7	5.3	5.9	4.8	5.1	6.2	6.1	6.1	5.3	6.1	4.9
Two	6.3	5.7	5.8	4.9	6.9	6.8	7.2	6.9	6.8	7.3	

15.23 Determine the p-value for the test in Exercise 15.22.

15.24 Some automobile dealers are able to sell a new car at a higher price than that advertised by adding extra charges that are often called "documentary fees." In 1990, Illinois became the first state to limit the amount of these extra charges. Suppose a state's attorney general wants to compare documentary fees charged by dealers in her state with those charged by dealers in a neighboring state. Random samples of 16 and 14 dealers were selected from the 2 states and produced the following values (all figures are in dollars).

Home State				Neighboring State			
125	188	190	175	185	150	99	110
190	135	180	195	149	125	120	139
125	125	120	225	175	145	165	
225	200	150	265	100	160	173	

Use the Mann-Whitney *U*-test to determine if there is sufficient evidence at the 5 percent level to conclude that a difference exists in the median documentary fees charged by dealers in the 2 states.

15.25 Determine the *p*-value for the test in Exercise 15.24.

15.26 To determine if a new additive improves the mileage performance of gasoline, 12 test runs were conducted with the additive, and 10 runs were made without it. The test results appear below, with all figures in miles per gallon (mpg).

With Additive	Without Additive
32.6	31.3
30.1	29.7
29.8	29.1
34.6	30.3
33.5	30.9
29.6	29.9
33.8	28.7
30.0	30.1
29.7	27.6
29.5	28.6
32.7	
33.1	

Is there sufficient evidence to conclude that the additive increases gasoline mileage? Use the Mann-Whitney *U*-test and a significance level of 5 percent.

15.27 Find the *p*-value for the test in Exercise 15.26.

15.28 Pellet stoves use wood pellets made from lumber industry waste as fuel. A stove manufacturer believes that its pellet stoves are more efficient than those of a competing brand. An independent testing laboratory measured the efficiency ratings for 10 of the manufacturer's stoves and for 12 of the competitor's stoves. The results are given on the following page.

Manufacturer	Competitor
80	74
83	75
79	79
85	80
79	72
82	74
77	80
81	77
84	78
76	78
	70
	76

Is there sufficient evidence at the 0.01 level to support the manufacturer's belief? Use the Mann-Whitney U-test.

15.29 Find the p-value for the test in Exercise 15.28.

15.30 To compare the protein content of two brands of animal feed, bags of each brand were randomly selected and analyzed. The percentages of protein for each sample are given below.

Brand 1	Brand 2
27.3	21.6
25.7	23.9
29.2	22.8
28.2	21.9
22.9	22.1
26.2	22.2
28.4	23.3
29.2	20.9
31.9	22.7
28.0	23.0
24.0	22.5
	22.4
	20.9
	21.1

Use the Mann-Whitney U-test to determine if the brands differ in their median protein content. Use a significance level of 0.05.

MINITAB Assignments

Ⓜ 15.31 Use MINITAB to apply the Mann-Whitney U-test to the samples in Exercise 15.30.

Ⓜ 15.32 An economist wanted to compare the hourly labor rates of television repairmen in two states. Service shops were randomly selected from each state, and the following hourly charges in dollars were obtained.

First State	Second State
50.00	45.00
48.00	47.00
48.00	41.00
47.00	49.00
46.00	41.50
49.00	45.00
51.50	42.50
48.00	44.00
49.50	49.00
47.50	46.00
45.00	58.00
50.00	55.00
60.00	50.00
54.50	
55.00	
49.50	

Use MINITAB to apply the Mann-Whitney *U*-test to determine whether a difference exists in the median hourly rates for these two states. Test with $\alpha = 0.05$.

15.4 The Kruskal-Wallis *H*-Test for a Completely Randomized Design

The Kruskal-Wallis *H*-test is a nonparametric alternative to the one-way analysis of variance for analyzing a completely randomized design. Chapter 13 discusses the one-way analysis of variance that is used to test the null hypothesis that k (2 or more) population means are equal. In addition to independent random samples, it assumes that the k populations have normal distributions with the same variance. The Kruskal-Wallis *H*-test is less restrictive. It assumes independent random samples, and like the Mann-Whitney *U*-test, it also assumes that the sampled populations have continuous distributions that are identical in shape. The Kruskal-Wallis *H*-procedure tests the null hypothesis that the k-distributions all have the same location. The alternative hypothesis is that at least one distribution is shifted relative to the others, and thus, the k populations do not all have the same central location (see Fig. 15.4).

In conducting the Kruskal-Wallis test, there are many similarities to the Mann-Whitney test in the preceding section. The samples are combined into a single group of $n = n_1 + n_2 + \cdots + n_k$ values, and the measurements are arranged in order of size from smallest to largest. The measurements are then ranked from 1 (smallest value) to n (largest value). In the event of ties, the tied measurements are assigned mean ranks in the same manner as described for the Mann-Whitney test. Next, the rank sum, R_i, is

Figure 15.4
$K = 3$ identical distributions with shifted locations.

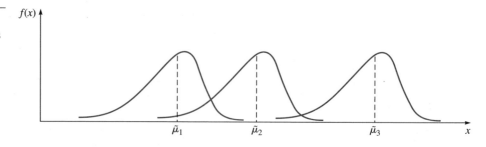

determined for each of the k samples, and a test statistic is calculated based on the values of the k rank sums. The statistic is

$$H = \frac{12}{n(n + 1)} \Sigma \frac{R_i^2}{n_i} - 3(n + 1).$$

It can be shown that if H_0 is true and the sample sizes are sufficiently large (each $n_i \geq 5$), the test statistic H has approximately a chi-square distribution with $df = k - 1$. H, moreover, measures the extent to which the ranks are randomly dispersed among the k samples. If there are great differences in the average ranks for the k samples, then H will be large, suggesting that the populations differ in their central locations. On the other hand, a small value of H indicates that the average ranks are nearly equal and thus supportive of the null hypothesis. Consequently, the rejection region is always an upper tail, and H_0 is rejected for $H > \chi_\alpha^2$. The Kruskal-Wallis H-test is summarized below.

Kruskal-Wallis H-Test for a Completely Randomized Design:
For testing the hypotheses

$$H_0: \tilde{\mu}_1 = \tilde{\mu}_2 = \cdots = \tilde{\mu}_k,$$

H_a: at least two populations have different medians,

the test statistic is

$$H = \frac{12}{n(n + 1)} \Sigma \frac{R_i^2}{n_i} - 3(n + 1). \tag{15.3}$$

The null hypothesis is rejected for $H > \chi_\alpha^2$.

Assumptions:
1. Independent random samples.
2. The sampled populations have continuous probability distributions with the same shape.
3. Each sample size n_i is at least 5.

Note:
1. The chi-square distribution has $df = k - 1$.
2. R_i is the rank sum of sample i, and $n = n_1 + n_2 + \cdots + n_k$.

Example 15.7

(© *Mark Burnett/Photo Researchers, Inc.*)

A commercial farm that grows and sells fish for consumption is testing three types of feed. Each food is fed for six weeks to one of three large pools of newly born fish of the same type. Eight fish are then randomly selected from each pool and weighed. Their weights in ounces are given below.

Diet 1	Diet 2	Diet 3
56	61	53
51	66	49
52	60	41
48	59	47
54	68	45
49	71	43
57	59	45
59	69	40

Use the Kruskal-Wallis *H*-test to determine whether one or more of the feeds tend to produce larger fish. Test at the 0.05 significance level.

Solution

Step 1: Hypotheses.

Let $\tilde{\mu}_1$, $\tilde{\mu}_2$, and $\tilde{\mu}_3$ denote the medians of the weight distributions for the three feeds.

$$H_0: \tilde{\mu}_1 = \tilde{\mu}_2 = \tilde{\mu}_3.$$

H_a: At least two of the median weights differ.

Step 2: Significance level.

$$\alpha = 0.05.$$

Step 3: Calculations.

Think of the 3 samples as being combined into one group of $n = n_1 + n_2 + n_3 = 24$ measurements, and then assign the smallest value a rank of 1, the next in size a rank of 2, and so on up to the largest, which gets a rank of 24 (ties are averaged as explained previously). The resulting assigned ranks are indicated in parentheses and given below.

Diet 1	Diet 2	Diet 3
56 (14)	61 (20)	53 (12)
51 (10)	66 (21)	49 (8.5)
52 (11)	60 (19)	41 (2)
48 (7)	59 (17)	47 (6)
54 (13)	68 (22)	45 (4.5)
49 (8.5)	71 (24)	43 (3)
57 (15)	59 (17)	45 (4.5)
59 (17)	69 (23)	40 (1)

$$R_1 = 95.5 \qquad R_2 = 163 \qquad R_3 = 41.5$$

$$n_1 = 8 \qquad n_2 = 8 \qquad n_3 = 8$$

For each feed the ranks are added, and the resulting rank sums are $R_1 = 95.5$, $R_2 = 163$, and $R_3 = 41.5$. (As a check, you can use the fact that the rank sums must add to $n(n + 1)/2 = 24(25)/2 = 300$.) The test statistic H equals the following.

$$H = \frac{12}{n(n + 1)} \Sigma \frac{R_i^2}{n_i} - 3(n + 1)$$

$$= \frac{12}{24(25)} \left(\frac{95.5^2}{8} + \frac{163^2}{8} + \frac{41.5^2}{8} \right) - 3(25)$$

$$= 18.53$$

Step 4: Rejection region.
H has approximately a chi-square distribution with $df = k - 1 = 2$. H_0 is rejected for $H > \chi_\alpha^2 = \chi_{.05}^2 = 5.99$. This value is obtained from Table 5.

Step 5: Conclusion.
Since $H = 18.53$ is in the rejection region, the null hypothesis is rejected. We therefore conclude that at least one of the feeds tends to produce larger fish.

Ⓜ **Example 15.8**   Use MINITAB to perform the Kruskal-Wallis H-test for the three samples of weights in Example 15.7.

Solution

The MINITAB command is **KRUSKAL-WALLIS** (or just **KRUS**, since only the first four letters of a command are needed). **KRUSKAL-WALLIS** requires a method of storing the data that differs from that for the analysis of variance command **AOVONEWAY**. All samples must be stacked in a single column. Consequently, we will enter in C1 the values of sample one, followed by the sample two values, and then the values of sample three. We also need to create a column C2 that identifies the sample associated with each value in C1.

```
SET C1
56 51 52 48 54 49 57 59
61 66 60 59 68 71 59 69
53 49 41 47 45 43 45 40
END
SET C2
1  1  1  1  1  1  1  1
2  2  2  2  2  2  2  2
3  3  3  3  3  3  3  3
END
```

A faster way of placing the indices 1, 2, and 3 in column C2 is to type the following.

```
SET C2
(1:3)8
END
```

Now type the following command.

```
KRUS C1 C2   # Samples are in C1 and indices are in C2
```

The output produced by MINITAB is displayed in Exhibit 15.3.

Exhibit 15.3

```
MTB > KRUS C1 C2 # Samples are in C1 and indices are in C2

LEVEL     NOBS    MEDIAN   AVE. RANK   Z VALUE
   1        8     53.00      11.9      -0.28
   2        8     63.50      20.4       3.86
   3        8     45.00       5.2      -3.58
OVERALL    24                12.5

H = 18.53   d.f. = 2   p = 0.000
H = 18.58   d.f. = 2   p = 0.000 (adj. for ties)
```

The value of the test statistic, $H = 18.53$, agrees with the value that we obtained in Example 15.7. Beside the *H*-value appears its *p*-value of 0.000, which indicates that the null hypothesis of equal medians must be rejected.

When there are ties in the data, MINITAB also computes an adjusted value of *H* (18.58 in Exhibit 15.3), which involves a slight modification of Formula 15.3.

Exhibit 15.3 also gives a *z*-value for the average rank of each sample. This is an indication of how that average rank differs from the overall mean rank (12.5) for all 24 observations.

Section 15.4 Exercises
The Kruskal-Wallis H-*Test for a Completely Randomized Design*

15.33 A completely randomized design produced the following sample results.

Sample 1	Sample 2	Sample 3
69	63	51
73	64	56
70	60	59
68	61	56
74	65	60
	67	

Use the Kruskal-Wallis *H*-test to determine if at least 2 of the sampled populations have different medians. Use $\alpha = 10$ percent.

15.34 Consider the following independent random samples.

Sample 1	Sample 2	Sample 3	Sample 4
17.9	12.3	11.1	13.3
14.2	13.8	10.1	13.9
15.7	10.9	14.2	10.8
12.7	12.7	10.3	16.5
14.9	10.2	12.1	13.3
14.0		10.5	

Test to determine if the sampled populations differ in location. Use the Kruskal-Wallis H-test and a significance level of 5 percent.

15.35 Exercise 15.30 considered a comparison of two brands of animal feed. The Mann-Whitney U-test was used on the following samples to determine if a difference exists in the median protein content of the two brands. Use the Kruskal-Wallis H-test and $\alpha = 0.05$ to perform the analysis.

Brand 1		Brand 2	
27.3	28.4	21.6	20.9
25.7	29.2	23.9	22.7
29.2	31.9	22.8	23.0
28.2	28.0	21.9	22.5
22.9	24.0	22.1	22.4
26.2		22.2	20.9
		23.3	21.1

15.36 A particular camcorder model is sold by mail order companies, camera shops, and general merchandise outlets. The manufacturer is interested in determining if the average selling price differs for these three sources. Seven stores of each type were randomly selected, and the following prices were obtained.

Type of Store	Selling Price in Dollars						
Mail order	899	929	900	979	925	950	959
Camera shop	995	935	950	979	979	995	929
General merch.	979	999	950	995	925	975	989

Use the Kruskal-Wallis H-test and $\alpha = 0.05$ to determine if a difference exists in the median prices for the types of stores.

15.37 Determine the approximate p-value for the test in Exercise 15.36.

15.38 A consumer products evaluation magazine tested three brands of flashlight batteries. Each brand was used in five flashlights, and the lights were left on until the batteries failed. The life lengths in hours are given below.

Brand	Length of Life in Hours				
A	7.3	6.9	5.8	7.9	8.2
B	6.7	7.1	6.0	6.5	5.9
C	7.5	8.3	7.9	8.4	8.3

Do the data provide sufficient evidence to indicate that at least two of the brands differ in their average length of life? Use the Kruskal-Wallis H-test and $\alpha = 0.01$.

15.39 A study was conducted to compare the four leading computer word processing programs. Each of four groups of students received training in one program for a four-week period. At the conclusion of the instruction, all participants were rated on their ability to perform a given task. Their ratings are given below.

Program 1	Program 2	Program 3	Program 4
24	29	24	25
22	27	38	23
21	21	25	24
20	25	24	25
21	22	29	23
	28		

Is there sufficient evidence to indicate differences in performance ratings for at least two of the four word processing programs? Use the Kruskal-Wallis *H*-test and a significance level of 5 percent.

15.40 Determine the approximate *p*-value for the test in Exercise 15.39.

15.41 In a state for which the price of milk is not regulated, 8 stores were randomly selected in the western, central, and eastern parts of the state. The prices charged by the 24 stores for a quart of milk appear below. Do the data provide sufficient evidence to conclude that there is a difference in the median prices for a quart of milk in the 3 geographical regions? Test at the 0.05 significance level, and use the Kruskal-Wallis *H*-test.

Region	Cost in Cents for a Quart of Milk							
Western	63	64	63	60	55	62	60	62
Central	67	62	69	68	65	65	65	66
Eastern	68	68	73	64	69	72	69	68

15.42 Radon, the second largest cause of lung cancer after smoking, is a radioactive gas produced by the natural decay of radium in the ground. Radon can seep into a home through openings in the foundation, and the EPA recommends that corrective measures be taken if levels reach 4 or more picocuries per liter (pc/l). To investigate radon levels in four public schools, an official took a sample of six readings at each school. The results, in pc/l, appear below.

School 1	School 2	School 3	School 4
1.7	5.3	5.1	1.2
1.3	4.2	4.1	2.9
0.8	3.9	5.2	1.5
1.5	5.7	3.3	2.3
0.9	3.8	2.6	2.1
1.1	5.2	4.9	1.8

Do the data suggest that radon levels tend to be higher at one or more of the schools? Use the Kruskal-Wallis *H*-test and a significance level of 5 percent.

15.43 Determine the approximate *p*-value for the test in Exercise 15.42.

MINITAB Assignments

Ⓜ **15.44** Use MINITAB to conduct the test in Exercise 15.42.

Ⓜ **15.45** A manufacturer of bond writing paper conducted a study to compare the quality of paper for 3 different sources of pulp used in the manufacturing process. For each type of pulp, 16 sheets of paper were randomly selected and assigned a quality rating. The results were as follows.

Pulp A		Pulp B		Pulp C	
78	74	73	62	79	91
69	89	65	80	87	90
75	71	78	76	84	87
89	80	76	72	98	92
76	88	73	56	87	85
78	84	65	69	92	78
65	82	69	61	91	86
78	91	74	71	87	92

Is there sufficient evidence to conclude that a difference exists in the average quality ratings for the 3 sources of pulp? Use MINITAB to apply the Kruskal-Wallis *H*-test and let $\alpha = 0.05$.

Ⓜ **15.46** With reference to Exercise 15.45, Use MINITAB to apply the Kruskal-Wallis *H*-test to determine if there is a difference in the average quality ratings for sources A and C. Test at the 0.05 significance level.

15.5   The Runs Test for Randomness

Throughout this book, whenever reference has been made to a sample for the purpose of making an inference, we have always assumed that the sample was random. The assumption of randomness is an integral part of the inferential process. In this section we will consider a very simple method for detecting a lack of randomness in a sequence of sample observations. It is based on the **theory of runs,** where a **run** is defined below.

> Consider a sequence of data, each of which is classified into one of two types. A **run** is a succession of occurrences of the same type. It is preceded and followed by either a different type or nothing at all.

To illustrate a run, consider a professor who constructs a quiz consisting of 12 true-false questions. The correct answers to questions one through twelve are as follows.

<u>T</u>   <u>F F F</u>   <u>T T</u>   <u>F</u>   <u>T T</u>   <u>F</u>   <u>T T</u>

This sequence of the two types of letters T and F contains seven runs, each of which is underlined. The theory of runs utilizes the number of runs in a sequence to check on the assumption of randomness in the data. A lack of randomness can be exhibited by too few runs that might indicate a tendency of the data to group together. For example, the smallest possible number of runs for a sequence containing seven T's and five F's is two. An example of this follows.

T T T T T T T   F F F F F

At the other extreme, a lack of randomness can be indicated by too many runs, which could suggest that the data have been altered or tampered with. For our illustration, the maximum number of possible runs is 11. One way that this occurs is in the following sequence.

T F T F T F T F T F T T

It is important to note that the test for randomness that we are about to describe is based on the order in which the observations are obtained and not on the frequency of occurrence of each type. For example, the sequence of true-false answers could consist of nine T's and three F's, and in spite of this imbalance, it might be considered random, depending on the number of runs in the sequence.

Runs Test for Randomness:

For testing the hypotheses

$$H_0: \text{the sequence of sample observations is random,}$$

$$H_a: \text{the sequence is not random,}$$

the test statistic is

$$x = \text{number of runs in the sequence.}$$

The null hypothesis is rejected if $x \le L_{\alpha/2}$ or if $x \ge R_{\alpha/2}$, where $L_{\alpha/2}$ and $R_{\alpha/2}$ are values from Table 7 that determine the left and right tails of the rejection region.

Note:
1. In Table 7, n_1 and n_2 denote the number of occurrences of the two types of elements in the sequence.
2. A blank space in the table indicates that n_1 or n_2 is not sufficiently large for a rejection region in that tail.

Example 15.9   Determine if there is sufficient evidence at the 5 percent significance level to indicate a lack of randomness in the following sequence of true-false answers:

T F F F T T F T T F T T

Solution

Step 1: Hypotheses.

H_0: The sequence of T's and F's is random.

H_a: The sequence of T's and F's is not random.

Step 2: Significance level.

$$\alpha = 0.05.$$

Step 3: Calculations.

<u>T</u> <u>F F F</u> <u>T T</u> <u>F</u> <u>T T</u> <u>F</u> <u>T T</u>

The arrangement contains $x = 7$ runs, consisting of $n_1 = 7$ T's and $n_2 = 5$ F's.

Step 4: Rejection region.
From Table 7, using the values of $n_1 = 7$, $n_2 = 5$, and a tail area of $\alpha/2 = 0.05/2 = 0.025$, we obtain $L_{\alpha/2} = L_{.025} = 3$ and $R_{\alpha/2} = R_{.025} = 11$. Therefore, the null hypothesis is rejected if $x \leq 3$ or $x \geq 11$.

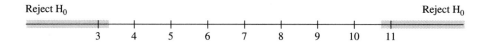

Reject H_0 Reject H_0

    3   4   5   6   7   8   9   10   11

Step 5: Conclusion.
Since $x = 7$ runs is not in the rejection region, H_0 is not rejected. Thus, there is insufficient evidence to indicate a lack of randomness in the sequence.

When n_1 or n_2 exceeds the range of values given in Table 7, an approximate z-statistic can be used. If both n_1 and n_2 are at least 10, the sampling distribution of the following statistic is approximately normal.

Large-Sample Runs Test for Randomness:
For testing the hypotheses

H_0: the sequence of sample observations is random,

H_a: the sequence is not random,

the test statistic is

$$z = \frac{x - \mu}{\sigma}. \qquad (15.4)$$

The null hypothesis is rejected if $x < -z_{\alpha/2}$ or if $x > z_{\alpha/2}$.

Assumptions:
$n_1 \geq 10$ and $n_2 \geq 10$.

Note:

1. x is the number of runs in the sequence, and

$$\mu = \frac{2n_1 n_2}{n_1 + n_2} + 1, \qquad \sigma = \sqrt{\frac{2n_1 n_2 (2n_1 n_2 - n_1 - n_2)}{(n_1 + n_2)^2 (n_1 + n_2 - 1)}}.$$

2. n_1 and n_2 denote the number of occurrences of the 2 types of elements in the sequence.

Example 15.10　The author of a mathematics text claims that the pronouns ''he'' and ''she'' are used randomly in the exercises. An examination of 40 consecutive problems in which these pronouns were used revealed the following sequence.

$$\begin{array}{cccccccccccccccccccc} \text{H} & \text{H} & \text{H} & \text{S} & \text{S} & \text{S} & \text{S} & \text{S} & \text{H} & \text{H} & \text{H} & \text{H} & \text{H} & \text{S} & \text{H} & \text{H} & \text{H} & \text{H} & \text{H} & \text{H} \\ \text{S} & \text{H} & \text{S} & \text{S} & \text{S} & \text{S} & \text{S} & \text{S} & \text{S} & \text{H} & \text{S} & \text{S} & \text{H} & \text{H} & \text{H} & \text{S} & \text{S} & \text{H} & \text{H} & \text{H} \end{array}$$

Is there sufficient evidence at the 0.05 level to conclude that the author is not using the pronouns randomly?

Solution

Step 1: Hypotheses.

H_0: The sequence of H's and S's is random.

H_a: The sequence is not random.

Step 2: Significance level.

$$\alpha = 0.05.$$

Step 3: Calculations.

$$\begin{array}{cccccccccccccccccccc} \text{H} & \text{H} & \text{H} & \text{S} & \text{S} & \text{S} & \text{S} & \text{S} & \text{H} & \text{H} & \text{H} & \text{H} & \text{H} & \text{S} & \text{H} & \text{H} & \text{H} & \text{H} & \text{H} & \text{H} \\ \underline{\text{S}} & \underline{\text{H}} & \underline{\text{S}} & \underline{\text{S}} & \underline{\text{S}} & \underline{\text{S}} & \underline{\text{S}} & \underline{\text{S}} & \underline{\text{S}} & \underline{\text{H}} & \underline{\text{S}} & \underline{\text{S}} & \underline{\text{H}} & \underline{\text{H}} & \underline{\text{H}} & \underline{\text{S}} & \underline{\text{S}} & \underline{\text{H}} & \underline{\text{H}} & \underline{\text{H}} \end{array}$$

The sequence contains $n_1 = 22$ H's and $n_2 = 18$ S's. The number of runs is $x = 13$.

$$\begin{aligned} \mu &= \frac{2n_1 n_2}{n_1 + n_2} + 1 \\ &= \frac{2(22)(18)}{22 + 18} + 1 \\ &= 20.8 \end{aligned}$$

$$\begin{aligned} \sigma &= \sqrt{\frac{2n_1 n_2 (2n_1 n_2 - n_1 - n_2)}{(n_1 + n_2)^2 (n_1 + n_2 - 1)}} \\ &= \sqrt{\frac{2(22)(18)[2(22)(18) - 22 - 18]}{(22 + 18)^2 (22 + 18 - 1)}} = 3.09 \end{aligned}$$

$$z = \frac{x - \mu}{\sigma}$$

$$= \frac{13 - 20.8}{3.09}$$

$$= -2.52$$

Step 4: Rejection region.
The rejection region is two tailed, and H_0 is rejected for $z < -z_{\alpha/2} = -z_{.025} = -1.96$ and $z > 1.96$.

Step 5: Conclusion.
From Step 3, $z = -2.52$, which is in the rejection region. Therefore, H_0 is rejected, and we conclude at the 5 percent significance level that the use of "he" and "she" is not random.

The runs test is not restricted to just a sequence of qualitative observations. It can be used to test the randomness of quantitative data by recording for each sample value whether it lies above (A) or below (B) the median. If a sample observation equals the median, it is disregarded.

Example 15.11

To illustrate the Mann-Whitney U-test in Example 15.5, two samples of weights of bags of potato chips were considered. The weights of brand two are repeated below.

8.13, 8.28, 8.31, 8.24, 8.17, 8.29, 8.10, 8.27, 8.26, 8.04, 8.21

If the weights were obtained in the order above, is there sufficient evidence at the 0.05 significance level to reject the assumption of randomness?

Solution

Step 1: Hypotheses.

$$H_0: \text{The sample is random.}$$

$$H_a: \text{The sample is not random.}$$

Step 2: Significance level.

$$\alpha = 0.05.$$

Step 3: Calculations.
The median of the sample is easily shown to be 8.24. The sample values are now classified as above (A) or below (B) 8.24 (a sample value of 8.24 is discarded).

8.13	8.28	8.31	8.24	8.17	8.29	8.10	8.27	8.26	8.04	8.21
B	A	A	—	B	A	B	A	A	B	B

The sequence of A's and B's contains $x = 7$ runs.

B A A B A B A A B B

Step 4: Rejection region.

Table 7 can be used with $n_1 = 5$ A's and $n_2 = 5$ B's. H_0 is rejected for $x \le L_{\alpha/2} = L_{.025} = 2$, and for $x \ge R_{\alpha/2} = R_{.025} = 10$.

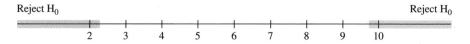

Reject H_0 Reject H_0

2   3   4   5   6   7   8   9   10

Step 5: Conclusion.

$x = 7$ runs is not in the rejection region. Thus, there is insufficient evidence at the 5 percent level to reject the assumption that the sample is random.

⑩Example 15.12

Use MINITAB to determine the number of runs above or below the median for the weights in Example 15.11.

Solution

First, store the data in column C1 and use the **MEDIAN** command to determine the median of the sample.

```
SET C1
8.13 8.28 8.31 8.24 8.17 8.29 8.10 8.27 8.26 8.04 8.21
END
MEDIAN C1
```

MINITAB will give the median value of 8.24. To obtain the number of runs above or below the median, type the following.

```
RUNS 8.24 C1     # 8.24 is the median of the data in C1
```

MINITAB will produce the output in Exhibit 15.4.

Exhibit 15.4

```
MTB > MEDIAN C1
    MEDIAN =      8.2400
MTB > RUNS 8.24 C1

    C1

    K =      8.2400

    THE OBSERVED NO. OF RUNS =    7
    THE EXPECTED NO. OF RUNS =    6.4545
      5 OBSERVATIONS ABOVE K    6 BELOW
  * N SMALL--FOLLOWING APPROX. MAY BE INVALID
                THE TEST IS SIGNIFICANT AT  0.7265
                CANNOT REJECT AT ALPHA = 0.05
```

The number of runs, 7, appears in Exhibit 15.4, and MINITAB also gives its approximate *p*-value based on a normal approximation. When there are too few observations to assure a good approximation, a warning message is printed, as has occurred in this illustration. The approximation is generally good if at least 10 observations are below the median and at least 10 are above it.

MINITAB's output for the **RUNS** test also gives the number of observations above the median (5 for the above example) and the number of observations at or below the median (6 in the above output). The latter is labeled ''BELOW'' in Exhibit 15.4.

Section 15.5 Exercises
The Runs Test for Randomness

For each of the following exercises that involve more than one row of data, the sequence of observations is row-wise.

15.47 Determine the number of runs in the following sequence of letters.

A B B B A A B A B B B B B A A B A B A A

15.48 Determine the number of runs in the following sequence.

X X X Y Y Y Y X X Y Y Y X X X Y X Y X Y

15.49 Determine the number of runs above or below the median for the following sequence of 11 numbers.

15   76   45   34   26   54   21   67   49   18   34

15.50 Construct a sequence of 15 elements consisting of the letters A and B, where the sequence contains 7 runs.

15.51 Construct a sequence of 13 numbers for which the median is 25 and there are 5 runs above or below the median.

15.52 A calculator with a random number key generated the following sequence of random digits.

9 5 6 7 2 9 5 7 0 2 4 6 3 5

Classify each digit as odd or even, and test if sufficient evidence exists to conclude a lack of randomness in the occurrence of even and odd integers. Use $\alpha = 5$ percent.

15.53 A coin was tossed 25 times, and the following sequence was observed.

H T T H H T T T H T T T T
H H T H T H T H H H T H

Do the data suggest that heads and tails are occurring in a nonrandom nanner? Use Table 7 and a significance level of 5 percent.

15.54 Use the large samples runs test to do Exercise 15.53.

15.55 A social club claims that it does not discriminate with regard to membership in its organization. During the last two years, the club has extended membership offers to 38 people, and the sexes of those who have received invitations are listed below in the order in which they were given.

M M M M M F M M M F M M M M F F M M M
M M F M M F M M M M F F F M M M M M F

Does there appear to be a lack of randomness in the sequence of sexes? Use the large samples run test and a significance level of 1 percent.

15.56 Two professors, A and B, are scheduled to teach a section of English literature during the same period next semester. The registrar's office claims that students are randomly assigned to one of the sections as they register for the course. Forty students signed up for the course, and the section assignments in the order of registration are given below.

A A A B A A B B B B B A A A A B B B B
A A A A B A A A B B B A B B B B A A A

Is there sufficient evidence at the 5 percent significance level to reject the claim that section assignments were made randomly? Use Table 7 to perform the test.

15.57 For Exercise 15.56, use the large samples runs test and a significance level of 0.05.

15.58 A self-service gasoline station sells two national brands of motor oil, B_1 and B_2. The sequence of sales for the last 30 oil purchases appears below.

B_1 B_2 B_1 B_1 B_2 B_1 B_2 B_1 B_2 B_1 B_2 B_1 B_2 B_2 B_1
B_2 B_1 B_2 B_1 B_2 B_2 B_1 B_1 B_1 B_2 B_2 B_1 B_1 B_2 B_1

Does there appear to be a lack of randomness in the order in which the brands are selected? Test at the 5 percent level.

15.59 Fifty hamburgers were purchased from a drive-in restaurant in the chronological order listed below. The following figures give the amount of fat in grams for each hamburger.

22.6	23.1	20.8	20.9	22.7	23.6	20.4	19.9	22.7	20.5
20.0	18.7	21.6	20.9	21.5	21.8	20.2	19.7	18.9	19.5
19.3	21.2	18.4	21.0	21.6	20.6	20.7	21.9	20.1	17.1
18.1	21.1	19.3	21.5	20.1	16.5	18.9	17.4	20.8	18.5
21.6	23.1	20.5	22.0	20.6	17.5	16.1	20.1	21.8	19.4

Do the data suggest nonrandomness in the fat contents? Test at the 5 percent level of significance (the median of the sample is 20.55).

MINITAB Assignments

Ⓜ 15.60 Use MINITAB to perform the test in Exercise 15.59.

Ⓜ 15.61 Use MINITAB to determine the number of runs above or below the median for the data in Exercise 15.52.

Ⓜ 15.62 The following are the ages of 70 adults who recently responded to an ad for life insurance designed for older citizens. The ages are listed in the order of response.

69	82	68	82	76	74	69	76	59	62
72	58	68	90	58	67	59	56	62	71
63	82	64	59	56	65	67	69	64	55
74	78	75	71	67	74	81	59	67	63
62	64	63	59	68	67	62	64	59	57
67	68	62	64	59	56	57	62	63	64
91	75	87	56	65	64	69	79	57	75

Use MINITAB to test if the ages occur in a nonrandom pattern. Use a 5 percent significance level.

15.6 Spearman's Rank Correlation Coefficient

The correlation coefficient that was introduced in Chapter 3 is called Pearson's product-moment correlation and is based on the assumption that x and y have a bivariate normal distribution. In the early part of this century, Charles Spearman developed a nonparametric correlation coefficient that is applicable under very general conditions. It is called **Spearman's rank correlation coefficient** and, interestingly, it is calculated with the same formula used for finding Pearson's product-moment correlation coefficient. Spearman's correlation coefficient, however, uses the ranks of the samples, rather than the actual measurements.

To illustrate the calculation of Spearman's rank correlation coefficient, r_s, suppose that a consumer testing group evaluated seven brands of peanut butter on the basis of several factors such as taste, texture, aroma, color, and so on. The total number of points assigned to each brand and its retail price are listed below.

Brand	Price	Rating
1	$1.69	65
2	$1.89	83
3	$1.85	89
4	$1.29	52
5	$1.49	79
6	$1.25	58
7	$1.45	67

We will calculate r_s to measure the strength of the relation between a brand's rating and its price. To determine the rank correlation coefficient, first replace the prices by their relative ranks, and then do the same for the ratings. Ties are handled by assigning mean ranks to tied values as described in Section 15.3. The two ranks for each brand are given below, where the ranks are denoted by x and y.

Brand	Price (x)	Rating (y)	xy	x^2	y^2
1	5	3	15	25	9
2	7	6	42	49	36
3	6	7	42	36	49
4	2	1	2	4	1
5	4	5	20	16	25
6	1	2	2	1	4
7	3	4	12	9	16
	28	28	135	140	140

The rank order correlation coefficient r_s can be calculated with the same formula that was used in Chapter 3 for r.

$$r_s = \frac{SS(xy)}{\sqrt{SS(x)SS(y)}} = \frac{\Sigma xy - \frac{(\Sigma x)(\Sigma y)}{n}}{\sqrt{\left[\Sigma x^2 - \frac{(\Sigma x)^2}{n}\right]\left[\Sigma y^2 - \frac{(\Sigma y)^2}{n}\right]}}$$

$$= \frac{135 - \frac{(28)(28)}{7}}{\sqrt{\left[140 - \frac{(28)^2}{7}\right]\left[140 - \frac{(28)^2}{7}\right]}} = \frac{23}{\sqrt{(28)(28)}} = 0.821$$

Note that $\Sigma x = \Sigma y$, $\Sigma x^2 = \Sigma y^2$, and $SS(x) = SS(y)$. When there are no ties in the values of each sample, these relations always hold. In fact, in this case there is a much simpler formula that can be used to calculate r_s. This is given on the next page, and it is recommended that you always use this formula except in situations for which there are several ties.

Spearman's Rank Correlation Coefficient:

$$r_s = \frac{SS(xy)}{\sqrt{SS(x)SS(y)}},$$

where

$$SS(xy) = \Sigma xy - \frac{(\Sigma x)(\Sigma y)}{n},$$

$$SS(x) = \Sigma x^2 - \frac{(\Sigma x)^2}{n},$$

$$SS(y) = \Sigma y^2 - \frac{(\Sigma y)^2}{n}.$$

When there are no ties in the values of each sample, the following shortcut formula is algebraically equivalent.

$$r_s = 1 - \frac{6\Sigma d_i^2}{n(n^2 - 1)},$$

where

$$d_i = x_i - y_i.$$

Note:
1. x_i is the rank of the *i*th observation in sample 1, y_i is the rank of the *i*th observation in sample 2, n is the number of pairs.
2. Unless there are several ties, it is recommended that the shortcut formula be used.

Example 15.13   Use the shortcut formula to calculate the rank correlation coefficient for the peanut butter ratings and prices.

Solution

The relative ranks for the prices and the ratings are copied below. We have also added a column of differences in the ranks and a column of their squares.

Brand	Price (x)	Rating (y)	$d = x - y$	d^2
1	5	3	2	4
2	7	6	1	1
3	6	7	-1	1
4	2	1	1	1
5	4	5	-1	1
6	1	2	-1	1
7	3	4	-1	1
				$\Sigma d^2 = 10$

$$r_s = 1 - \frac{6\Sigma d^2}{n(n^2 - 1)} = 1 - \frac{6(10)}{7(49 - 1)} = 0.821$$

The rank correlation coefficient r_s can be used as the test statistic for testing the null hypothesis that there is no association between two variables. Letting ρ_s denote the true rank correlation between the variables, the null hypothesis is $H_0: \rho_s = 0$, and this is rejected if the value of r_s differs significantly from 0. Critical values that define the rejection region have been tabulated, and some of these values are given in Table 8 of Appendix A. The hypothesis test is described below and is illustrated in Example 15.14.

Spearman's Test for Rank Correlation:

For testing the hypothesis

$$H_0: \rho_s = 0 \quad \text{(there is no association between the variables)}$$

the test statistic is

$$r_s = 1 - \frac{6\Sigma d_i^2}{n(n^2 - 1)}. \tag{15.5}$$

The rejection region is given by the following.

For $H_a: \rho_s < 0$, H_0 is rejected when $r_s < -r_\alpha$;

For $H_a: \rho_s > 0$, H_0 is rejected when $r_s > r_\alpha$;

For $H_a: \rho_s \neq 0$, H_0 is rejected when $r_s < -r_{\alpha/2}$ or $r_s > r_{\alpha/2}$.

Assumption:
The sample of pairs is random.

Note:
1. The number of pairs is n, and d_i is the difference between ranks for the two sample values in the ith pair.
2. The value of r_α or $r_{\alpha/2}$ that determines the rejection region is found from Table 8.

Example 15.14   In a 1988 special report on "America's Best Colleges," *U.S. News & World Report* gave the following tuition charges and the average annual pay (1987–88) for full professors at the following large universities. With $\alpha = 5$ percent, use Spearman's test to determine if the sample correlation coefficient is significantly different from zero.

School	Tuition	Avg. Pay
Calif. Inst. of Tech.	$11,789	$69,900
Columbia University	$12,878	$64,800
Georgetown University	$11,990	$64,300
Harvard University	$13,665	$73,200
Princeton University	$13,380	$67,800
Stanford University	$12,564	$70,800
University of Chicago	$12,200	$64,200
Univ. of Cal./Berkeley	$ 6,037	$64,200
Univ. of Pennsylvania	$12,750	$64,300
Yale University	$12,960	$67,700

Solution

Step 1: Hypotheses.

$$H_0: \rho_s = 0 \quad \text{(tuition and salary are not associated)}$$

$$H_a: \rho_s \neq 0 \quad \text{(tuition and salary are associated)}$$

Step 2: Significance level.

$$\alpha = 0.05.$$

Step 3: Calculations.
The 10 tuitions are replaced by their ranks, and the same is done for the salaries.

School	Tuition(x)	Pay(y)	d	d^2
Calif. Inst. of Tech.	2	8	−6	36
Columbia University	7	5	2	4
Georgetown University	3	3.5	−0.5	0.25
Harvard University	10	10	0	0
Princeton University	9	7	2	4
Stanford University	5	9	−4	16
University of Chicago	4	1.5	2.5	6.25
Univ. of Cal./Berkeley	1	1.5	−0.5	0.25
Univ. of Pennsylvania	6	3.5	2.5	6.25
Yale University	8	6	2	4
				$\Sigma d^2 = 77$

$$r_s = 1 - \frac{6\Sigma d^2}{n(n^2 - 1)} = 1 - \frac{6(77)}{10(100 - 1)} = 0.533$$

Step 4: Rejection region.
The rejection region is two tailed because the relation in H_a is $\neq$. Table 8 is used with $n = 10$ and a tail area of $\alpha/2 = 0.025$. The value of $r_{.025}$ is 0.648. Therefore, H_0 is rejected for $r < -0.648$ and for $r > 0.648$.

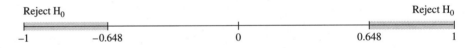

Reject H_0 Reject H_0

−1   −0.648   0   0.648   1

Step 5: Conclusion.
Since $r_s = 0.533$ is not in the rejection region, the null hypothesis is not rejected. There is insufficient evidence at the 0.05 level to conclude that there is an association between salary and tuition.

Ⓜ**Example 15.15**

MINITAB does not have an explicit command for calculating Spearman's rank correlation coefficient. However, it can easily be determined by using the **RANK** and **CORRELATION** commands. We will illustrate by using the tuition and salaries data presented in Example 15.14. We begin by entering the data in columns C1 and C2.

```
READ C1 C2
11789 69900
12878 64800
11990 64300
13665 73200
13380 67800
12564 70800
```

```
12200 64200
 6037 64200
12750 64300
12960 67700
END
```

We now need to obtain the relative ranks for the numbers in C1 and C2. The **RANK** command accomplishes this.

```
RANK C1 C3    # ranks nos. in C1 and puts ranks in C3
RANK C2 C4
```

The ranks that have been created in columns C3 and C4 are shown in Exhibit 15.5. With these columns of ranks, Spearman's rank correlation can be obtained by applying the **CORRELATION** command to C3 and C4. The results appear in Exhibit 15.5. Notice that the rank correlation of 0.530 differs slightly from the value of 0.533 that was obtained in Example 15.14. This is because there are ties in the column of salaries, and in Example 15.14 the shortcut formula was used. Since MINITAB uses the exact formula, there is a small discrepancy in the results.

Exhibit 15.5

```
MTB > PRINT C3 C4

ROW    C3      C4

 1      2     8.0
 2      7     5.0
 3      3     3.5
 4     10    10.0
 5      9     7.0
 6      5     9.0
 7      4     1.5
 8      1     1.5
 9      6     3.5
10      8     6.0

MTB > CORRELATION C3 C4

Correlation of C3 and C4 = 0.530
```

Section 15.6 Exercises

Spearman's Rank Correlation Coefficient

15.63 Use Table 8 to determine the rejection region for testing the hypotheses

$$H_0: \rho_s = 0$$

$$H_a: \rho_s \neq 0$$

where $\alpha = 0.05$ and $n = 24$.

15.64 Use Table 8 to determine the rejection region for detecting a positive rank correlation at the significance level $\alpha = 0.01$. Assume there are 20 pairs of data.

15.65 Use Table 8 to determine the rejection region for detecting a negative rank correlation at the significance level $\alpha = 0.05$, where $n = 13$ points.

15.66 Calculate Spearman's rank correlation coefficient for the following data.

x	37	32	36	31	35
y	88	82	85	80	84

15.67 The table below gives the populations of the 8 most populous metropolitan areas in the United States in 1990. Also given is each area's population rank in 1980 (The *Wall Street Journal*, March 3, 1990). Rank the 1990 populations from high (assign a rank of 1) to low (assign a rank of 8), and calculate the correlation coefficient of the population ranks for 1980 and 1990.

Metropolitan Area	1990 Population (in thousands)	1980 Population Rank
L.A./Long Beach	8,771	2
New York	8,625	1
Chicago	6,308	3
Philadelphia	4,973	4
Detroit	4,409	5
Washington, DC	3,710	7
Houston	3,509	8
Boston	2,837	6

15.68 Do the following data provide sufficient evidence of an association between x and y? Use Spearman's rank correlation coefficient, and test at the 10 percent significance level.

x	y
185	65
437	32
287	57
930	21
487	43
431	39
846	25

15.69 The following sample gives the tread depth (in hundredths of an inch) and the number of miles of usage (in thousands) for a sample of 10 tires of the same brand. In Exercise 14.20, the reader was asked to calculate Pearson's correlation coefficient r for the data ($r = -0.909$). Calculate Spearman's rank correlation coefficient r_s.

Miles (x)	39	37	35	15	25	38	18	60	36	40
Tread (y)	10	15	16	30	21	14	34	3	10	14

15.70 Use the results of Exercise 15.69 to determine if there is sufficient evidence to conclude that mileage and tread depth are negatively correlated. Test with $\alpha = 5$ percent.

15.71 Two computer magazines performed independent evaluations of 9 personal computer systems. Each magazine ranked the systems from best to worst, and their rankings appear in the table below. Is there sufficient evidence of a positive correlation between the rankings of the 2 magazines? Test at the 5 percent significance level.

Computer System	First Magazine's Ratings	Second Magazine's Ratings
1	6	7
2	2	3
3	8	9
4	3	2
5	1	1
6	4	5
7	9	8
8	5	4
9	7	6

15.72 Determine the approximate p-value for the test in Exercise 15.71.

15.73 A professor conducted a study to determine if there is an association between the final examination scores in her course and the times required to complete the exam. Exam scores and times (in minutes) appear below for 15 students.

Student	Exam Score	Time
1	86	108
2	93	100
3	73	115
4	78	113
5	54	118
6	93	99
7	69	110
8	78	109
9	84	111
10	82	117
11	41	120
12	67	116
13	98	89
14	74	112
15	71	110

Use Spearman's rank correlation coefficient to determine if there is an association between the exam score and the time required to complete the exam. Test at the 1 percent significance level.

15.74 Determine the approximate p-value for the test in Exercise 15.73.

MINITAB Assignments

Ⓜ 15.75 Use MINITAB to calculate the rank correlation coefficient for the data in Exercise 15.73.

Ⓜ 15.76 The chapter opener referred to a study conducted by *MONEY* magazine (March, 1990) in which 50 tax professionals were hired to complete a 1040 Federal income tax return for a hypothetical family. Only 2 of the 50 professionals determined the correct tax due. The table below gives the fee charged by each preparer and the amount by which their result differed from the correct tax (all figures are in dollars).

Fee	Error	Fee	Error	Fee	Error	Fee	Error
990	2232	300	53	1015	1484	750	4228
795	2019	450	189	600	1659	2500	4307
750	1665	950	256	1100	1831	276	5064
1950	1400	800	457	422	1852	900	5188
400	1345	1450	494	550	2051	280	5454
960	1111	1425	609	1200	2074	4000	5473
1300	875	650	614	2000	2098	750	5518
850	651	770	737	1500	2198	970	5973
1450	618	750	808	1100	2788	1150	8103
1360	215	975	912	1100	2867	400	8288
1285	0	720	1124	1750	3079	520	9178
750	0	960	1300	1350	3240		
271	54	450	1407	640	3338		

Assess if the fee charged is related to performance by using MINITAB to calculate the rank correlation coefficient between the fee and the size of the error.

Looking Back

Chapter 15 discussed **nonparametric tests** that can be used in place of their parametric counterparts that were considered in previous chapters. Appealing features of nonparametric tests are their less restrictive assumptions and simpler calculations. Generally, they are slightly less powerful than corresponding parametric tests when the underlying assumptions of the latter hold. However, in situations for which the assumptions are not true, a parametric test is often more powerful.

The following nonparametric tests of hypotheses were discussed in this chapter.

1. The One-Sample Sign Test for a Population Median $\tilde{\mu}$.

$$H_0: \tilde{\mu} = \tilde{\mu}_0 \quad (\tilde{\mu}_0 \text{ is a constant})$$

The test statistic is

$$x = \text{the number of sample values that exceed } \tilde{\mu}_0.$$

A large-sample test statistic is

$$z = \frac{x - 0.5n}{0.5\sqrt{n}},$$

where n is the number of sample values retained after discarding those equal to $\tilde{\mu}_0$.

2. **The Two-Sample Sign Test: Paired Samples.**

$$H_0: \tilde{\mu}_D = 0 \quad (\tilde{\mu}_D \text{ is the median of the population of differences})$$

The test statistic is

$$x = \text{the number of pairs with a positive difference.}$$

A large-sample test statistic is

$$z = \frac{x - 0.5n}{0.5\sqrt{n}},$$

where n is the number of pairs with nonzero differences.

3. **The Mann-Whitney U-Test: Independent Samples.**

$$H_0: (\tilde{\mu}_1 - \tilde{\mu}_2) = 0$$

The test statistic is

$$z = \frac{U_1 - \dfrac{n_1 n_2}{2}}{\sqrt{\dfrac{n_1 n_2 (n_1 + n_2 + 1)}{12}}},$$

where

$$U_1 = n_1 n_2 + \frac{n_2(n_2 + 1)}{2} - R_2,$$

R_2 is the rank sum for sample two,
n_1 and n_2 are the sample sizes.

4. The Kruskal-Wallis *H*-Test for a Completely Randomized Design.

$$H_0: \tilde{\mu}_1 = \tilde{\mu}_2 = \cdots = \tilde{\mu}_k$$

The test statistic is

$$H = \frac{12}{n(n+1)} \Sigma \frac{R_i^2}{n_i} - 3(n+1),$$

where R_i is the rank sum of sample i, and $n = n_1 + \cdots + n_k$.

5. The Runs Test for Randomness.

H_0: The sequence of sample observations is random.

The test statistic is

$$x = \text{the number of runs in the sequence.}$$

A large sample test statistic is

$$z = \frac{x - \mu}{\sigma},$$

where

$$\mu = \frac{2n_1 n_2}{n_1 + n_2} + 1, \qquad \sigma = \sqrt{\frac{2n_1 n_2(2n_1 n_2 - n_1 - n_2)}{(n_1 + n_2)^2(n_1 + n_2 - 1)}}$$

n_1 and n_2 denote the number of occurrences of the two types in the sample.

6. Spearman's Test for Rank Correlation.

$$H_0: \rho_s = 0 \quad \text{(there is no association between the variables)}$$

The test statistic is

$$r_s = 1 - \frac{6\Sigma d_i^2}{n(n^2 - 1)},$$

where n is the number of pairs, and d_i is the difference between ranks for the two sample values in the ith pair.

Key Words

In reviewing this chapter, you should be able to define, explain, and illustrate each of the following.

nonparametric test *(page 603)*

one-sample sign test *(page 604)*

two-sample sign test *(page 608)*

Mann-Whitney *U*-test *(page 615)*

rank sum *(page 616)*

Kruskal-Wallis *H*-test *(page 623)*

run *(page 630)*

runs test for randomness *(page 630)*

Ⓜ *MINITAB Commands*

```
READ _ _ (page 609)
END (page 609)
LET _ (page 609)
STEST _ _; (page 610)
ALTERNATIVE _.
SET _ (page 618)
MANN-WHITNEY _ _; (page 619)
ALTERNATIVE _.
```

```
KRUSKAL-WALLIS _ _ (page 626)
MEDIAN _ (page 635)
RUNS _ _ (page 635)
RANK _ _ (page 643)
CORRELATION _ _ (page 643)
```

Review Exercises

15.77 A movie theater shows its featured film at 7:00 P.M. and 9:30 P.M. Random samples of 14 patrons at each showing were selected and asked their age. The results were the following.

Showing	Patrons' Ages													
7:00 P.M.	57	49	41	56	39	40	37	26	19	68	49	73	28	44
9:30 P.M.	23	20	32	43	21	27	25	52	40	26	34	38	25	21

Use the Mann-Whitney *U*-test to determine if there is a difference in the median age of patrons for the two showings. Test at the 5 percent significance level.

15.78 During a recent broadcast, a radio talk show received 28 calls. The sexes of the callers, in order of occurrence, are given below.

F M M F F F F M F M F F M M M M M F M F F F F M M M M M

Use Table 7 to determine if there is a lack of randomness with regard to the sex of the callers. Let $\alpha = 0.05$.

15.79 Use the large samples runs test to solve Exercise 15.78.

15.80 Eleven secretaries participated in a one-day workshop designed to increase their typing speeds. The participants' speeds (words per minute) were measured before and after the workshop. They are given on the next page.

Secretary	Speed Before the Program	Speed After the Program
1	59	64
2	43	51
3	58	55
4	47	59
5	37	48
6	62	62
7	41	45
8	50	52
9	43	53
10	49	52
11	63	65

Use a nonparametric test to determine if there is sufficient evidence at the 0.05 level to conclude that the course tends to increase one's typing speed.

15.81   Determine the p-value for the test in Exercise 15.80.

15.82   A Ferris wheel ride is supposed to last at least 5 minutes. Thirty-six operations of the ride were timed, and the following times (in seconds) were obtained.

283	274	296	301	294	288	302	275	297
291	306	316	285	296	289	295	300	291
305	289	298	287	281	295	291	290	316
275	296	284	303	295	303	290	278	299

Determine if there is sufficient evidence at the 0.01 level to conclude that the median duration of the ride is less than 5 minutes.

15.83   Determine the p-value for the test in Exercise 15.82.

15.84   A company interviewed 10 recent college graduates for an engineering position. Each applicant was assigned a composite score based on a written exam and personal interviews. The scores and college grade point averages of the candidates are given in the following table.

Applicant	Interview Score	GPA
1	47	3.12
2	43	3.59
3	37	2.98
4	46	3.71
5	30	2.76
6	31	3.21
7	44	3.34
8	34	2.95
9	40	3.06
10	35	2.88

Is there sufficient evidence to indicate that the interview score is positively correlated with grade point average? Use a nonparametric test and $\alpha = 0.01$.

15.85 Determine the approximate p-value for the test in Exercise 15.84.

15.86 A completely randomized design produced the following sample results.

Sample 1	Sample 2	Sample 3	Sample 4
65	59	72	56
76	60	75	54
68	62	67	53
63	58	74	51
64	61	71	55
59			58

Use the Kruskal-Wallis H-test to determine if at least two of the sampled populations differ in location. Test with $\alpha = 1$ percent.

15.87 A state conducts a daily lottery for which a 3-digit number is selected. The selected numbers for 30 consecutive drawings are given below (in the order row 1, then row 2).

711 356 189 347 982 597 735 916 069 901 735 914 036 300 918
131 016 733 868 971 350 071 415 555 783 027 271 244 612 085

Use Table 7 to determine if the data suggest nonrandomness in the numbers drawn. Use $\alpha = 5$ percent (the sample median is 485).

15.88 For Exercise 15.87, use the large-sample runs test and a significance level of 5 percent.

15.89 Dangerous chemicals from industrial wastes linked to cancer and other diseases can enter the food chain through their presence in lake sediments. The amounts of DDT were determined for samples of trout from four lakes and are given below.

Lake	Levels of DDT in parts per million					
One	1.7	1.4	1.9	1.1	2.1	1.8
Two	0.3	0.7	0.5	0.1	1.1	0.9
Three	2.7	1.9	2.0	1.5	2.6	
Four	1.2	3.1	1.9	3.7	2.8	3.5

Do the data provide sufficient evidence at the 0.05 level to conclude that there is a difference in the median DDT levels for the four lakes?

15.90 Determine the approximate p-value for the test in Exercise 15.89.

MINITAB Assignments

Ⓜ 15.91 Refer to Exercise 15.82, and use MINITAB to determine if there is sufficient evidence at the 0.05 level to conclude that the median duration of the ride is less than 5 minutes.

Ⓜ 15.92 Refer to Exercise 15.84, and use MINITAB to calculate the rank correlation coefficient between interview score and grade point average.

Ⓜ 15.93 Use MINITAB to determine the number of runs above or below the median in Exercise 15.87.

Ⓜ 15.94 A furniture manufacturer wanted to compare drying times for 4 brands of stains. Each stain was applied to 10 chairs, and the drying times in minutes appear below.

Brand 1	Brand 2	Brand 3	Brand 4
80.6	91.6	90.5	86.7
81.3	83.5	98.5	75.4
82.8	83.4	97.5	79.7
81.5	88.6	99.9	76.5
80.4	96.7	96.9	75.7
79.7	84.8	90.5	84.7
82.3	88.4	96.7	74.5
81.7	89.5	93.8	83.3
80.6	84.4	97.8	84.2
81.5	85.1	96.8	75.3

Is there sufficient evidence to conclude that a difference exists in the median drying times for the 4 brands? Use MINITAB to apply a nonparametric test, and let $\alpha = 0.05$.

Ⓜ 15.95 Refer to Exercise 15.94, and use MINITAB to apply the Mann-Whitney U-test to determine whether a difference exists in the median drying times for brands 1 and 3. Test at the 0.01 significance level.

Ⓜ 15.96 A marketing firm conducted a study to determine if there is a difference in the average price of two brands of frozen pizza. Thirty-three food stores were randomly selected, and the prices of the two brands were recorded. They are given below, where the figures are in dollars.

Store	Brand A	Brand B	Store	Brand A	Brand B
1	3.69	3.59	18	3.76	3.70
2	4.19	4.03	19	3.85	3.79
3	3.75	3.75	20	3.64	3.68
4	3.50	3.43	21	3.65	3.57
5	3.65	3.55	22	3.76	3.67
6	3.73	3.69	23	4.09	3.98
7	3.89	3.95	24	3.94	3.90
8	3.93	3.75	25	3.85	3.79

Store	Brand A	Brand B	Store	Brand A	Brand B
9	3.95	3.88	26	3.69	3.69
10	4.25	4.00	27	3.49	3.79
11	3.41	3.38	28	3.76	3.58
12	3.68	3.60	29	3.84	3.76
13	3.86	3.84	30	3.96	3.87
14	3.83	3.79	31	3.58	3.50
15	3.72	3.75	32	3.74	3.71
16	3.99	3.99	33	3.87	3.76
17	3.85	3.80			

Use MINITAB to perform a nonparametric test to determine if one of the brands tends to be priced higher than the other. Test at the 0.01 significance level.

Appendix A

Table 1 Binomial Probabilities

$n = 1$

The Tabulated Values are the Cumulative Probabilities
$P(x \leq k)$ for $k = 0$ through 1.

k	p = .05	p = .10	p = .20	p = .30	p = .40	p = .50	p = .60	p = .70	p = .80	p = .90	p = .95
0	0.950	0.900	0.800	0.700	0.600	0.500	0.400	0.300	0.200	0.100	0.050
1	1.000	1.000	1.000	1.000	1.000	1.000	1.000	1.000	1.000	1.000	1.000

$n = 2$

The Tabulated Values are the Cumulative Probabilities
$P(x \leq k)$ for $k = 0$ through 2.

k	p = .05	p = .10	p = .20	p = .30	p = .40	p = .50	p = .60	p = .70	p = .80	p = .90	p = .95
0	0.902	0.810	0.640	0.490	0.360	0.250	0.160	0.090	0.040	0.010	0.003
1	0.998	0.990	0.960	0.910	0.840	0.750	0.640	0.510	0.360	0.190	0.098
2	1.000	1.000	1.000	1.000	1.000	1.000	1.000	1.000	1.000	1.000	1.000

$n = 3$

The Tabulated Values are the Cumulative Probabilities
$P(x \leq k)$ for $k = 0$ through 3.

k	p = .05	p = .10	p = .20	p = .30	p = .40	p = .50	p = .60	p = .70	p = .80	p = .90	p = .95
0	0.857	0.729	0.512	0.343	0.216	0.125	0.064	0.027	0.008	0.001	0.000
1	0.993	0.972	0.896	0.784	0.648	0.500	0.352	0.216	0.104	0.028	0.007
2	1.000	0.999	0.992	0.973	0.936	0.875	0.784	0.657	0.488	0.271	0.143
3	1.000	1.000	1.000	1.000	1.000	1.000	1.000	1.000	1.000	1.000	1.000

$n = 4$

The Tabulated Values are the Cumulative Probabilities
$P(x \leq k)$ for $k = 0$ through 4.

k	p = .05	p = .10	p = .20	p = .30	p = .40	p = .50	p = .60	p = .70	p = .80	p = .90	p = .95
0	0.815	0.656	0.410	0.240	0.130	0.062	0.026	0.008	0.002	0.000	0.000
1	0.986	0.948	0.819	0.652	0.475	0.312	0.179	0.084	0.027	0.004	0.000
2	1.000	0.996	0.973	0.916	0.821	0.687	0.525	0.348	0.181	0.052	0.014
3	1.000	1.000	0.998	0.992	0.974	0.937	0.870	0.760	0.590	0.344	0.185
4	1.000	1.000	1.000	1.000	1.000	1.000	1.000	1.000	1.000	1.000	1.000

Table 1   Binomial Probabilities (continued)

n = 5

n = 5

The Tabulated Values are the Cumulative Probabilities
$P(x \leq k)$ for $k = 0$ through 5.

k	p = .05	p = .10	p = .20	p = .30	p = .40	p = .50	p = .60	p = .70	p = .80	p = .90	p = .95
0	0.774	0.590	0.328	0.168	0.078	0.031	0.010	0.002	0.000	0.000	0.000
1	0.977	0.919	0.737	0.528	0.337	0.187	0.087	0.031	0.007	0.000	0.000
2	0.999	0.991	0.942	0.837	0.683	0.500	0.317	0.163	0.058	0.009	0.001
3	1.000	1.000	0.993	0.969	0.913	0.812	0.663	0.472	0.263	0.081	0.023
4	1.000	1.000	1.000	0.998	0.990	0.969	0.922	0.832	0.672	0.410	0.226
5	1.000	1.000	1.000	1.000	1.000	1.000	1.000	1.000	1.000	1.000	1.000

n = 6

n = 6

The Tabulated Values are the Cumulative Probabilities
$P(x \leq k)$ for $k = 0$ through 6.

k	p = .05	p = .10	p = .20	p = .30	p = .40	p = .50	p = .60	p = .70	p = .80	p = .90	p = .95
0	0.735	0.531	0.262	0.118	0.047	0.016	0.004	0.001	0.000	0.000	0.000
1	0.967	0.886	0.655	0.420	0.233	0.109	0.041	0.011	0.002	0.000	0.000
2	0.998	0.984	0.901	0.744	0.544	0.344	0.179	0.070	0.017	0.001	0.000
3	1.000	0.999	0.983	0.930	0.821	0.656	0.456	0.256	0.099	0.016	0.002
4	1.000	1.000	0.998	0.989	0.959	0.891	0.767	0.580	0.345	0.114	0.033
5	1.000	1.000	1.000	0.999	0.996	0.984	0.953	0.882	0.738	0.469	0.265
6	1.000	1.000	1.000	1.000	1.000	1.000	1.000	1.000	1.000	1.000	1.000

n = 7

n = 7

The Tabulated Values are the Cumulative Probabilities
$P(x \leq k)$ for $k = 0$ through 7.

k	p = .05	p = .10	p = .20	p = .30	p = .40	p = .50	p = .60	p = .70	p = .80	p = .90	p = .95
0	0.698	0.478	0.210	0.082	0.028	0.008	0.002	0.000	0.000	0.000	0.000
1	0.956	0.850	0.577	0.329	0.159	0.062	0.019	0.004	0.000	0.000	0.000
2	0.996	0.974	0.852	0.647	0.420	0.227	0.096	0.029	0.005	0.000	0.000
3	1.000	0.997	0.967	0.874	0.710	0.500	0.290	0.126	0.033	0.003	0.000
4	1.000	1.000	0.995	0.971	0.904	0.773	0.580	0.353	0.148	0.026	0.004
5	1.000	1.000	1.000	0.996	0.981	0.937	0.841	0.671	0.423	0.150	0.044
6	1.000	1.000	1.000	1.000	0.998	0.992	0.972	0.918	0.790	0.522	0.302
7	1.000	1.000	1.000	1.000	1.000	1.000	1.000	1.000	1.000	1.000	1.000

Table 1   Binomial Probabilities *(continued)*

$n = 8$

The Tabulated Values are the Cumulative Probabilities
$P(x \leq k)$ for $k = 0$ through 8.

k	p = .05	p = .10	p = .20	p = .30	p = .40	p = .50	p = .60	p = .70	p = .80	p = .90	p = .95
0	0.663	0.430	0.168	0.058	0.017	0.004	0.001	0.000	0.000	0.000	0.000
1	0.943	0.813	0.503	0.255	0.106	0.035	0.009	0.001	0.000	0.000	0.000
2	0.994	0.962	0.797	0.552	0.315	0.145	0.050	0.011	0.001	0.000	0.000
3	1.000	0.995	0.944	0.806	0.594	0.363	0.174	0.058	0.010	0.000	0.000
4	1.000	1.000	0.990	0.942	0.826	0.637	0.406	0.194	0.056	0.005	0.000
5	1.000	1.000	0.999	0.989	0.950	0.855	0.685	0.448	0.203	0.038	0.006
6	1.000	1.000	1.000	0.999	0.991	0.965	0.894	0.745	0.497	0.187	0.057
7	1.000	1.000	1.000	1.000	0.999	0.996	0.983	0.942	0.832	0.570	0.337
8	1.000	1.000	1.000	1.000	1.000	1.000	1.000	1.000	1.000	1.000	1.000

$n = 9$

The Tabulated Values are the Cumulative Probabilities
$P(x \leq k)$ for $k = 0$ through 9.

k	p = .05	p = .10	p = .20	p = .30	p = .40	p = .50	p = .60	p = .70	p = .80	p = .90	p = .95
0	0.630	0.387	0.134	0.040	0.010	0.002	0.000	0.000	0.000	0.000	0.000
1	0.929	0.775	0.436	0.196	0.071	0.020	0.004	0.000	0.000	0.000	0.000
2	0.992	0.947	0.738	0.463	0.232	0.090	0.025	0.004	0.000	0.000	0.000
3	0.999	0.992	0.914	0.730	0.483	0.254	0.099	0.025	0.003	0.000	0.000
4	1.000	0.999	0.980	0.901	0.733	0.500	0.267	0.099	0.020	0.001	0.000
5	1.000	1.000	0.997	0.975	0.901	0.746	0.517	0.270	0.086	0.008	0.001
6	1.000	1.000	1.000	0.996	0.975	0.910	0.768	0.537	0.262	0.053	0.008
7	1.000	1.000	1.000	1.000	0.996	0.980	0.929	0.804	0.564	0.225	0.071
8	1.000	1.000	1.000	1.000	1.000	0.998	0.990	0.960	0.866	0.613	0.370
9	1.000	1.000	1.000	1.000	1.000	1.000	1.000	1.000	1.000	1.000	1.000

$n = 10$

The Tabulated Values are the Cumulative Probabilities
$P(x \leq k)$ for $k = 0$ through 10.

k	p = .05	p = .10	p = .20	p = .30	p = .40	p = .50	p = .60	p = .70	p = .80	p = .90	p = .95
0	0.599	0.349	0.107	0.028	0.006	0.001	0.000	0.000	0.000	0.000	0.000
1	0.914	0.736	0.376	0.149	0.046	0.011	0.002	0.000	0.000	0.000	0.000
2	0.988	0.930	0.678	0.383	0.167	0.055	0.012	0.002	0.000	0.000	0.000
3	0.999	0.987	0.879	0.650	0.382	0.172	0.055	0.011	0.001	0.000	0.000
4	1.000	0.998	0.967	0.850	0.633	0.377	0.166	0.047	0.006	0.000	0.000
5	1.000	1.000	0.994	0.953	0.834	0.623	0.367	0.150	0.033	0.002	0.000
6	1.000	1.000	0.999	0.989	0.945	0.828	0.618	0.350	0.121	0.013	0.001
7	1.000	1.000	1.000	0.998	0.988	0.945	0.833	0.617	0.322	0.070	0.012
8	1.000	1.000	1.000	1.000	0.998	0.989	0.954	0.851	0.624	0.264	0.086
9	1.000	1.000	1.000	1.000	1.000	0.999	0.994	0.972	0.893	0.651	0.401
10	1.000	1.000	1.000	1.000	1.000	1.000	1.000	1.000	1.000	1.000	1.000

Table 1   Binomial Probabilities *(continued)*

n = 15   n = 15

The Tabulated Values are the Cumulative Probabilities
$P(x \le k)$ for $k = 0$ through 15.

k	p = .05	p = .10	p = .20	p = .30	p = .40	p = .50	p = .60	p = .70	p = .80	p = .90	p = .95
0	0.463	0.206	0.035	0.005	0.000	0.000	0.000	0.000	0.000	0.000	0.000
1	0.829	0.549	0.167	0.035	0.005	0.000	0.000	0.000	0.000	0.000	0.000
2	0.964	0.816	0.398	0.127	0.027	0.004	0.000	0.000	0.000	0.000	0.000
3	0.995	0.944	0.648	0.297	0.091	0.018	0.002	0.000	0.000	0.000	0.000
4	0.999	0.987	0.836	0.515	0.217	0.059	0.009	0.001	0.000	0.000	0.000
5	1.000	0.998	0.939	0.722	0.403	0.151	0.034	0.004	0.000	0.000	0.000
6	1.000	1.000	0.982	0.869	0.610	0.304	0.095	0.015	0.001	0.000	0.000
7	1.000	1.000	0.996	0.950	0.787	0.500	0.213	0.050	0.004	0.000	0.000
8	1.000	1.000	0.999	0.985	0.905	0.696	0.390	0.131	0.018	0.000	0.000
9	1.000	1.000	1.000	0.996	0.966	0.849	0.597	0.278	0.061	0.002	0.000
10	1.000	1.000	1.000	0.999	0.991	0.941	0.783	0.485	0.164	0.013	0.001
11	1.000	1.000	1.000	1.000	0.998	0.982	0.909	0.703	0.352	0.056	0.005
12	1.000	1.000	1.000	1.000	1.000	0.996	0.973	0.873	0.602	0.184	0.036
13	1.000	1.000	1.000	1.000	1.000	1.000	0.995	0.965	0.833	0.451	0.171
14	1.000	1.000	1.000	1.000	1.000	1.000	1.000	0.995	0.965	0.794	0.537
15	1.000	1.000	1.000	1.000	1.000	1.000	1.000	1.000	1.000	1.000	1.000

n = 20   n = 20

The Tabulated Values are the Cumulative Probabilities
$P(x \le k)$ for $k = 0$ through 20.

k	p = .05	p = .10	p = .20	p = .30	p = .40	p = .50	p = .60	p = .70	p = .80	p = .90	p = .95
0	0.358	0.122	0.012	0.001	0.000	0.000	0.000	0.000	0.000	0.000	0.000
1	0.736	0.392	0.069	0.008	0.001	0.000	0.000	0.000	0.000	0.000	0.000
2	0.925	0.677	0.206	0.035	0.004	0.000	0.000	0.000	0.000	0.000	0.000
3	0.984	0.867	0.411	0.107	0.016	0.001	0.000	0.000	0.000	0.000	0.000
4	0.997	0.957	0.630	0.238	0.051	0.006	0.000	0.000	0.000	0.000	0.000
5	1.000	0.989	0.804	0.416	0.126	0.021	0.002	0.000	0.000	0.000	0.000
6	1.000	0.998	0.913	0.608	0.250	0.058	0.006	0.000	0.000	0.000	0.000
7	1.000	1.000	0.968	0.772	0.416	0.132	0.021	0.001	0.000	0.000	0.000
8	1.000	1.000	0.990	0.887	0.596	0.252	0.057	0.005	0.000	0.000	0.000
9	1.000	1.000	0.997	0.952	0.755	0.412	0.128	0.017	0.001	0.000	0.000
10	1.000	1.000	0.999	0.983	0.872	0.588	0.245	0.048	0.003	0.000	0.000
11	1.000	1.000	1.000	0.995	0.943	0.748	0.404	0.113	0.010	0.000	0.000
12	1.000	1.000	1.000	0.999	0.979	0.868	0.584	0.228	0.032	0.000	0.000
13	1.000	1.000	1.000	1.000	0.994	0.942	0.750	0.392	0.087	0.002	0.000
14	1.000	1.000	1.000	1.000	0.998	0.979	0.874	0.584	0.196	0.011	0.000
15	1.000	1.000	1.000	1.000	1.000	0.994	0.949	0.762	0.370	0.043	0.003
16	1.000	1.000	1.000	1.000	1.000	0.999	0.984	0.893	0.589	0.133	0.016
17	1.000	1.000	1.000	1.000	1.000	1.000	0.996	0.965	0.794	0.323	0.075
18	1.000	1.000	1.000	1.000	1.000	1.000	0.999	0.992	0.931	0.608	0.264
19	1.000	1.000	1.000	1.000	1.000	1.000	1.000	0.999	0.988	0.878	0.642
20	1.000	1.000	1.000	1.000	1.000	1.000	1.000	1.000	1.000	1.000	1.000

Table 2 Values of $e^{-\mu}$

μ	$e^{-\mu}$	μ	$e^{-\mu}$	μ	$e^{-\mu}$	μ	$e^{-\mu}$	μ	$e^{-\mu}$
0.00	1.000000	2.40	0.090718	4.80	0.008230	7.20	0.000747	9.60	0.000068
0.05	0.951229	2.45	0.086294	4.85	0.007828	7.25	0.000710	9.65	0.000064
0.10	0.904837	2.50	0.082085	4.90	0.007447	7.30	0.000676	9.70	0.000061
0.15	0.860708	2.55	0.078082	4.95	0.007083	7.35	0.000643	9.75	0.000058
0.20	0.818731	2.60	0.074274	5.00	0.006738	7.40	0.000611	9.80	0.000055
0.25	0.778801	2.65	0.070651	5.05	0.006409	7.45	0.000581	9.85	0.000053
0.30	0.740818	2.70	0.067206	5.10	0.006097	7.50	0.000553	9.90	0.000050
0.35	0.704688	2.75	0.063928	5.15	0.005799	7.55	0.000526	9.95	0.000048
0.40	0.670320	2.80	0.060810	5.20	0.005517	7.60	0.000500	10.00	0.000045
0.45	0.637628	2.85	0.057844	5.25	0.005248	7.65	0.000476	10.05	0.000043
0.50	0.606531	2.90	0.055023	5.30	0.004992	7.70	0.000453	10.10	0.000041
0.55	0.576950	2.95	0.052340	5.35	0.004748	7.75	0.000431	10.15	0.000039
0.60	0.548812	3.00	0.049787	5.40	0.004517	7.80	0.000410	10.20	0.000037
0.65	0.522046	3.05	0.047359	5.45	0.004296	7.85	0.000390	10.25	0.000035
0.70	0.496585	3.10	0.045049	5.50	0.004087	7.90	0.000371	10.30	0.000034
0.75	0.472367	3.15	0.042852	5.55	0.003887	7.95	0.000353	10.35	0.000032
0.80	0.449329	3.20	0.040762	5.60	0.003698	8.00	0.000335	10.40	0.000030
0.85	0.427415	3.25	0.038774	5.65	0.003518	8.05	0.000319	10.45	0.000029
0.90	0.406570	3.30	0.036883	5.70	0.003346	8.10	0.000304	10.50	0.000028
0.95	0.386741	3.35	0.035084	5.75	0.003183	8.15	0.000289	10.55	0.000026
1.00	0.367879	3.40	0.033373	5.80	0.003028	8.20	0.000275	10.60	0.000025
1.05	0.349938	3.45	0.031746	5.85	0.002880	8.25	0.000261	10.65	0.000024
1.10	0.332871	3.50	0.030197	5.90	0.002739	8.30	0.000249	10.70	0.000023
1.15	0.316637	3.55	0.028725	5.95	0.002606	8.35	0.000236	10.75	0.000021
1.20	0.301194	3.60	0.027324	6.00	0.002479	8.40	0.000225	10.80	0.000020
1.25	0.286505	3.65	0.025991	6.05	0.002358	8.45	0.000214	10.85	0.000019
1.30	0.272532	3.70	0.024724	6.10	0.002243	8.50	0.000203	10.90	0.000018
1.35	0.259240	3.75	0.023518	6.15	0.002133	8.55	0.000194	10.95	0.000018
1.40	0.246597	3.80	0.022371	6.20	0.002029	8.60	0.000184	11.00	0.000017
1.45	0.234570	3.85	0.021280	6.25	0.001930	8.65	0.000175	11.05	0.000016
1.50	0.223130	3.90	0.020242	6.30	0.001836	8.70	0.000167	11.10	0.000015
1.55	0.212248	3.95	0.019255	6.35	0.001747	8.75	0.000158	11.15	0.000014
1.60	0.201897	4.00	0.018316	6.40	0.001662	8.80	0.000151	11.20	0.000014
1.65	0.192050	4.05	0.017422	6.45	0.001581	8.85	0.000143	11.25	0.000013
1.70	0.182684	4.10	0.016573	6.50	0.001503	8.90	0.000136	11.30	0.000012
1.75	0.173774	4.15	0.015764	6.55	0.001430	8.95	0.000130	11.35	0.000012
1.80	0.165299	4.20	0.014996	6.60	0.001360	9.00	0.000123	11.40	0.000011
1.85	0.157237	4.25	0.014264	6.65	0.001294	9.05	0.000117	11.45	0.000011
1.90	0.149569	4.30	0.013569	6.70	0.001231	9.10	0.000112	11.50	0.000010
1.95	0.142274	4.35	0.012907	6.75	0.001171	9.15	0.000106	11.55	0.000010
2.00	0.135335	4.40	0.012277	6.80	0.001114	9.20	0.000101	11.60	0.000009
2.05	0.128735	4.45	0.011679	6.85	0.001059	9.25	0.000096	11.65	0.000009
2.10	0.122456	4.50	0.011109	6.90	0.001008	9.30	0.000091	11.70	0.000008
2.15	0.116484	4.55	0.010567	6.95	0.000959	9.35	0.000087	11.75	0.000008
2.20	0.110803	4.60	0.010052	7.00	0.000912	9.40	0.000083	11.80	0.000008
2.25	0.105399	4.65	0.009562	7.05	0.000867	9.45	0.000079	11.85	0.000007
2.30	0.100259	4.70	0.009095	7.10	0.000825	9.50	0.000075	11.90	0.000007
2.35	0.095369	4.75	0.008652	7.15	0.000785	9.55	0.000071	11.95	0.000006

Table 3A   Standard Normal Curve Areas (negative z-values)

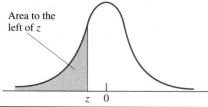

Area to the left of z

Each table value is the cumulative area to the left of the specified z-value.

z   0

			The Second Decimal Digit of z						
z	.00	.01	.02	.03	.04	.05	.06	.07	.08	.09
−3.4	.0003	.0003	.0003	.0003	.0003	.0003	.0003	.0003	.0003	.0002
−3.3	.0005	.0005	.0005	.0004	.0004	.0004	.0004	.0004	.0004	.0003
−3.2	.0007	.0007	.0006	.0006	.0006	.0006	.0006	.0005	.0005	.0005
−3.1	.0010	.0009	.0009	.0009	.0008	.0008	.0008	.0008	.0007	.0007
−3.0	.0013	.0013	.0013	.0012	.0012	.0011	.0011	.0011	.0010	.0010
−2.9	.0019	.0018	.0018	.0017	.0016	.0016	.0015	.0015	.0014	.0014
−2.8	.0026	.0025	.0024	.0023	.0023	.0022	.0021	.0021	.0020	.0019
−2.7	.0035	.0034	.0033	.0032	.0031	.0030	.0029	.0028	.0027	.0026
−2.6	.0047	.0045	.0044	.0043	.0041	.0040	.0039	.0038	.0037	.0036
−2.5	.0062	.0060	.0059	.0057	.0055	.0054	.0052	.0051	.0049	.0048
−2.4	.0082	.0080	.0078	.0075	.0073	.0071	.0069	.0068	.0066	.0064
−2.3	.0107	.0104	.0102	.0099	.0096	.0094	.0091	.0089	.0087	.0084
−2.2	.0139	.0136	.0132	.0129	.0125	.0122	.0119	.0116	.0113	.0110
−2.1	.0179	.0174	.0170	.0166	.0162	.0158	.0154	.0150	.0146	.0143
−2.0	.0228	.0222	.0217	.0212	.0207	.0202	.0197	.0192	.0188	.0183
−1.9	.0287	.0281	.0274	.0268	.0262	.0256	.0250	.0244	.0239	.0233
−1.8	.0359	.0351	.0344	.0336	.0329	.0322	.0314	.0307	.0301	.0294
−1.7	.0446	.0436	.0427	.0418	.0409	.0401	.0392	.0384	.0375	.0367
−1.6	.0548	.0537	.0526	.0516	.0505	.0495	.0485	.0475	.0465	.0455
−1.5	.0668	.0655	.0643	.0630	.0618	.0606	.0594	.0582	.0571	.0559
−1.4	.0808	.0793	.0778	.0764	.0749	.0735	.0721	.0708	.0694	.0681
−1.3	.0968	.0951	.0934	.0918	.0901	.0885	.0869	.0853	.0838	.0823
−1.2	.1151	.1131	.1112	.1093	.1075	.1056	.1038	.1020	.1003	.0985
−1.1	.1357	.1335	.1314	.1292	.1271	.1251	.1230	.1210	.1190	.1170
−1.0	.1587	.1562	.1539	.1515	.1492	.1469	.1446	.1423	.1401	.1379
−0.9	.1841	.1814	.1788	.1762	.1736	.1711	.1685	.1660	.1635	.1611
−0.8	.2119	.2090	.2061	.2033	.2005	.1977	.1949	.1922	.1894	.1867
−0.7	.2420	.2389	.2358	.2327	.2296	.2266	.2236	.2206	.2177	.2148
−0.6	.2743	.2709	.2676	.2643	.2611	.2578	.2546	.2514	.2483	.2451
−0.5	.3085	.3050	.3015	.2981	.2946	.2912	.2877	.2843	.2810	.2776
−0.4	.3446	.3409	.3372	.3336	.3300	.3264	.3228	.3192	.3156	.3121
−0.3	.3821	.3783	.3745	.3707	.3669	.3632	.3594	.3557	.3520	.3483
−0.2	.4207	.4168	.4129	.4090	.4052	.4013	.3974	.3936	.3897	.3859
−0.1	.4602	.4562	.4522	.4483	.4443	.4404	.4364	.4325	.4286	.4247
−0.0	.5000	.4960	.4920	.4880	.4840	.4801	.4761	.4721	.4681	.4641

Table 3B   Standard Normal Curve Areas (positive z-values)

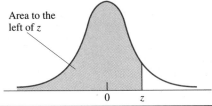

Area to the left of z

Each table value is the cumulative area to the left of the specified z-value.

0   z

				The Second Decimal Digit of z					
z	.00	.01	.02	.03	.04	.05	.06	.07	.08	.09
0.0	.5000	.5040	.5080	.5120	.5160	.5199	.5239	.5279	.5319	.5359
0.1	.5398	.5438	.5478	.5517	.5557	.5596	.5636	.5675	.5714	.5753
0.2	.5793	.5832	.5871	.5910	.5948	.5987	.6026	.6064	.6103	.6141
0.3	.6179	.6217	.6255	.6293	.6331	.6368	.6406	.6443	.6480	.6517
0.4	.6554	.6591	.6628	.6664	.6700	.6736	.6772	.6808	.6844	.6879
0.5	.6915	.6950	.6985	.7019	.7054	.7088	.7123	.7157	.7190	.7224
0.6	.7257	.7291	.7324	.7357	.7389	.7422	.7454	.7486	.7517	.7549
0.7	.7580	.7611	.7642	.7673	.7704	.7734	.7764	.7794	.7823	.7852
0.8	.7881	.7910	.7939	.7967	.7995	.8023	.8051	.8078	.8106	.8133
0.9	.8159	.8186	.8212	.8238	.8264	.8289	.8315	.8340	.8365	.8389
1.0	.8413	.8438	.8461	.8485	.8508	.8531	.8554	.8577	.8599	.8621
1.1	.8643	.8665	.8686	.8708	.8729	.8749	.8770	.8790	.8810	.8830
1.2	.8849	.8869	.8888	.8907	.8925	.8944	.8962	.8980	.8997	.9015
1.3	.9032	.9049	.9066	.9082	.9099	.9115	.9131	.9147	.9162	.9177
1.4	.9192	.9207	.9222	.9236	.9251	.9265	.9279	.9292	.9306	.9319
1.5	.9332	.9345	.9357	.9370	.9382	.9394	.9406	.9418	.9429	.9441
1.6	.9452	.9463	.9474	.9484	.9495	.9505	.9515	.9525	.9535	.9545
1.7	.9554	.9564	.9573	.9582	.9591	.9599	.9608	.9616	.9625	.9633
1.8	.9641	.9649	.9656	.9664	.9671	.9678	.9686	.9693	.9699	.9706
1.9	.9713	.9719	.9726	.9732	.9738	.9744	.9750	.9756	.9761	.9767
2.0	.9772	.9778	.9783	.9788	.9793	.9798	.9803	.9808	.9812	.9817
2.1	.9821	.9826	.9830	.9834	.9838	.9842	.9846	.9850	.9854	.9857
2.2	.9861	.9864	.9868	.9871	.9875	.9878	.9881	.9884	.9887	.9890
2.3	.9893	.9896	.9898	.9901	.9904	.9906	.9909	.9911	.9913	.9916
2.4	.9918	.9920	.9922	.9925	.9927	.9929	.9931	.9932	.9934	.9936
2.5	.9938	.9940	.9941	.9943	.9945	.9946	.9948	.9949	.9951	.9952
2.6	.9953	.9955	.9956	.9957	.9959	.9960	.9961	.9962	.9963	.9964
2.7	.9965	.9966	.9967	.9968	.9969	.9970	.9971	.9972	.9973	.9974
2.8	.9974	.9975	.9976	.9977	.9977	.9978	.9979	.9979	.9980	.9981
2.9	.9981	.9982	.9982	.9983	.9984	.9984	.9985	.9985	.9986	.9986
3.0	.9987	.9987	.9987	.9988	.9988	.9989	.9989	.9989	.9990	.9990
3.1	.9990	.9991	.9991	.9991	.9992	.9992	.9992	.9992	.9993	.9993
3.2	.9993	.9993	.9994	.9994	.9994	.9994	.9994	.9995	.9995	.9995
3.3	.9995	.9995	.9995	.9996	.9996	.9996	.9996	.9996	.9996	.9997
3.4	.9997	.9997	.9997	.9997	.9997	.9997	.9997	.9997	.9997	.9998

Table 4   Student's *t*-Values for Specified Tail Areas

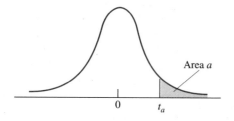

Each table entry is the *t*-value whose right-tail area equals the column heading, and whose degrees of freedom (*df*) equal the row number.

Degrees of Freedom (*df*)	Amount of Area in One Tail				
	.100	.050	.025	.010	.005
1	3.078	6.314	12.706	31.821	63.657
2	1.886	2.920	4.303	6.965	9.925
3	1.638	2.353	3.182	4.541	5.841
4	1.533	2.132	2.776	3.747	4.604
5	1.476	2.015	2.571	3.365	4.032
6	1.440	1.943	2.447	3.143	3.707
7	1.415	1.895	2.365	2.998	3.499
8	1.397	1.860	2.306	2.896	3.355
9	1.383	1.833	2.262	2.821	3.250
10	1.372	1.812	2.228	2.764	3.169
11	1.363	1.796	2.201	2.718	3.106
12	1.356	1.782	2.179	2.681	3.055
13	1.350	1.771	2.160	2.650	3.012
14	1.345	1.761	2.145	2.624	2.977
15	1.341	1.753	2.131	2.602	2.947
16	1.337	1.746	2.120	2.583	2.921
17	1.333	1.740	2.110	2.567	2.898
18	1.330	1.734	2.101	2.552	2.878
19	1.328	1.729	2.093	2.539	2.861
20	1.325	1.725	2.086	2.528	2.845
21	1.323	1.721	2.080	2.518	2.831
22	1.321	1.717	2.074	2.508	2.819
23	1.319	1.714	2.069	2.500	2.807
24	1.318	1.711	2.064	2.492	2.797
25	1.316	1.708	2.060	2.485	2.787
26	1.315	1.706	2.056	2.479	2.779
27	1.314	1.703	2.052	2.473	2.771
28	1.313	1.701	2.048	2.467	2.763
29	1.311	1.699	2.045	2.462	2.756
Infinity	1.282	1.645	1.960	2.326	2.576

Table 5 Chi-Square Values for Specified Areas

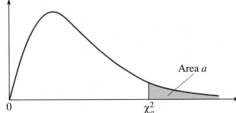

Each table entry is the χ^2-value whose area to the right equals the column heading, and whose degrees of freedom (*df*) equal the row number.

Area *a*

χ_a^2

Amount of Area to the Right of the Table Value

df	.995	.990	.975	.950	.900	.100	.050	.025	.010	.005
1	0.00	0.00	0.00	0.00	0.02	2.71	3.84	5.02	6.63	7.88
2	0.01	0.02	0.05	0.10	0.21	4.61	5.99	7.38	9.21	10.60
3	0.07	0.11	0.22	0.35	0.58	6.25	7.81	9.35	11.34	12.84
4	0.21	0.30	0.48	0.71	1.06	7.78	9.49	11.14	13.28	14.86
5	0.41	0.55	0.83	1.15	1.61	9.24	11.07	12.83	15.09	16.75
6	0.68	0.87	1.24	1.64	2.20	10.64	12.59	14.45	16.81	18.55
7	0.99	1.24	1.69	2.17	2.83	12.02	14.07	16.01	18.48	20.28
8	1.34	1.65	2.18	2.73	3.49	13.36	15.51	17.53	20.09	21.95
9	1.73	2.09	2.70	3.33	4.17	14.68	16.92	19.02	21.67	23.59
10	2.16	2.56	3.25	3.94	4.87	15.99	18.31	20.48	23.21	25.19
11	2.60	3.05	3.82	4.57	5.58	17.28	19.68	21.92	24.72	26.76
12	3.07	3.57	4.40	5.23	6.30	18.55	21.03	23.34	26.22	28.30
13	3.57	4.11	5.01	5.89	7.04	19.81	22.36	24.74	27.69	29.82
14	4.07	4.66	5.63	6.57	7.79	21.06	23.68	26.12	29.14	31.32
15	4.60	5.23	6.26	7.26	8.55	22.31	25.00	27.49	30.58	32.80
16	5.14	5.81	6.91	7.96	9.31	23.54	26.30	28.85	32.00	34.27
17	5.70	6.41	7.56	8.67	10.09	24.77	27.59	30.19	33.41	35.72
18	6.26	7.01	8.23	9.39	10.86	25.99	28.87	31.53	34.81	37.16
19	6.84	7.63	8.91	10.12	11.65	27.20	30.14	32.85	36.19	38.58
20	7.43	8.26	9.59	10.85	12.44	28.41	31.41	34.17	37.57	40.00
21	8.03	8.90	10.28	11.59	13.24	29.62	32.67	35.48	38.93	41.40
22	8.64	9.54	10.98	12.34	14.04	30.81	33.92	36.78	40.29	42.80
23	9.26	10.20	11.69	13.09	14.85	32.01	35.17	38.08	41.64	44.18
24	9.89	10.86	12.40	13.85	15.66	33.20	36.42	39.36	42.98	45.56
25	10.52	11.52	13.12	14.61	16.47	34.38	37.65	40.65	44.31	46.93
26	11.16	12.20	13.84	15.38	17.29	35.56	38.89	41.92	45.64	48.29
27	11.81	12.88	14.57	16.15	18.11	36.74	40.11	43.19	46.96	49.65
28	12.46	13.56	15.31	16.93	18.94	37.92	41.34	44.46	48.28	50.99
29	13.12	14.26	16.05	17.71	19.77	39.09	42.56	45.72	49.59	52.34
30	13.79	14.95	16.79	18.49	20.60	40.26	43.77	46.98	50.89	53.67
40	20.71	22.16	24.43	26.51	29.05	51.81	55.76	59.34	63.69	66.77
50	27.99	29.71	32.36	34.76	37.69	63.17	67.50	71.42	76.15	79.49
60	35.53	37.48	40.48	43.19	46.46	74.40	79.08	83.30	88.38	91.95
70	43.28	45.44	48.76	51.74	55.33	85.53	90.53	95.02	100.42	104.21
80	51.17	53.54	57.15	60.39	64.28	96.58	101.88	106.63	112.33	116.32
90	59.20	61.75	65.65	69.13	73.29	107.56	113.14	118.14	124.11	128.30
100	67.33	70.06	74.22	77.93	82.36	118.50	124.34	129.56	135.81	140.18

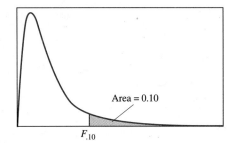

Area = 0.10

$F_{.10}$

Each table entry is the F-value that has an area of 0.10 to the right and whose degrees of freedom are given by the column and row numbers

Table 6A F-Distribution*

Right-Hand Tail Area of 0.10

ddf	Numerator df (ndf)								
	1	2	3	4	5	6	7	8	9
1	39.86	49.50	53.59	55.83	57.24	58.20	58.91	59.44	59.86
2	8.53	9.00	9.16	9.24	9.29	9.33	9.35	9.37	9.38
3	5.54	5.46	5.39	5.34	5.31	5.28	5.27	5.25	5.24
4	4.54	4.32	4.19	4.11	4.05	4.01	3.98	3.95	3.94
5	4.06	3.78	3.62	3.52	3.45	3.40	3.37	3.34	3.32
6	3.78	3.46	3.29	3.18	3.11	3.05	3.01	2.98	2.96
7	3.59	3.26	3.07	2.96	2.88	2.83	2.78	2.75	2.72
8	3.46	3.11	2.92	2.81	2.73	2.67	2.62	2.59	2.56
9	3.36	3.01	2.81	2.69	2.61	2.55	2.51	2.47	2.44
10	3.29	2.92	2.73	2.61	2.52	2.46	2.41	2.38	2.35
11	3.23	2.86	2.66	2.54	2.45	2.39	2.34	2.30	2.27
12	3.18	2.81	2.61	2.48	2.39	2.33	2.28	2.24	2.21
13	3.14	2.76	2.56	2.43	2.35	2.28	2.23	2.20	2.16
14	3.10	2.73	2.52	2.39	2.31	2.24	2.19	2.15	2.12
15	3.07	2.70	2.49	2.36	2.27	2.21	2.16	2.12	2.09
16	3.05	2.67	2.46	2.33	2.24	2.18	2.13	2.09	2.06
17	3.03	2.64	2.44	2.31	2.22	2.15	2.10	2.06	2.03
18	3.01	2.62	2.42	2.29	2.20	2.13	2.08	2.04	2.00
19	2.99	2.61	2.40	2.27	2.18	2.11	2.06	2.02	1.98
20	2.97	2.59	2.38	2.25	2.16	2.09	2.04	2.00	1.96
21	2.96	2.57	2.36	2.23	2.14	2.08	2.02	1.98	1.95
22	2.95	2.56	2.35	2.22	2.13	2.06	2.01	1.97	1.93
23	2.94	2.55	2.34	2.21	2.11	2.05	1.99	1.95	1.92
24	2.93	2.54	2.33	2.19	2.10	2.04	1.98	1.94	1.91
25	2.92	2.53	2.32	2.18	2.09	2.02	1.97	1.93	1.89
26	2.91	2.52	2.31	2.17	2.08	2.01	1.96	1.92	1.88
27	2.90	2.51	2.30	2.17	2.07	2.00	1.95	1.91	1.87
28	2.89	2.50	2.29	2.16	2.06	2.00	1.94	1.90	1.87
29	2.89	2.50	2.28	2.15	2.06	1.99	1.93	1.89	1.86
30	2.88	2.49	2.28	2.14	2.05	1.98	1.93	1.88	1.85
40	2.84	2.44	2.23	2.09	2.00	1.93	1.87	1.83	1.79
60	2.79	2.39	2.18	2.04	1.95	1.87	1.82	1.77	1.74
120	2.75	2.35	2.13	1.99	1.90	1.82	1.77	1.72	1.68
∞	2.71	2.30	2.08	1.94	1.85	1.77	1.72	1.67	1.63

Denominator df

* Reprinted, with permission of the *Biometrika* trustees, from Merrington, M. and C. M. Thompson. "Tables of percentage points of the inverted beta (F) distribution." *Biometrika* 33(1943).

Table 6A   *F*-Distribution (*continued*)

Right-Hand Tail Area of 0.10

Numerator *df* (*ndf*)

ddf	10	12	15	20	24	30	40	60	120	∞
1	60.19	60.71	61.22	61.74	62.00	62.26	62.53	62.79	63.06	63.33
2	9.39	9.41	9.42	9.44	9.45	9.46	9.47	9.47	9.48	9.49
3	5.23	5.22	5.20	5.18	5.18	5.17	5.16	5.15	5.14	5.13
4	3.92	3.90	3.87	3.84	3.83	3.82	3.80	3.79	3.78	3.76
5	3.30	3.27	3.24	3.21	3.19	3.17	3.16	3.14	3.12	3.10
6	2.94	2.90	2.87	2.84	2.82	2.80	2.78	2.76	2.74	2.72
7	2.70	2.67	2.63	2.59	2.58	2.56	2.54	2.51	2.49	2.47
8	2.54	2.50	2.46	2.42	2.40	2.38	2.36	2.34	2.32	2.29
9	2.42	2.38	2.34	2.30	2.28	2.25	2.23	2.21	2.18	2.16
10	2.32	2.28	2.24	2.20	2.18	2.16	2.13	2.11	2.08	2.06
11	2.25	2.21	2.17	2.12	2.10	2.08	2.05	2.03	2.00	1.97
12	2.19	2.15	2.10	2.06	2.04	2.01	1.99	1.96	1.93	1.90
13	2.14	2.10	2.05	2.01	1.98	1.96	1.93	1.90	1.88	1.85
14	2.10	2.05	2.01	1.96	1.94	1.91	1.89	1.86	1.83	1.80
15	2.06	2.02	1.97	1.92	1.90	1.87	1.85	1.82	1.79	1.76
16	2.03	1.99	1.94	1.89	1.87	1.84	1.81	1.78	1.75	1.72
17	2.00	1.96	1.91	1.86	1.84	1.81	1.78	1.75	1.72	1.69
18	1.98	1.93	1.89	1.84	1.81	1.78	1.75	1.72	1.69	1.66
19	1.96	1.91	1.86	1.81	1.79	1.76	1.73	1.70	1.67	1.63
20	1.94	1.89	1.84	1.79	1.77	1.74	1.71	1.68	1.64	1.61
21	1.92	1.87	1.83	1.78	1.75	1.72	1.69	1.66	1.62	1.59
22	1.90	1.86	1.81	1.76	1.73	1.70	1.67	1.64	1.60	1.57
23	1.89	1.84	1.80	1.74	1.72	1.69	1.66	1.62	1.59	1.55
24	1.88	1.83	1.78	1.73	1.70	1.67	1.64	1.61	1.57	1.53
25	1.87	1.82	1.77	1.72	1.69	1.66	1.63	1.59	1.56	1.52
26	1.86	1.81	1.76	1.71	1.68	1.65	1.61	1.58	1.54	1.50
27	1.85	1.80	1.75	1.70	1.67	1.64	1.60	1.57	1.53	1.49
28	1.84	1.79	1.74	1.69	1.66	1.63	1.59	1.56	1.52	1.48
29	1.83	1.78	1.73	1.68	1.65	1.62	1.58	1.55	1.51	1.47
30	1.82	1.77	1.72	1.67	1.64	1.61	1.57	1.54	1.50	1.46
40	1.76	1.71	1.66	1.61	1.57	1.54	1.51	1.47	1.42	1.38
60	1.71	1.66	1.60	1.54	1.51	1.48	1.44	1.40	1.35	1.29
120	1.65	1.60	1.55	1.48	1.45	1.41	1.37	1.32	1.26	1.19
∞	1.60	1.55	1.49	1.42	1.38	1.34	1.30	1.24	1.17	1.00

Denominator *df*

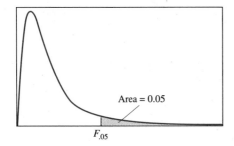

Area = 0.05

$F_{.05}$

Each table entry is the F-value that has an area of 0.05 to the right and whose degrees of freedom are given by the column and row numbers.

Table 6B   F-Distribution

Right-Hand Tail Area of 0.05

ddf	\multicolumn{9}{c}{Numerator df (ndf)}
	1	2	3	4	5	6	7	8	9
1	161.4	199.5	215.7	224.6	230.2	234.0	236.8	238.9	240.5
2	18.51	19.00	19.16	19.25	19.30	19.33	19.35	19.37	19.38
3	10.13	9.55	9.28	9.12	9.01	8.94	8.89	8.85	8.81
4	7.71	6.94	6.59	6.39	6.26	6.16	6.09	6.04	6.00
5	6.61	5.79	5.41	5.19	5.05	4.95	4.88	4.82	4.77
6	5.99	5.14	4.76	4.53	4.39	4.28	4.21	4.15	4.10
7	5.59	4.74	4.35	4.12	3.97	3.87	3.79	3.73	3.68
8	5.32	4.46	4.07	3.84	3.69	3.58	3.50	3.44	3.39
9	5.12	4.26	3.86	3.63	3.48	3.37	3.29	3.23	3.18
10	4.96	4.10	3.71	3.48	3.33	3.22	3.14	3.07	3.02
11	4.84	3.98	3.59	3.36	3.20	3.09	3.01	2.95	2.90
12	4.75	3.89	3.49	3.26	3.11	3.00	2.91	2.85	2.80
13	4.67	3.81	3.41	3.18	3.03	2.92	2.83	2.77	2.71
14	4.60	3.74	3.34	3.11	2.96	2.85	2.76	2.70	2.65
15	4.54	3.68	3.29	3.06	2.90	2.79	2.71	2.64	2.59
16	4.49	3.63	3.24	3.01	2.85	2.74	2.66	2.59	2.54
17	4.45	3.59	3.20	2.96	2.81	2.70	2.61	2.55	2.49
18	4.41	3.55	3.16	2.93	2.77	2.66	2.58	2.51	2.46
19	4.38	3.52	3.13	2.90	2.74	2.63	2.54	2.48	2.42
20	4.35	3.49	3.10	2.87	2.71	2.60	2.51	2.45	2.39
21	4.32	3.47	3.07	2.84	2.68	2.57	2.49	2.42	2.37
22	4.30	3.44	3.05	2.82	2.66	2.55	2.46	2.40	2.34
23	4.28	3.42	3.03	2.80	2.64	2.53	2.44	2.37	2.32
24	4.26	3.40	3.01	2.78	2.62	2.51	2.42	2.36	2.30
25	4.24	3.39	2.99	2.76	2.60	2.49	2.40	2.34	2.28
26	4.23	3.37	2.98	2.74	2.59	2.47	2.39	2.32	2.27
27	4.21	3.35	2.96	2.73	2.57	2.46	2.37	2.31	2.25
28	4.20	3.34	2.95	2.71	2.56	2.45	2.36	2.29	2.24
29	4.18	3.33	2.93	2.70	2.55	2.43	2.35	2.28	2.22
30	4.17	3.32	2.92	2.69	2.53	2.42	2.33	2.27	2.21
40	4.08	3.23	2.84	2.61	2.45	2.34	2.25	2.18	2.12
60	4.00	3.15	2.76	2.53	2.37	2.25	2.17	2.10	2.04
120	3.92	3.07	2.68	2.45	2.29	2.17	2.09	2.02	1.96
∞	3.84	3.00	2.60	2.37	2.21	2.10	2.01	1.94	1.88

D
e
n
o
m
i
n
a
t
o
r

d
f

Table 6B   *F*-Distribution *(continued)*

Right-Hand Tail Area of 0.05

					Numerator *df (ndf)*					
ddf	10	12	15	20	24	30	40	60	120	∞
1	241.9	243.9	245.9	248.0	249.1	250.1	251.1	252.2	253.3	254.3
2	19.40	19.41	19.43	19.45	19.45	19.46	19.47	19.48	19.49	19.50
3	8.79	8.74	8.70	8.66	8.64	8.62	8.59	8.57	8.55	8.53
4	5.96	5.91	5.86	5.80	5.77	5.75	5.72	5.69	5.66	5.63
5	4.74	4.68	4.62	4.56	4.53	4.50	4.46	4.43	4.40	4.36
6	4.06	4.00	3.94	3.87	3.84	3.81	3.77	3.74	3.70	3.67
7	3.64	3.57	3.51	3.44	3.41	3.38	3.34	3.30	3.27	3.23
8	3.35	3.28	3.22	3.15	3.12	3.08	3.04	3.01	2.97	2.93
9	3.14	3.07	3.01	2.94	2.90	2.86	2.83	2.79	2.75	2.71
10	2.98	2.91	2.85	2.77	2.74	2.70	2.66	2.62	2.58	2.54
11	2.85	2.79	2.72	2.65	2.61	2.57	2.53	2.49	2.45	2.40
12	2.75	2.69	2.62	2.54	2.51	2.47	2.43	2.38	2.34	2.30
13	2.67	2.60	2.53	2.46	2.42	2.38	2.34	2.30	2.25	2.21
14	2.60	2.53	2.46	2.39	2.35	2.31	2.27	2.22	2.18	2.13
15	2.54	2.48	2.40	2.33	2.29	2.25	2.20	2.16	2.11	2.07
16	2.49	2.42	2.35	2.28	2.24	2.19	2.15	2.11	2.06	2.01
17	2.45	2.38	2.31	2.23	2.19	2.15	2.10	2.06	2.01	1.96
18	2.41	2.34	2.27	2.19	2.15	2.11	2.06	2.02	1.97	1.92
19	2.38	2.31	2.23	2.16	2.11	2.07	2.03	1.98	1.93	1.88
20	2.35	2.28	2.20	2.12	2.08	2.04	1.99	1.95	1.90	1.84
21	2.32	2.25	2.18	2.10	2.05	2.01	1.96	1.92	1.87	1.81
22	2.30	2.23	2.15	2.07	2.03	1.98	1.94	1.89	1.84	1.78
23	2.27	2.20	2.13	2.05	2.01	1.96	1.91	1.86	1.81	1.76
24	2.25	2.18	2.11	2.03	1.98	1.94	1.89	1.84	1.79	1.73
25	2.24	2.16	2.09	2.01	1.96	1.92	1.87	1.82	1.77	1.71
26	2.22	2.15	2.07	1.99	1.95	1.90	1.85	1.80	1.75	1.69
27	2.20	2.13	2.06	1.97	1.93	1.88	1.84	1.79	1.73	1.67
28	2.19	2.12	2.04	1.96	1.91	1.87	1.82	1.77	1.71	1.65
29	2.18	2.10	2.03	1.94	1.90	1.85	1.81	1.75	1.70	1.64
30	2.16	2.09	2.01	1.93	1.89	1.84	1.79	1.74	1.68	1.62
40	2.08	2.00	1.92	1.84	1.79	1.74	1.69	1.64	1.58	1.51
60	1.99	1.92	1.84	1.75	1.70	1.65	1.59	1.53	1.47	1.39
120	1.91	1.83	1.75	1.66	1.61	1.55	1.50	1.43	1.35	1.25
∞	1.83	1.75	1.67	1.57	1.52	1.46	1.39	1.32	1.22	1.00

(Left margin label, reading vertically: D e n o m i n a t o r *d f*)

Adapted from Newmark J. *Statistics and Probability in Modern Life.* 5th ed. Saunders College Publishing, Philadelphia, 1992, p.A.14.

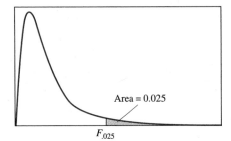

Area = 0.025

$F_{.025}$

Each table entry is the F-value that has an area of 0.025 to the right and whose degrees of freedom are given by the column and row numbers.

Table 6C F-Distribution

Right-Hand Tail Area of 0.025

| ddf | \multicolumn{9}{c}{Numerator df (ndf)} |
	1	2	3	4	5	6	7	8	9
1	647.8	799.5	864.2	899.6	921.8	937.1	948.2	956.7	963.3
2	38.51	39.00	39.17	39.25	39.30	39.33	39.36	39.37	39.39
3	17.44	16.04	15.44	15.10	14.88	14.73	14.62	14.54	14.47
4	12.22	10.65	9.98	9.60	9.36	9.20	9.07	8.98	8.90
5	10.01	8.43	7.76	7.39	7.15	6.98	6.85	6.76	6.68
6	8.81	7.26	6.60	6.23	5.99	5.82	5.70	5.60	5.52
7	8.07	6.54	5.89	5.52	5.29	5.12	4.99	4.90	4.82
8	7.57	6.06	5.42	5.05	4.82	4.65	4.53	4.43	4.36
9	7.21	5.71	5.08	4.72	4.48	4.32	4.20	4.10	4.03
10	6.94	5.46	4.83	4.47	4.24	4.07	3.95	3.85	3.78
11	6.72	5.26	4.63	4.28	4.04	3.88	3.76	3.66	3.59
12	6.55	5.10	4.47	4.12	3.89	3.73	3.61	3.51	3.44
13	6.41	4.97	4.35	4.00	3.77	3.60	3.48	3.39	3.31
14	6.30	4.86	4.24	3.89	3.66	3.50	3.38	3.29	3.21
15	6.20	4.77	4.15	3.80	3.58	3.41	3.29	3.20	3.12
16	6.12	4.69	4.08	3.73	3.50	3.34	3.22	3.12	3.05
17	6.04	4.62	4.01	3.66	3.44	3.28	3.16	3.06	2.98
18	5.98	4.56	3.95	3.61	3.38	3.22	3.10	3.01	2.93
19	5.92	4.51	3.90	3.56	3.33	3.17	3.05	2.96	2.88
20	5.87	4.46	3.86	3.51	3.29	3.13	3.01	2.91	2.84
21	5.83	4.42	3.82	3.48	3.25	3.09	2.97	2.87	2.80
22	5.79	4.38	3.78	3.44	3.22	3.05	2.93	2.84	2.76
23	5.75	4.35	3.75	3.41	3.18	3.02	2.90	2.81	2.73
24	5.72	4.32	3.72	3.38	3.15	2.99	2.87	2.78	2.70
25	5.69	4.29	3.69	3.35	3.13	2.97	2.85	2.75	2.68
26	5.66	4.27	3.67	3.33	3.10	2.94	2.82	2.73	2.65
27	5.63	4.24	3.65	3.31	3.08	2.92	2.80	2.71	2.63
28	5.61	4.22	3.63	3.29	3.06	2.90	2.78	2.69	2.61
29	5.59	4.20	3.61	3.27	3.04	2.88	2.76	2.67	2.59
30	5.57	4.18	3.59	3.25	3.03	2.87	2.75	2.65	2.57
40	5.42	4.05	3.46	3.13	2.90	2.74	2.62	2.53	2.45
60	5.29	3.93	3.34	3.01	2.79	2.63	2.51	2.41	2.33
120	5.15	3.80	3.23	2.89	2.67	2.52	2.39	2.30	2.22
∞	5.02	3.69	3.12	2.79	2.57	2.41	2.29	2.19	2.11

Denominator df

Table 6C *F*-**Distribution** (*continued*)

Right-Hand Tail Area of 0.025

ddf	Numerator *df* (*ndf*)									
	10	**12**	**15**	**20**	**24**	**30**	**40**	**60**	**120**	**∞**
1	968.6	976.7	984.9	993.1	997.2	1001	1006	1010	1014	1018
2	39.40	39.41	39.43	39.45	39.46	39.46	39.47	39.48	39.49	39.50
3	14.42	14.34	14.25	14.17	14.12	14.08	14.04	13.99	13.95	13.90
4	8.84	8.75	8.66	8.56	8.51	8.46	8.41	8.36	8.31	8.26
5	6.62	6.52	6.43	6.33	6.28	6.23	6.18	6.12	6.07	6.02
6	5.46	5.37	5.27	5.17	5.12	5.07	5.01	4.96	4.90	4.85
7	4.76	4.67	4.57	4.47	4.42	4.36	4.31	4.25	4.20	4.14
8	4.30	4.20	4.10	4.00	3.95	3.89	3.84	3.78	3.73	3.67
9	3.96	3.87	3.77	3.67	3.61	3.56	3.51	3.45	3.39	3.33
10	3.72	3.62	3.52	3.42	3.37	3.31	3.26	3.20	3.14	3.08
11	3.53	3.43	3.33	3.23	3.17	3.12	3.06	3.00	2.94	2.88
12	3.37	3.28	3.18	3.07	3.02	2.96	2.91	2.85	2.79	2.72
13	3.25	3.15	3.05	2.95	2.89	2.84	2.78	2.72	2.66	2.60
14	3.15	3.05	2.95	2.84	2.79	2.73	2.67	2.61	2.55	2.49
15	3.06	2.96	2.86	2.76	2.70	2.64	2.59	2.52	2.46	2.40
16	2.99	2.89	2.79	2.68	2.63	2.57	2.51	2.45	2.38	2.32
17	2.92	2.82	2.72	2.62	2.56	2.50	2.44	2.38	2.32	2.25
18	2.87	2.77	2.67	2.56	2.50	2.44	2.38	2.32	2.26	2.19
19	2.82	2.72	2.62	2.51	2.45	2.39	2.33	2.27	2.20	2.13
20	2.77	2.68	2.57	2.46	2.41	2.35	2.29	2.22	2.16	2.09
21	2.73	2.64	2.53	2.42	2.37	2.31	2.25	2.18	2.11	2.04
22	2.70	2.60	2.50	2.39	2.33	2.27	2.21	2.14	2.08	2.00
23	2.67	2.57	2.47	2.36	2.30	2.24	2.18	2.11	2.04	1.97
24	2.64	2.54	2.44	2.33	2.27	2.21	2.15	2.08	2.01	1.94
25	2.61	2.51	2.41	2.30	2.24	2.18	2.12	2.05	1.98	1.91
26	2.59	2.49	2.39	2.28	2.22	2.16	2.09	2.03	1.95	1.88
27	2.57	2.47	2.36	2.25	2.19	2.13	2.07	2.00	1.93	1.85
28	2.55	2.45	2.34	2.23	2.17	2.11	2.05	1.98	1.91	1.83
29	2.53	2.43	2.32	2.21	2.15	2.09	2.03	1.96	1.89	1.81
30	2.51	2.41	2.31	2.20	2.14	2.07	2.01	1.94	1.87	1.79
40	2.39	2.29	2.18	2.07	2.01	1.94	1.88	1.80	1.72	1.64
60	2.27	2.17	2.06	1.94	1.88	1.82	1.74	1.67	1.58	1.48
120	2.16	2.05	1.94	1.82	1.76	1.69	1.61	1.53	1.43	1.31
∞	2.05	1.94	1.83	1.71	1.64	1.57	1.48	1.39	1.27	1.00

Denominator *df*

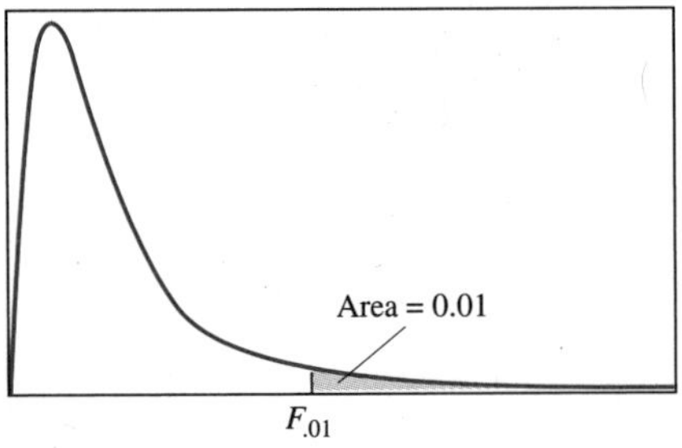

Each table entry is the F-value that has an area of 0.01 to the right and whose degrees of freedom are given by the column and row numbers.

Table 6D *F*-Distribution

Right-Hand Tail Area of 0.01

Numerator *df* (*ndf*)

Denominator *df*

ddf	1	2	3	4	5	6	7	8	9
1	4052	4999.5	5403	5625	5764	5859	5928	5982	6022
2	98.50	99.00	99.17	99.25	99.30	99.33	99.36	99.37	99.39
3	34.12	30.82	29.46	28.71	28.24	27.91	27.67	27.49	27.35
4	21.20	18.00	16.69	15.98	15.52	15.21	14.98	14.80	14.66
5	16.26	13.27	12.06	11.39	10.97	10.67	10.46	10.29	10.16
6	13.75	10.92	9.78	9.15	8.75	8.47	8.26	8.10	7.98
7	12.25	9.55	8.45	7.85	7.46	7.19	6.99	6.84	6.72
8	11.26	8.65	7.59	7.01	6.63	6.37	6.18	6.03	5.91
9	10.56	8.02	6.99	6.42	6.06	5.80	5.61	5.47	5.35
10	10.04	7.56	6.55	5.99	5.64	5.39	5.20	5.06	4.94
11	9.65	7.21	6.22	5.67	5.32	5.07	4.89	4.74	4.63
12	9.33	6.93	5.95	5.41	5.06	4.82	4.64	4.50	4.39
13	9.07	6.70	5.74	5.21	4.86	4.62	4.44	4.30	4.19
14	8.86	6.51	5.56	5.04	4.69	4.46	4.28	4.14	4.03
15	8.68	6.36	5.42	4.89	4.56	4.32	4.14	4.00	3.89
16	8.53	6.23	5.29	4.77	4.44	4.20	4.03	3.89	3.78
17	8.40	6.11	5.18	4.67	4.34	4.10	3.93	3.79	3.68
18	8.29	6.01	5.09	4.58	4.25	4.01	3.84	3.71	3.60
19	8.18	5.93	5.01	4.50	4.17	3.94	3.77	3.63	3.52
20	8.10	5.85	4.94	4.43	4.10	3.87	3.70	3.56	3.46
21	8.02	5.78	4.87	4.37	4.04	3.81	3.64	3.51	3.40
22	7.95	5.72	4.82	4.31	3.99	3.76	3.59	3.45	3.35
23	7.88	5.66	4.76	4.26	3.94	3.71	3.54	3.41	3.30
24	7.82	5.61	4.72	4.22	3.90	3.67	3.50	3.36	3.26
25	7.77	5.57	4.68	4.18	3.85	3.63	3.46	3.32	3.22
26	7.72	5.53	4.64	4.14	3.82	3.59	3.42	3.29	3.18
27	7.68	5.49	4.60	4.11	3.78	3.56	3.39	3.26	3.15
28	7.64	5.45	4.57	4.07	3.75	3.53	3.36	3.23	3.12
29	7.60	5.42	4.54	4.04	3.73	3.50	3.33	3.20	3.09
30	7.56	5.39	4.51	4.02	3.70	3.47	3.30	3.17	3.07
40	7.31	5.18	4.31	3.83	3.51	3.29	3.12	2.99	2.89
60	7.08	4.98	4.13	3.65	3.34	3.12	2.95	2.82	2.72
120	6.85	4.79	3.95	3.48	3.17	2.96	2.79	2.66	2.56
∞	6.63	4.61	3.78	3.32	3.02	2.80	2.64	2.51	2.41

Table 6D   F-Distribution (continued)

Right-Hand Tail Area of 0.01

ddf	\multicolumn{10}{c}{Numerator df (ndf)}									
	10	12	15	20	24	30	40	60	120	∞
1	6056	6106	6157	6209	6235	6261	6287	6313	6339	6366
2	99.40	99.42	99.43	99.45	99.46	99.47	99.47	99.48	99.49	99.50
3	27.23	27.05	26.87	26.69	26.60	26.50	26.41	26.32	26.22	26.13
4	14.55	14.37	14.20	14.02	13.93	13.84	13.75	13.65	13.56	13.46
5	10.05	9.89	9.72	9.55	9.47	9.38	9.29	9.20	9.11	9.02
6	7.87	7.72	7.56	7.40	7.31	7.23	7.14	7.06	6.97	6.88
7	6.62	6.47	6.31	6.16	6.07	5.99	5.91	5.82	5.74	5.65
8	5.81	5.67	5.52	5.36	5.28	5.20	5.12	5.03	4.95	4.86
9	5.26	5.11	4.96	4.81	4.73	4.65	4.57	4.48	4.40	4.31
10	4.85	4.71	4.56	4.41	4.33	4.25	4.17	4.08	4.00	3.91
11	4.54	4.40	4.25	4.10	4.02	3.94	3.86	3.78	3.69	3.60
12	4.30	4.16	4.01	3.86	3.78	3.70	3.62	3.54	3.45	3.36
13	4.10	3.96	3.82	3.66	3.59	3.51	3.43	3.34	3.25	3.17
14	3.94	3.80	3.66	3.51	3.43	3.35	3.27	3.18	3.09	3.00
15	3.80	3.67	3.52	3.37	3.29	3.21	3.13	3.05	2.96	2.87
16	3.69	3.55	3.41	3.26	3.18	3.10	3.02	2.93	2.84	2.75
17	3.59	3.46	3.31	3.16	3.08	3.00	2.92	2.83	2.75	2.65
18	3.51	3.37	3.23	3.08	3.00	2.92	2.84	2.75	2.66	2.57
19	3.43	3.30	3.15	3.00	2.92	2.84	2.76	2.67	2.58	2.49
20	3.37	3.23	3.09	2.94	2.86	2.78	2.69	2.61	2.52	2.42
21	3.31	3.17	3.03	2.88	2.80	2.72	2.64	2.55	2.46	2.36
22	3.26	3.12	2.98	2.83	2.75	2.67	2.58	2.50	2.40	2.31
23	3.21	3.07	2.93	2.78	2.70	2.62	2.54	2.45	2.35	2.26
24	3.17	3.03	2.89	2.74	2.66	2.58	2.49	2.40	2.31	2.21
25	3.13	2.99	2.85	2.70	2.62	2.54	2.45	2.36	2.27	2.17
26	3.09	2.96	2.81	2.66	2.58	2.50	2.42	2.33	2.23	2.13
27	3.06	2.93	2.78	2.63	2.55	2.47	2.38	2.29	2.20	2.10
28	3.03	2.90	2.75	2.60	2.52	2.44	2.35	2.26	2.17	2.06
29	3.00	2.87	2.73	2.57	2.49	2.41	2.33	2.23	2.14	2.03
30	2.98	2.84	2.70	2.55	2.47	2.39	2.30	2.21	2.11	2.01
40	2.80	2.66	2.52	2.37	2.29	2.20	2.11	2.02	1.92	1.80
60	2.63	2.50	2.35	2.20	2.12	2.03	1.94	1.84	1.73	1.60
120	2.47	2.34	2.19	2.03	1.95	1.86	1.76	1.66	1.53	1.38
∞	2.32	2.18	2.04	1.88	1.79	1.70	1.59	1.47	1.32	1.00

Denominator df

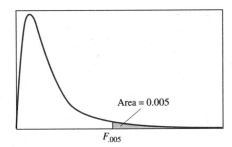

Each table entry is the F-value that has an area of 0.005 to the right and whose degrees of freedom are given by the column and row numbers.

Area = 0.005

$F_{.005}$

Table 6E F-Distribution

Right-Hand Tail Area of 0.005

Numerator df (ndf)

ddf	1	2	3	4	5	6	7	8	9
1	16211	20000	21615	22500	23056	23437	23715	23925	24091
2	198.5	199.0	199.2	199.2	199.3	199.3	199.4	199.4	199.4
3	55.55	49.80	47.47	46.19	45.39	44.84	44.43	44.13	43.88
4	31.33	26.28	24.26	23.15	22.46	21.97	21.62	21.35	21.14
5	22.78	18.31	16.53	15.56	14.94	14.51	14.20	13.96	13.77
6	18.63	14.54	12.92	12.03	11.46	11.07	10.79	10.57	10.39
7	16.24	12.40	10.88	10.05	9.52	9.16	8.89	8.68	8.51
8	14.69	11.04	9.60	8.81	8.30	7.95	7.69	7.50	7.34
9	13.61	10.11	8.72	7.96	7.47	7.13	6.88	6.69	6.54
10	12.83	9.43	8.08	7.34	6.87	6.54	6.30	6.12	5.97
11	12.23	8.91	7.60	6.88	6.42	6.10	5.86	5.68	5.54
12	11.75	8.51	7.23	6.52	6.07	5.76	5.52	5.35	5.20
13	11.37	8.19	6.93	6.23	5.79	5.48	5.25	5.08	4.94
14	11.06	7.92	6.68	6.00	5.56	5.26	5.03	4.86	4.72
15	10.80	7.70	6.48	5.80	5.37	5.07	4.85	4.67	4.54
16	10.58	7.51	6.30	5.64	5.21	4.91	4.69	4.52	4.38
17	10.38	7.35	6.16	5.50	5.07	4.78	4.56	4.39	4.25
18	10.22	7.21	6.03	5.37	4.96	4.66	4.44	4.28	4.14
19	10.17	7.09	5.92	5.27	4.85	4.56	4.34	4.18	4.04
20	9.94	6.99	5.82	5.17	4.76	4.47	4.26	4.09	3.96
21	9.83	6.89	5.73	5.09	4.68	4.39	4.18	4.01	3.88
22	9.73	6.81	5.65	5.02	4.61	4.32	4.11	3.94	3.81
23	9.63	6.73	5.58	4.95	4.54	4.26	4.05	3.88	3.75
24	9.55	6.66	5.52	4.89	4.49	4.20	3.99	3.83	3.69
25	9.48	6.60	5.46	4.84	4.43	4.15	3.94	3.78	3.64
26	9.41	6.54	5.41	4.79	4.38	4.10	3.89	3.73	3.60
27	9.34	6.49	5.36	4.74	4.34	4.06	3.85	3.69	3.56
28	9.28	6.44	5.32	4.70	4.30	4.02	3.81	3.65	3.52
29	9.23	6.40	5.28	4.66	4.26	3.98	3.77	3.61	3.48
30	9.18	6.35	5.24	4.62	4.23	3.95	3.74	3.58	3.45
40	8.83	6.07	4.98	4.37	3.99	3.71	3.51	3.35	3.22
60	8.49	5.79	4.73	4.14	3.76	3.49	3.29	3.13	3.01
120	8.18	5.54	4.50	3.92	3.55	3.28	3.09	2.93	2.81
∞	7.88	5.30	4.28	3.72	3.35	3.09	2.90	2.74	2.62

Denominator df

Table 6E ***F*-Distribution *(continued)***

Right-Hand Tail Area of 0.005

Numerator *df (ndf)*

ddf	10	12	15	20	24	30	40	60	120	∞
1	24224	24426	24630	24836	24940	25044	25148	25253	25359	25465
2	199.4	199.4	199.4	199.4	199.5	199.5	199.5	199.5	199.5	199.5
3	43.69	43.39	43.08	42.78	42.62	42.47	42.31	42.15	41.99	41.83
4	20.97	20.70	20.44	20.17	20.03	19.89	19.75	19.61	19.47	19.32
5	13.62	13.38	13.15	12.90	12.78	12.66	12.53	12.40	12.27	12.14
6	10.25	10.03	9.81	9.59	9.47	9.36	9.24	9.12	9.00	8.88
7	8.38	8.18	7.97	7.75	7.65	7.53	7.42	7.31	7.19	7.08
8	7.21	7.01	6.81	6.61	6.50	6.40	6.29	6.18	6.06	5.95
9	6.42	6.23	6.03	5.83	5.73	5.62	5.52	5.41	5.30	5.19
10	5.85	5.66	5.47	5.27	5.17	5.07	4.97	4.86	4.75	4.64
11	5.42	5.24	5.05	4.86	4.76	4.65	4.55	4.44	4.34	4.23
12	5.09	4.91	4.72	4.53	4.43	4.33	4.23	4.12	4.01	3.90
13	4.82	4.64	4.46	4.27	4.17	4.07	3.97	3.87	3.76	3.65
14	4.60	4.43	4.25	4.06	3.96	3.86	3.76	3.66	3.55	3.44
15	4.42	4.25	4.07	3.88	3.79	3.69	3.58	3.48	3.37	3.26
16	4.27	4.10	3.92	3.73	3.64	3.54	3.44	3.33	3.22	3.11
17	4.14	3.97	3.79	3.61	3.51	3.41	3.31	3.21	3.10	2.98
18	4.03	3.86	3.68	3.50	3.40	3.30	3.20	3.10	2.99	2.87
19	3.93	3.76	3.59	3.40	3.31	3.21	3.11	3.00	2.89	2.78
20	3.85	3.68	3.50	3.32	3.22	3.12	3.02	2.92	2.81	2.69
21	3.77	3.60	3.43	3.24	3.15	3.05	2.95	2.84	2.73	2.61
22	3.70	3.54	3.36	3.18	3.08	2.98	2.88	2.77	2.66	2.55
23	3.64	3.47	3.30	3.12	3.02	2.92	2.82	2.71	2.60	2.48
24	3.59	3.42	3.25	3.06	2.97	2.87	2.77	2.66	2.55	2.43
25	3.54	3.37	3.20	3.01	2.92	2.82	2.72	2.61	2.50	2.38
26	3.49	3.33	3.15	2.97	2.87	2.77	2.67	2.56	2.45	2.33
27	3.45	3.28	3.11	2.93	2.83	2.73	2.63	2.52	2.41	2.29
28	3.41	3.25	3.07	2.89	2.79	2.69	2.59	2.48	2.37	2.25
29	3.38	3.21	3.04	2.86	2.76	2.66	2.56	2.45	2.33	2.21
30	3.34	3.18	3.01	2.82	2.73	2.63	2.52	2.42	2.30	2.18
40	3.12	2.95	2.78	2.60	2.50	2.40	2.30	2.18	2.06	1.93
60	2.90	2.74	2.57	2.39	2.29	2.19	2.08	1.96	1.83	1.69
120	2.71	2.54	2.37	2.19	2.09	1.98	1.87	1.75	1.61	1.43
∞	2.52	2.36	2.19	2.00	1.90	1.79	1.67	1.53	1.36	1.00

Denominator *df*

Table 7 Critical Values for the Number of Runs in a Two-Tailed Test with $\alpha = 0.05$

Each Cell Contains the Critical Values $L_{.025}$ and $R_{.025}$

The Larger of n_1 and n_2

	5	6	7	8	9	10	11	12	13	14	15	16	17	18	19	20
2								2/6	2/6	2/6	2/6	2/6	2/6	2/6	2/6	2/6
3		2/8	2/8	2/8	2/8	2/8	2/8	2/8	2/8	2/8	3/8	3/8	3/8	3/8	3/8	3/8
4	2/9	2/9	2/10	3/10	3/10	3/10	3/10	3/10	3/10	3/10	3/10	4/10	4/10	4/10	4/10	4/10
5	2/10	3/10	3/11	3/11	3/12	3/12	4/12	4/12	4/12	4/12	4/12	4/12	4/12	5/12	5/12	5/12
6		3/11	3/12	3/12	4/13	4/13	4/13	4/13	5/14	5/14	5/14	5/14	5/14	5/14	6/14	6/14
7			3/13	4/13	4/14	5/14	5/14	5/14	5/15	5/15	6/15	6/16	6/16	6/16	6/16	6/16
8				4/14	5/14	5/15	5/15	6/16	6/16	6/16	6/16	6/17	7/17	7/17	7/17	7/17
9					5/15	5/16	6/16	6/16	6/17	7/17	7/18	7/18	7/18	8/18	8/18	8/18
10						6/16	6/17	7/17	7/18	7/18	7/18	8/19	8/19	8/19	8/20	9/20
11							7/17	7/18	7/19	8/19	8/19	8/20	9/20	9/20	9/21	9/21
12								7/19	8/19	8/20	8/20	9/21	9/21	9/21	10/22	10/22
13									8/20	9/20	9/21	9/21	10/22	10/22	10/23	10/23
14										9/21	9/22	10/22	10/23	10/23	11/23	11/24
15											10/22	10/23	11/23	11/24	11/24	12/25
16												11/23	11/24	11/25	12/25	12/25
17													11/25	12/25	12/26	13/26
18														12/26	13/26	13/27
19															13/27	13/27
20																14/28

The Smaller of n_1 and n_2

Table values are for a two-tailed test with $\alpha = 0.05$.

From C. Eisenhart and F. Swed, "Tables for testing randomness of grouping in a sequence of alternatives." *The Annals of Statistics, 14(1943)*, 66–87. Reprinted by permission.

Table 8   Critical Values of Spearman's Rank Correlation Coefficient

Sample Size n	Amount of Probability in One Tail			
	0.05	0.025	0.01	0.005
5	0.900	—	—	—
6	0.829	0.886	0.943	—
7	0.714	0.786	0.893	0.929
8	0.643	0.738	0.833	0.881
9	0.600	0.683	0.783	0.833
10	0.564	0.648	0.745	0.794
11	0.523	0.623	0.736	0.818
12	0.497	0.591	0.703	0.780
13	0.475	0.566	0.673	0.745
14	0.457	0.545	0.646	0.716
15	0.441	0.525	0.623	0.689
16	0.425	0.507	0.601	0.666
17	0.412	0.490	0.582	0.645
18	0.399	0.476	0.564	0.625
19	0.388	0.462	0.549	0.608
20	0.377	0.450	0.534	0.591
21	0.368	0.438	0.521	0.576
22	0.359	0.428	0.508	0.562
23	0.351	0.418	0.496	0.549
24	0.343	0.409	0.485	0.537
25	0.336	0.400	0.475	0.526
26	0.329	0.392	0.465	0.515
27	0.323	0.385	0.456	0.505
28	0.317	0.377	0.448	0.496
29	0.311	0.370	0.440	0.487
30	0.305	0.364	0.432	0.478

Adapted from Newmark J. *Statistics and Probability in Modern Life.* 5th ed. Saunders College Publishing, Philadelphia, 1992, p.A.19.

Appendix B
Data Disk

A computer diskette is available, free of charge, to instructors who adopt this book. The disk contains the data sets for 384 exercises, including all data sets for the MINITAB problems. Data sets are stored as ASCII files, and they may be freely duplicated for student use at adopting institutions.

All data files are stored on the diskette in a directory called DATA. The name of each data file identifies the chapter and the problem number. A file name is of the form **CxPy.DAT,** where x is the chapter number and y is the problem number. For example, the data for Problem 11.124 (problem 124 in Chapter 11) is stored with the file name C11P124.DAT. The data files can be imported into MINITAB by using the **READ** and **SET** commands. To illustrate, consider Problem 11.124 that involves two samples. To store samples 1 and 2 in columns C1 and C2, insert the data disk in your disk drive B and type the following at the MINITAB prompt MTB >.

READ 'B:\DATA\C11P124' C1 C2

If drive A is used, replace B in the above command with A. Because each file includes MINITAB's default file extension DAT, it does not have to be specified in the **READ** command.

The **SET** command must be used for data files that pertain to only one sample. For example, Problem 2.19 involves a single sample. To store the values of the sample in column C1, type the following command.

SET 'B:\DATA\C2P19' C1

The data disk includes the data files listed below and a text file named README.DOC that contains these instructions for importing the files into MINI-TAB.

Chapter 1:

C1P13.DAT	C1P14.DAT	C1P19.DAT	C1P20.DAT	C1P21.DAT	C1P22.DAT
C1P23.DAT	C1P24.DAT	C1P25.DAT			

Chapter 2:

C2P1.DAT	C2P2.DAT	C2P3.DAT	C2P4.DAT	C2P14.DAT	C2P15.DAT
C2P16.DAT	C2P17.DAT	C2P18.DAT	C2P19.DAT	C2P20.DAT	C2P21.DAT
C2P26.DAT	C2P27.DAT	C2P28.DAT	C2P29.DAT	C2P30.DAT	C2P31.DAT
C2P32.DAT	C2P33.DAT	C2P34.DAT	C2P35.DAT	C2P36.DAT	C2P37.DAT
C2P38.DAT	C2P39.DAT	C2P69.DAT	C2P70.DAT	C2P71.DAT	C2P83.DAT
C2P84.DAT	C2P85.DAT	C2P86.DAT	C2P87.DAT	C2P97.DAT	C2P98.DAT
C2P99.DAT	C2P105.DAT	C2P112.DAT	C2P113.DAT	C2P114.DAT	C2P115.DAT
C2P116.DAT	C2P117.DAT	C2P124.DAT	C2P125.DAT	C2P126.DAT	C2P127.DAT
C2P132.DAT	C2P133.DAT	C2P138.DAT	C2P139.DAT	C2P140.DAT	C2P141.DAT

Chapter 3:

C3P23.DAT	C3P30.DAT	C3P32.DAT	C3P33.DAT	C3P34.DAT	C3P35.DAT
C3P36.DAT	C3P37.DAT	C3P38.DAT	C3P39.DAT	C3P50.DAT	C3P51.DAT
C3P52.DAT	C3P53.DAT	C3P54.DAT	C3P55.DAT	C3P56.DAT	C3P57.DAT
C3P58.DAT	C3P59.DAT	C3P68.DAT	C3P69.DAT	C3P70.DAT	C3P71.DAT
C3P72.DAT	C3P73.DAT	C3P74.DAT			

Chapter 5:

C5P81.DAT	C5P117.DAT	C5P118.DAT

Chapter 9:

C9P18.DAT	C9P19.DAT	C9P20.DAT	C9P21.DAT	C9P71.DAT	C9P74.DAT
C9P75.DAT	C9P76.DAT	C9P77.DAT	C9P152.DAT	C9P153.DAT	C9P159.DAT
C9P160.DAT	C9P167.DAT	C9P168.DAT	C9P179.DAT	C9P180.DAT	C9P181.DAT
C9P186.DAT	C9P187.DAT	C9P188.DAT	C9P189.DAT	C9P190.DAT	

Chapter 10:

C10P28.DAT	C10P29.DAT	C10P44.DAT	C10P45.DAT	C10P48.DAT	C10P49.DAT
C10P71.DAT	C10P75.DAT	C10P89.DAT	C10P94.DAT	C10P111.DAT	C10P112.DAT
C10P122.DAT	C10P123.DAT	C10P124.DAT	C10P128.DAT		

Chapter 11:

C11P26.DAT	C11P27.DAT	C11P28.DAT	C11P31.DAT	C11P32.DAT	C11P33.DAT
C11P34.DAT	C11P35.DAT	C11P36.DAT	C11P37.DAT	C11P38.DAT	C11P39.DAT
C11P40.DAT	C11P41.DAT	C11P42.DAT	C11P44.DAT	C11P45.DAT	C11P46.DAT
C11P47.DAT	C11P48.DAT	C11P49.DAT	C11P50.DAT	C11P51.DAT	C11P103.DAT
C11P107.DAT	C11P108.DAT	C11P121.DAT	C11P122.DAT	C11P123.DAT	C11P124.DAT
C11P125.DAT	C11P126.DAT				

Chapter 12:

C12P19.DAT	C12P20.DAT	C12P21.DAT	C12P22.DAT	C12P23.DAT	C12P24.DAT
C12P25.DAT	C12P26.DAT	C12P27.DAT	C12P28.DAT	C12P29.DAT	C12P30.DAT
C12P31.DAT	C12P32.DAT	C12P33.DAT	C12P34.DAT	C12P35.DAT	C12P39.DAT
C12P45.DAT	C12P46.DAT	C12P47.DAT	C12P48.DAT	C12P49.DAT	C12P50.DAT
C12P51.DAT	C12P52.DAT	C12P53.DAT	C12P54.DAT	C12P55.DAT	C12P60.DAT
C12P61.DAT	C12P64.DAT	C12P65.DAT	C12P73.DAT	C12P74.DAT	

Chapter 13:

C13P7.DAT	C13P8.DAT	C13P9.DAT	C13P10.DAT	C13P11.DAT	C13P14.DAT
C13P15.DAT	C13P16.DAT	C13P17.DAT	C13P18.DAT	C13P19.DAT	C13P20.DAT
C13P21.DAT	C13P22.DAT	C13P23.DAT	C13P24.DAT	C13P25.DAT	C13P26.DAT
C13P27.DAT	C13P28.DAT	C13P29.DAT	C13P30.DAT	C13P31.DAT	C13P32.DAT
C13P33.DAT	C13P34.DAT	C13P35.DAT	C13P36.DAT	C13P37.DAT	C13P40.DAT
C13P41.DAT	C13P42.DAT	C13P43.DAT	C13P44.DAT	C13P45.DAT	C13P46.DAT
C13P47.DAT	C13P48.DAT	C13P49.DAT	C13P55.DAT	C13P57.DAT	C13P58.DAT
C13P59.DAT	C13P60.DAT	C13P61.DAT	C13P62.DAT	C13P63.DAT	C13P64.DAT
C13P65.DAT	C13P66.DAT	C13P67.DAT	C13P68.DAT		

Chapter 14:

C14P6.DAT	C14P7.DAT	C14P8.DAT	C14P9.DAT	C14P12.DAT	C14P13.DAT
C14P14.DAT	C14P15.DAT	C14P16.DAT	C14P17.DAT	C14P18.DAT	C14P19.DAT
C14P20.DAT	C14P21.DAT	C14P22.DAT	C14P25.DAT	C14P26.DAT	C14P27.DAT
C14P28.DAT	C14P29.DAT	C14P30.DAT	C14P31.DAT	C14P32.DAT	C14P33.DAT
C14P34.DAT	C14P35.DAT	C14P39.DAT	C14P40.DAT	C14P41.DAT	C14P42.DAT
C14P43.DAT	C14P44.DAT	C14P45.DAT	C14P46.DAT	C14P47.DAT	C14P48.DAT
C14P49.DAT	C14P50.DAT	C14P51.DAT	C14P52.DAT	C14P55.DAT	C14P56.DAT
C14P57.DAT	C14P58.DAT	C14P59.DAT	C14P60.DAT	C14P61.DAT	C14P62.DAT
C14P63.DAT	C14P64.DAT	C14P65.DAT	C14P66.DAT		

Chapter 15:

C15P1.DAT	C15P2.DAT	C15P3.DAT	C15P4.DAT	C15P5.DAT	C15P6.DAT
C15P7.DAT	C15P8.DAT	C15P9.DAT	C15P10.DAT	C15P11.DAT	C15P12.DAT
C15P13.DAT	C15P14.DAT	C15P15.DAT	C15P16.DAT	C15P17.DAT	C15P18.DAT
C15P19.DAT	C15P20.DAT	C15P21.DAT	C15P22.DAT	C15P23.DAT	C15P24.DAT
C15P25.DAT	C15P26.DAT	C15P27.DAT	C15P28.DAT	C15P29.DAT	C15P30.DAT
C15P31.DAT	C15P32.DAT	C15P33.DAT	C15P34.DAT	C15P35.DAT	C15P36.DAT
C15P37.DAT	C15P38.DAT	C15P39.DAT	C15P40.DAT	C15P41.DAT	C15P42.DAT
C15P43.DAT	C15P44.DAT	C15P45.DAT	C15P46.DAT	C15P49.DAT	C15P52.DAT
C15P59.DAT	C15P60.DAT	C15P61.DAT	C15P62.DAT	C15P66.DAT	C15P67.DAT
C15P68.DAT	C15P69.DAT	C15P70.DAT	C15P71.DAT	C15P72.DAT	C15P73.DAT
C15P74.DAT	C15P75.DAT	C15P76.DAT	C15P77.DAT	C15P80.DAT	C15P81.DAT
C15P82.DAT	C15P83.DAT	C15P84.DAT	C15P85.DAT	C15P86.DAT	C15P87.DAT
C15P88.DAT	C15P89.DAT	C15P90.DAT	C15P91.DAT	C15P92.DAT	C15P93.DAT
C15P94.DAT	C15P95.DAT	C15P96.DAT			

Appendix C
Bibliography

Bell, E. T. *Men of Mathematics,* Simon and Schuster, New York, 1937.

Cochran, W. G. *Sampling Techniques,* Wiley, New York, 1963.

Cochran, W. G. and Cox, G. M. *Experimental Designs,* 2d ed. Wiley, New York, 1957.

Devore, J. and Peck, R. *Statistics: The Exploration and Analysis of Data,* West, St. Paul, MN, 1986.

Draper, N. R. and Smith, H. *Applied Regression Analysis,* 2d ed. Wiley, New York, 1981.

Feller, W. *An Introduction to Probability Theory and Its Applications,* Vol. 1, 3d ed. Wiley, New York, 1968.

Freedman et al. *Statistics,* 2d ed. Norton, New York, 1991.

Freund, J. E. *Mathematical Statistics,* 5th ed. Prentice Hall, Englewood Cliffs, NJ, 1992.

Freund, J. E. and Simon, G. A. *Statistics: A First Course,* 5th ed. Prentice Hall, Englewood Cliffs, NJ, 1991.

Hacking, I. *The Emergence of Probability,* Cambridge University Press, New York, 1975.

Hald, A. *A History of Probability and Statistics and Their Applications before 1750,* Wiley, New York, 1990.

Hicks, C. R. *Fundamental Concepts in the Design of Experiments,* 3d ed. Holt, Rinehart, and Winston, New York, 1982.

Hogg, R. V. and Craig, A. T. *Introduction to Mathematical Statistics,* 4th ed. Macmillan, New York, 1986.

Johnson, R. *Elementary Statistics,* 5th ed. PWS-KENT, Boston, 1988.

Khazanie, R. *Elementary Statistics in a World of Applications,* 3d ed. Scott, Foresman, Glenview, IL, 1990.

Kish, L. *Survey Sampling,* Wiley, New York, 1965.

Larsen, R. J. and Marx, M. L. *Statistics,* Prentice Hall, Englewood Cliffs, NJ, 1990.

McClave, J. T. and Dietrich, F. H. *Statistics,* 5th ed. Dellen, San Francisco, 1991.

Mendenhall, W. and Beaver, R. *Introduction to Probability and Statistics,* 8th ed. PWS-KENT, Boston, 1991.

Mendenhall, W., Wackerly, D., and Scheaffer, R. L. *Mathematical Statistics with Applications,* 4th ed. PWS-KENT, Boston, 1990.

Meyers, R. H. *Classical and Modern Regression with Applications,* 2d ed. PWS-KENT, Boston, 1990.

Minitab Inc., *MINITAB Reference Manual,* Release 8, PC Version, Minitab, Inc., State College, PA, 1991.

Mood, A. M., Graybill, F. A., and Boes, D. C. *Introduction to the Theory of Statistics,* 3d ed. McGraw-Hill, New York, 1974.

Moore, D. S. and McCabe, G. P. *Introduction to the Practice of Statistics,* Freeman, New York, 1989.

Mosteller, F. and Rourke, R. E. *Sturdy Statistics,* Addison-Wesley, Reading, MA, 1973.

National Bureau of Standards Handbook 91, *Experimental Statistics,* U.S. Government Printing Office, Washington, D.C., 1963.

Neter, J., Wasserman, W., and Whitmore, G. A. *Applied Statistics,* 3d ed. Allyn and Bacon, Boston, 1987.

Newmark, J. *Statistics and Probability in Modern Life,* 5th ed. Saunders College Publishing, Philadelphia, 1992.

Ott, L. *An Introduction to Statistical Methods and Data Analysis,* 3d ed. PWS-KENT, Boston, 1988.

Salvia, A. A. *Introduction to Statistics,* Saunders College Publishing, Philadelphia, 1990.

Scheaffer, R. L. and Farber, E. *The Student Edition of MINITAB, Release 8,* Addison-Wesley, Reading MA, 1992.

Scheaffer, R. L., Mendenhall, W., and Ott, L. *Elementary Survey Sampling,* 4th ed. PWS-KENT, Boston, 1990.

Snedecor, G. W. and Cochran, W. G. *Statistical Methods,* 8th ed. Iowa State University Press, Ames, IA, 1989.

Stigler, S. M. *The History of Statistics: The Measurement of Uncertainty Before 1900,* Belkamp Press of Harvard University, Cambridge, MA, 1986.

Tanur, J. M. et al. eds. *Statistics: A Guide to the Unknown,* 3d ed. Wadsworth/Brooks-Cole, Belmont, CA, 1989.

Triola, M. F. *Elementary Statistics,* 4th ed. Addison-Wesley, Reading, MA, 1989.

Tukey, J. W. *Exploratory Data Analysis,* Addison-Wesley, Reading, MA, 1977.

U.S. Bureau of the Census, *Statistical Abstract of the United States: 1991,* U.S. Government Printing Office, Washington, D.C., 1991.

Velleman, P. F. and Hoaglin, D. C. *Applications, Basics, and Computing of Exploratory Data Analysis,* PWS-KENT, Boston, 1981.

Walpole, R. E. and Myers, R. H. *Probability and Statistics for Engineers and Scientists,* 3d ed. Macmillan, New York, 1985.

Weiss, N. A. *Elementary Statistics,* Addison-Wesley, Reading, MA, 1989.

Weiss, N. A. and Hassett, M. J. *Introductory Statistics,* 3d ed. Addison-Wesley, Reading, MA, 1991.

Williams, B. *A Sampler on Sampling,* Wiley, New York, 1978.

Answers to Odd-Numbered Section Exercises and All Review Exercises

Chapter 1
Sections 1.1–1.3

1.1 Sample

1.3 a. Descriptive
 b. Inferential
 c. Descriptive
 d. Descriptive
 e. Inferential

1.5 a. The 113 faculty would be considered a sample if one were interested in making inferences about all college faculty in the United States. (Other answers are possible.)
 b. If one were only interested in describing faculty at this particular college, then the 113 faculty would be considered a population.

1.7 a. The population of interest is all people who will actually vote in the mayoral election.
 b. It is impossible to know in advance who will vote in the election.

1.9 a. The population of interest is all sardines in the catch.
 b. The parameter of interest is the average weight of sardines in the catch.
 c. The statistic used is the sample average, and its value is 38.9.
 d. They probably were caught in the same location, and this might contain sardines that were mostly small or mostly large.

1.11 a. The population of interest is all homes in the city.
 b. The parameter of interest is the percentage of city homes with an in-sink garbage disposal.
 c. The statistic used was the percentage of the 880 homes with an in-sink garbage disposal, and its value was 35 percent.

Ⓜ 1.13 SUM = 4535.0

Ⓜ 1.15

ROW	F	C
1	0	-17.7778
2	10	-12.2222
3	20	-6.6667
4	30	-1.1111
5	40	4.4444
6	50	10.0000
7	60	15.5556
8	70	21.1111
9	80	26.6667
10	90	32.2222
11	100	37.7778

Ⓜ 1.17

ROW	N	CUBE	CUBE RT
1	1	1	1.00000
2	2	8	1.25992
3	3	27	1.44225
4	4	64	1.58740
5	5	125	1.70998
6	6	216	1.81712
7	7	343	1.91293
8	8	512	2.00000
9	9	729	2.08008
10	10	1000	2.15443
11	11	1331	2.22398
12	12	1728	2.28943
13	13	2197	2.35133
14	14	2744	2.41014
15	15	3375	2.46621
16	16	4096	2.51984
17	17	4913	2.57128
18	18	5832	2.62074
19	19	6859	2.66840
20	20	8000	2.71442
21	21	9261	2.75892
22	22	10648	2.80204
23	23	12167	2.84387
24	24	13824	2.88450
25	25	15625	2.92402

Ⓜ 1.19

ROW	STUDENT	FE
1	1	73
2	2	85
3	3	72
4	4	69
5	5	87
6	6	94
7	7	86
8	8	81
9	9	58
10	10	98

Ⓜ 1.21

ROW	STUDENT	FE
1	9	58
2	4	69
3	3	72
4	1	73
5	8	81
6	2	85
7	7	86
8	5	87
9	6	94
10	10	98

Ⓜ 1.23

ROW	STUDENT	E1	E2	E3	FE	AVG
1	1	69	78	54	73	69.4
2	2	89	87	72	85	83.6
3	3	75	81	47	69	68.2
4	4	91	86	58	72	75.8
5	5	93	91	69	87	85.4
6	6	99	95	83	94	93.0
7	7	72	82	68	86	78.8
8	8	84	89	70	81	81.0
9	9	65	69	39	58	57.8
10	10	94	94	82	98	93.2

Ⓜ 1.25

ROW	AMOUNT	RANK
1	8.13	8
2	23.64	48
3	14.53	30
4	5.98	1
5	15.89	33
6	18.01	39
7	22.10	46
8	8.22	9
9	11.93	22
10	10.06	14
11	17.23	37
12	9.84	13
13	19.25	42
14	17.95	38
15	11.85	21
16	15.85	32
17	7.04	3
18	11.53	19
19	12.66	24
20	24.45	49
21	19.18	41
22	13.83	28
23	18.62	40
24	6.58	2
25	13.37	27
26	7.56	5
27	13.16	26
28	9.65	12
29	14.46	29
30	21.16	45
31	11.34	18
32	29.48	50
33	10.55	16
34	8.01	7
35	23.34	47
36	10.24	15
37	14.74	31
38	10.99	17
39	11.71	20
40	8.72	10
41	13.09	25
42	7.70	6
43	8.94	11
44	16.64	36
45	12.37	23
46	16.29	34
47	19.81	43
48	7.46	4
49	16.53	35
50	19.83	44

Chapter 2
Section 2.1

2.1

Percentage	Frequency
30–39	1
40–49	2
50–59	4
60–69	2
70–79	6
80–89	1

2.3

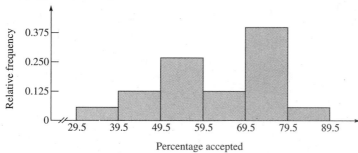

2.5   **Class**

20–27
28–35
36–43
44–51
52–59

2.7   **Class**

99.5–139.5
139.5–179.5
179.5–219.5
219.5–259.5
259.5–299.5
299.5–339.5

2.9   **Class**

69.5– 78.5
78.5– 87.5
87.5– 96.5
96.5–105.5
105.5–114.5
114.5–123.5
123.5–132.5
132.5–141.5

2.11 8, 8, 8, 8, 8, 8, 8, 8, 8, 8, 9, 9, 9, 9, 9, 9, 10, 10, 11, 11

2.13 a. 5   f. 39
      b. 44.5–49.5 g. 37
      c. 24.5   h. 24
      d. 69.5   i. 97
      e. 55

2.15

Unemployment Rate	Relative Frequency
2.0– 2.9	0.02
3.0– 3.9	0.12
4.0– 4.9	0.16
5.0– 5.9	0.16
6.0– 6.9	0.16
7.0– 7.9	0.16
8.0– 8.9	0.14
9.0– 9.9	0
10.0–10.9	0.06
11.0–11.9	0
12.0–12.9	0.02

(Other answers are possible.)

2.17

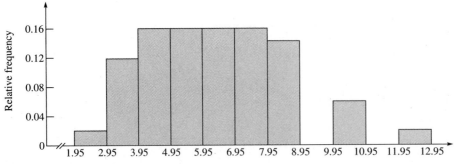

(Other answers are possible.)

 2.19 `Histogram of SCORES   N = 79`

```
   Midpoint    Count
      470.0        2   **
      490.0        6   ******
      510.0        4   ****
      530.0        7   *******
      550.0       13   *************
      570.0       13   *************
      590.0       16   ****************
      610.0        7   *******
      630.0        1   *
      650.0        1   *
      670.0        2   **
      690.0        5   *****
      710.0        1   *
      730.0        1   *
```

 2.21 `Histogram of 1980    N = 32`

```
   Midpoint    Count
        60       10   **********
        70        5   *****
        80        8   ********
        90        4   ****
       100        1   *
       110        2   **
       120        2   **
```

`Histogram of 1987   N = 32`

```
   Midpoint    Count
        80        2   **
        90        6   ******
       100        3   ***
       110        5   *****
       120        2   **
       130        5   *****
       140        2   **
       150        0
       160        2   **
       170        1   *
       180        3   ***
       190        1   *
```

Section 2.2

2.23 20, 20.5, 21.7, 21.8, 21.8, 21.8, 21.8, 22, 22, 22, 22, 22.2, 22.2, 22.5, 22.5, 22.7, 23, 23.6

2.25 $4.25, $4.25, $4.25, $4.40, $4.40, $4.40, $4.45, $4.50, $4.50, $4.50, $4.50, $4.55, $4.60, $4.65, $4.75, $4.75, $4.75, $5.00, $5.00, $5.00, $5.00

2.27

Stem	Leaf
0	5 9 8 6
1	7 3 9 1 0 6 0 5 7 4 2 1 7 4 3
2	6 6 6 9 7
3	9 5 1 7
4	5 9 7 7
5	0
6	6
7	
8	
9	
10	4
11	
12	
13	
14	
15	
16	
17	7
18	
19	4

2.29

2.31

```
 ..:   : :.:..  .:.:..  .::: .. .   .   . ...   .
+++++++++++++++++++++++++++++++++++++++++++++++++- Age
20        30         40         50         60        70
```

2.33

Stem	Leaf
2	1 3 9 7 7 2 9 3
3	5 5 7 9 1 6 7 5 2 4 8 1 0 7
4	3 2 8 1 2 3 3 1 5 6 1
5	3 8 9 0 6
6	1 4

2.35

Stem	Leaf		Stem	Leaf
3.2	9		3.2	9
3.3	9		3.3	9
3.4	5 1		3.4	1 5
3.5	3 0		3.5	0 3
3.6	7 0 4 2		3.6	0 2 4 7
3.7	1 4 5 3		3.7	1 3 4 5
3.8	8		3.8	8
3.9	0		3.9	0
4.0	0 0		4.0	0 0

Ⓜ 2.37
```
Stem-and-leaf of GPA      N  = 47
Leaf Unit = 0.010

     4    28 0479
     9    29 44459
    15    30 112689
    22    31 4567899
   (5)    32 14689
    20    33 1266799
    13    34 11
    11    35 147
     8    36 167
     5    37 257
     2    38 7
     1    39 7
```

Ⓜ 2.39
```
Stem-and-leaf of SAT      N  = 79
Leaf Unit = 10

     8    4 77889999
   (53)   5 0111222333344444444444455566666666677777888888888999999
    18    6 0000011247788889
     2    7 12
```

Section 2.3

2.41

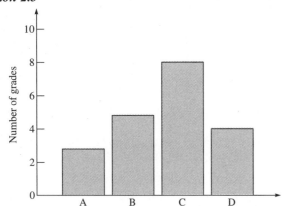

2.43

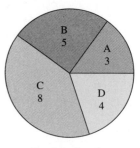

Number of grades

2.45

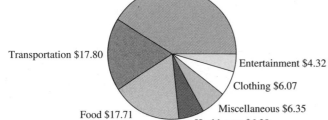

Average expenditure per $100.00

2.47

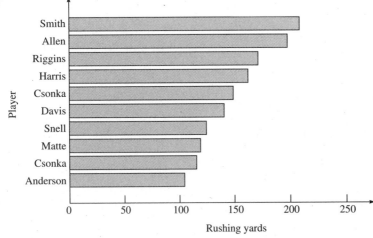

2.49

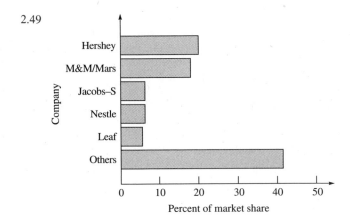

2.51

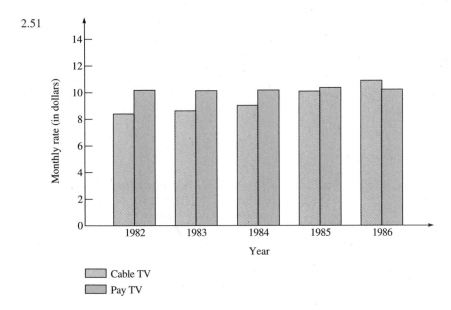

Section 2.4

2.53 $7.50

2.55 a. 5.2
 b. 6
 c. 8

2.57 a. 43.7
 b. The data would be regarded as a sample if it were used to estimate the mean age of all faculty at the school. It would be denoted by $\bar{x}$ and would be a statistic.
 c. The data would be considered a population if it were only used to determine the mean age of the math faculty at this school. It would be denoted by μ and would be a parameter.

2.59 a. $75,500
 b. $53,000
 c. Median
 d. Mean

2.61 Yes, because the total weight was 30,732 pounds and this exceeds the maximum.

2.63 77.96

2.65 2.73

2.67 33.5

Ⓜ 2.69 a. **MEAN** = **44.800**
 b. If the number of motorists in each state were known, then the percentage of motorists in the 50 states who use their seat belts could be found. It

would equal the weighted mean of the 50 state percentages, where the weights would be the number of motorists in each of the states.

Ⓜ 2.71 MEAN   =   74.043

MEDIAN =   74.100

Section 2.5
2.73 a. 9
b. 13.5
c. 3.7
2.75 a. $\Sigma(x - \bar{x})^2 = 61.7$
b. $\Sigma x^2 - (\Sigma x)^2/n = 61.7$
c. 3.2
2.77 Range $= 10.71$, $s = 4.730$
2.79 a. Each sample has a mean of 5 and a range of 10.
b. Sample A
c. $s_A = 4.1$, $s_B = 3.1$
d. The standard deviation is the better measure of variability. Each sample value is used in calculating the standard deviation, while only the largest and smallest values are used to calculate the range.
2.81 a. 0.002
b. 2. If 2 is divided by 1,000, it will give the standard deviation of the original sample.
2.83 $s = 2.2$
Ⓜ 2.85 MEAN   =   3.2734
MEDIAN =   3.2400
ST.DEV. =   0.29066
Ⓜ 2.87 MEAN   =   78.791
MEAN   =   122.73

MEDIAN =   75.400
MEDIAN =   115.30

ST.DEV. =   17.943
ST.DEV. =   31.700

Section 2.6
2.89 Data sets with mound-shaped distributions
2.91 a. Approximately 68 percent
b. Approximately 95 percent
c. Approximately 100 percent

2.93 The percentage is at least ($\geq$) 93.75 percent.
2.95 a. Between 26.4 and 29.6 ounces
b. Between 25.6 and 30.4 ounces
c. Approximately 16 percent
2.97 a. $35/50 = 70$ percent
b. $47/50 = 94$ percent
c. $50/50 = 100$ percent
d. The percentages given by the Empirical rule are approximately 68, 95, and 100 percent for Parts a, b, and c, respectively. These values are very close to the actual percentages obtained.
e. The percentages given by Chebyshev's theorem for Parts b and c, respectively, are at least 75 and at least 89 percent. These values are very conservative and are considerably less than the actual percentages obtained.
2.99 10.75

Section 2.7
2.101

Price	z-score
50	-1.000
45	-1.625
60	0.250

2.103 40 cents
2.105 a. 0.95
b. 0.72
c. The nonresident tuition at Temple University
2.107 a. Approximately 50 percent
b. Approximately 75 percent
c. Approximately 25 percent
2.109 a. 21.00, 31.00, 37.50, 44.50, and 64.00
b. 43.00
c. 13.50
2.111 5-number summary: 0, 200, 400, 600, and 1,000
IQR $= 400$
Range $= 1,000$

Ⓜ 2.113

ROW	GPA	Z-SCORE	ROW	GPA	Z-SCORE
1	3.29	0.05710	25	3.36	0.29793
2	3.67	1.36446	26	3.37	0.33233
3	3.28	0.02269	27	3.39	0.40114
4	2.89	-1.31908	28	3.75	1.63970
5	3.01	-0.90623	29	3.41	0.46995
6	2.94	-1.14706	30	3.97	2.39659
7	3.31	0.12590	31	2.87	-1.38789
8	2.80	-1.62872	32	2.99	-0.97503
9	3.54	0.91720	33	3.02	-0.87182
10	3.15	-0.42456	34	3.09	-0.63099
11	3.08	-0.66540	35	3.87	2.05255
12	3.01	-0.90623	36	3.14	-0.45897
13	3.17	-0.35576	37	3.18	-0.32135
14	3.36	0.29793	38	3.24	-0.11493
15	3.57	1.02042	39	3.41	0.46995
16	3.39	0.40114	40	3.61	1.15803
17	3.32	0.16031	41	2.94	-1.14706
18	3.19	-0.28695	42	3.06	-0.73420
19	2.95	-1.11265	43	2.94	-1.14706
20	3.19	-0.28695	44	3.66	1.33006
21	2.84	-1.49110	45	3.77	1.70850
22	3.16	-0.39016	46	3.72	1.53648
23	3.21	-0.21814	47	3.51	0.81399
24	3.26	-0.04612			

Ⓜ 2.115

```
                 ---------------
      ----------I     +    I------------------    *
                 ---------------
 ----+---------+---------+---------+---------+---------+--GPA
    2.75      3.00      3.25      3.50      3.75      4.00
```

Review Exercises

2.117 a. 13.35
b. 13.3
c. 3.3
d. 1.03
e.

Stem	Leaf
11	9
12	8 9 0
13	0 6 8
14	3 0
15	2

2.118 $\bar{x} = 15.6$; $s = 10.46$
2.119 215 pounds
2.120 $s = 5.1$
2.121 The z-value for milk is 3.25, and the z-value for ground beef is 3.08. Therefore, milk is relatively more overpriced.
2.122 37.8 years
2.123 42.0 hours

2.124

Stem	Leaf
2	5 7
3	9 8 0 2 4
4	3 2 1 8 0 1 9 8 2 6 6 2 0 0 0 2 7 8 2 3
5	2 1 5 2 1 9 1 2
6	1

2.125

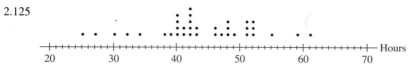

2.126

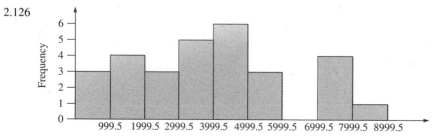

Miles

(Other answers are possible.)

2.127 a.

Class	Frequency
0	3
1	9
2	18
3	9
4	8
5	1

b. $\bar{x} = 2.3$; $s = 1.2$

2.128 Meritorious (M)

2.129

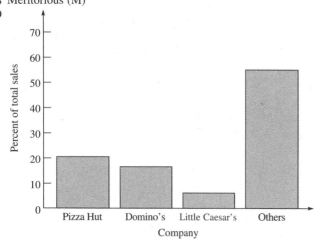

2.130

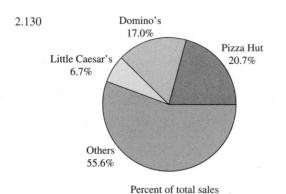

Domino's 17.0%

Pizza Hut 20.7%

Little Caesar's 6.7%

Others 55.6%

Percent of total sales

2.131

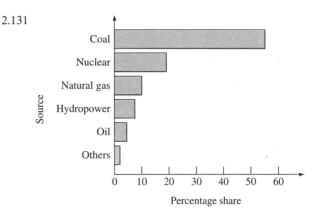

Percentage share

2.132 a.

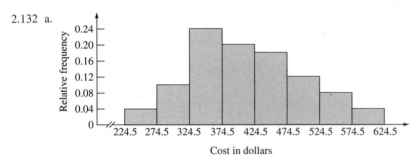

The distribution is slightly skewed to the right.

b. 48/50 = 96 percent of the data are within 2 standard deviations of the mean, and 100 percent are within 3 standard deviations.

c. The percentages given by the Empirical rule are approximately 95 and 100 percent for 2 and 3 standard deviations, respectively. These values are very close to the actual percentages obtained.

d. The percentages given by Chebyshev's theorem for 2 and 3 standard deviations, respectively, are at least 75 and at least 89 percent. These values are very conservative and are considerably less than the actual percentages obtained.

2.133 $90

2.134 a. Approximately 68 percent
     b. Approximately 2.5 percent
     c. Approximately 2.5 percent

2.135 At least 84 percent of the scores will be between 54 and 94.

2.136 5-number summary: 79.60, 94.95, 115.30, 139.70, and 186.20
     $R = 106.6$
     $IQR = 44.75$

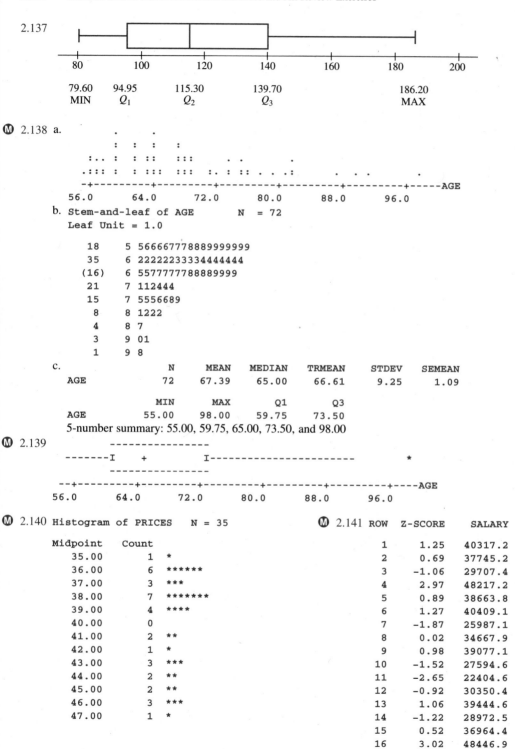

2.137

```
79.60      94.95         115.30           139.70                        186.20
MIN         Q₁            Q₂                Q₃                           MAX
```

Ⓜ 2.138 a.

```
              .          .

         :   :   :    :
    :.. :   : ::   :::       .  .              .
  .:::  :   : :::  :::  :. : ::  . . . .:     .  . .              .
  -+---------+---------+---------+---------+---------+-----AGE
  56.0      64.0      72.0      80.0      88.0      96.0
```

b. Stem-and-leaf of AGE   N = 72
   Leaf Unit = 1.0

```
   18     5 566667778889999999
   35     6 22222233334444444
  (16)    6 5577777788889999
   21     7 112444
   15     7 5556689
    8     8 1222
    4     8 7
    3     9 01
    1     9 8
```

c.
	N	MEAN	MEDIAN	TRMEAN	STDEV	SEMEAN
AGE	72	67.39	65.00	66.61	9.25	1.09

	MIN	MAX	Q1	Q3
AGE	55.00	98.00	59.75	73.50

5-number summary: 55.00, 59.75, 65.00, 73.50, and 98.00

Ⓜ 2.139

```
                  ----------------
     -------I     +       I----------------------           *
                  ----------------
  --+---------+---------+---------+---------+---------+----AGE
  56.0      64.0      72.0      80.0      88.0      96.0
```

Ⓜ 2.140 Histogram of PRICES   N = 35

Midpoint	Count	
35.00	1	*
36.00	6	******
37.00	3	***
38.00	7	*******
39.00	4	****
40.00	0	
41.00	2	**
42.00	1	*
43.00	3	***
44.00	2	**
45.00	2	**
46.00	3	***
47.00	1	*

Ⓜ 2.141

ROW	Z-SCORE	SALARY
1	1.25	40317.2
2	0.69	37745.2
3	-1.06	29707.4
4	2.97	48217.2
5	0.89	38663.8
6	1.27	40409.1
7	-1.87	25987.1
8	0.02	34667.9
9	0.98	39077.1
10	-1.52	27594.6
11	-2.65	22404.6
12	-0.92	30350.4
13	1.06	39444.6
14	-1.22	28972.5
15	0.52	36964.4
16	3.02	48446.9

Chapter 3
Section 3.1
3.1 $b_0 = 3, b_1 = 8$
3.3 $b_0 = 0, b_1 = 4$
3.5 $b_0 = 10, b_1 = 0$
3.7 $b_0 = -8, b_1 = 4/3$
3.9 a. $y = 3 + 8x$

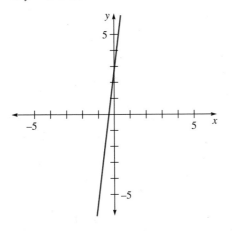

b. $5x - 4y = 20$

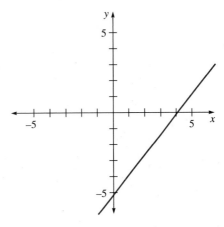

c. $y = 4x$

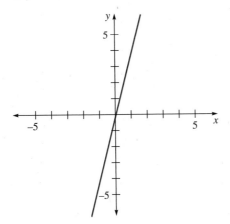

d. $2x + 3y = 10$

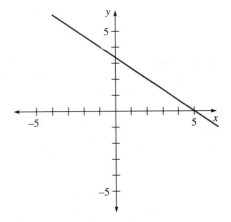

3.11 $b_1 = -3/7$
3.13 $y = 7 + 10x$

3.15 $y = 4 - x$

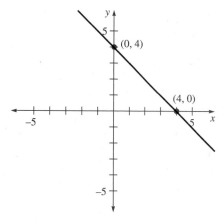

3.17 a. $y = 35.00 + 0.30x$
      b. $b_1 = 0.30$, $b_0 = 35.00$
      c. $30.00
3.19 a. 1.5
      b. 1,500
3.21 $SS(x) = 26.8$, $SS(y) = 36.8$, $SS(xy) = 30.2$
3.23 $SS(x) = 1,464.1$, $SS(y) = 790.1$, $SS(xy) = -1,004.1$

Sections 3.2 and 3.3

3.25 a.

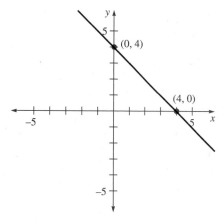

      b. Yes
3.27 $y = 1.6 + 1.7x$
3.29 $y = 0.7829 + 2.2541x$
3.31 a. $y = 0.1 + 1.2x$
      b.

x	y	Pred. y	Error	(Error)²
0	1	0.1	0.9	0.81
1	0	1.3	-1.3	1.69
3	4	3.7	0.3	0.09
4	5	4.9	0.1	0.01
			0	2.60

c. SSE = 2.60. The error column has a sum of 0.

d.

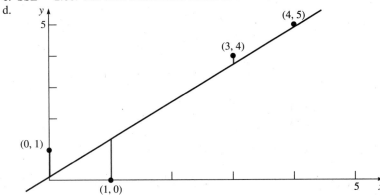

3.33 a.

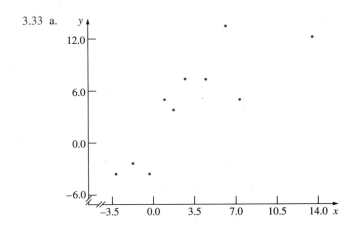

b. $y = 0.913 + 1.021x$

c. 2.955 percent

Ⓜ 3.35

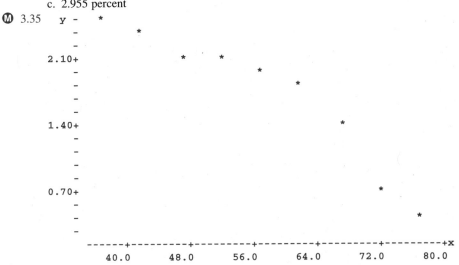

3.37

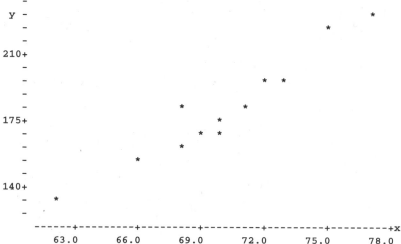

 3.39 The regression equation is
y = -289 + 6.70 x

Section 3.4

3.41 Positively correlated

3.43 Positively correlated

3.45 Positively correlated

3.47 Positively correlated

3.49 $r = 0.96$

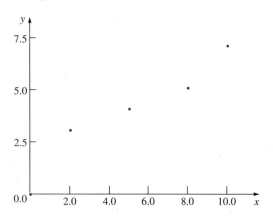

3.51 $r = -0.93$. Tread depth (y) tends to decrease as the number of miles of usage (x) increases.

3.53 $r = 0.21$. The linear relationship is very weak between the number of wins and the batting average.

3.55 $r = -0.95$. As a person's age increases, the estimated years of added life decrease almost linearly.

 3.57 Correlation of x and y = 0.214

Ⓜ 3.59 a.

```
        75+
            -
   SALES  -                                              *
            -                                         *
            -
        60+                              *         *
            -                          *     *
            -                    *
            -
            -              *
        45+          *     *
            -       *
            -    *           *
            -  *
            -     *
        30+
            -
          ------+---------+---------+---------+---------+---------+DJIA
              1500      1750      2000      2250      2500      2750
```

b. The regression equation is
   SALES = - 2.39 + 0.0270 DJIA

c. Correlation of DJIA and SALES = 0.960

Review Exercises

3.60 $b_0 = 5, b_1 = -2.5$

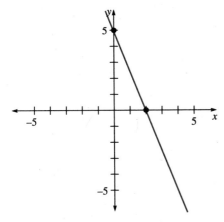

3.61 $y = 1/6 - 7x/3$
3.62 a. $y = 135 + 75x$
   b. $b_1 = 75, b_0 = 135$
3.63 10

3.64

3.65 $y = 6.306 - 1.038x$
3.66 $r = -0.96$. The data points lie near a straight line with negative slope.
3.67 $r = 0.92, b_1 = 1.2, SS(x) = 10, SS(y) = 17$, and $0.92 = 1.2\sqrt{10/17}$.

3.68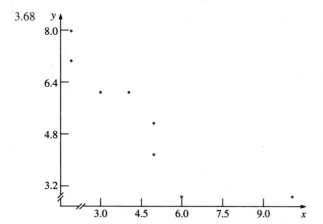

b. $y = 8.076 - 0.611x$
c. $r = -0.87$. The subtest score tends to decrease as the time after injury increases.
d. $y = 3.8 \approx 4$

3.69 a. $r = 0.199$
b. No

3.70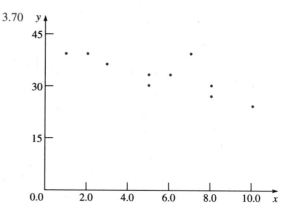

b. $y = 40.678 - 1.396x$
c. $r = -0.81$. The installation time tends to decrease as the experience of a worker increases.

Ⓜ 3.71 a.

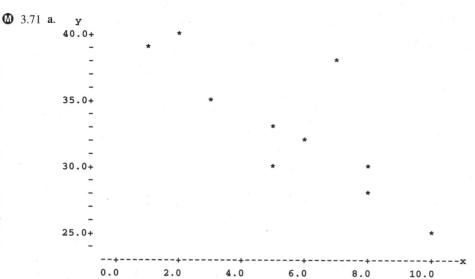

b. The regression equation is
$y = 40.7 - 1.40\ x$
c. Correlation of x and y $= -0.809$

Ⓜ 3.72

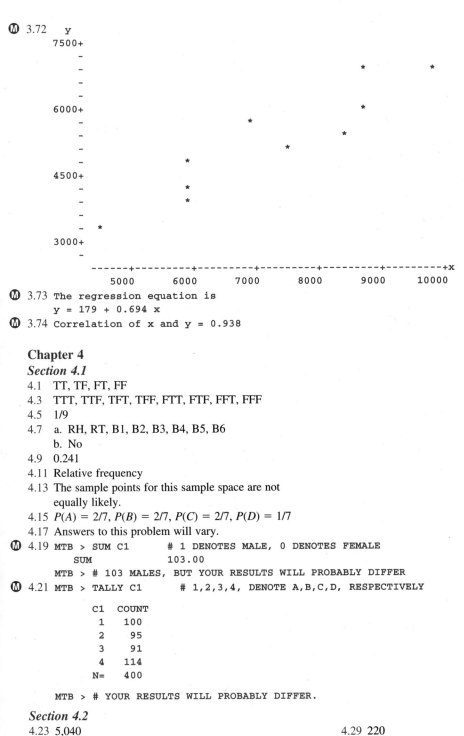

Ⓜ 3.73 The regression equation is
y = 179 + 0.694 x

Ⓜ 3.74 Correlation of x and y = 0.938

Chapter 4
Section 4.1
4.1 TT, TF, FT, FF
4.3 TTT, TTF, TFT, TFF, FTT, FTF, FFT, FFF
4.5 1/9
4.7 a. RH, RT, B1, B2, B3, B4, B5, B6
     b. No
4.9 0.241
4.11 Relative frequency
4.13 The sample points for this sample space are not
     equally likely.
4.15 $P(A) = 2/7$, $P(B) = 2/7$, $P(C) = 2/7$, $P(D) = 1/7$
4.17 Answers to this problem will vary.

Ⓜ 4.19
```
MTB > SUM C1        # 1 DENOTES MALE, 0 DENOTES FEMALE
    SUM            103.00
MTB > # 103 MALES, BUT YOUR RESULTS WILL PROBABLY DIFFER
```

Ⓜ 4.21
```
MTB > TALLY C1     # 1,2,3,4, DENOTE A,B,C,D, RESPECTIVELY

        C1   COUNT
         1    100
         2     95
         3     91
         4    114
        N=    400

MTB > # YOUR RESULTS WILL PROBABLY DIFFER.
```

Section 4.2
4.23 5,040
4.25 100
4.27 120

4.29 220
4.31 88
4.33 1

4.35 1,024

4.37 32,768

4.39

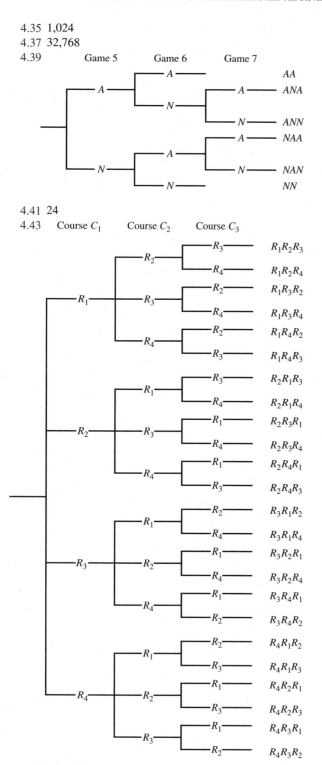

4.41 24

4.43

4.45 24

4.47 362,880

4.49 2,598,960

4.51 0.000009234

4.53 3,838,380

4.55 The first element in the ordered arrangement can be selected in n ways, then the second can be selected in $(n - 1)$ ways, then the third in $(n - 2)$ ways, . . . and lastly, the kth element can be selected in $[n - (k - 1)] = (n - k + 1)$ ways. Thus, by the Multiplication rule, the number of possible arrangements is

$$P(n, k) = n(n - 1)(n - 2) \cdots (n - k + 1)$$
$$= n(n - 1)(n - 2) \cdots (n - k + 1) \cdot \frac{(n - k)!}{(n - k)!}$$
$$= n!/(n - k)!$$

Section 4.3

4.57 Mutually exclusive

4.59 Not mutually exclusive

4.61 Not mutually exclusive

4.63 Not mutually exclusive

4.65 0.562

4.67 a. Selecting a green card
      b. Selecting a card that is either red or black
         (Other answers are possible for Parts a and b.)

4.69 0.94

4.71 81 percent

4.73 0.55

4.75 0.71

4.77 0.769

Section 4.4

4.79 Dependent

4.81 Independent

4.83 Independent

4.85 Dependent

4.87 Dependent

4.89 a. The probability that a company will show a profit for the year if December sales are good.
      b. The probability of good December sales if the company shows a profit for the year.
      c. The probability of good December sales and a profitable year.

4.91 $P(A|B) = 0.5$, $P(B|A) = 0.4$

4.93 0.86

4.95 a. 0.0714

b. 0.00476

4.97 a. 0.610

b. 0.165

c. 0.271

d. 0.105

4.99 0.76

4.101 29.9 percent

Sections 4.5 and 4.6

4.103 The birth of a wire-haired fox terrier is a female.

4.105 At least one component in an optical scanner does not function properly.

4.107 All 7 department faculty are married.

4.109 All in a class of 10 took the exam.

4.111 Fewer than 8 cards in a 13-card bridge hand are red.

4.113 0.306

4.115 0.9999

4.117 0.012

4.119 a. 0.00107

b. 5.06×10^{-15}

4.121 a. 0.45

b. 0.90

c. 0.30

d. 0.65

e. 0.417

f. 0.75

4.123 0.575

4.125 0.913

Review Exercises

4.127 0.6

4.128 0.52

4.129 a. 10

b. 105

c. 105

d. 1,000

4.130 792

4.131 0.348

4.132 $P(B|A) = 0.59$, $P(A|B) = 0.875$

4.133 Events A and B are not independent and they are not mutually exclusive.

4.134 a. 1/9

b. 5/9

4.135 0.127

4.136 $(S, 1)$ $(S, 2)$ $(S, 5)$ $(\bar{S}, 1)$, $(\bar{S}, 2)$ $(\bar{S}, 5)$

4.137 a. Not mutually exclusive

b. Not mutually exclusive

c. Mutually exclusive

d. Not mutually exclusive

4.138 24

4.139 a. 0.60

b. 0.25

c. 0.85

d. 0.556

4.140 720

4.141 48

4.142 0.94

4.143 240

4.144 a. Yes

b. No

c. Yes

4.145 28

4.146 1,260

4.147 a. 15,504

b. 5,940

c. 0.383

4.148 0.999996

4.149 0.375

4.150 a. 0.225

b. 0.28

4.151 0.73

4.152 0.936

4.153 a. 0.0081

b. 0.7599

c. 0.9919

4.154 0.176

4.155 0.00104

4.156 0.75

4.157 0.017

4.158 0.353

4.159 0.507

4.160 0.253

Ⓜ 4.161

```
         C1   COUNT
          1    28
          2    28
          3    32
          4    33
          5    20
          6    39
         N=   180
MTB > # REL. FREQ. of ONE = 28/180 = 0.156

         C1   COUNT
          1    58
          2    64
          3    58
          4    63
          5    56
          6    61
         N=   360
MTB > # REL. FREQ. OF ONE = 58/360 = 0.161

         C1   COUNT
          1    88
          2   102
          3    86
          4    98
          5    79
          6    87
         N=   540
MTB > # REL. FREQ. OF ONE = 88/540 = 0.163

         C1   COUNT
          1   121
          2   124
          3   116
          4   116
          5   112
          6   131
         N=   720
MTB > # REL. FREQ. OF ONE = 121/720 = 0.168
MTB > # 0.168 IS CLOSE TO THE THEORETICAL PROBABILITY OF 1/6
MTB > # YOUR RESULTS WILL PROBABLY DIFFER
```

Ⓜ 4.162

```
MTB > # 1 = CLUB, 2 = DIAMOND, 3 = HEART, 4 = SPADE
MTB > TALLY C1

         C1   COUNT
          1     2
          2     8
          3    13
          4     7
         N=    30

MTB > # YOUR RESULTS WILL PROBABLY DIFFER.
```

Ⓜ 4.163 MTB > TALLY C1

```
              C1    COUNT
               1      11
               2      15
               3      14
              N=      40
```

MTB > # YOUR RESULTS WILL PROBABLY DIFFER.

Ⓜ 4.164 MTB > PRINT C2

```
         C2
              8     1     3
```

MTB > # YOUR RESULTS WILL PROBABLY DIFFER.

Ⓜ 4.165 MTB > PRINT C2

```
    C2
        12   50   61    1   14   48   30    5   45   59    2   17   42
        19    4   53   35   27   22   31   13    6   25   51   34   10
        64   55   47   38   39   60
```

MTB > # YOUR RESULTS WILL PROBABLY DIFFER.

Chapter 5
Section 5.1
5.1 Continuous

5.3 Continuous

5.5 Discrete

5.7 Continuous

5.9 Continuous

5.11 Continuous

5.13 Continuous

5.15 Discrete

5.17 Discrete

5.19 Continuous

5.21 Continuous

5.23 Continuous

5.25 Discrete

5.27 Discrete

5.29 X is the number of copies sold;
0, 1, 2, . . . , 725,000;
discrete.

5.31 X is the length (in miles) of coastline that has been damaged by the oil spill;
$0 \leq x \leq 13$;
continuous.

5.33 X is the number of students in Spanish I who will enroll in Spanish II;
0, 1, 2, . . . , 97;
discrete.

5.35 X is the number of activations of the switch that are required to make it fail;
1, 2, 3, . . . ;
discrete.

Section 5.2
5.37 0.27

5.39

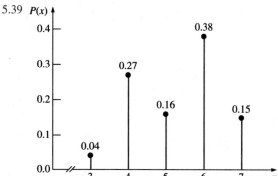

5.41

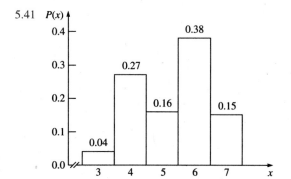

5.43 Because all the probabilities are nonnegative and sum to 1, it is a valid probability distribution.

5.45 It is a valid probability distribution because each probability is nonnegative and the sum of the probabilities is 1.

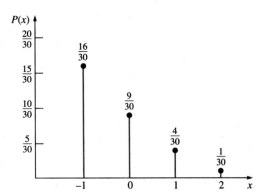

5.47 It is a valid probability distribution because each probability is nonnegative and the sum of the probabilities is 1.

x	0	1	2
$P(x)$	0.09	0.42	0.49

5.49 a. 0.648
     b. 0.352
5.51 a. 1H, 2H, 3H, 4H, 5H, 6H, 1T, 2T, 3T, 4T, 5T, 6T
     b. 1/12 to each sample point
     c. 1H $x = 2$ 4H $x = 5$ 1T $x = 1$ 4T $x = 4$
        2H $x = 3$ 5H $x = 6$ 2T $x = 2$ 5T $x = 5$
        3H $x = 4$ 6H $x = 7$ 3T $x = 3$ 6T $x = 6$
     d.

x	1	2	3	4	5	6	7
$P(x)$	1/12	2/12	2/12	2/12	2/12	2/12	1/12

5.53

x	0	1	2
$P(x)$	0.0324	0.2952	0.6724

5.55

x	0	1	2
$P(x)$	0.1	0.6	0.3

Ⓜ 5.57
```
Histogram of C3    N = 162
Each * represents 2 obs.

Mispoint   Count
    0        59    ******************************
    1        39    *******************
    2        36    ******************
    3        22    ***********
    4         6    ***

MTB > # YOUR RESULTS WILL PROBABLY DIFFER.
```
Ⓜ 5.59
```
Histogram of C3    N = 600
Each * represents 10 obs.

Midpoint   COUNT
    0       205    ********************
    1       395    ****************************************

MTB > # 0 DENOTES A TAIL AND 1 DENOTES A HEAD.
MTB > # YOUR RESULTS WILL PROBABLY DIFFER.
```

Sections 5.3 and 5.4
5.61 $\mu = 2.3$, $\sigma^2 = 2.01$, $\sigma = 1.42$
5.63 $\mu = 0.66$, $\sigma^2 = 13.36$, $\sigma = 3.66$
5.65 $\mu = 3.375$, $\sigma = 0.74$

5.67 $\mu = 1.4$, $\sigma = 0.65$
5.69 $\mu = 6$, $\sigma = 2.19$
5.71 $P(1.62 \le x \le 10.38) = 1$

5.73 $\mu = 1.77$

5.75 $\sigma = 0.90$

5.77 $145

Ⓜ **5.79** `MTB > SUM C3    # THIS IS THE MEAN`
      `SUM       =       52.667`

Ⓜ **5.81** `MTB > SUM C3    # THIS IS THE MEAN`
      `SUM       =       71.420`
     `MTB > PRINT K1   # THIS IS THE VARIANCE`
     `K1          114.344`

Review Exercises

5.83 $\mu = 0.91$

5.84 a. X is the number of monitors in the lot of 100 that operate satisfactorily.
   b. Discrete
   c. 0, 1, 2, . . . , 100

5.85 a. X is the amount of beverage in the cup.
   b. Continuous
   c. $0 \le x \le 8$

5.86 Continuous

5.87 Discrete

5.88 Discrete

5.89 Continuous

5.90 Discrete

5.91 Continuous

5.92 Continuous

5.93 Continuous

5.94 Continuous

5.95 Discrete

5.96 0.44

5.97 It is a valid probability distribution because each probability is nonnegative and the sum of the probabilities is 1.

5.98 It is not a valid probability distribution because the sum of the probabilities does not equal 1.

5.99 It is a valid probability distribution because each probability is nonnegative and the sum of the probabilities is 1.

5.100 $P(x) = x^2/55$ for $x = 1, 2, 3, 4, 5$

5.101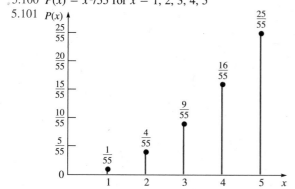

5.102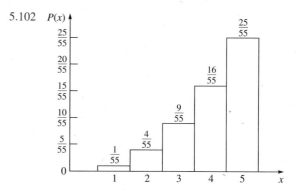

5.103 a. 0.8192
   b. 0.0272

5.104 $\mu = 0.8, \sigma = 0.8$

5.105

x	0	1	2
$P(x)$	1/16	6/16	9/16

5.106 a. AA, AB, AC, AD, BA, BB, BC, BD, CA, CB, CC, CD, DA, DB, DC, DD
   b. 1/16 to each sample point
   c.

AA $x = 0$	BA $x = 1$	CA $x = 0$	DA $x = 0$
AB $x = 1$	BB $x = 2$	CB $x = 1$	DB $x = 1$
AC $x = 0$	BC $x = 1$	CC $x = 0$	DC $x = 0$
AD $x = 0$	BD $x = 1$	CD $x = 0$	DD $x = 0$

   d.

x	0	1	2
$P(x)$	9/16	6/16	1/16

   e. $\mu = 0.5, \sigma = 0.61$

5.107

x	0	1	2	3
$P(x)$	0.008	0.096	0.384	0.512

5.108 $\mu = 5.8, \sigma^2 = 8.96, \sigma = 2.99$

5.109 $\mu = 1.43, \sigma^2 = 1.1451, \sigma = 1.07$

5.110 $\mu = 4.5, \sigma = 1.36$

5.111 $P(1.78 \le x \le 7.22) = 1$

5.112 $\mu = 11/9, \sigma = 0.79$

5.113 $\mu = 8.61$

5.114 $\sigma = 0.69$

5.115 $6,250

Ⓜ 5.116 `Histogram of C3    N = 360`
`Each * represents 2 obs.`

```
Midpoint    Count
       1       73    ***************************************
       2       42    *********************
       3       69    **********************************
       4       49    *************************
       5       89    **********************************************
       6       38    *******************
```

`MTB > # RESULTS FOR THIS PROBLEM WILL VARY.`

Ⓜ 5.117 `MTB > SUM C3    # THIS IS THE MEAN`
`    SUM    =      172.50`

Ⓜ 5.118 `MTB > PRINT K1   # THIS IS THE VARIANCE`
`K1        1022.74`
`MTB > PRINT K2   # THIS IS THE STD.DEV.`
`K2         31.9804`

Ⓜ 5.119 `MTB > SUM C3    # THIS IS THE MEAN`
`    SUM    =      67.000`

Ⓜ 5.120 `MTB > PRINT K2   # THIS IS THE STD.DEV.`
`K2         23.6854`

Ⓜ 5.121 `MTB > HIST C3    # PART C`
`Histogram of C3    N = 400`
`Each * represents 5 obs.`

```
Midpoint    Count
       0       27    ******
       1      104    ********************
       2      151    ******************************
       3       95    ******************
       4       23    *****
```

`MTB > MEAN C3    # PART D`
`    MEAN    =      1.9575`
`MTB > STDEV C3   # PART D`
`    ST.DEV. =      0.99909`
`MTB > # RESULTS FOR THIS PROBLEM WILL VARY.`

Chapter 6
Section 6.1

6.1 This is a binomial experiment in which a success can be defined as the tossing of a head on a single throw; $n = 7$ and $p = 0.4$.

6.3 Because the population is extremely large, this can be regarded as a binomial experiment where a success is the selection of a household with a personal computer; $n = 500$ and $p = 0.15$.

6.5 This is not a binomial experiment; the trials are not independent.

6.7 This is a binomial experiment where a success is answering a question correctly; $n = 10$ and $p = 0.5$.

6.9 This is a binomial experiment where a success is the use of a charge card by a customer; $n = 3$ and $p = 0.70$.

6.11 Because the population is extremely large, this can be regarded as a binomial experiment where a success is the selection of a household with a microwave oven; $n = 20$ and $p = 0.75$.

Section 6.2

6.13 0.001
6.15 0.165
6.17 0.896
6.19 0.017
6.21 0.954
6.23 0.121
6.25 0.167
6.27 0.667

6.29 0.788

6.31 0.336

6.33 0.328

6.35 a. 0.96875

     b. 0.969

6.37 a. 0.378

     b. 0.642

6.39 a. 0.188

     b. 0.836

     c. 0.352

     d. 0.669

6.41 a. 0.879

     b. 0.114

     c. 0.121

Ⓜ 6.43
```
          K   P( X LESS OR = K)
       30.00              0.3805
 MTB > # ANSWER = 1 - 0.3805 = 0.6195
```

Ⓜ 6.45
```
          K           P( X = K)
        7.00              0.1467
 MTB > # ANSWER (A) = 0.1467

          K   P( X LESS OR = K)
        6.00              0.6065
 MTB > # ANSWER (B) = 0.6065
```

Section 6.3

6.47 $\mu = 90$, $\sigma = 3$

6.49 $\mu = 30$, $\sigma = 5$

6.51 $\mu = 200$, $\sigma = 10$

6.53 $\mu = 12$, $\sigma = 3$

6.55 $\mu = 30$, $\sigma = 3.87$

6.57 $\mu = 60$, $\sigma = 3.87$

6.59 $\sigma = 1$

6.61 $\mu = 42.67$; this does not have to be an integer because it is a long-run average value.

6.63 a. $\mu = 21$

     b. $\sigma = 3.97$

     c. No, because the number actually awarded, 9, has a z-value of -3.02, placing it more than 3 standard deviations below the mean. By Chebyshev's theorem, the probability of this occurrence is small.

6.65 a. $\mu = 200$

     b. $\sigma = 10$

     c. 180 to 220

Section 6.4

6.67 0.238

6.69 0.643

6.71 0.083

6.73 2.6×10^{-7}

6.75 a. 0.716

     b. 0.268

6.77 0.00174

6.79 $\mu = 1.5$, $\sigma^2 = 0.583$, $\sigma = 0.764$

6.81 0.804

Section 6.5

6.83 0.195

6.85 0.186

6.87 0.525

6.89 0.001

6.91 0.045

6.93 0.393

6.95 a. 0.008

     b. 0.857

6.97 a. $\sigma^2 = 4$, $\sigma = 2$

     b. $\sigma^2 = 5$, $\sigma = 2.236$

     c. $\sigma^2 = 3.4$, $\sigma = 1.844$

     d. $\sigma^2 = 1.2$, $\sigma = 1.095$

     e. $\sigma^2 = 1.6$, $\sigma = 1.265$

Ⓜ 6.99
```
          K           P( X = K)
        0.00              0.9512
 MTB > # ANSWER = 1 - 0.9512 = 0.0488
```

Ⓜ 6.101
```
          K   P( X LESS OR = K)
       13.00              0.6309
 MTB > # ANSWER (A) = 1 - 0.6309 = 0.3691

          K   P( X LESS OR = K)
       13.00              0.6278
 MTB > # ANSWER (B) = 1 - 0.6278 = 0.3722
```

Review Exercises

6.103 0.088

6.104 0.857

6.105 0.379

6.106 0.086

6.107 0.722

6.108 0.175

6.109 0.387

6.110 0.929

6.111 0.441

6.112 $\mu = 750$

6.113 0.262

6.114 a. 0.237

      b. 0.001

6.115 0.900

6.116 0.670

6.117 $\mu = 680$, $\sigma = 14.75$

6.118 a. 0.263

      b. 0.264

6.119 $\mu = 5,000$ and $\sigma = 50$. As few as 4,800 heads is very unlikely because it is 4 standard deviations below the mean ($z = -4$).

6.120 $n = 448$, $p = 1/8$

6.121 $p = 0.5$

Ⓜ 6.122

K	P(X = K)
29.00	0.0884
30.00	0.0216

MTB > # ANSWER = 0.0884 + 0.0216 = 0.1100

Ⓜ 6.123

K	P(X LESS OR = K)
2.00	0.6988

MTB > # ANSWER = 1 - 0.6988 = 0.3012

Ⓜ 6.124

K	P(X LESS OR = K)
7.00	0.0316

Ⓜ 6.125

K	P(X LESS OR = K)
7.00	0.2202

Chapter 7
Section 7.2
7.1 0.3907
7.3 0.4986
7.5 0.5636
7.7 0.0604
7.9 0.3256
7.11 0.9306
7.13 0.1762
7.15 0.0146
7.17 0.9515

7.19 0.9911
7.21 0.0934
7.23 0.1210
7.25 0.9978
7.27 0.3078
7.29 0.2088
7.31 0.9592
7.33 0.1911
7.35 0.0288
7.37 0.9974; Empirical rule: ≈ 1; Chebyshev's theorem: ≥ 0.8889
7.39 2.26
7.41 0.26
7.43 -2.72
7.45 2.88
7.47 0.53
7.49 0.36
7.51 -0.67
7.53 0
7.55 1.645
7.57 2.33
7.59 -1.96

Ⓜ 7.61

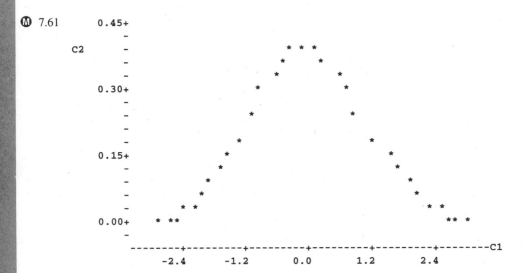

MTB > # YOUR RESULTS MAY DIFFER.

 7.63 MTB > # PART A

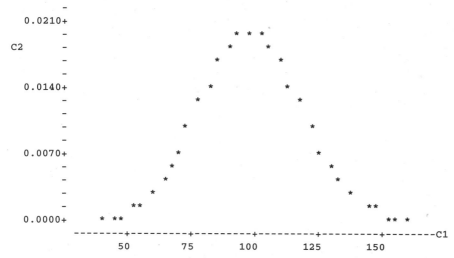

MTB > # PART B

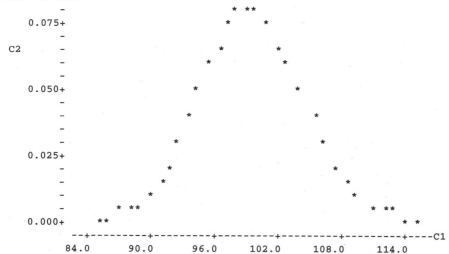

MTB > # YOUR RESULTS MAY DIFFER FOR PARTS A AND B.

Section 7.3

7.65	2	7.87	0.3278
7.67	−1	7.89	318.5
7.69	−1.5	7.91	256.5
7.71	−3.25	7.93	0.0668
7.73	0.8413	7.95	10.56 percent
7.75	0.6915	7.97	82.04
7.77	0.1832	7.99	8.332
7.79	0.4825	7.101	161.68
7.81	0.2401	7.103	a. 11.51 percent
7.83	0.8745		b. 78.74 percent
7.85	0.9842	7.105	5.168

Ⓜ 7.107   13.5000   0.3002
        MTB > # ANSWER A = 0.3002

            17.0000   0.6215
        MTB > # ANSWER B = 1 - 0.6215 = 0.3785

            10.0000   0.0874
            20.0000   0.8470
        MTB > # ANSWER C = 0.8470 - 0.0874 = 0.7596

Ⓜ 7.109   0.0500   8.7916
        MTB > # ANSWER = 8.7916

Section 7.4

7.111 a. 0.1762
       b. 0.176
       c. 0.1742
7.113 0.2052, compared to 0.207 from Table 1.
7.115 0.1190
7.117 0.0618
7.119 0.1611
7.121 0.9708
Ⓜ 7.123   K   P(X LESS OR = K)
            4.00   0.2680
        MTB > # ANSWER A = 0.2680

            -0.5000   0.0023
            4.5000   0.2569
        MTB > # ANSWER B = 0.2569 - 0.0023 = 0.2546

Review Exercises

7.125 0.9850
7.126 0.0256
7.127 a. 0.1472
       b. 0.8050
       c. 0.1472
7.128 0.9876, compared to at least 0.84 from Chebyshev's
       theorem.

7.129 − 1.92
7.130 1.08
7.131 a. 0.52
       b. − 1.04
7.132 a. 2.05
       b. 1.41
7.133 156.7
7.134 a. 0.9878
       b. 0.8931
7.135 0.0028
7.136 3.07 pounds
7.137 0.0239
7.138 0.0559
7.139 0.7287
7.140 38.20 percent
7.141 70.09
7.142 96.29 percent
7.143 0.9554
7.144 0.0918
Ⓜ 7.145   K   P(X LESS OR = K)
            24.00   0.0573
        MTB > # ANSWER A = 1 - 0.0573 = 0.9427

            24.5000   0.0562
        MTB > # ANSWER B = 1 - 0.0562 = 0.9438
Ⓜ 7.146   43.3000   0.0947
        MTB > # ANSWER A = 0.0947

            0.7500   52.8376
        MTB > # Answer B = 52.8376
Ⓜ 7.147   3.5700   0.9998
        MTB > # ANSWER A = 0.9998

            0.9100   1.3408
        MTB > # ANSWER B = 1.3408

Ⓜ 7.148   -0.6200   0.2676
        MTB > # ANSWER A = 0.2676

            1.8600   0.9686
        MTB > # ANSWER B = 1 - 0.9686 = 0.0314

            1.1100   0.8665
            -1.6400   0.0505
        MTB > # ANSWER C = 0.8665 - 0.0505 = 0.8160

Ⓜ 7.149   0.8300   165.7362
        MTB > # ANSWER = 165.7362
Ⓜ 7.150   0.9600   1.7507
        MTB > # ANSWER = 1.7507

7.151

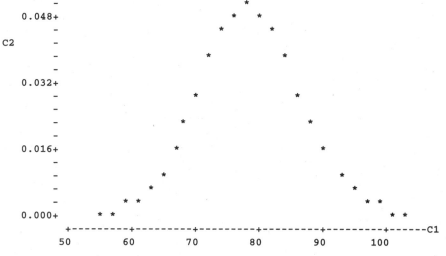

```
MTB > # YOUR RESULTS MAY DIFFER.
```

 7.152

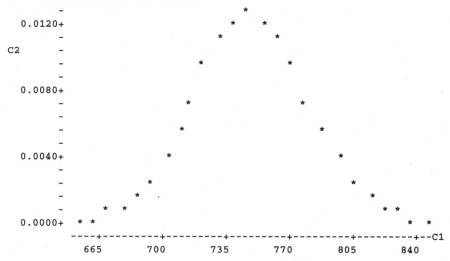

```
MTB > # YOUR RESULTS MAY DIFFER.
```

Chapter 8
Section 8.1

8.1   **56**

8.3   **45**

8.5   **75,287,520**

8.7   **1/12,650**

8.9   a. **6**
     b. *WX, WY, WZ, XY, XZ, YZ*
     c. **1/6**

8.11 In each case the probability is 0.5.

8.13 a. Assign the integers from 1 to 7 to the 7 faculty. Then use MINITAB to randomly select 2 of these numbers. The 2 faculty associated with the selected integers constitute the random sample. (There are other possible procedures.)
     b. 1/21

Ⓜ **8.15** MTB > PRINT C2
```
    C2
      62    99    31    12    43    53    97    86    93    72    44
      55    39    87   107    29    65    48    21    92    79    64
      94    61    20
    MTB > # RESULTS FOR THIS PROBLEM WILL VARY.
```
Ⓜ **8.17** MTB > PRINT C2
```
    C2
     516   562   630   647   660   665   593   669   735   536   719
     563   682   529   582   657   573   706   521   561
    MTB > # RESULTS FOR THIS PROBLEM WILL VARY.
```

Section 8.2

8.19 Stratified random sample

8.21 Cluster sample

8.23 Systematic sample

Section 8.3

8.25 a,b.

Sample	Mean	Sample	Mean
(1, 1)	1	(5, 1)	3
(1, 3)	2	(5, 3)	4
(1, 5)	3	(5, 5)	5
(1, 7)	4	(5, 7)	6
(3, 1)	2	(7, 1)	4
(3, 3)	3	(7, 3)	5
(3, 5)	4	(7, 5)	6
(3, 7)	5	(7, 7)	7

c.

$\bar{x}$	1	2	3	4	5	6	7
$P(\bar{x})$	1/16	2/16	3/16	4/16	3/16	2/16	1/16

d.

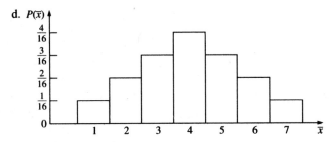

8.27 a,b.

Sample	Mean	Sample	Mean	Sample	Mean	Sample	Mean	Sample	Mean
(1, 1)	1.0	(2, 1)	1.5	(3, 1)	2.0	(4, 1)	2.5	(5, 1)	3.0
(1, 2)	1.5	(2, 2)	2.0	(3, 2)	2.5	(4, 2)	3.0	(5, 2)	3.5
(1, 3)	2.0	(2, 3)	2.5	(3, 3)	3.0	(4, 3)	3.5	(5, 3)	4.0
(1, 4)	2.5	(2, 4)	3.0	(3, 4)	3.5	(4, 4)	4.0	(5, 4)	4.5
(1, 5)	3.0	(2, 5)	3.5	(3, 5)	4.0	(4, 5)	4.5	(5, 5)	5.0

c.

$\bar{x}$	1.0	1.5	2.0	2.5	3.0	3.5	4.0	4.5	5.0
$P(\bar{x})$	1/25	2/25	3/25	4/25	5/25	4/25	3/25	2/25	1/25

d.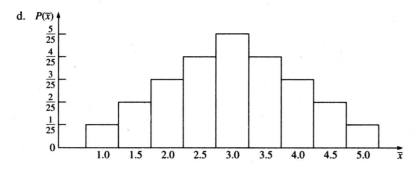

8.29

Sample	Range	Sample	Range	Sample	Range	Sample	Range	Sample	Range
(1, 1)	0	(2, 1)	1	(3, 1)	2	(4, 1)	3	(5, 1)	4
(1, 2)	1	(2, 2)	0	(3, 2)	1	(4, 2)	2	(5, 2)	3
(1, 3)	2	(2, 3)	1	(3, 3)	0	(4, 3)	1	(5, 3)	2
(1, 4)	3	(2, 4)	2	(3, 4)	1	(4, 4)	0	(5, 4)	1
(1, 5)	4	(2, 5)	3	(3, 5)	2	(4, 5)	1	(5, 5)	0

R	0	1	2	3	4
$P(R)$	5/25	8/25	6/25	4/25	2/25

Ⓜ 8.31

```
Histogram of MEANS    N = 100
Midpoint    Count
    1.0        7   *******
    1.5        8   ********
    2.0       13   *************
    2.5       14   **************
    3.0       23   ***********************
    3.5       16   ****************
    4.0       13   *************
    4.5        4   ****
    5.0        2   **

MTB > # RESULTS FOR THIS PROBLEM WILL VARY.
```

Section 8.4

8.33 C

8.35 a. 1.5

   b. 2

8.37 576

8.39 a. Approximately normal

   b. An estimate of 79

   c. An estimate of 3

8.41 a. $\mu = \Sigma x P(x) = 3$; $\sigma = \sqrt{\Sigma x^2 P(x) - \mu^2} = 1$

   b. $\mu_{\bar{x}} = \Sigma \bar{x} P(\bar{x}) = 3$; $\sigma_{\bar{x}} = \sqrt{9.25 - 3^2} = 0.5$

   c. $\mu_{\bar{x}} = \mu = 3$; $\sigma_{\bar{x}} = \sigma/\sqrt{n} = 0.5$

Ⓜ 8.43 Histogram of C2   N = 100
```
     Midpoint    Count
           43       1   *
           44       1   *
           45       6   ******
           46       4   ****
           47       9   *********
           48       9   *********
           49      12   ************
           50      17   *****************
           51      15   ***************
           52      14   **************
           53       6   ******
           54       4   ****
           55       2   **

     MEAN      =       49.704
     ST.DEV. =         2.5268
MTB > # THEORETICAL MEAN = 50
MTB > # THEORETICAL STDEV = 12/SQRT(25) = 2.4
MTB > # THEORETICAL SHAPE IS THAT OF A NORMAL CURVE
MTB > # RESULTS FOR THIS PROBLEM WILL VARY.
```

Ⓜ 8.45 Histogram of C2   N = 100
```
     Midpoint    Count
          2.5       1   *
          2.6       2   **
          2.7       6   ******
          2.8      11   ***********
          2.9      24   ************************
          3.0      21   *********************
          3.1      12   ************
          3.2      16   ****************
          3.3       6   ******
          3.4       1   *

     MEAN      =        2.9890
     St.DEV. =          0.17991
MTB > # THEORETICAL MEAN = 3
MTB > # THEORETICAL STDEV = SQRT(2)/SQRT(60) = 0.18
MTB > # THEORETICAL SHAPE IS APPROXIMATELY A NORMAL CURVE
MTB > # RESULTS FOR THIS PROBLEM WILL VARY.
```

Section 8.5

8.47 0.0228

8.49 0.7642

8.51 0.2236

8.53 0.0062

8.55 0.0784

Review Exercises

8.57 a. 0.0013
     b. 0.0228
     c. 0.8931

8.58 a. 0.6554
     b. 0.0808
     c. 0.9370

8.59 0.1112

8.60 a. 2
     b. 1.1

8.61 It will decrease to 70.7 percent of its original value.

8.62 The expected shape would be approximately normal. The expected mean and standard deviation would be approximately 88.5 and 1.2, respectively.

8.63 2,118,760

8.64 Cluster sampling

8.65 Systematic sampling

8.66 Stratified random sampling

8.67 Random sampling

8.68 0.9772

8.69 0.9750

8.70 0.2112

8.71 The probability is only 0.0008, making the claim unlikely.

8.72 a. 21

b. AB, AC, AD, AE, AF, AG, BC, BD, BE, BF, BG, CD, CE, CF, CG, DE, DF, DG, EF, EG, FG

c. Assign the integers from 1 to 7 to the 7 puppies. Then use MINITAB to randomly select 2 of these numbers. The 2 puppies associated with the selected integers constitute the random sample. (There are other possible procedures.)

d. 1/21

8.73 a. $\mu = 5.5$, $\sigma = 1.708$

b,c.

Sample	Mean	Sample	Mean	Sample	Mean	Sample	Mean
(3, 3)	3.0	(4, 6)	5.0	(6, 3)	4.5	(7, 6)	6.5
(3, 4)	3.5	(4, 7)	5.5	(6, 4)	5.0	(7, 7)	7.0
(3, 5)	4.0	(4, 8)	6.0	(6, 5)	5.5	(7, 8)	7.5
(3, 6)	4.5	(5, 3)	4.0	(6, 6)	6.0	(8, 3)	5.5
(3, 7)	5.0	(5, 4)	4.5	(6, 7)	6.5	(8, 4)	6.0
(3, 8)	5.5	(5, 5)	5.0	(6, 8)	7.0	(8, 5)	6.5
(4, 3)	3.5	(5, 6)	5.5	(7, 3)	5.0	(8, 6)	7.0
(4, 4)	4.0	(5, 7)	6.0	(7, 4)	5.5	(8, 7)	7.5
(4, 5)	4.5	(5, 8)	6.5	(7, 5)	6.0	(8, 8)	8.0

d.

$\bar{x}$	3.0	3.5	4.0	4.5	5.0	5.5	6.0	6.5	7.0	7.5	8.0
$P(\bar{x})$	1/36	2/36	3/36	4/36	5/36	6/36	5/36	4/36	3/36	2/36	1/36

e. $\mu_{\bar{x}} = \Sigma \bar{x} P(\bar{x}) = 5.5$; $\sigma_{\bar{x}} = \sqrt{31.7083 - 5.5^2} = 1.208$

f. $\mu_{\bar{x}} = \mu = 5.5$; $\sigma_{\bar{x}} = 1.708/\sqrt{2} = 1.208$

Ⓜ 8.74
```
Histogram of C2    N = 100
Midpoint   Count
     3.0      6    ******
     3.5      9    *********
     4.0      7    *******
     4.5      3    ***
     5.0     13    *************
     5.5     28    ****************************
     6.0     11    ***********
     6.5      5    *****
     7.0      7    *******
     7.5      7    *******
     8.0      4    ****
    MEAN    =      5.4200
    ST.DEV. =      1.3156
MTB > # THEORETICAL MEAN IS 5.5
MTB > # THEORETICAL STDEV IS 1.708/SQRT(2) = 1.208
MTB > # RESULTS FOR THIS PROBLEM WILL VARY.
```

Ⓜ 8.75 Histogram of C2   N = 100
```
      Midpoint    Count
           4.8        4    ****
           5.0        6    ******
           5.2       16    ****************
           5.4       25    *************************
           5.6       27    ***************************
           5.8       15    ***************
           6.0        3    ***
           6.2        3    ***
           6.4        1    *
        MEAN    =        5.4856
        ST.DEV. =        0.30205
     MTB > # THEORETICAL MEAN IS 5.5
     MTB > # THEORETICAL STDEV IS 1.708/SQRT(32) = 0.302
     MTB > # RESULTS FOR THIS PROBLEM WILL VARY.
```
Ⓜ 8.76 Histogram of C2   N = 100
```
      Mispoint    Count
           450        1    *
           460        0
           470       12    ************
           480       15    ***************
           490       16    ****************
           500       21    *********************
           510       18    ******************
           520       12    ************
           530        3    ***
           540        2    **

        MEAN    =        497.24
        ST.DEV. =        17.781
     MTB > # THEORETICAL SHAPE IS THAT OF A NORMAL CURVE
     MTB > # THEORETICAL MEAN IS 500
     MTB > # THEORETICAL STDEV IS 75/SQRT(20) = 16.77
     MTB > # RESULTS FOR THIS PROBLEM WILL VARY.
```
Ⓜ 8.77 C2
```
       71   50   87   93   69   47   62    9   72    3   40
       25   33    5   18   48   61  100   21   70   16   23
       85   82   56   27   10   95   65   36   20    7   12
       81   13   64    8   11   63    4   43   44   74   58
       92   22   54   80   73   53
     MTB > # PATIENTS WITH THESE NUMBERS WILL RECEIVE SPECIAL DIET.
     MTB > # RESULTS FOR THIS PROBLEM WILL VARY.
```
Ⓜ 8.78 C2
```
      126  278  204  297    2   42  110  183  252  239  188
      184  271  301  288   78  240  295  291  255  208  147
        3   98  114  159   65  283   72   11
     MTB > # EMPLOYEES WITH THESE NUMBERS WILL RECEIVE PRIZES.
     MTB > # RESULTS FOR THIS PROBLEM WILL VARY.
```

Chapter 9

Sections 9.1 and 9.2

9.1 a. 1.70
     b. 1.88
     c. 2.05
     d. 2.17

9.3 A point estimate is a single value; a confidence interval estimate consists of an interval of values with an associated level of confidence.

9.5 75.9 ± 1.32

9.7 a. 25.4 ± 3.92
     b. 25.4 ± 5.15
     c. It increases.

9.9 a. 98.84 ± 1.96
     b. 98.84 ± 3.92
     c. The width is directly proportional to the standard deviation.

9.11 13.36 ± 0.102

9.13 850 ± 18.2

9.15 $\$577 \pm \$14.66$

9.17 97.8 ± 0.74

Ⓜ 9.19 THE ASSUMED SIGMA =0.355

	N	MEAN	STDEV	SE MEAN	99.0 PERCENT C.I.
C1	41	9.8506	0.3550	0.0554	(9.7075, 9.9937)

Ⓜ 9.21 THE ASSUMED SIGMA =1.60

	N	MEAN	STDEV	SE MEAN	90.0 PERCENT C.I.
C1	40	19.990	1.645	0.253	(19.573, 20.407)

Section 9.3

9.23 1.319
9.25 2.977
9.27 2.201
9.29 2.462
9.31 1.645
9.33 1.706
9.35 2.602
9.37 1.638
9.39 -1.782
9.41 -2.326
9.43 -1.333
9.45 0.01
9.47 0.90
9.49 0.025
9.51 0.945
9.53 2.718
9.55 -1.638
9.57 2.779

Ⓜ 9.59   2.2200   0.9844
          MTB > # ANSWER A = 1 - 0.9844 = 0.0156

          0.6800   0.4707
          MTB > # ANSWER B = 0.4707

Section 9.4

9.61 a. 1.345
      b. 1.721
      c. 2.763
      d. 2.306

9.63 575 ± 53.1

9.65 25 ± 2.9

9.67 28.9 ± 1.50

9.69 34.3 ± 1.78

9.71 8.88 ± 0.16

9.73 $\$115 \pm \$13.85$

Ⓜ 9.75

	N	MEAN	STDEV	SE MEAN	95.0 PERCENT C.I.
C1	15	16.900	1.326	0.342	(16.166, 17.634)

Ⓜ 9.77
```
     C2      -
             -                                                          *
             -
         1.2+                                              2
             -
             -                                      *
             -
             -                               3
         0.0+                          2
             -
             -              2
             -        2
             -
        -1.2+
             -    2
             -
             -
             -
             ------+---------+---------+---------+---------+---------+C1
               15.20     16.00     16.80     17.60     18.40     19.20
```

Correlation of C1 and C2 = 0.987
MTB > # NORMALITY ASSUMPTION APPEARS REASONABLE

Section 9.5
9.79 0.30 ± 0.038
9.81 0.15 ± 0.024
9.83 0.65 ± 0.027
9.85 92% ± 3.8%
9.87 0.14 ± 0.057
9.89 16% ± 1.9%
9.91 0.219 ± 0.008

Section 9.6
9.93   260
9.95   547
9.97   228
9.99   271
9.101 203
9.103 865
9.105 2,401

Section 9.7
9.107 33.41
9.109 29.82
9.111 40.65
9.113 67.50
9.115 59.20
9.117 5.58
9.119 53.54
9.121 6.57
9.123 14.68
9.125 32.00
9.127 32.67

9.129 0.975
9.131 0.99
9.133 0.875
9.135 21.67
9.137 13.79

Ⓜ 9.139   0.8500   51.4746
            0.1500   32.6255
MTB > # ANSWER 1 = 51.4746, ANSWER 2 = 32.6255

Ⓜ 9.141   0.3500   16.1089
MTB > # ANSWER = 16.1089

Section 9.8
9.143 $9.77 < \sigma^2 < 29.10$
9.145 $3.13 < \sigma < 5.39$
9.147 $1.75 < \sigma^2 < 123.81$
9.149 $0.314 < \sigma^2 < 1.023$
9.151 $0.168 < \sigma < 0.400$
9.153 $0.107 < \sigma < 0.283$
9.155 $1.59 < \sigma^2 < 15.09$

Review Exercises
9.157 478.7 ± 12.49
9.158 $529.43 < \sigma^2 < 1,749.25$
9.159 6.3 ± 0.18
9.160 $0.188 < \sigma < 0.424$
9.161 1,002.8 ± 1.04
9.162 $2.92 < \sigma < 4.07$
9.163 423
9.164 139

9.165 **1,037**

9.166 **944**

9.167 **$0.0850 < \sigma^2 < 0.2064$**

9.168 **$0.291 < \sigma < 0.454$**

9.169 **419**

9.170 **52.8 ± 0.64**

9.171 **0.27 ± 0.042**

9.172 **$45\% \pm 2.8\%$**

9.173 **42**

9.174 **2,401**

9.175 **a. 2.069**

     b. 2.764

9.176 **a. 50.99**

     b. 23.68

9.177 **37.69**

9.178 **1.699**

9.179 **$3,503 \pm 435$**

Ⓜ 9.180

```
           N     MEAN    STDEV   SE MEAN   99.0 PERCENT C.I.
C1         8   3503.00   351.61   124.31   ( 3067.89, 3938.11)
```

Ⓜ 9.181

```
C2    -
      -
      -
 1.0+                                                    *
      -
      -                                          *
      -
      -                                        *
      -
 0.0+                                  2
      -
      -                          *
      -
      -                  *
      -
-1.0+
      -
      -       *
      -
    --+---------+---------+---------+---------+---------+----C1
    2750      3000      3250      3500      3750      4000
```

Correlation of C1 and C2 = 0.967

MTB > # THE NORMALITY ASSUMPTION APPEARS REASONABLE.

Ⓜ 9.182

```
       -
0.090+                   *  *  *
       -                       *
C2     -                *          *
       -
       -             *          *
0.060+
       -                          *
       -        *                   *
       -
       -
0.030+     *
       -   *
       -
       -  *                       *
       -  *                      *  *
       - *                        * * * *
0.000+ *
     +---------+---------+---------+---------+---------+------C1
     0.0       5.0      10.0      15.0      20.0      25.0
```

MTB > # YOUR PLOT MAY VARY IN APPEARANCE.

Ⓜ 9.183

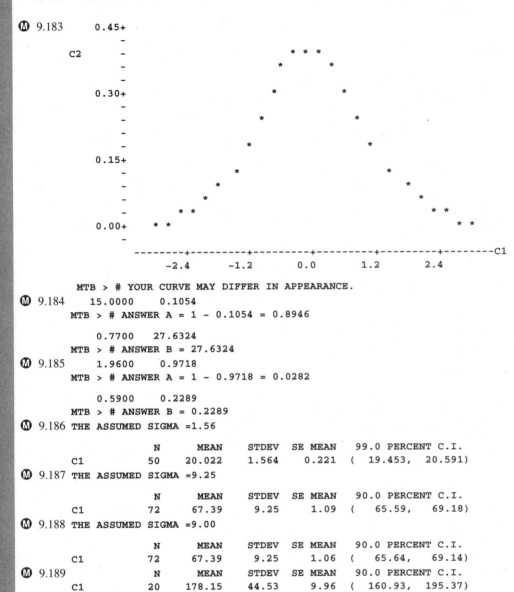

```
        0.45+
            -
    C2      -                         *   *   *
            -                      *           *
            -
        0.30+                   *               *
            -
            -                *                   *
            -
            -             *                       *
        0.15+
            -          *                             *
            -        *                                 *
            -     *                                      *
            -   *                                          *
        0.00+  *  *                                           *   *
            --------+---------+---------+---------+---------+--------C1
                  -2.4      -1.2       0.0       1.2       2.4
```

        MTB > # YOUR CURVE MAY DIFFER IN APPEARANCE.
Ⓜ 9.184   15.0000   0.1054
        MTB > # ANSWER A = 1 - 0.1054 = 0.8946

            0.7700   27.6324
        MTB > # ANSWER B = 27.6324
Ⓜ 9.185   1.9600   0.9718
        MTB > # ANSWER A = 1 - 0.9718 = 0.0282

            0.5900   0.2289
        MTB > # ANSWER B = 0.2289
Ⓜ 9.186 THE ASSUMED SIGMA =1.56

                    N   MEAN   STDEV   SE MEAN   99.0 PERCENT C.I.
        C1   50   20.022   1.564   0.221 (19.453, 20.591)
Ⓜ 9.187 THE ASSUMED SIGMA =9.25

                    N   MEAN   STDEV   SE MEAN   90.0 PERCENT C.I.
        C1   72   67.39   9.25   1.09   (65.59,   69.18)
Ⓜ 9.188 THE ASSUMED SIGMA =9.00

                    N   MEAN   STDEV   SE MEAN   90.0 PERCENT C.I.
        C1   72   67.39   9.25   1.06   (65.64,   69.14)
Ⓜ 9.189   N   MEAN   STDEV   SE MEAN   90.0 PERCENT C.I.
        C1   20   178.15   44.53   9.96   (160.93, 195.37)
```
```

Ⓜ 9.190   C2

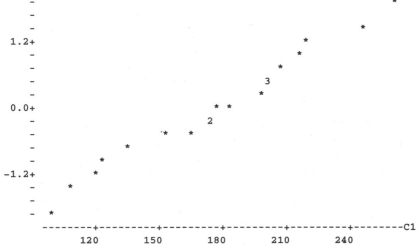

```
      C2
         -
         -
         -
                                                                          *
    1.2+
         -
         -                                                         *
         -
         -                                                    *
         -                                                  *
         -                                               *
         -                                          3
         -                                        *
    0.0+
         -                                  *   *
         -                               2
         -
         -                         '*      *
         -                    *
         -
         -                  *
         -
   -1.2+               *
         -
         -          *
         -
         -
       -     *
         --------+---------+---------+---------+---------+--------C1
              120       150       180       210       240
```

MTB > CORR C1 C2
Correlation of C1 and C2 = 0.987
MTB > # THE NORMALITY ASSUMPTION APPEARS REASONABLE.

Chapter 10
Section 10.1
10.1   a. Type I
       b. Type II
       c. Type I
       d. Type II
10.3   By decreasing the value of α, the value of β is increased, thus increasing the probability of committing a Type II error.
10.5   a. Type II
       b. Type I
10.7   a. Type I
       b. Type II
10.9   a. A Type I error will occur if it is concluded that the restaurant is selling underweight hamburgers, when in fact it is not.
       b. A Type II error will occur if it is concluded that the restaurant is not selling underweight hamburgers, when in fact it is.
       c. Type I
10.11 a. $H_0: \mu = 66$
          $H_a: \mu > 66$
       b. $H_0: \mu = 66$
          $H_a: \mu < 66$
10.13 $H_0: \mu = 950$
       $H_a: \mu > 950$

Section 10.2
10.15 a. $z < -2.326$
       b. $z < -2.576, z > 2.576$
       c. $z > 2.326$
10.17 a. $H_0: \mu = 95$
          $H_a: \mu \neq 95$
       b. $\alpha = 0.01$, which is the probability of committing a Type I error.
       c. $z = -2.50$
       d. $z < -2.576, z > 2.576$
       e. Fail to reject H_0
10.19 $H_0: \mu = 80,000$
       $H_a: \mu > 80,000$
       $z = 1.87$
       RR: $z > 1.645$
       Reject H_0 (yes)
10.21 $H_0: \mu = 150$
       $H_a: \mu > 150$
       $z = 1.97$
       RR: $z > 1.645$
       Reject H_0 (yes)
10.23 $H_0: \mu = 573$
       $H_a: \mu \neq 573$
       $z = 2.36$
       RR: $z < -1.96, z > 1.96$
       Reject H_0 (yes)

10.25 $H_0: \mu = 1,600$
      $H_a: \mu \neq 1,600$
      $z = -1.73$
      RR: $z < -1.96, z > 1.96$
      Fail to reject H_0 (no)

10.27 $H_0: \mu = 5,000$
      $H_a: \mu > 5,000$
      $z = 1.75$
      RR: $z > 1.645$
      Reject H_0 (yes)

Ⓜ 10.29 TEST OF MU = 19.900 VS MU G.T. 19.900
         THE ASSUMED SIGMA = 1.56

	N	MEAN	STDEV	SE MEAN	Z	P VALUE
C1	50	20.022	1.564	0.221	0.55	0.29

         MTB > # FAIL TO REJECT Ho. INSUFFICIENT EVIDENCE THAT MEAN EXCEEDS 19.9.

Section 10.3

10.31 $t > 2.602$

10.33 $t < -1.812, t > 1.812$

10.35 $t < -1.860$

10.37 $H_0: \mu = 590$
      $H_a: \mu \neq 590$
      $t = -1.23$
      RR: $t < -2.977, t > 2.977$
      Fail to reject H_0

10.39 $H_0: \mu = 25$
      $H_a: \mu \neq 25$
      $t = -5.48$
      RR: $t < -4.032, t > 4.032$
      Reject H_0 (yes)

10.41 $H_0: \mu = 8.4$
      $H_a: \mu > 8.4$
      $t = 2.65$
      RR: $t > 1.703$
      Reject H_0 (yes)

10.43 $H_0: \mu = 24.8$
      $H_a: \mu \neq 24.8$
      $t = -1.99$
      RR: $t < -2.069, t > 2.069$
      Fail to reject H_0

10.45 $H_0: \mu = 6.5$
      $H_a: \mu < 6.5$
      $t = -2.45$
      RR: $t < -1.833$
      Reject H_0 (yes)

10.47 $H_0: \mu = 50$
      $H_a: \mu \neq 50$
      $t = 3.21$
      RR: $t < -2.064, t > 2.064$
      Reject H_0 (yes)

Ⓜ 10.49 TEST OF MU = 38.000 VS MU N.E. 38.000

	N	MEAN	STDEV	SE MEAN	T	P VALUE
C1	23	27.783	9.010	1.879	-5.44	0.0000

         MTB > # REJECT Ho. SUFFICIENT EVIDENCE THAT A CHANGE HAS OCCURRED.

Section 10.4

10.51 $H_0: p = 0.40$
      $H_a: p < 0.40$
      $z = -1.64$
      RR: $z < -2.326$
      Fail to reject H_0.

10.53 $H_0: p = 0.70$
      $H_a: p \neq 0.70$
      $z = 1.75$
      RR: $z < -1.96, z > 1.96$
      Fail to reject H_0 (no)

10.55 $H_0: p = 0.50$
$H_a: p > 0.50$
$z = 1.58$
RR: $z > 1.645$
Fail to reject H_0 (no)

10.57 $H_0: p = 0.90$
$H_a: p < 0.90$
$z = -3.22$
RR: $z < -2.326$
Reject H_0 (yes)

10.59 $H_0: p = 0.37$
$H_a: p \neq 0.37$
$z = 0.95$
RR: $z < -2.576, z > 2.576$
Fail to reject H_0 (no)

10.61 $H_0: p = 0.5$
$H_a: p \neq 0.5$
$z = 2.21$
RR: $z < -1.96, z > 1.96$
Reject H_0 (yes)

Section 10.5

10.63 a. $\chi^2 < 10.12$
b. $\chi^2 > 30.14$
c. $\chi^2 < 8.91, \chi^2 > 32.85$

10.65 $H_0: \sigma^2 = 28$
$H_a: \sigma^2 < 28$
$\chi^2 = 16.46$
RR: $\chi^2 < 10.85$
Fail to reject H_0

10.67 $H_0: \sigma^2 = 36$
$H_a: \sigma^2 \neq 36$
$\chi^2 = 31.50$
RR: $\chi^2 < 10.12, \chi^2 > 30.14$
Reject H_0

10.69 $H_0: \sigma^2 = 5.76$
$H_a: \sigma^2 \neq 5.76$
$\chi^2 = 13.45$
RR: $\chi^2 < 0.83, \chi^2 > 12.83$
Reject H_0 (yes)

10.71 $H_0: \sigma^2 = 0.16$
$H_a: \sigma^2 > 0.16$
$\chi^2 = 24.63$
RR: $\chi^2 > 16.92$
Reject H_0 (yes)

10.73 $H_0: \sigma^2 = 4$
$H_a: \sigma^2 \neq 4$
$\chi^2 = 202.30$
RR: $\chi^2 < 51.74, \chi^2 > 90.53$
Reject H_0 (yes)

10.75 $H_0: \sigma^2 = 0.0625$
$H_a: \sigma^2 > 0.0625$
$\chi^2 = 80.67$
RR: $\chi^2 > 55.76$
Reject H_0 (yes)

Section 10.6

10.77 a. No
b. Yes
c. No
d. Yes
e. No
f. Yes

10.79 0.0485

10.81 0.3270

10.83 p-value < 0.005

10.85 $0.05 < p$-value < 0.10

10.87 $H_0: \mu = 150$
$H_a: \mu > 150$
$\alpha = 0.05$
$z = 1.97$
p-value $= 0.0244$
Reject H_0 (yes)

10.89 $H_0: \mu = 6.5$
$H_a: \mu < 6.5$
$\alpha = 0.05$
$t = -2.45$
$0.01 < p$-value < 0.025
Reject H_0 (yes)

10.91 $H_0: p = 0.10$
$H_a: p < 0.10$
$\alpha = 0.05$
$z = -1.03$
p-value $= 0.1515$
Fail to reject H_0 (no)

10.93 $H_0: \sigma^2 = 0.0225$
$H_a: \sigma^2 < 0.0225$
$\alpha = 0.05$
$\chi^2 = 21.78$
p-value > 0.10
Fail to reject H_0 (no)

Ⓜ 10.95   -1.2700   0.1020
MTB > # P-VALUE = 0.1020

Ⓜ 10.97   22.1400   0.9242
MTB > # P-VALUE = 1 - 0.9242 = 0.0758

Review Exercises

10.98 $z > 2.326$

10.99 $z < -1.645$

10.100 $t < -2.228, t > 2.228$

10.101 $t < -2.508$

10.102 $z < -1.645$, $z > 1.645$

10.103 $z < -1.282$

10.104 $z < -1.96$, $z > 1.96$

10.105 $\chi^2 > 40.11$

10.106 $\chi^2 < 13.85$, $\chi^2 > 36.42$

10.107 a. Type II
 b. Type I
 c. $z = -1.16$
   RR: $z < -1.645$
   Fail to reject H_0
 d. p-value $= 0.1230$

10.108 $H_0: p = 1/3$
 $H_a: p < 1/3$
 $z = -1.15$
 RR: $z < -1.645$
 Fail to reject H_0 (no)

10.109 $H_0: \mu = 50$
 $H_a: \mu > 50$
 $z = 2.18$
 RR: $z > 1.645$
 Reject H_0 (yes)

10.110 $H_0: \mu = 50$
 $H_a: \mu > 50$
 $\alpha = 0.05$
 $z = 2.18$
 p-value $= 0.0146$
 Reject H_0 (yes)

10.111 a. $H_0: \mu = 40.3$
   $H_a: \mu < 40.3$
   $t = -2.49$
   RR: $t < -1.729$
   Reject H_0 (yes)
 b. Random sample, normal population

10.112 $0.01 < p$-value < 0.025

10.113 $H_0: \sigma^2 = 0.36$
 $H_a: \sigma^2 > 0.36$
 $\chi^2 = 21.30$
 RR: $\chi^2 > 23.68$
 Fail to reject H_0 (no)

10.114 $0.05 < p$-value < 0.10

10.115 $H_0: \mu = 66$
 $H_a: \mu \ne 66$
 $z = -2.80$
 RR: $z < -2.576$, $z > 2.576$
 Reject H_0

10.116 p-value $= 0.0052$

10.117 $H_0: p = 0.35$
 $H_a: p \ne 0.35$
 $z = 1.28$
 RR: $z < -1.96$, $z > 1.96$
 Fail to reject H_0 (no)

10.118 p-value $= 0.2006$

10.119 $H_0: \mu = 53$
 $H_a: \mu > 53$
 $t = 7.90$
 RR: $t > 2.567$
 Reject H_0 (yes)

10.120 $H_0: \sigma^2 = 0.04$
 $H_a: \sigma^2 \ne 0.04$
 $\chi^2 = 14.50$
 RR: $\chi^2 < 0.21$, $\chi^2 > 14.86$
 Fail to reject H_0 (no)

10.121 $0.01 < p$-value < 0.02

Ⓜ 10.122 TEST OF MU = 15.000 VS MU L.T. 15.000

	N	MEAN	STDEV	SE MEAN	T	P VALUE
C1	15	14.427	0.473	0.122	-4.70	0.0002

MTB > # REJECT Ho. THERE IS SUFFICIENT EVIDENCE.

Ⓜ 10.123 TEST OF MU = 800.000 VS MU G.T. 800.000

	N	MEAN	STDEV	SE MEAN	T	P VALUE
C1	13	844.382	483.763	134.172	0.33	0.37

MTB > # FAIL TO REJECT Ho. THERE IS NOT SUFFICIENT EVIDENCE.

Ⓜ 10.124 TEST OF MU = 75.000 VS MU N.E. 75.000
THE ASSUMED SIGMA = 2.22

	N	MEAN	STDEV	SE MEAN	Z	P VALUE
C1	50	73.938	2.217	0.314	-3.38	0.0007

MTB > # REJECT Ho. THERE IS SUFFICIENT EVIDENCE.

Ⓜ 10.125   1.7800   0.9625
MTB > # P-VALUE = 1 - 0.9625 = 0.0375

ⓜ 10.126   2.0100   0.9709
        MTB > # P-VALUE = 2(1 - 0.9709) = 0.0582

ⓜ 10.127   18.5300   0.3259
        MTB > # P-VALUE = 1 - 0.3259 = 0.6741

ⓜ 10.128 TEST OF MU = 80.000 VS MU G.T. 80.000
        THE ASSUMED SIGMA = 6.11

	N	MEAN	STDEV	SE MEAN	Z	P VALUE
C1	36	81.111	6.112	1.019	1.09	0.14

        MTB > # FAIL TO REJECT Ho. THERE IS NOT SUFFICIENT EVIDENCE.

Chapter 11

Section 11.1

11.1 H_0: $(\mu_1 - \mu_2) = 0$
      H_a: $(\mu_1 - \mu_2) > 0$
      $z = 2.38$
      RR: $z > 1.645$
      Reject H_0

11.3 H_0: $(\mu_1 - \mu_2) = 10$
      H_a: $(\mu_1 - \mu_2) \neq 10$
      $z = 2.92$
      RR: $z < -2.576$, $z > 2.576$
      Reject H_0

11.5 H_0: $(\mu_1 - \mu_2) = 5$
      H_a: $(\mu_1 - \mu_2) > 5$
      $z = 1.06$
      RR: $z > 1.645$
      Fail to reject H_0

11.7 H_0: $(\mu_1 - \mu_2) = 0$
      H_a: $(\mu_1 - \mu_2) \neq 0$
      $z = 2.72$
      RR: $z < -1.96$, $z > 1.96$
      Reject H_0 (yes)

11.9 a. H_0: $(\mu_1 - \mu_2) = 0$
        H_a: $(\mu_1 - \mu_2) > 0$
        $z = 1.82$
        RR: $z > 1.645$
        Reject H_0 (yes)
      b. p-value = 0.0344

11.11 H_0: $(\mu_1 - \mu_2) = 0$
        H_a: $(\mu_1 - \mu_2) \neq 0$
        $z = 18.41$
        RR: $z < -2.576$, $z > 2.576$
        Reject H_0 (yes)

11.13 -1.7 ± 0.75

11.15 6.1 ± 0.87

Sections 11.2 and 11.3

11.17 3.6 ± 2.044

11.19 7 ± 11.890

11.21 0.72 ± 0.575

11.23 0.78 ± 0.504

11.25 $0.025 < p\text{-value} < 0.05$

11.27 p-value > 0.10

11.29 If $n_1 = n_2 = n$, then

$$s_p^2 = \frac{(n-1)s_1^2 + (n-1)s_2^2}{(n-1) + (n-1)}$$

$$= \frac{(n-1)(s_1^2 + s_2^2)}{2(n-1)}$$

$$= \frac{s_1^2 + s_2^2}{2}$$

ⓜ 11.31 TWOSAMPLE T FOR C1 VS C2

	N	MEAN	STDEV	SE MEAN
C1	12	38.29	1.83	0.53
C2	10	35.00	2.84	0.90

    95 PCT CI FOR MU C1 - MU C2: (1.20, 5.38)
    TTEST MU C1 = MU C2 (VS NE): T= 3.29 P=0.0037 DF= 20
    POOLED STDEV =   2.34
    MTB > # SINCE P-VALUE = 0.0037 < 0.05, REJECT Ho. THERE IS A DIFFERENCE.

Ⓜ 11.33 TWOSAMPLE T FOR C1 VS C2

```
              N      MEAN    STDEV   SE MEAN
    C1   11   5.500   0.588    0.18
    C2   10   6.460   0.771    0.24

    95 PCT CI FOR MU C1 - MU C2: (-1.60, -0.32)
    TTEST MU C1 = MU C2 (VS NE): T= -3.19  P=0.0058  DF=  16
    MTB > # SINCE P-VALUE = 0.0058 < 0.05, REJECT Ho. DIFFERENCE IS SIGNIF.
```

Section 11.4

11.35 H_0: $\mu_D = 0$
H_a: $\mu_D < 0$
$t = -2.40$
RR: $t < -2.132$
Reject H_0

11.37 H_0: $\mu_D = 7$
H_a: $\mu_D \neq 7$
$t = 1.28$
RR: $t < -4.032$, $t > 4.032$
Fail to reject H_0

11.39 H_0: $\mu_D = 0$
H_a: $\mu_D > 0$
$t = 2.71$
RR: $t > 2.015$
Reject H_0 (yes)

11.41 H_0: $\mu_D = 0$
H_a: $\mu_D > 0$
$t = 2.31$
RR: $t > 1.895$
Reject H_0 (yes)

11.43 H_0: $\mu_D = 0$
H_a: $\mu_D > 0$
$t = 7.14$
RR: $t > 1.796$
Reject H_0 (yes)

11.45 4.1 ± 2.273

11.47 2.5 ± 1.637

ⓂⓂ 11.49 TEST OF MU = 0.000 VS MU G.T. 0.000

```
              N      MEAN    STDEV   SE MEAN      T     P VALUE
    C3        36    1.694    3.161    0.527    3.22    0.0014
    MTB > # SINCE P-VALUE = 0.0014 < 0.05, REJECT Ho. HERB IS EFFECTIVE.
```

ⓂⓂ 11.51 TEST OF MU = 0.000 VS MU G.T. 0.000

```
              N      MEAN    STDEV   SE MEAN      T     P VALUE
    C3        32   43.944   20.691    3.658   12.01    0.0000
    MTB > # SINCE P-VALUE = 0.0000 < 0.01, REJECT Ho. INCREASE HAS OCCURRED.
```

Section 11.5

11.53 0.10 ± 0.055

11.55 0.08 ± 0.054

11.57 p-value $= 0.1118$

11.59 H_0: $(p_1 - p_2) = 0$
H_a: $(p_1 - p_2) > 0$
$z = 8.28$
RR: $z > 1.645$
Reject H_0 (yes)

11.61 -0.153 ± 0.077

11.63 p-value $= 0.2236$

11.65 H_0: $(p_1 - p_2) = 0$
H_a: $(p_1 - p_2) \neq 0$
$z = 1.40$
RR: $z < -1.96$, $z > 1.96$
Fail to reject H_0 (no)

Section 11.6

11.67 **3.89**

11.69 **2.84**

11.71 **2.51**

11.73 2.32
11.75 6.84
11.77 0.10
11.79 0.95
11.81 0.04
11.83 3.73
11.85 4.00

Ⓜ 11.87   1.1200   0.6275
        MTB > # ANSWER A = 0.6275

             0.9300   1.8980
        MTB > # ANSWER B = 1.8980

Ⓜ 11.89

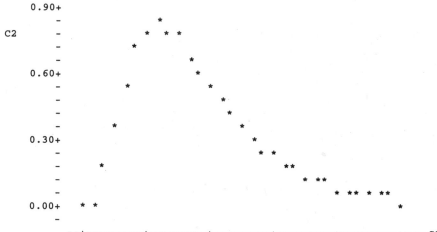

MTB > # YOUR GRAPH MAY DIFFER

Section 11.7

11.91 $H_0: \sigma_1^2 = \sigma_2^2$
       $H_a: \sigma_1^2 > \sigma_2^2$
       $F = 4.39$
       RR: $F > 2.53$
       Reject H_0

11.93 $0.005 < p\text{-value} < 0.01$

11.95 $H_0: \sigma_1^2 = \sigma_2^2$
       $H_a: \sigma_1^2 \neq \sigma_2^2$
       $F = 4.82$
       RR: $F > 4.65$
       Reject H_0

11.97 $H_0: \sigma_1^2 = \sigma_2^2$
       $H_a: \sigma_1^2 > \sigma_2^2$
       $F = 2.61$
       RR: $F > 2.53$
       Reject H_0 (yes)

11.99 $H_0: \sigma_1^2 = \sigma_2^2$
       $H_a: \sigma_1^2 \neq \sigma_2^2$
       $F = 1.28$
       RR: $F > 2.46$
       Fail to reject H_0 (no)

Ⓜ 11.101 a. *p*-value > 0.20

b.   1.2800   0.6752
      MTB > # ANSWER B = 2(1 - 0.6752) = 0.6496

Ⓜ 11.103 MTB > PRINT K1 K2
      K1   0.362381
      K2   0.184000
      MTB > # PART A: VAR(1) = 0.362381, VAR(2) = 0.184

      MTB > # PARTS B AND C FOLLOW.
      MTB > PRINT K3
      K3   1.96947
      MTB > # F = 1.96947

         1.9695   0.7630
      MTB > # P-VALUE = 2(1 - 0.763) = 0.474
      MTB > # SINCE P-VALUE > 0.05, FAIL TO REJECT Ho
      MTB > # DIFFERENCE IS NOT STATISTICALLY SIGNIFICANT

Review Exercises

11.104 $H_0: (p_1 - p_2) = 0$
$H_a: (p_1 - p_2) \neq 0$
$z = 4.72$
RR: $z < -2.576$, $z > 2.576$
Reject H_0

11.105 a. $H_0: (p_1 - p_2) = 0$
$H_a: (p_1 - p_2) > 0$
$z = 2.54$
RR: $z > 1.645$
Reject H_0 (yes)
b. *p*-value $= 0.0055$

11.106 0.048 ± 0.032

11.107 $H_0: \mu_D = 0$
$H_a: \mu_D < 0$
$t = -3.29$
RR: $t < -2.132$
Reject H_0

11.108 -4.6 ± 3.886

11.109 $H_0: \sigma_1^2 = \sigma_2^2$
$H_a: \sigma_1^2 \neq \sigma_2^2$
$F = 1.36$
RR: $F > 2.90$
Fail to reject H_0 (no)

11.110 $H_0: (\mu_1 - \mu_2) = 0$
$H_a: (\mu_1 - \mu_2) \neq 0$
$t = 1.68$
RR: $t < -2.086$, $t > 2.086$
Fail to reject H_0 (no)

11.111 13 ± 13.379

11.112 a. $H_0: \sigma_1^2 = \sigma_2^2$
$H_a: \sigma_1^2 \neq \sigma_2^2$
$F = 3.17$
RR: $F > 2.53$
Reject H_0 (yes)

b. $H_0: (\mu_1 - \mu_2) = 0$
$H_a: (\mu_1 - \mu_2) \neq 0$
$t = 1.94$
RR: $t < -2.064$, $t > 2.064$
Fail to reject H_0 (no)

11.113 $0.05 < $ *p*-value < 0.10

11.114 0.7 ± 0.386

11.115 0.15 ± 0.082

11.116 $H_0: (\mu_1 - \mu_2) = 0$
$H_a: (\mu_1 - \mu_2) > 0$
$z = 2.15$
RR: $z > 1.645$
Reject H_0 (yes)

11.117 *p*-value $= 0.0158$

11.118 $H_0: \sigma_1^2 = \sigma_2^2$
$H_a: \sigma_1^2 \neq \sigma_2^2$
$F = 11.60$
RR: $F > 2.70$
Reject H_0 (yes)

11.119 $H_0: (\mu_1 - \mu_2) = 20$
$H_a: (\mu_1 - \mu_2) > 20$
$t = 3.88$
RR: $t > 1.860$
Reject H_0

11.120 36.7 ± 14.45

11.121 $H_0: \mu_D = 0$
$H_a: \mu_D > 0$
$t = 4.69$
RR: $t > 2.602$
Reject H_0 (yes)

Ⓜ 11.122 TEST OF MU = 0.000 VS MU G.T. 0.000

	N	MEAN	STDEV	SE MEAN	T	P VALUE
C3	16	7.625	6.500	1.625	4.69	0.0001

MTB > # SINCE P-VALUE < 0.01, REJECT Ho. TREATMENT IS EFFECTIVE.

Ⓜ 11.123

	N	MEAN	STDEV	SE MEAN	99.0 PERCENT C.I.
C3	16	7.62	6.50	1.62	(2.84, 12.41)

Ⓜ 11.124 K1   10.8071
      K2   0.629925
      MTB > # PART A: VAR(1) = 10.8071, VAR(2) = 0.629925
      MTB > # PARTS B AND C FOLLOW.

      K3   17.1562
      MTB > # F VALUE = 17.1562

         17.1562   1.0000
      MTB > # P-VALUE $\approx$ 2(1 - 1.0000) $\approx$ 0
      MTB > # REJECT Ho SINCE P-VALUE < 0.05
      MTB > # THE DIFFERENCE IS STATISTICALLY SIGNIFICANT.

Ⓜ 11.125 TWOSAMPLE T FOR C1 VS C2

	N	MEAN	STDEV	SE MEAN
C1	10	36.96	3.29	1.0
C2	12	32.442	0.794	0.23

      95 PCT CI FOR MU C1 - MU C2: (2.1, 6.93)
      TTEST MU C1 = MU C2 (VS NE): T= 4.24 P=0.0022 DF= 9
      MTB > # SINCE P-VALUE < 0.01, REJECT Ho. BRANDS DIFFER.

Ⓜ 11.126 TWOSAMPLE T FOR C1 VS C2

	N	MEAN	STDEV	SE MEAN
C1	10	36.96	3.29	1.0
C2	12	32.442	0.794	0.23

      90 PCT CI FOR MU C1 - MU C2: (2.6, 6.47)
      TTEST MU C1 = MU C2 (VS NE): T= 4.24 P=0.0022 DF= 9

Ⓜ 11.127
```
      0.90+
         -                        **  *
   C2    -                  *          *
         -
         -              *            *
      0.60+                            *
         -
         -                       *
         -          *              *
         -                          *
      0.30+   .                          *
         -          *                  *  *
         -                             **
         -                          *  **
         -      *                   * ** * **
      0.00+   *  *                              *
         -
         --+---------+---------+---------+---------+---------+----C1
          0.00      0.60      1.20      1.80      2.40      3.00
```
      MTB > # YOUR GRAPH MAY DIFFER.

Ⓜ 11.128   1.2300   0.7001
      MTB > # ANSWER A = 0.7001

            0.8300   1.4843
      MTB > # ANSWER B = 1.4843

Chapter 12
Section 12.2

12.1 $H_0: p_1 = 0.30, p_2 = 0.10, p_3 = 0.40, p_4 = 0.20$
  $H_a:$ Not all of the above are true.
  $\chi^2 = 5.66$
  RR: $\chi^2 > 6.25$
  Fail to reject H_0 (no)

12.3 p-value > 0.10

12.5 p-value < 0.005

12.7 $H_0: p_1 = 1/3, p_2 = 1/3, p_3 = 1/3$
  $H_a:$ Not all of the above are true.
  $\chi^2 = 2.60$
  RR: $\chi^2 > 4.61$
  Fail to reject H_0

12.9 p-value < 0.005

12.11 $0.05 < p$-value < 0.10

12.13 $H_0: p_1 = 0.15, p_2 = 0.24, p_3 = 0.32, p_4 = 0.20,$
  $p_5 = 0.09$
  $H_a:$ Not all of the above are true.
  $\chi^2 = 5.01$
  RR: $\chi^2 > 9.49$
  Fail to reject H_0 (no)

Ⓜ 12.15   8.5800   0.9275
  MTB > # P-VALUE = (1 - 0.9275) = 0.0725

Ⓜ 12.17   5.0100   0.7137
  MTB > # P-VALUE = (1 - 0.7137) = 0.2863

Section 12.3

12.19 $H_0:$ The row and column classifications are
  independent.
  $H_a:$ They are dependent.
  $\chi^2 = 7.55$
  RR: $\chi^2 > 5.99$
  Reject H_0

12.21 $0.05 < p$-value < 0.10

12.23 $H_0:$ Damage is independent of the presence of an
  extinguisher.
  $H_a:$ They are dependent.
  $\chi^2 = 32.82$
  RR: $\chi^2 > 11.34$
  Reject H_0 (yes)

12.25 $H_0:$ Month of injury is independent of gender.
  $H_a:$ They are dependent.
  $\chi^2 = 11.68$
  RR: $\chi^2 > 19.68$
  Fail to reject H_0 (no)

12.27 $H_0:$ Completion semester is independent of GPA.
  $H_a:$ They are dependent.
  $\chi^2 = 10.47$
  RR: $\chi^2 > 9.49$
  Reject H_0 (yes)

12.29 $H_0:$ The 2 methods of classification are independent.
  $H_a:$ They are dependent.
  $\chi^2 = 19.59$
  RR: $\chi^2 > 5.99$
  Reject H_0 (yes)

Ⓜ 12.31 Expected counts are printed below observed counts

	C1	C2	C3	Total
1	26	66	40	132
	38.30	61.55	32.15	
2	30	24	7	61
	17.70	28.45	14.85	
Total	56	90	47	193

ChiSq =   3.950 + 0.321 + 1.919 +
          8.548 + 0.695 + 4.153 = 19.588

df = 2
MTB > # SINCE 19.588 IS IN RR (> 5.99), REJECT Ho.
      # YES, THERE IS A DEPENDENCY.

Ⓜ 12.33 Expected counts are printed below observed counts

	C1	C2	Total
1	189	291	480
	183.22	296.78	
2	35	62	97
	37.02	59.98	
3	29	30	59
	22.52	36.48	
4	17	32	49
	18.70	30.30	
5	13	33	46
	17.56	28.44	
6	5	13	18
	6.87	11.13	
7	4	12	16
	6.11	9.89	
Total	292	473	765

```
ChiSq =  0.183 +  0.113 +
         0.111 +  0.068 +
         1.864 +  1.151 +
         0.155 +  0.096 +
         1.183 +  0.731 +
         0.509 +  0.314 +
         0.727 +  0.449 = 7.654
df = 6
MTB > # SINCE 7.654 IS NOT IN RR (> 12.59), FAIL TO REJECT Ho.
MTB > # NO. INSUFFICIENT EVIDENCE OF A RELATIONSHIP.
```

Ⓜ 12.35   94.6800   1.0000

```
MTB > # P-VALUE = (1 - 1.000) ≈ 0.0000
```

Section 12.4

12.37 H_0: $p_1 = p_2 = p_3$
H_a: Not all proportions are equal.
$\chi^2 = 4.88$
RR: $\chi^2 > 5.99$
Fail to reject H_0 (no)

12.39 H_0: $p_1 = p_2 = p_3 = p_4$
H_a: Not all probabilities are equal.
$\chi^2 = 4.10$
RR: $\chi^2 > 6.25$
Fail to reject H_0 (no)

12.41 $0.005 < p\text{-value} < 0.01$

12.43 H_0: $p_1 = p_2$
H_a: $p_1 \neq p_2$
$z = 2.65$
RR: $z < -1.96$, $z > 1.96$
Reject H_0 (yes)

12.45 H_0: $p_1 = p_2 = p_3 = p_4 = p_5$
H_a: Not all probabilities are equal.
$\chi^2 = 10.17$
RR: $\chi^2 > 9.49$
Reject H_0 (yes)

12.47 H_0: $p_1 = p_2 = p_3$
H_a: Not all probabilities are equal.
$\chi^2 = 28.56$
RR: $\chi^2 > 9.21$
Reject H_0 (yes)

Ⓜ 12.49 Expected counts are printed below observed counts

	ME	NH	MA	RI	CT	VT	Total
1	375	321	315	362	358	347	2078
	346.33	346.33	346.33	346.33	346.33	346.33	
2	125	179	185	138	142	153	922
	153.67	153.67	153.67	153.67	153.67	153.67	
Total	500	500	500	500	500	500	3000

```
ChiSq =  2.373 +  1.853 +  2.835 +  0.709 +  0.393 +  0.001 +
         5.348 +  4.176 +  6.389 +  1.597 +  0.886 +  0.003 = 26.563
df = 5
MTB > # SINCE 26.563 IS IN RR (> 11.07), REJECT Ho.
MTB > # A DIFFERENCE IN PROPORTIONS EXISTS.
```

Ⓜ 12.51 Expected counts are printed below observed counts

	ME	NH	VT	Total
1	375	321	347	1043
	347.67	347.67	347.67	
2	125	179	153	457
	152.33	152.33	152.33	
Total	500	500	500	1500

```
ChiSq =  2.149 +  2.045 +  0.001 +
         4.904 +  4.668 +  0.003 = 13.771
df = 2
MTB > # SINCE 13.771 IS IN RR (> 5.99), REJECT Ho.
MTB > # A DIFFERENCE IN PROPORTIONS EXISTS.
```

Review Exercises

12.53 H_0: The 2 methods of classification are independent.
H_a: They are dependent.
$\chi^2 = 42.47$
RR: $\chi^2 > 9.49$
Reject H_0 (yes)

12.54 p-value < 0.005

12.55 There are 2 cells with very small expected counts (much less than 5).

12.56 H_0: $p_1 = p_2 = p_3 = p_4 = p_5 = p_6 = p_7$
H_a: Not all proportions are equal.
$\chi^2 = 43.98$
RR: $\chi^2 > 12.59$
Reject H_0 (yes)

12.57 H_0: $p_1 = p_2 = p_3$
H_a: Not all proportions are equal.
$\chi^2 = 4.58$
RR: $\chi^2 > 5.99$
Fail to reject H_0 (no)

12.58 p-value < 0.005

12.59 p-value > 0.10

12.60 H_0: Color preference is independent of size.
H_a: They are dependent.
$\chi^2 = 3.15$
RR: $\chi^2 > 7.81$
Fail to reject H_0

12.61 p-value > 0.10

12.62 H_0: $p_1 = 0.20$, $p_2 = 0.20$, $p_3 = 0.20$, $p_4 = 0.20$, $p_5 = 0.20$
H_a: Not all of the above are true.
$\chi^2 = 12.21$
RR: $\chi^2 > 13.28$
Fail to reject H_0 (no)

12.63 $0.01 < p$-value < 0.025

12.64 H_0: Mortgage selection is independent of income level.
H_a: They are dependent.
$\chi^2 = 34.54$
RR: $\chi^2 > 12.59$
Reject H_0

12.65 p-value < 0.005

12.66 H_0: $p_1 = p_2 = p_3 = p_4$
H_a: Not all proportions are equal.
$\chi^2 = 34.35$
RR: $\chi^2 > 11.34$
Reject H_0 (yes)

12.67 H_0: $p_1 = p_2 = p_3$
H_a: Not all proportions are equal.
$\chi^2 = 25.68$
RR: $\chi^2 > 5.99$
Reject H_0 (yes)
12.68 p-value < 0.005

12.69 H_0: $p_1 = 0.24$, $p_2 = 0.38$, $p_3 = 0.21$, $p_4 = 0.17$
H_a: Not all of the above are true.
$\chi^2 = 20.15$
RR: $\chi^2 > 7.81$
Reject H_0 (yes)
12.70 p-value < 0.005

Ⓜ 12.71 Expected counts are printed below observed counts

	C1	C2	C3	Total
1	697	603	571	1871
	623.67	623.67	623.67	
2	503	597	629	1729
	576.33	576.33	576.33	
Total	1200	1200	1200	3600

ChiSq = 8.623 + 0.685 + 4.448 +
         9.331 + 0.741 + 4.813 = 28.640
df = 2
MTB > # SINCE 28.64 IS IN RR (> 9.21), REJECT Ho.
MTB > # SUFF. EVIDENCE THAT THE 3 PROPORTIONS ARE NOT ALL EQUAL.

Ⓜ 12.72   28.6400   1.0000
MTB > # P-VALUE = (1 - 1.0000) ≈ 0.0000

Ⓜ 12.73 Expected counts are printed below observed counts

	C1	C2	C3	C4	Total
1	27	108	79	35	249
	23.10	106.27	77.01	42.61	
2	18	99	71	48	236
	21.90	100.73	72.99	40.39	
Total	45	207	150	83	485

ChiSq = 0.657 + 0.028 + 0.051 + 1.360 +
         0.694 + 0.030 + 0.054 + 1.435 = 4.309
df = 3
MTB > # SINCE 4.309 IS NOT IN RR (> 7.81), FAIL TO REJECT Ho.
MTB > # INSUFFICIENT EVIDENCE THAT THEY ARE DEPENDENT.

Ⓜ 12.74   4.3090   0.7700
MTB > # P-VALUE = (1 - 0.7700) = 0.2300

Ⓜ 12.75 Expected counts are printed below observed counts

	C1	C2	C3	Total
1	60	47	32	139
	46.33	46.33	46.33	
2	340	353	368	1061
	353.67	353.67	353.67	
Total	400	400	400	1200

ChiSq = 4.031 + 0.010 + 4.434 +
         0.528 + 0.001 + 0.581 = 9.585
df = 2
MTB > # SINCE 9.585 IS IN RR (> 5.99), REJECT Ho.
MTB > # SUFF. EVIDENCE THAT THE PROPORTIONS ARE NOT ALL EQUAL.

Ⓜ 12.76   9.5850   0.9917
MTB > # P-VALUE = 1 - 0.9917 = 0.0083

Chapter 13

Sections 13.1 and 13.2

13.1   42.5714

13.3   37.48

13.5   a. 11.6
     b. 1,012.8
     c. 16.7647
     d.

Source	df	SS	MS	F
Treatments	2	1,012.8	506.4	30.21
Error	17	285.0	16.7647	
Total	19	1,297.8		

13.7   a. 336
     b. 336
     c. 46
     d. 46
     e.

Source	df	SS	MS	F
Treatments	2	336	168	32.87
Error	9	46	5.111	
Total	11	382		

13.9

Source	df	SS	MS	F
Treatments	2	157.11	78.55	9.67
Error	10	81.20	8.12	
Total	12	238.31		

13.11

Source	df	SS	MS	F
Treatments	4	752.58	188.14	9.91
Error	19	360.75	18.99	
Total	23	1,113.33		

13.13   a.

Source	df	SS	MS	F
Treatments	7	21.7	3.1	1.82
Error	28	47.6	1.7	
Total	35	69.3		

     b. 8
     c. 36

Sections 13.3 and 13.4

13.15   a.

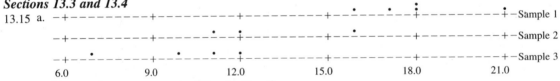

The sample means appear to differ significantly.

     b. $H_0: \mu_1 = \mu_2 = \mu_3$
       H_a: Not all the means are equal.

Source	df	SS	MS	F
Treatments	2	146.92	73.46	15.74
Error	9	42.00	4.667	
Total	11	188.92		

RR: $F > 4.26$
Reject H_0

13.17   a. 18 ± 2.19
     b. 5 ± 3.57

13.19 p-value < 0.005

13.21 $H_0: \mu_1 = \mu_2$
     $H_a: \mu_1 \neq \mu_2$

Source	df	SS	MS	F
Treatments	1	59.10	59.10	10.82
Error	20	109.23	5.462	
Total	21	168.33		

RR: $F > 4.35$
Reject H_0

13.23 $\$3.29 \pm \$2.85$

13.25 p-value < 0.005

13.27 2.475 ± 0.852

13.29 4.2 ± 0.644

13.31 $H_0: \mu_1 = \mu_2 = \mu_3$
H_a: Not all the means are equal.

Source	df	SS	MS	F
Treatments	2	11.66	5.83	2.37
Error	20	49.21	2.46	
Total	22	60.87		

RR: $F > 3.49$
Fail to reject H_0 (no)

13.33 -1.68 ± 1.40
13.35 p-value < 0.005

13.37 $H_0: \mu_1 = \mu_3 = \mu_4$
H_a: Not all the means are equal.

Source	df	SS	MS	F
Treatments	2	3.210	1.605	3.53
Error	14	6.365	0.455	
Total	16	9.575		

RR: $F > 3.74$
Fail to reject H_0 (no)

13.39 $-\$1,554 \pm \$858$

Ⓜ 13.41 MTB > # FROM THE MINITAB OUTPUT, MEAN (2) = 87.60, MEAN(4) = 79.60,
MTB > # MSE = 12.9 WITH DF = 36. FROM THESE RESULTS, WE OBTAIN:
MTB > # 95% CI FOR MU(2) - MU(4) IS 8.00 +/- 3.15

Ⓜ 13.43 ANALYSIS OF VARIANCE

SOURCE	DF	SS	MS	F	p
FACTOR	3	51.123	17.041	29.82	0.000
ERROR	20	11.430	0.571		
TOTAL	23	62.553			

Ⓜ 13.45 ANALYSIS OF VARIANCE

SOURCE	DF	SS	MS	F	p
FACTOR	3	14.149	4.716	12.68	0.000
ERROR	19	7.065	0.372		
TOTAL	22	21.215			

Review Exercises

13.46 $H_0: \mu_1 = \mu_2 = \mu_3$
H_a: Not all the means are equal.

Source	df	SS	MS	F
Treatments	2	5,942	2,971	3.74
Error	18	14,286.57	793.70	
Total	20	20,228.57		

RR: $F > 6.01$
Fail to reject H_0

13.47 $0.025 < p$-value < 0.05
13.48 $\$934.43 \pm \$18.46$
13.49 $-\$38.71 \pm \$26.11$
13.50 19.13
13.51 If $n_1 = n_2 = \cdots = n_k = n$, then

$$MSE = \frac{(n-1)s_1^2 + (n-1)s_2^2 + \cdots + (n-1)s_k^2}{(n-1) + (n-1) + \cdots + (n-1)}$$

$$= \frac{(n-1)(s_1^2 + s_2^2 + \cdots + s_k^2)}{k(n-1)}$$

$$= \frac{s_1^2 + s_2^2 + \cdots + s_k^2}{k}$$

13.52 a. 991.2
b. 21.423
c.

Source	df	SS	MS	F
Treatments	3	991.2	330.4	15.42
Error	26	557	21.423	
Total	29	1,548.2		

d. $H_0: \mu_1 = \mu_2 = \mu_3 = \mu_4$
H_a: Not all the means are equal.
$F = 15.42$
RR: $F > 2.98$
Reject H_0

13.53 39 ± 2.497
13.54 12 ± 4.404

13.55 a.

4.0   8.0   12.0

There appears to be a significant difference in the sample means.

b. $H_0: \mu_1 = \mu_2 = \mu_3 = \mu_4$
H_a: Not all the means are equal.

Source	df	SS	MS	F
Treatments	3	75.05	25.017	8.55
Error	14	40.95	2.925	
Total	17	116		

RR: $F > 3.34$
Reject H_0

13.56 $H_0: \mu_1 = \mu_2 = \mu_3 = \mu_4 = \mu_5 = \mu_6 = \mu_7 = \mu_8$
H_a: Not all the means are equal.

Source	df	SS	MS	F
Treatments	7	493	70.43	6.80
Error	22	228	10.364	
Total	29	721		

RR: $F > 3.59$
Reject H_0

13.57 $H_0: \mu_1 = \mu_2 = \mu_3$
H_a: Not all the means are equal.

Source	df	SS	MS	F
Treatments	2	244.33	122.17	17.99
Error	21	142.63	6.792	
Total	23	386.96		

RR: $F > 3.47$
Reject H_0 (yes)

13.58 p-value < 0.005
13.59 65.875 ± 1.917
13.60 7.75 ± 2.710

13.61 $H_0: \mu_1 = \mu_2 = \mu_3 = \mu_4$
H_a: Not all the means are equal.

Source	df	SS	MS	F
Treatments	3	107.28	35.76	2.92
Error	17	208.53	12.266	
Total	20	315.81		

RR: $F > 3.20$
Fail to reject H_0 (no)

13.62 $0.05 < p$-value < 0.10
13.63 6.4 ± 3.85
13.64 $H_0: \mu_1 = \mu_2 = \mu_3$
H_a: Not all the means are equal.

Source	df	SS	MS	F
Treatments	2	6.729	3.364	7.90
Error	12	5.108	0.426	
Total	14	11.837		

RR: $F > 3.89$
Reject H_0 (yes)

Ⓜ 13.65 ANALYSIS OF VARIANCE

SOURCE	DF	SS	MS	F	p
FACTOR	2	206.10	103.05	19.39	0.000
ERROR	30	159.41	5.31		
TOTAL	32	365.52			

```
                                     INDIVIDUAL 95 PCT CI'S FOR MEAN
                                     BASED ON POOLED STDEV
    LEVEL    N     MEAN    STDEV  ------+---------+---------+---------+
    C1      12    38.292   1.827                   (----*-----)
    C2      10    35.000   2.838   (-----*-----)
    C3      11    41.273   2.240                         (-----*-----)
                                  ------+---------+---------+---------+
    POOLED STDEV =    2.305         35.0      37.5      40.0      42.5
    MTB > # REJECT Ho SINCE P-VALUE < 0.01. SUFF. EVIDENCE OF A DIFFERENCE.
    MTB > # PART B: IT APPEARS THAT MU(2) DIFFERS FROM MU(1) AND MU(3).
          # ALSO MU(1) MAY DIFFER FROM MU(3).
```

Ⓜ 13.66

```
    MTB > # FROM THE MINITAB OUTPUT, MEAN(1) = 38.29, MEAN(3) = 41.27,
    MTB > # MSE = 5.31 WITH DF = 30.  FROM THESE RESULTS, WE OBTAIN:
    MTB > # 95% CI FOR MU(1) - MU(3) IS -2.98 +/- 1.89
```

Ⓜ 13.67 ANALYSIS OF VARIANCE

SOURCE	DF	SS	MS	F	p
FACTOR	2	2556.8	1278.4	29.90	0.000
ERROR	45	1924.2	42.8		
TOTAL	47	4481.0			

```
                                     INDIVIDUAL 95 PCT CI'S FOR MEAN
                                     BASED ON POOLED STDEV
    LEVEL    N     MEAN    STDEV  -----+---------+---------+---------+-
    C1      16    79.187   7.626               (----*----)
    C2      16    70.000   6.673   (----*----)
    C3      16    87.875   5.058                         (----*---)
                                  -----+---------+---------+---------+-
    POOLED STDEV =    6.539         70.0      77.0      84.0      91.0
    MTB > # REJECT Ho SINCE P-VALUE < 0.05. SUFF. EVIDENCE OF A DIFFERENCE.
```

Ⓜ 13.68 ANALYSIS OF VARIANCE

SOURCE	DF	SS	MS	F	p
FACTOR	1	603.8	603.8	14.42	0.001
ERROR	30	1256.2	41.9		
TOTAL	31	1860.0			

```
                                     INDIVIDUAL 95 PCT CI'S FOR MEAN
                                     BASED ON POOLED STDEV
    LEVEL    N     MEAN    STDEV  ---------+---------+---------+-------
    C1      16    79.187   7.626   (-----*------)
    C3      16    87.875   5.058                    (------*-----)
                                  ---------+---------+---------+-------
    POOLED STDEV =    6.471         80.0      85.0      90.0
    MTB > # REJECT Ho SINCE P-VALUE < 0.05. SUFF. EVIDENCE OF A DIFFERENCE.
```

Chapter 14
Sections 14.1 and 14.2

14.1 A deterministic model assumes that the phenomenon of interest can be predicted precisely, while a probabilistic model allows for variation in the predictions.

14.3 A probabilistic model produces different values of y for the same x because it contains a random error component. A deterministic model has no random component and thus gives the same y-value for a given value of x.

14.5 $\hat{y} = 50 + 45x$
     Deterministic model

14.7

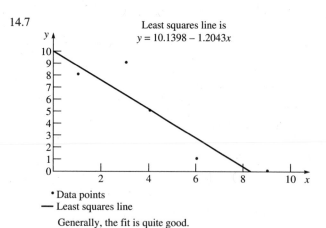

Least squares line is
$y = 10.1398 - 1.2043x$

• Data points
— Least squares line

Generally, the fit is quite good.

14.9 a. and b.

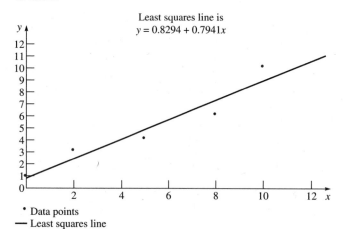

Least squares line is
$y = 0.8294 + 0.7941x$

• Data points
— Least squares line

The line fits the points well.

c. 3.918
d. 1.306
e. 1.143

14.11 SSE = 888.5
$s^2 = 9.066$

14.13

Least squares line is
$y = -10.7944 + 0.04859x$

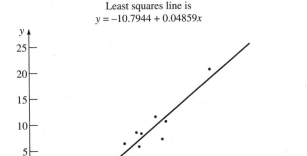

• Data points
— Least squares line

14.15 a. and b.

Least squares line is
$y = 4.5674 - 0.050267x$

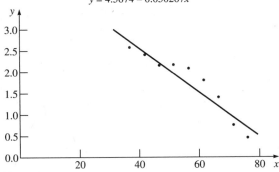

• Data points
— Least squares line

  c. SSE = 0.4298
   $s^2 = 0.0614$

Sections 14.3 and 14.4

14.17 a. -0.910
  b. 0.827
  c. The least squares line accounts for 82.7 percent of
   the total variation in the y-values.

14.19 1.127 ± 0.437

14.21 $H_0: \beta_1 = 0$
  $H_a: \beta_1 < 0.$
  $t = -7.37$
  RR: $t < -2.896$
  Reject H_0 (yes)

14.23 0.086
  The least squares line accounts for 8.6 percent of the
  total variation in the y-values.

14.25 a. -0.872
  b. y tends to decrease as x increases.
  c. 0.760
  d. The least squares line accounts for 76.0 percent of
   the total variation in the y-values.

14.27 p-value < 0.01

14.29 a. 0.825
  The least squares line accounts for 82.5 percent of
  the total variation in the y-values.
  b. $H_0: \beta_1 = 0$
  $H_a: \beta_1 \neq 0$
  $t = 5.32$
  RR: $t < -3.707, t > 3.707$
  Reject H_0 (yes)

14.31 0.0486 ± 0.0224

14.33 p-value < 0.01

Section 14.5

14.35 a.

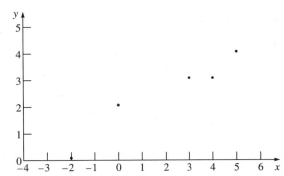

• Data points

  b. $\hat{y} = 1.4 + 0.5x$
  c. 0.924
  The least squares line accounts for 92.4 percent of
  the total variation in the y-values.
  d. $H_0: \beta_1 = 0$
  $H_a: \beta_1 \neq 0$
  $t = 6.04$
  RR: $t < -3.182, t > 3.182$
  Reject H_0 (yes)
  e. 2.9 ± 0.736
  f. 2.9 ± 1.704
  g. Widths are 1.472 and 3.408

14.37 a. 291.7 ± 2.967
       b. Narrower because 21 is closer to $\bar{x}$.
       c. 20
14.39 76.71 ± 3.75
14.41 $\$8.65 \pm \$5.74$
14.43 a.

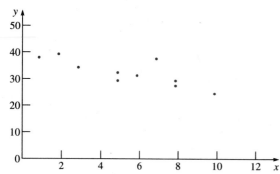

• Data points

       b. $\hat{y} = 40.678 - 1.396x$
       c. 0.654
          The least squares line accounts for 65.4 percent of
          the total variation in the y-values.
       d. $H_0: \beta_1 = 0$
          $H_a: \beta_1 \neq 0$
          $t = -3.89$
          RR: $t < -2.306, t > 2.306$
          Reject H_0
       e. 33.70 ± 7.51

Section 14.6
Ⓜ 14.45 a. $\hat{y} = 4.57 - 0.0503x$
         b. 0.898
            The least squares line accounts for 89.8 percent of
            the total variation in the y-values.
         c. $t = -7.86$
            p-value $= 0.000$
            Yes
         d. For $x = 50$, we are 95 percent confident that the
            long run average years of added life is between
            1.8318 and 2.2763.

Ⓜ 14.47 a. $\hat{y} = -10.8 + 0.0486x$
         b. $r^2 = 0.825$, which suggests that the fit is quite good.
         c. $t = 5.32$
            p-value $= 0.002$
            Yes
         d. For a randomly selected year when $x = \$400$, the
            probability is 0.95 that y is between $\$2.904$ and
            $\$14.38$.
         e. The interval is too wide to be of much practical
            value.
Ⓜ 14.49 a. 82.5 percent
            The least squares line accounts for 82.5 percent of
            the total variation in the y-values.
         b. $r^2 = 137.34/166.45 = 0.825$

Ⓜ 14.51 MTB > # PART A

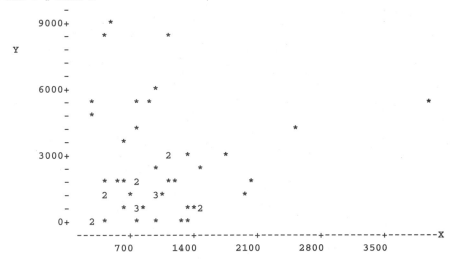

```
MTB > # PART B
MTB > # FEE AND ERROR SIZE DO NOT APPEAR TO BE CORRELATED.

Correlation of X and Y = 0.072

MTB > PRINT K1 # COEFFICIENT OF DETERMINATION
K1        0.00518400
MTB > # PART C

Predictor      Coef      Stdev    t-ratio        p
Constant      2110.8      612.4      3.45     0.001
X             0.2558     0.5126      0.50     0.620

MTB > # P-VALUE FOR SLOPE IS 0.62 > 0.05. FAIL TO REJECT Ho.
MTB > # CAN NOT CONCLUDE THAT X AND Y ARE LINEARLY RELATED.
```

Review Exercises

14.52 **a.**

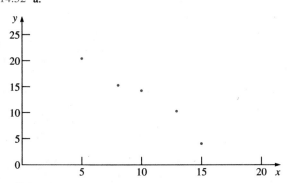

• Data points

  b. $\hat{y} = 27.640 - 1.4745x$
  c. $r = -0.976$
  $\quad r^2 = 0.953$

  d. $H_0: \beta_1 = 0$
  $H_a: \beta_1 \neq 0$
  $t = -7.84$
  RR: $t < -3.182$, $t > 3.182$
  Reject H_0 (yes)
  e. $s = 1.490$
  f. -1.4745 ± 0.4424
  g. 11.420 ± 2.174
  h. 11.420 ± 5.216
14.53 **a.** $y = 135 + 75x$
  b. The model is deterministic because, for a given value of x, it always yields the same y-value.
14.54 **a.** 240
  b. 6.3158
  c. 2.5131

14.55 a.

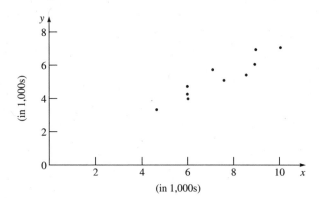

• Data points

b. $\hat{y} = 179.26 + 0.69367x$

c. 0.879

The least squares line accounts for 87.9 percent of the total variation in the y-values.

14.56 $H_0: \beta_1 = 0$
$H_a: \beta_1 \neq 0$
$t = 7.62$
RR: $t < -2.306, t > 2.306$
Reject H_0
p-value < 0.01

14.57 $4{,}341 \pm 420$

14.58 $4{,}341 \pm 1{,}138$

14.59 a.

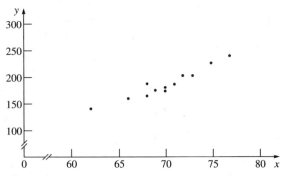

• Data points

b. $\hat{y} = -288.59 + 6.70278x$

c. 0.922

The least squares line accounts for 92.2 percent of the total variation in the y-values.

d. $H_0: \beta_1 = 0$
$H_a: \beta_1 \neq 0$
$t = 10.85$
RR: $t < -2.228, t > 2.228$
Reject H_0

e. 180.61 ± 19.06

14.60 a.

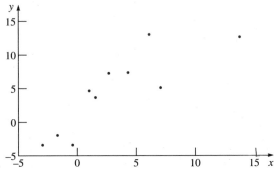

• Data points

b. $\hat{y} = 0.91285 + 1.02105x$

c. 0.848

d. $H_0: \beta_1 = 0$
$H_a: \beta_1 \neq 0$
$t = 4.53$
RR: $t < -2.306, t > 2.306$
Reject H_0

e. 71.9%

14.61 $2.95\% \pm 2.53\%$

14.62 $2.95\% \pm 8.15\%$

Ⓜ 14.63 MTB > # PART A

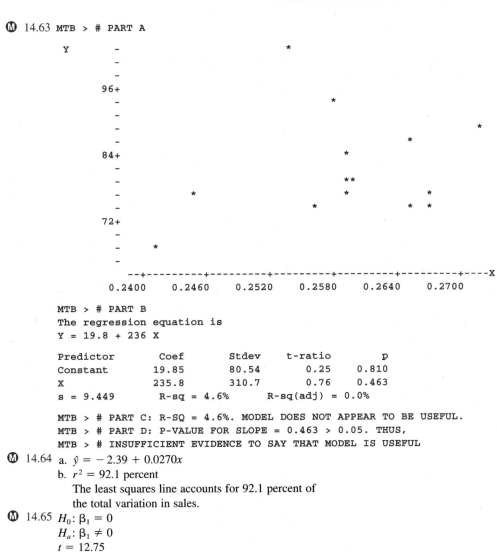

```
Y      -                              *
       -
       -
       -
  96+  -                                 *
       -
       -                                            *
       -                                       *
       -
  84+  -                                  *
       -
       -                                     **
       -               *                     *        *
       -                               *           *  *
  72+  -
       -
       -       *
       --+---------+---------+---------+---------+---------+----X
       0.2400   0.2460   0.2520   0.2580   0.2640   0.2700
```

MTB > # PART B
The regression equation is
Y = 19.8 + 236 X

Predictor	Coef	Stdev	t-ratio	p
Constant	19.85	80.54	0.25	0.810
X	235.8	310.7	0.76	0.463

s = 9.449   R-sq = 4.6%   R-sq(adj) = 0.0%

MTB > # PART C: R-SQ = 4.6%. MODEL DOES NOT APPEAR TO BE USEFUL.
MTB > # PART D: P-VALUE FOR SLOPE = 0.463 > 0.05. THUS,
MTB > # INSUFFICIENT EVIDENCE TO SAY THAT MODEL IS USEFUL

Ⓜ 14.64 a. $\hat{y} = -2.39 + 0.0270x$
b. $r^2 = 92.1$ percent
   The least squares line accounts for 92.1 percent of
   the total variation in sales.

Ⓜ 14.65 $H_0: \beta_1 = 0$
$H_a: \beta_1 \neq 0$
$t = 12.75$
p-value $= 0.000$
Reject H_0

Ⓜ 14.66 For weeks when $x = 2,300$, we are 95 percent confident that the long run average weekly sales is between $57,441 and $62,099.

Chapter 15
Sections 15.1 and 15.2

15.1 $H_0: \tilde{\mu}_D = 0$
  $H_a: \tilde{\mu}_D < 0$
  $x = 2$
  RR: 0, 1 ($\alpha = 0.035$, $n = 8$)
  Fail to reject H_0 (no)

15.3 $H_0: \tilde{\mu} = 40$
  $H_a: \tilde{\mu} < 40$
  $x = 5$
  RR: 0,1,2,3,4,5,6 ($\alpha = 0.058$, $n = 20$)
  Reject H_0 (yes)

15.5 $H_0: \tilde{\mu} = 15$
  $H_a: \tilde{\mu} < 15$
  $x = 3$
  RR: 0,1,2,3,4 ($\alpha = 0.059$, $n = 15$)
  Reject H_0 (yes)

15.7 p-value = 0.377

15.9 $H_0: \tilde{\mu}_D = 0$
  $H_a: \tilde{\mu}_D > 0$
  $x = 5$
  RR: 5,6 ($\alpha = 0.109$, $n = 6$)
  Reject H_0 (yes)

15.11 p-value = 0.0055

15.13 p-value = 0.0238

15.15 p-value = 0.0625

Ⓜ 15.17 SIGN TEST OF MEDIAN = 80.00 VERSUS G.T.   80.00

```
              N   BELOW   EQUAL   ABOVE   P-VALUE    MEDIAN
C1           36     15       1      20    0.2498     82.50
MTB > # SINCE P-VALUE = 0.2498 > 0.05, FAIL TO REJECT Ho.
MTB > # INSUFFICIENT EVIDENCE THAT MEDIAN EXCEEDS 80.
```

Ⓜ 15.19 SIGN TEST OF MEDIAN = 0.00000 VERSUS G.T.   0.00000

```
              N   BELOW   EQUAL   ABOVE   P-VALUE    MEDIAN
BEF-AFTR     36     10       1      25    0.0083     2.000
MTB > # SINCE P-VALUE = 0.0083 < 0.01, REJECT Ho.
MTB > # SUFFICIENT EVIDENCE THAT HERB IS EFFECTIVE IN LOWERING B.P.
```

Section 15.3

15.21 $H_0: \tilde{\mu}_1 - \tilde{\mu}_2 = 0$
  $H_a: \tilde{\mu}_1 - \tilde{\mu}_2 \neq 0$
  $z = 0.92$
  RR: $z < -1.96$, $z > 1.96$
  Fail to reject H_0

15.23 p-value = 0.0102

15.25 p-value = 0.0198

15.27 p-value = 0.0239

15.29 p-value = 0.0028

Ⓜ 15.31 Mann-Whitney Confidence Interval and Test

```
C1          N =  11     Median =      28.000
C2          N =  14     Median =      22.300
Point estimate for ETA1-ETA2 is      5.350
95.4 pct c.i. for ETA1-ETA2 is (3.501,6.699)
W = 217.0
Test of ETA1 = ETA2  vs.  ETA1  n.e. ETA2 is significant at 0.0001
The test is significant at 0.0001 (adjusted for ties)

MTB > # SINCE P-VALUE = 0.0001 < 0.05, REJECT Ho.
MTB > # SUFFICIENT EVIDENCE THAT THE MEDIANS DIFFER.
```

Section 15.4

15.33 H_0: $\tilde{\mu}_1 = \tilde{\mu}_2 = \tilde{\mu}_3$
H_a: At least 2 populations have different medians.
$H = 13.11$
RR: $\chi^2 > 4.61$
Reject H_0

15.35 H_0: $\tilde{\mu}_1 = \tilde{\mu}_2$
H_a: $\tilde{\mu}_1 \neq \tilde{\mu}_2$
$H = 16.41$
RR: $\chi^2 > 3.84$
Reject H_0

15.37 $0.05 < p\text{-value} < 0.10$

15.39 H_0: $\tilde{\mu}_1 = \tilde{\mu}_2 = \tilde{\mu}_3 = \tilde{\mu}_4$
H_a: At least 2 populations have different medians.
$H = 8.18$
RR: $\chi^2 > 7.81$
Reject H_0 (yes)

15.41 H_0: $\tilde{\mu}_1 = \tilde{\mu}_2 = \tilde{\mu}_3$
H_a: At least 2 populations have different medians.
$H = 15.53$
RR: $\chi^2 > 5.99$
Reject H_0 (yes)

15.43 $p\text{-value} < 0.005$

Ⓜ 15.45

LEVEL	NOBS	MEDIAN	AVE. RANK	Z VALUE
1	16	78.00	24.4	-0.02
2	16	71.50	11.6	-4.52
3	16	87.00	37.5	4.54
OVERALL	48		24.5	

```
H = 27.33  d.f. = 2  p = 0.000
H = 27.41  d.f. = 2  p = 0.000 (adj. for ties)

MTB > # SINCE P-VALUE = 0.000 < 0.05, REJECT Ho.
MTB > # THERE IS SUFFICIENT EVIDENCE THAT A DIFFERENCE EXISTS.
```

Section 15.5

15.47 11

15.49 7

15.51 1, 2, 26, 3, 4, 25, 27, 28, 29, 30, 31, 5, 6
(Other answers are possible.)

15.53 H_0: The sequence of observations is random.
H_a: The sequence is not random.
$x = 15$
RR: $x \leq 8, x \geq 19$
Fail to reject H_0 (no)

15.55 H_0: The sequence of observations is random.
H_a: The sequence is not random.
$z = -0.74$
RR: $z < -2.576, z > 2.576$
Fail to reject H_0 (no)

15.57 H_0: The sequence of observations is random.
H_a: The sequence is not random.
$z = -2.56$
RR: $z < -1.96, z > 1.96$
Reject H_0 (yes)

15.59 H_0: The sequence of observations is random.
H_a: The sequence is not random.
$z = -1.14$
RR: $z < -1.96, z > 1.96$
Fail to reject H_0 (no)

Ⓜ 15.61

```
THE OBSERVED NO. OF RUNS =   10
THE EXPECTED NO. OF RUNS =    7.8571
   6 OBSERVATIONS ABOVE K      8 BELOW
```

Section 15.6

15.63 $r_s < -0.409, r_s > 0.409$

15.65 $r_s < -0.475$

15.67 $r_s = 0.905$

15.69 $r_s = -0.866$

15.71 H_0: $\rho_s = 0$
H_a: $\rho_s > 0$
$r_s = 0.933$
RR: $r_s > 0.600$
Reject H_0 (yes)

15.73 H_0: $\rho_s = 0$
H_a: $\rho_s \neq 0$
$r_s = -0.779$
RR: $r_s < -0.689, r_s > 0.689$
Reject H_0 (yes)

Ⓜ 15.75 Correlation of C3 and C4 = -0.779

Review Exercises

15.77 $H_0: \tilde{\mu}_1 - \tilde{\mu}_2 = 0$
$H_a: \tilde{\mu}_1 - \tilde{\mu}_2 \neq 0$
$z = 2.57$
RR: $z < -1.96, z > 1.96$
Reject H_0 (yes)

15.78 H_0: The sequence of observations is random.
H_a: The sequence is not random.
$x = 12$
RR: $x \leq 9, x \geq 21$
Fail to reject H_0

15.79 H_0: The sequence of observations is random.
H_a: The sequence is not random.
$z = -1.13$
RR: $z < -1.96, z > 1.96$
Fail to reject H_0

15.80 $H_0: \tilde{\mu}_D = 0$
$H_a: \tilde{\mu}_D > 0$
$x = 9$
RR: 8,9,10 ($\alpha = 0.055, n = 10$)
Reject H_0 (yes)

15.81 p-value $= 0.011$

15.82 $H_0: \tilde{\mu} = 300$
$H_a: \tilde{\mu} < 300$
$z = -3.21$
RR: $z < -2.326$
Reject H_0 (yes)

15.83 p-value $= 0.0007$

15.84 $H_0: \rho_s = 0$
$H_a: \rho_s > 0$
$r_s = 0.685$
RR: $r_s > 0.745$
Fail to reject H_0 (no)

15.85 $0.01 < p$-value < 0.025

15.86 $H_0: \tilde{\mu}_1 = \tilde{\mu}_2 = \tilde{\mu}_3 = \tilde{\mu}_4$
H_a: At least 2 populations have different medians.
$H = 17.33$
RR: $\chi^2 > 11.34$
Reject H_0 (yes)

15.87 H_0: The sequence of observations is random.
H_a: The sequence is not random.
$x = 14$
RR: $x \leq 10, x \geq 22$
Fail to reject H_0

15.88 H_0: The sequence of observations is random.
H_a: The sequence is not random.
$z = -0.74$
RR: $z < -1.96, z > 1.96$
Fail to reject H_0

15.89 $H_0: \tilde{\mu}_1 = \tilde{\mu}_2 = \tilde{\mu}_3 = \tilde{\mu}_4$
H_a: At least 2 populations have different medians.
$H = 15.21$
RR: $\chi^2 > 7.81$
Reject H_0 (yes)

15.90 p-value < 0.005

Ⓜ 15.91

```
SIGN TEST OF MEDIAN = 300.0 VERSUS  L.T.   300.0
                 N   BELOW  EQUAL  ABOVE    P-VALUE     MEDIAN
C1              36     27     1      8      0.0009      294.5
MTB > # SINCE P-VALUE = 0.0009 < 0.05, REJECT Ho.
MTB > # SUFFICIENT EVIDENCE THAT MEDIAN IS LESS THAN 5 MINUTES.
```

Ⓜ 15.92 Correlation of C3 and C4 = 0.685

Ⓜ 15.93

```
MEDIAN =       485.00

    K =    485.0000

    THE OBSERVED NO. OF RUNS =   14
    THE EXPECTED NO. OF RUNS =   16.0000
    15 OBSERVATIONS ABOVE K    15 BELOW
```

Ⓜ 15.94

```
LEVEL     NOBS    MEDIAN   AVE. RANK    Z VALUE
  1        10     81.40      11.4       -2.83
  2        10     86.75      25.0        1.42
  3        10     96.85      35.0        4.51
  4        10     78.10      10.6       -3.11
OVERALL    40                20.5

H = 30.03  d.f. = 3  p = 0.000
H = 30.04  d.f. = 3  p = 0.000 (adj. for ties)
MTB > # SINCE P-VALUE = 0.000 < 0.05, REJECT Ho.
MTB > # SUFFICIENT EVIDENCE THAT THE MEDIANS ARE NOT ALL EQUAL.
```

(M) 15.95 `Mann-Whitney Confidence Interval and Test`

```
C1          N =  10     Median =      81.400
C3          N =  10     Median =      96.850
Point estimate for ETA1-ETA2 is      -15.500
95.5 pct c.i. for ETA1-ETA2 is (-17.000,-12.301)
W = 55.0
Test of ETA1 = ETA2  vs.  ETA1 n.e. ETA2 is significant at 0.0002
The test is significant at 0.0002 (adjusted for ties)
MTB > # SINCE P-VALUE = 0.0002 < 0.01, REJECT Ho.
MTB > # SUFFICIENT EVIDENCE THAT MEDIANS 1 AND 3 DIFFER.
```

(M) 15.96 `SIGN TEST OF MEDIAN = 0.00000 VERSUS  N.E.  0.00000`

```
              N  BELOW  EQUAL  ABOVE   P-VALUE     MEDIAN
C3           33     4     3      26    0.0001     0.06000
MTB > # SINCE P-VALUE = 0.0001 < 0.01, REJECT Ho.
MTB > # SUFFICIENT EVIDENCE THAT THE AVERAGE PRICES DIFFER.
```

INDEX

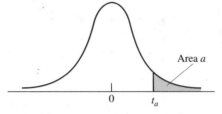

Area a

0   t_a

Each table entry is the t-value whose right-tail area equals the column heading, and whose degrees of freedom (df) equal the row number.

Student's t-Values for Specified Tail Areas

Degrees of Freedom (df)	Amount of Area in One Tail				
	.100	.050	.025	.010	.005
1	3.078	6.314	12.706	31.821	63.657
2	1.886	2.920	4.303	6.965	9.925
3	1.638	2.353	3.182	4.541	5.841
4	1.533	2.132	2.776	3.747	4.604
5	1.476	2.015	2.571	3.365	4.032
6	1.440	1.943	2.447	3.143	3.707
7	1.415	1.895	2.365	2.998	3.499
8	1.397	1.860	2.306	2.896	3.355
9	1.383	1.833	2.262	2.821	3.250
10	1.372	1.812	2.228	2.764	3.169
11	1.363	1.796	2.201	2.718	3.106
12	1.356	1.782	2.179	2.681	3.055
13	1.350	1.771	2.160	2.650	3.012
14	1.345	1.761	2.145	2.624	2.977
15	1.341	1.753	2.131	2.602	2.947
16	1.337	1.746	2.120	2.583	2.921
17	1.333	1.740	2.110	2.567	2.898
18	1.330	1.734	2.101	2.552	2.878
19	1.328	1.729	2.093	2.539	2.861
20	1.325	1.725	2.086	2.528	2.845
21	1.323	1.721	2.080	2.518	2.831
22	1.321	1.717	2.074	2.508	2.819
23	1.319	1.714	2.069	2.500	2.807
24	1.318	1.711	2.064	2.492	2.797
25	1.316	1.708	2.060	2.485	2.787
26	1.315	1.706	2.056	2.479	2.779
27	1.314	1.703	2.052	2.473	2.771
28	1.313	1.701	2.048	2.467	2.763
29	1.311	1.699	2.045	2.462	2.756
Infinity	1.282	1.645	1.960	2.326	2.576